Heterocyclic
N-Oxides

Authors

Angelo Albini, D. Chem.
Associate Professor
Department of Organic Chemistry
University of Pavia
Pavia, Italy

and

Silvio Pietra, D. Chem.
Professor
Department of Organic Chemistry
University of Pavia
Pavia, Italy

CRC Press
Boca Raton   Ann Arbor   Boston

Library of Congress Cataloging-in-Publication Data

Albini, Angelo.
    Heterocyclic *N*-oxides / authors, Angelo Albini, Silvio Pietra.
         p.   cm.
    Includes bibliographical references and index.
    ISBN 0-8493-4552-9
    1. Heterocyclic compounds. 2. Organo nitrogen compounds. 3. Nitrogen oxides.
  I. Pietra, Silvio. II. Title.
  QD401.A33   1991
  547′.59—dc20   90-2525
                                                            CIP

Direct all inquiries to CRC Press, Inc., 2000 Corporate Blvd., N.W., Boca Raton, Florida, 33431.

© 1991 by CRC Press, Inc.

International Standard Book Number 0-8493-4552-9

Library of Congress Card Number 90-2525
Printed in the United States

PREFACE

Heterocyclic *N*-oxides have been the subject of intensive investigations in the last decades. Two excellent treatises[1,2] summarized the state of the art at the end of the 1960s. Since then, research has progressed at a remarkable pace, as shown by the average publication of 300 papers a year (taking into account only those concerned with the chemical and physical properties of these compounds, and the papers and patents devoted to applications are at least twice as much). Furthermore, the advancement has not only been quantitative, but also qualitative, and new lines of research have been opened.

Thus, it appeared to be worthwhile to prepare a new monograph on this subject. We attempted to review the chemistry of *N*-oxides under all its aspects, synthesis, physical properties, reactions, and applications; to include tables and representative procedures; and yet to maintain the size of the book within reasonable limits. It was not deemed feasible to include all relevant references, but we tried to attain a large coverage, with particular emphasis on the most recent advances. We made no attempt to give a historical perspective. According to the publisher's habit, references are given with full title, and the comment in the text is accordingly reduced. We hope that the format chosen gives both sufficient information directly from the text and convenient access to the primary literature. Literature coverage is to the end of 1988, and partially through 1989.

The acceptance of sharing this task by my former mentor, Professor Silvio Pietra, weighed greatly in my decision to carry out this project. Together we did the planning, collection, and arrangement of the references, and he personally prepared the major part of the section on synthesis. A merciless illness prevented him from seeing the work completed. I hope this book as it is now presented is not unworthy of him.

This monograph is dedicated to Professor Pietra, in deep appreciation for the many things he taught me during many years of collaboration. I always will affectionately remember him and be inspired by his style in the lab and in life.

Obviously, the responsibility for any mistake is mine. Every practitioner of chemistry is surely aggravated by the many mistakes and misquotations commonly found in the printed literature and, in particular, in review papers. One cannot avoid being anxious when presenting yet another review, being aware that a sizeable number of mistakes must have surely survived checking. I will be very grateful to anyone who kindly points out to me any mistakes, misquotations, or omissions.

Invaluable help in the preparation of this work was given by Dr. E. Fasani, who devised and executed the drawings. I am also greatly indebted to Dr. V. Regondi for conducting the computer literature search, and to Prof. G. Spinolo and to Drs. M. Mella, I. Baldini and R. Giardino for helping in arranging it. Finally, I am grateful to Emeritus Professor M. Colonna and to Prof. I. Degani for their encouragement. My family helped me with their love and understanding in these years.

REFERENCES

1. **Ochiai, E.,** *Aromatic Amine Oxides,* Elsevier, Amsterdam, 1967.
2. **Katritzky, A. R. and Lagowski, J. M.,** *Chemistry of Heterocyclic N-Oxides,* Academic Press, London, 1971.

A. Albini

AUTHORS

Angelo Albini, D. Chem., was born in Milan, Italy in 1946. He studied chemistry at the University of Pavia, Pavia, Italy, presenting a thesis on the photochemistry of phenazine.

After a period with the pharmaceutical company Lepetit (1971) and postdoctoral work at Pavia (1972 to 1973) and at the Max Plank Institut für Strahlenchemie at Mülheim, West Germany (1973 to 1975), he joined the Faculty at Pavia as assistant and then associate professor. He was later visiting professor at the University of Western Ontario (1977 to 1978), at the University of Odense, Denmark (1983), and again at Mülheim (1983).

His research interests are in the field of aromatic and heterocyclic compounds, in particular, photochemistry of heterocyclic N-oxides, thermal and photochemical substitution on heterocylic compounds, heterocyclic azides, electron transfer reactions of aromatic compounds, and metal complexes of nitrogen heterocycles. He has authored more than 80 research papers and presented several contributions to Symposia in Photochemistry and in Heterocylic Chemistry.

Silvio Pietra, D. Chem., born in 1923, had been Professor of Chemistry, the University of Pavia, Italy, for about 40 years until his untimely death in January 1990, and Head of the Department of Organic Chemistry for several years.

He received his degree in Pavia in 1946, and apart from a postdoctoral stay at Harvard, he spent the whole of his career in that university as Assistant Professor (1950) and Professor (1973) of Chemistry.

A member of the Italian Chemical Society and the European Photochemical Association, he has been the author of about 90 research papers. His interest in recent years has centered on the photochemistry of heterocycles and the photodegradation of dyes.

TABLE OF CONTENTS

Chapter 1

INTRODUCTION

I. SCOPE

N-oxides are characterized by the presence of a so-called "donative" (or "coordinate-covalent") bond between nitrogen and oxygen, viz. of a covalent bond formed by the overlap of the nonbonding electron pair on the nitrogen with an empty orbital on the oxygen atom. This is usually graphically indicated by an arrow or by formal charges (formulae **1** and **2**, the symbols used in the former are usually preferred in the present text). The nitrogen atom can be sp^3 (as in aliphatic amine oxides **3**), sp^2 (as in nitrones **4** and in azoxy derivatives **5**), or sp (as in nitrile oxides **6**) hybridized. In the first two groups of compounds, the *N*-oxide function can be part of a ring, as in cyclic amine *N*-oxides **7** and cyclic nitrones **8**, and this obviously holds for aromatic rings, as in the case of pyridine 1-oxide **9**, the prototype of heteroaromatic *N*-oxides (or, less properly, aromatic amine oxides). This last group of compounds happens to display a peculiar chemistry, and the present discussion is centered around them, with only marginal reference to nonaromatic cyclic derivatives such as **7** and **8**.

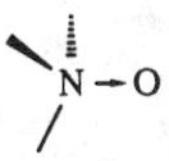

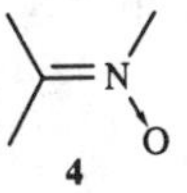

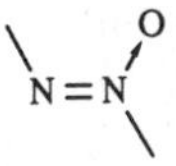

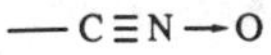

Nomenclature presents no problems, since these compounds are indicated by adding *oxide* after the name of the heterocycle, preceded by a number referring to the position involved (less precisely, but very commonly, by the letter *N*, or in the case of polyaza heterocycles, indications such as *N*, *N*′, *N*^{a}, or *N*^{1}). The prefix *oxido*, or, improperly, *oxy* or *oxo*, is sometimes used when the whole heterocyclic moiety is considered as a substituent, e.g., 1-oxidopyridin-2-yldiazomethane, **10**.

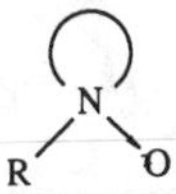

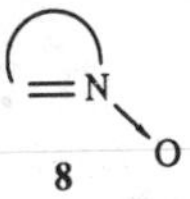

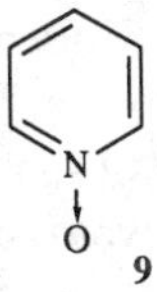

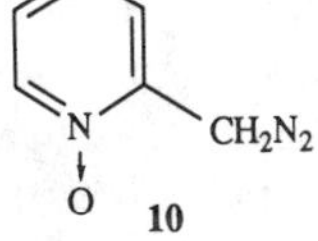

Newkome and Paudler appropriately stated that, although 40% of the papers published in organic chemistry are concerned with heterocyclic chemistry, organic chemistry textbooks often discuss the latter subject only sketchily, and this tends to give students the impression that heterocyclic chemistry is a discipline that is distinct from organic chemistry, whereas the chemistry of heterocycles "encompasses most, if not all, of the general reactions of organic chemistry."[1]

The same considerations (except for the percentage of published papers, which is considerable but surely less than 40%) apply to the case of *N*-oxides considered in the context of the general framework of heterocyclic chemistry. Many books on heterocycles include only a passing note on these derivatives, although practically all types of syntheses and reactions are represented, and indeed a discussion of *N*-oxide chemistry in direct comparison with what is observed with the corresponding nonoxidized substrates helps in understanding many mechanistic schemes.

The effect of the $\geqslant$N$\rightarrow$O function in pyridine 1-oxides and congeners is quite different from the effect of the nitrogen atom in azines, from which these compounds are formally derived. As discussed in Chapter 4, *N*-oxides do not fit well into the normal classification in π-deficient and π-excessive heterocyles, as they undergo both processes that are typical of the first (e.g., nucleophilic substitution) and of the latter (e.g., electrophilic substitution, addition) classes, as well as processes that are attributed to the $\geqslant$N$\rightarrow$O function itself, with modifications due to the aromaticity (e.g., cycloaddition, photorearrangement, and, in a sense, also the typical deoxidative substitution).

Growing recognition of the importance of *N*-oxides is demonstrated by the greater attention given to these compounds in recent books on heterocyclic chemistry,[2] as well as chapters devoted to them in books on single heterocycles, e.g., in the Weissberger series.[3-14]

In the following chapters, physical properties, synthetic methods, reactions, and uses of these compounds are considered. Exhaustive reference to the literature would have been incompatible with the planned size of the book, but it has been attempted to give a large coverage, in particular with regard to the most characteristic properties of *N*-oxides. Tabular surveys, as well as detailed experimental procedures, are meant to convey an idea of the typical experimental conditions, rather than an extensive tabulation of the individual reported reactions and products, since the applicability of mild methods is what makes the chemistry of heterocyclic *N*-oxides so interesting. According to the publisher's custom, literature references are reported by full title. Therefore, applications of a reaction to various derivatives are only briefly mentioned in the text, or are even simply quoted through the reference number, since the title of the paper allows one, at least in most cases, to understand the scope of the method.

II. HISTORICAL BACKGROUND

Several heterocyclic *N*-oxides have been prepared already in the 19th century by cyclization methods from oximes, nitroso, and nitro derivatives.[15-22] Identification of this class of compounds was not unambigous, however, and there is a considerable confusion in the earlier literature about the formulae; as an example, *N*-oxides were in some cases confused for oxoheterocycles,[15] and in other cases, though the oxygen atom was recognized as bonded to the nitrogen atom, it was considered as a part of the heterocyclic ring (see the formulae proposed for the heterocyclic ring in phenazine 5-oxide **11**[17] and the phenoxazine dye resazurin **12**).[22]

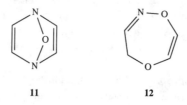

11   **12**

In the third decade of this century, Meisenheimer[23-24] showed that these compounds are obtained by peracid *N*-oxidation of nitrogen heterocycles (and that the products thus obtained are identic to those from cyclization processes), and in this way, the correct formula was definitely assessed and *N*-oxides were recognized as a new class of organic compounds.[23-24] In the following years, several reactions were discovered, and the nature of the N-O bond elucidated. Thus, Meisenheimer recognized that heteroaromatic *N*-oxides can not be resolved,[25] and Linton found that the dipole moment of these compounds is much lower that for their aliphatic counterpart, supporting the notion that there is back-donation from the oxygen atom to the ring.[26] Impressed by the last observation, both Ochiai[27-28] and, later but independently, den Hertog[29] surmised that electrophilic substitution on these substrates should be favored. This proved to be case and gave impetus to lively research in the field, and to the discovery of several other reactions, in addition to electrophilic substitution. Again during the 1940s, another line of thought, the analogy between heterocyclic *N*-oxides and nitrones, led Colonna to further fruitful investigations.[30]

The development of the subject during the 1950s and the 1960s is demonstrated by the publication of several reviews[31,32] and of two authoritative books, the first by Ochiai (1967)[33] and the latter by Katritzky and Lagowski (1971).[34]

The pace of research has certainly not slackened in the last two decades, with possibly a shift of attention from the general reactions of heterocycles (aromatic substitution) to those that are more typical of *N*-oxides, i.e., deoxidative substitution, cycloaddition, and photochemical reactivity.

III. OCCURRENCE

While a relatively large amount of saturated heterocyclic *N*-oxides are present in nature, only a few aromatic *N*-oxides have been obtained from natural sources, in most cases hydroxylated derivatives, e.g., the lethal toxin of the mushroom *Cortinensis orellanus*, orellanine (a bipyridine dioxide);[35] some pyrazine oxides, e.g., aspergillic acid and related hydroxamic acids,[36] as well as emimycin;[37] quinoline derivatives such as the "Pyo" substances of *Pseudomonas aeruginosa*,[38] and phenazines such as the antibiotics iodinin and mixin.[39] Canthin-6-one 3-oxide, a β-carboline derivative, has been isolated from *Ailanthus altissima*[40] (Figure 1).

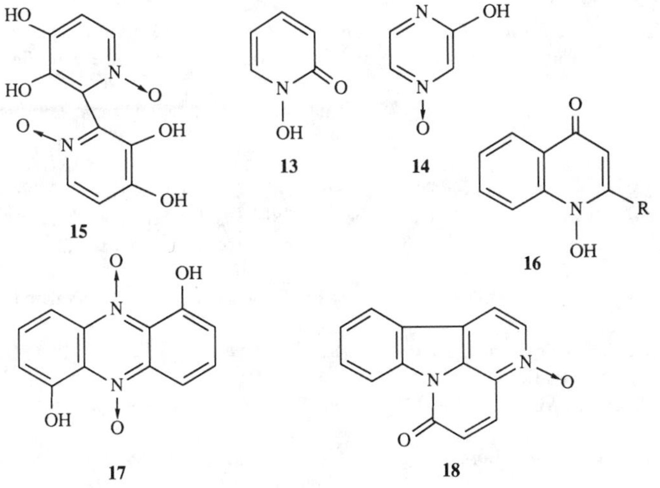

FIGURE 1. Heteroaromatic *N*-oxides from natural sources. (**13**) Aspergillic acid, (**14**) emimycin, (**15**) orellanine, (**16**) "Pyo" substances from *Pseudomonas aeruginosa*, (**17**) iodinin, (**18**) canthin-6-one 3-oxide.

However, *N*-oxidation is a quite general process during the metabolization of nitrogen heterocycles, and several *N*-oxides have been obtained as drug metabolites.[41]

REFERENCES

1. **Newkome, G.R. and Paudler, W.W.,** *Contemporary Heterocyclic Chemistry,* Wiley, New York, 1982, 1.
2. **Katritzky, A.R. and Rees, C.W. Eds.,** *Comprehensive Heterocyclic Chemistry,* Pergamon Press, Oxford, 1984.
3. **Abramovitch, R.A. and Smith, E.M.,** Pyridine-1-oxides, in *Pyridine and Its Derivatives,* Abramovitch, R.A., Ed., Wiley, New York, 1974, chap. 4.
4. **Jones, G. and Baty, D.J.,** Quinoline *N*-oxides, in *Quinolines,* Jones, G., Ed., Wiley, New York, 1982, Part II, chap. 3.
5. **Itai, T.,** Pyridazine *N*-oxides, in *Pyridazines,* Castle, R.N., Ed., Wiley, New York, 1973, chap. 8.
6. **Brown, D.J.,** The *N*-alkylated pyrimidines and pyrimidine *N*-oxides, in *The Pyrimidines,* Wiley, New York, 1962, chap. 10 (Suppl. I, 1970; Suppl. II, 1985).
7. **Barlin, G.B.,** *The Pyrazines,* Wiley, New York, 1982.
8. **Singerman, G.Y.,** Cinnoline *N*-oxides, in *Condensed Pyridazines,* Castle, R. N., Ed., Wiley, New York, 1973, chap. 1J.
9. **Cheeseman, G.W.H. and Cookson, R.F.,** Quinoxaline mono and di-*N*-oxides, in *Condensed Pyrazines,* Wiley, New York, 1979, chap. 4.
10. **Patel, N.R.,** Phthalazine *N*-oxides, in *Condensed Pyridazines,* Castle, R. N., Ed., Wiley, New York, 1973, chap. 2C.
11. **Armarego, W.L.F.,** Quinazoline *N*-oxides, in *Fused Pyrimidines,* Brown, D.J., Ed., Wiley, New York, 1967, Part I, chap. 9.
12. **Lister, J.H.,** Purine *N*-oxides, in *Fused Pyrimidines,* Brown, D.J., Ed., Wiley, New York, 1971, Part II, chap. 11.
13. **Brown, D.J.,** Alkoxypteridines, *N*-alkylpteridones and pteridine *N*-oxides, in *Fused Pyrimidines,* Brown, D.J., Ed., Wiley, New York, 1988, Part III, chap. 7.
14. **Smith, D.M.,** Benzimidazole-*N*-oxides, in *Benzimidazole and Congeneric Tricyclic Compounds,* Preston, P.N., Ed., Wiley, New York, 1981, Part I, chap. 2; Spande, T.F., Hydroxyindole, indole alcohols and indolethiols, in *Indoles,* Houlihan, W.J., Ed., Wiley, New York, 1979, Vol. 3, chap. 8.
15. **Ost, H.,** Action of hydroxylamine and ethylamine on comanic acid, *J. Prakt. Chem.,* 29, 378, 1884.
16. **Heller, G. and Sourlis, A.,** Reduction of nitroderivatives with zinc dust and acetic acid, *Ber. Dtsch. Chem. Ges.,* 41, 2692, 1908.
17. **Wohl, A. and Aue, W.,** Action of nitrobenzene on aniline in the presence of alkali, *Ber. Dtsch. Chem. Ges.,* 34, 2442, 1901.
18. **Zinin, N.,** On some derivatives of azoxybenzide, *Ann. Chem.,* 114, 217, 1860.
19. **Werner, A. and Stiasny, E.,** On nitroderivatives of azo-, azoxy-, and hydrazo-benzene, *Ber. Dtsch. Chem. Ges.,* 32, 3256, 1899.
20. **Täuber, E.,** On diphenylazone, a new ring-closed, nitrogen containing compound, *Ber. Dtsch. Chem. Ges.,* 24, 3081, 1891.
21. **Weselesky, P.,** On the azoderivatives of resorcine, *Ber. Dtsch. Chem. Ges.,* 4, 613, 1871.
22. **Nietzki, R.,** Synthesis of Weselesky's Resorcine Blue, *Ber. Dtsch. Chem. Ges.,* 24, 3366, 1891.
23. **Meisenheimer, J. and Statz, E.,** On quinaldine oxide, *Ber. Dtsch. Chem. Ges.,* 58, 2324, 1925.
24. **Meisenheimer, J. and Statz, E.,** On pyridine, quinoline, and isoquinoline *N*-oxide, *Ber. Dtsch. Chem. Ges.,* 59, 1848, 1926.
25. **Meisenheimer, J.,** Attempts to resolve quinoline oxide. Stereochemistry of tervalent nitrogen, *Ber. Dtsch. Chem. Ges.,* 66, 985, 1933.
26. **Linton, E.P.,** The dipole moment of amine oxides, *J. Am. Chem. Soc.,* 62, 1945, 1940.
27. **Ochiai, E.,** A new classification of tertiary amine oxides, *Proc. Imp. Acad. (Tokyo),* 19, 307, 1943.
28. **Ochiai, E., Ishikawa, M., and Katada, M.,** New conception of tertiary amine *N*-oxides, *Yakugaku Zasshi,* 63, 307, 1943.
29. **den Hertog, H.J. and Overhoff, J.,** Pyridine *N*-oxide as an intermediate for the preparation of 2- and 4-substituted pyridines, *Rec. Trav. Chim. Pays-Bas,* 69, 468, 1950.
30. **Colonna, M.,** Analogies between aldonitrones and *N*-oxides of pyridine bases, *Boll. Sci. Fac. Chim. Ind. Bologna,* 134, 1940.
31. **Katritzky, A.R.,** The chemistry of the aromatic heterocyclic *N*-oxides, *Q. Rev.,* 10, 395, 1956.
32. **Ochiai, E.,** Recent Japanese work on the chemistry of pyridine 1-oxide and related compounds, *J. Org. Chem.,* 18, 534, 1953.
33. **Ochiai, E.,** *Aromatic Amine Oxides,* Elsevier, Amsterdam, 1967.
34. **Katritzky, A.R. and Lagowski, J.M.,** *Chemistry of Heterocyclic N-Oxides,* Academic Press, London, 1971.
35. a. **Antkowiak, W.Z. and Gessner, W.P.,** The structure of orellanine and orelline, *Tetrahedron Lett.,* 1931, 1979; b. **Tiecco, M.,** New reactions of pyridines and total synthesis of the fungal toxin orellanine, *Bull. Soc. Chim. Belg.,* 95, 1009, 1986.

36. a. **Dutcher, J. D.,** Aspergillic acid, an antibiotic substance produced by *Aspergillus flavus*. I. General properties; formation of desoxyaspergillic acid; structure conclusions, *J. Biol. Chem.,* 171, 321, 1947; b. **Neilands, J.B.,** Hydroxamic acids in nature, *Science,* 156, 1443, 1967.

37. **Terao, M.,** Structure of emimycin, *J. Antibiotics (Tokyo),* Sec. A., 16, 182, 1963.

38. a. **Clemo, G.R. and Daglish, A.F.,** The constitution of the pigment of *Chromobacterium iodinum, J. Chem. Soc.,* 1481, 1950; b. **Sigg, H.P. and Toth, A.,** On the structure of phenazine *N*-oxide, *Helv. Chim. Acta.,* 50, 716, 1967.

39. **Budzikiewicz, H., Schaller, U., Korth, H., and Pulverer, G.,** Alkylquinolines and their *N*-oxides from *Pseudomonas aeruginosa, Monatsh. Chem.,* 110, 947, 1979.

40. **Ohmoto, T., Tanaka, R., and Nikaido, T.,** Studies on the constituents of *Ailanthus altissima* Swingle. On the alkaloidal constituents, *Chem. Pharm. Bull.,* 24, 1532, 1976.

41. **Gorrod, J.W. and Damani, L.A.,** The metabolic *N*-oxidation of 3-substituted pyridines in various animals species in vivo, *Eur. J. Drug Metab. Pharmacokinet.,* 5, 53, 1980.

Chapter 2

PHYSICAL PROPERTIES

I. MOLECULAR STRUCTURE

A. ELECTRONIC STRUCTURE AND MOLECULAR DIMENSIONS

Heteroaromatic *N*-oxides contain a donative bond from nitrogen to oxygen and thus a formal positive charge on the first atom (see, e.g., formula **1** for the prototype of the series, pyridine 1-oxide). Mesomeric formulae **2a—c** show that the charge is also distributed on the ring. On the other hand, as always happens when the atom that acts as the σ-donor has a π orbital available for electron acceptance, back-donation from the oxygen takes place and the additional mesomeric formulae **3a—c** must be considered. As will be shown in the following, this fact dominates both the physical and chemical properties of *N*-oxides (Table 1).

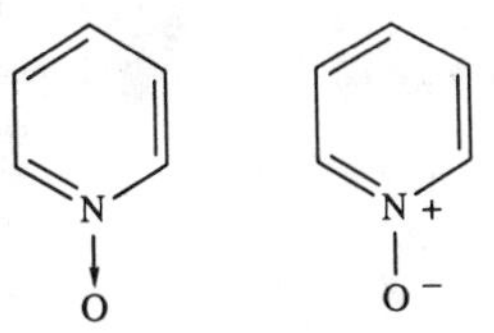

The first consequence affects the dimension of the molecule. Microwave and and electron-diffraction measurements show that the six-membered ring in pyridine 1-oxide is more regular and similar to benzene than pyridine, **4.**[1-4] The angle C_2-N-C_6 is very near to 120° and, contrary to the case of compound **4,** the N-C_2 bond is not shortened; on the contrary, all the bonds in the ring are quite similar in length. The N-O bond is relatively short, 1.28 Å. The steric requirement of the ≥N→O group was evaluated using an NMR study of the hindering to rotation in some pyridinophane 1-oxides, and it was found that it is comparable to that of the aromatic C-F group. In a Stuart-Briegleb model of pyridine 1-oxide, the calotte representing the quaternized nitrogen is very similar to an aromatic carbon, and the ≥N→O group extends ca. 2.63 Å from the vertex of the hexagon.[5]

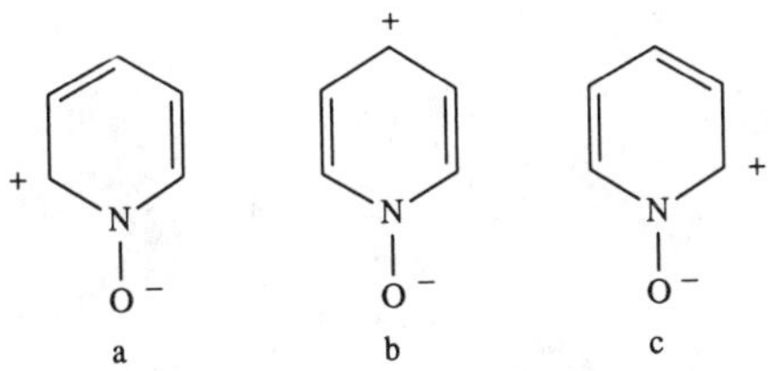

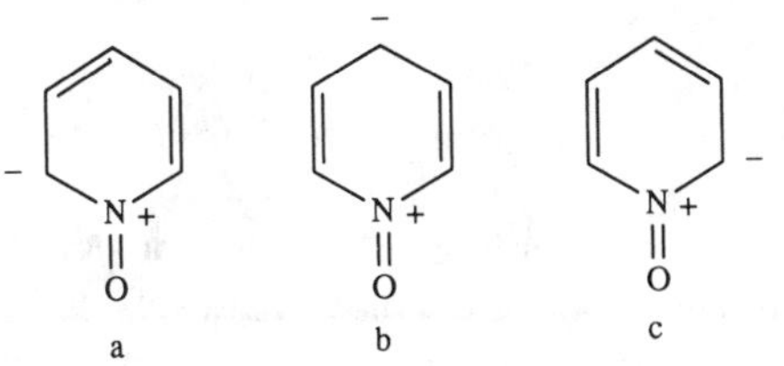

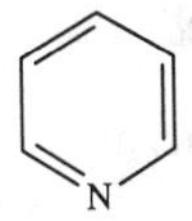

Various spectroscopic evidence (e.g., N-O stretching frequencies,[6] molecular dipole moments,[7,8] [1]H, ^{13}C, ^{14}C, and ^{15}N, as well as [17]O chemical shifts[9-14] and crystal structure determinations; see Section IIB) support the partial double-bond character of the N-O bond, evaluated at ca. 15—20%. This is, of course, influenced by the substituents, e.g., strong electron-withdrawing groups in position 4 favor canonical forms **3,** and thus a quinoidal structure, while with electron-donating substituents, formula **2** is favored instead, and the N-O bond becomes longer.[15,16] As predicted by this simple model, the π character of the N-O bond increases in pyrazine and pyridazine 1-oxides, but not in pyrimidine oxide.[16] A partial double-bond

TABLE 1
Molecular Dimensions for Pyridine 1-Oxide

	Bond length, Å			Angle, degrees	
	a	b		a	b
N-O	1.278 ± 0.01	1.290 ± 0.015	C_2NC_6	119.8 ± 0.3	120.9 ± 1.8
$N-C_2$	1.362 ± 0.003	1.380 ± 0.011	NC_2C_3	120.7 ± 0.2	118.6
C_2-C_3	1.389 ± 0.005	1.381 ± 0.009	$C_2C_3C_4$	120.6	124.6
C_3-C_4	1.395	1.393 ± 0.008	$C_3C_4C_5$	117.6	114.1 ± 2.5

[a] From microwave measurements.[1]
[b] From gas-phase electron diffraction measurements.[3]

TABLE 2
N-O π-Bond Character, Formal Charge on the Oxygen Atom, and N-O Bond Length in Some 4-Substituted Pyridine 1-Oxides

Substituent	Occupancy[a]	π-Bond order[a]	Formal charge on oxygen[a]	N-O bond length, Å[b]
$PhCH_2O$	1.849	0.168	–0.78	
Me	1.836	0.182	–0.80	1.405
Cl	1.823	0.197	–0.78	1.262
H	1.778	0.247	–0.73	1.29
NO_2	1.719	0.312	–0.67	1.281

[a] From ^{17}O NQR measurements.[13]
[b] From gas-phase electron diffraction measurements.[15]

character is present also in the *N*-oxides with different ring dimensions, e.g., in furoxans[17] (Table 2).

The double role of the *N*-oxide function that can act both as a π-electron donor and as a π-electron acceptor leads to an enhanced effect of both electron-withdrawing and -donating substituents on the electronic structure and spectroscopic properties,[8] and linear free energy correlations must take into account this twofold mesomeric effect.[18]

B. MOLECULAR ORBITALS AND PHOTOELECTRON SPECTROSCOPY

The first important characteristic of heteroaromatic *N*-oxides is, of course, their aromaticity. For example, the aromatic character of pyridine 1-oxide can be appreciated through the *aromatic index* calculated by Bird on the basis of the regularity of bond lengths,[19] as well as through the cyclic conjugation theory;[20] according to Devanneaux and Labarre, the attachment of the oxygen atom increases the aromaticity and delocalization in the pyridine ring.[21]

Many theoretical calculation have been carried out on these molecules and have been correlated with chemical and/or spectroscopic properties.[22-31] Recent results show that molecular orbital calculations with a different level of sophistication (e.g., *ab initio,* HAM/3, INDO-, CNDO-, and PPP-type) give, in almost every case, the correct order, at least for the high-lying MOs,[22] and these correlate satisfactorily with the experimental values from photoelectron spectroscopy (PES).[32]

The most interesting result from the He I or II PES spectra of pyridine 1-oxide is that, while the second band at 9.22 eV corresponds to a σ orbital centered on the oxygen atom (the main contribution is from a p orbital lying in the plane of the molecule), and occurs at a value similar to that found in aliphatic *N*-oxides, the first band is observed at a much lower potential (8.38 eV) and corresponds to a π orbital with a large coefficient on the oxygen atom (p orbital perpendicular to the molecular plane).[32-34] Such a structure of the HOMO is a constant feature of aromatic *N*-oxides with the expected variations along the series, i.e., the ionization potential

TABLE 3
Vertical Ionization Potentials (eV) of N-Oxides

N-oxide	Ionization potential, type of orbital					Ref.
Pyridine	8.38,π	9.22,σ_O	10.18,π	11.59,π	13.0,π	33
Pyrazine	9.17,π	9.9,10.3,σ_N,σ_O	10.68,π	12.72,π	13.85,π	35
Quinoline	8.00,π	9.01,σ_O	9.50,π	9.75,π	11.35,π	36
Isoquinoline	7.98,π	9.03,σ_O	9.21,π	10.18,π	10.88,π	36
Acridine	7.45,π	8.77,σ_O	9.04,π	9.3,π	9.63,π	36
Phenazine	8.10,π	9.27,π	9.46,σ_O			37

decreases with increasing conjugation (Table 3) and the localization of the HOMO on the oxygen atom also decreases, e.g., from the coefficient 0.68 in pyridine 1-oxide to 0.46 in phenazine 5-oxide.[35-37] This completely contrasts with what observed in the corresponding azines (e.g., in pyridine, the highest lying π orbital is almost degererate with the σ_N orbital and is much more stabilized, ca. 9.8 eV), while both the observed potential and the localization of the two highest occupied MOs in N-oxides are similar to those of nitrones.[40] The peculiarity of the HOMO, a relatively high-lying orbital of the π type, that is strongly localized outside the ring, somewhat detracts from the aromatic character of N-oxides, at least with respect to azines (Figure 1, Table 3).

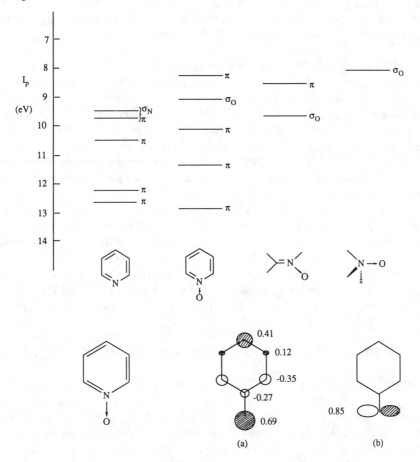

FIGURE 1. Energy (from photoelectron spectra) of the highest lying molecular orbitals in pyridine, pyridine 1-oxide, aliphatic nitrones, and amines. Atomic coefficient for the HOMO (a) and NHOMO (b) in pyridine 1-oxide.

While theoretical calculations adequately reproduce the spectroscopic properties of *N*-oxides, attempts to predict the chemical reactivity (in particular, the preferred position of attack in aromatic substitutions) on the basis of MO data (e.g., total charge, π-electron density, localization energy, frontier MO characteristics, and correlation with the hardness or softness of the reagents)[23,38,39] have met only limited success. Part of the trouble is due to the fact that different chemical species (e.g., the neutral or H-bonded or protonated *N*-oxide, or the corresponding *O*-alkyl or -acyl species; see Chapter 4) are involved in the different reactions.

It can at least be stated that the characteristics of the frontier MOs enable the prediction of the gross features of the reactivity, i.e., the high-lying HOMO explains the enhanced (with respect to azines) reactivity with electrophiles, and in particular the electrophilic substitution, mainly at position γ (the ring-position with the largest coefficient, followed by the α position), as well as the easy interaction with dipolarophiles, in view of the large coefficients on the nitrone moiety (however, the loss of aromaticity that would be implied prevents, in most cases, irreversible cycloaddition). Furthermore, since a low-lying LUMO is maintained, the easy nucleophilic substitution of azines is retained. Finally, the HOMO/LUMO geometric difference requires an extensive electron redistribution on excitation and is reflected in the high quantum yield photoreactivity of *N*-oxides. A detailed mechanistic discussion is deferred to Chapter 4.

It has been shown that core ionization energies (measured by X-ray photoelectron spectroscopy [ESCA]) are sensitive to the molecular structure, just as the external orbitals, e.g., the ionization potential of N_{1s} orbitals undergo changes of the same magnitude as π orbitals in substituted pyridine 1-oxides.[41] The most important results from ESCA spectroscopy refer to the parameters of cyclic azo dioxides (e.g., diazetine dioxide and benzo[*c*]cinnoline dioxide)[42a,42b] and of furoxans, as compared to the open-chain dinitroso[42c] derivatives.

C. POLARITY AND ENERGY OF THE N-O BOND

The dipole moments of many heteroaromatic *N*-oxides have been derived either from measurements of dielectric constants in solution, or, more recently, from microwave spectra.[2,7,8,43] As already pointed out by Linton in 1940,[7] the difference in dipole moment between pyridine 1-oxide and pyridine (2.02 D) is much smaller than the difference between trimethylamine *N*-oxide and trimethylamine (4.37 D), and this implies the contribution of canonical structures such as **3a—c**, which partially counterbalance charge separation. The above-mentioned dual nature of the *N*-oxide function leads to an increased effect of both electron-donating and -withdrawing substituents on the dipole moment with respect to carbocyclic aromatics and azines. The difference between the dipole moment of pyridine 1-oxide and that of pyridine-boron trihalide complexes has been compared with the difference between trimethylamine *N*-oxide and trimethylamine-boron trihalide, and the contribution of mesomeric formulae such as **3** has been evaluated.[44]

The bond dissociation enthalpy for the N-O bond in pyridine 1-oxide is 301.7 ± 2.8 kJM[-1] (72.06 kcalM[-1]), a value higher than that derived for both open-chain[46] and cyclic nitrones[47a] as well as furoxans[47b,47c] (247—268 kJM[-1]) and approaching that of azoxybenzene (321.5 kJM[-1]).[46]

II. PHYSICAL AND SPECTROSCOPIC PROPERTIES

A. ASPECT, SOLUBILITY

The lower members of the series, e.g., pyridine and quinoline 1-oxides, are colorless solids that are freely soluble in water; more conjugated molecules, e.g., acridine or phenazine *N*-oxides, are only sparingly water soluble. The enthalpies of dissolution of pyridine 1-oxide have been found to not correlate with the polarity of the solvents.[47d]

Anhydrous and deliquescent crystals of pyridine 1-oxide are obtained by recrystallization from diethyl ether or by sublimation under reduced pressure. Quinoline and isoquinoline *N*-

oxides crystallize both in the anhydrous state (again hygroscopic) and as hydrates. Phenazine 5-oxide is a high-melting solid, which, like the previous compounds, sublimes undecomposed (however, partial thermal deoxygenation is a frequent phenomenon with N-oxides, see Section IID and Chapter 4). Azabenzene and naphthalene N-oxides usually yield stable crystalline hydrochlorides; while with higher members, such as phenazine 5-oxides, the hydrochlorides are moisture sensitive (Table 4).

B. CRYSTALLOGRAPHIC STUDIES

In the crystalline state, pyridine 1-oxide is present in two independent forms, with considerable deformation with respect to the gas-phase structure.[48] The packing is much less compact than in trimethylamine N-oxide, and the energy gained through the alternation of the dipole moment is less, as reflected in the melting point of 66°C (to be compared with 207°C for the aliphatic derivative).

Some molecular parameters obtained from crystallographic studies are reported in Table 5. The N-O distance in heteroaromatic N-oxides changes only slightly in the corresponding salts and metal complexes.[49,50] In several cases it has been shown that the N-oxide function in the crystalline state is involved in an intramolecular (e.g., 8-hydroxyquinoline)[51] or intermolecular (e.g., 2-hydroxymethylpyridine 1-oxide)[52] hydrogen bond, again with little effect on the bond length. However, this parameter is strongly dependent on the substituent, and thus on the π-bond order (see 4-nitropyridine 1-oxide[53] and compare with Section IB). The situation is, of course, different in five-membered derivatives[57-60] and in cyclic nitrones.[61]

Examination of the physical properties of some pyridine 1-oxides in the liquid state shows only a limited ordering through dipole orientation (see Section IIIB2 for self-association in solution)[62] (Table 5).

C. NUCLEAR MAGNETIC RESONANCE AND QUADRUPOLE RESONANCE

The NMR spectra of heterocyclic N-oxides for all the nuclei concerned give both useful information on the electronic structure and indications for identification.

The resonance of the protons in α and γ to the $\geqslant$N→O function are less deshielded than in the corresponding azines (typically by 0.2—0.5 ppm), while the protons in ß and those in condensed rings show little difference from the azines, except for the protons in *peri*, which are strongly deshielded, typically by 0.6 ppm or more.[9,63,64] These are diagnostic characteristics, together with the effect of acids, that cause a downfield shift of protons in α and γ (however, these remain at higher fields than in protonated azines, in accordance with the fact that the $\geqslant$N$^+$-OH group is less electron withdrawing than the $\geqslant$N$^+$-H group) and the return to normal values of protons in *peri*. The anisotropic effect of the group extends to protons outside the ring, provided that these are sufficiently near to the oxygen atom, e.g., to α-alkyl groups;[65,66] in rigid systems, e.g., in the quinolinophanes **5,** the *syn* proton is easily distinguished from its downfield shift (as much as 1.5 ppm) in passing from the azine to the N-oxide.[66]

The H_α-H_β coupling constant is more similar to that observed in benzene than to that of azines,[9,63,64] again probably an indication of the smaller modification of the aromatic ring caused by the N-oxide group as compared to the aza substitution of the parent compounds.

As expected, observed resonances are strongly solvent dependent, in particular in a protic solvent where a hydrogen bond is formed. With several pyrazines and pyrimidines, it has been noticed that benzene causes a larger shift in the N-oxides than in the azines, indicative of a more strongly associated π-complex.[67] The effect of lanthanide shift reagents has been extensively investigated (the observed association constant can be both larger and smaller than in the azines) and its origin has been discussed.[68-71] 4-Picoline 1-oxide has been considered to be as a convenient substrate for revealing the role of a contact (i.e., due to spin delocalization) term in the shift, since the sign of the observed effect is then the same as in stable metal complexes.[71]

The effect of the 2- and 3-pyridyl 1-oxide moieties on an adjacent methylene group shows that heterocyclic *N*-oxides are to be considered as strongly electron-withdrawing substituents, more so than the parent compounds[72] (Tables 6 and 7).

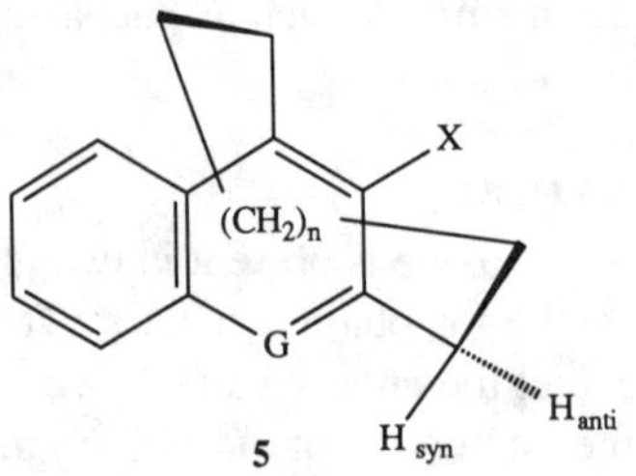

5

$G = N, N \longrightarrow O$

In the ^{13}C spectra, both the α and γ carbons and the carbons in *peri* are shielded with respect to the azines.[10-13,73] The hypothesis has been made that the last effect is due to a steric interaction between the *peri*-C-H bond and the solvent shell of the strongly polar $\geqslant N \rightarrow O$ group, rather than to magnetic anisotropy, a rationalization in accordance with the deshielding observed in the proton and the shielding observed in the carbon spectra.[11] The shielding of the α-carbon is observed not only in azine *N*-oxides, but also in other derivatives, e.g., the *N*-oxides of benzodiazepines.[74]

A large amount of data are also available on ^{15}N[12,75,76] and ^{17}O NMR[16,77-81] spectra as well as ^{14}N and ^{17}O nuclear quadrupole resonance.[13,14,82] The ranges of the observed nitrogen and oxygen chemical shifts are characteristic for the *N*-oxide function, and, in the latter case, substituents cause a large change, as do protic and acidic solvents. Substituents chemical shifts for pyridine 1-oxides have been correlated with other physical constants and with substituent constants, obtaining a more quantitative picture of the dual role (π-donor and π-acceptor) of the $\geqslant N \rightarrow O$ group and of the N-O bond order.[12,18,76-79,81] Localization of the *N*-oxide group in polyazines by means of nitrogen chemical shifts through comparison with the results from additivity rules has proved to be effective in some cases,[76a] but in others the small shifts observed have proven to be unreliable for unambigous assignment.[76b] In accordance with the above-mentioned effects in the proton spectra, steric effects (marked deshielding by 2-alkyl groups and by *peri* protons, e.g., in 4-azaphenanthrene 4-oxide) are noticed in ^{17}O spectra.[80]

The conformational properties of some derivatives of 2-aminopyridine 1-oxide have been studied by ^{1}H NMR[83] (Tables 8 and 9).

D. MASS SPECTRA

Fragmentation under electron impact of *N*-oxides again indicates their stability due to their aromatic character. The molecular ion is often the main peak observed, and is virtually always so under low-potential conditions;[84,85] this contrasts with aliphatic *N*-oxides, in which molecular peaks are weak and important fragments arise from water elimination and, when possible, Cope rearrangement.[84,86] Deoxygenation followed by fragmentation identical to that observed from the corresponding base has been proposed as a diagnostic property for heteroaromatic *N*-oxides.[87] However, later studies showed that this is a complex phenomenon, attributable in large part to thermal deoxygenation in the source (and therefore is not apparent if the sample is vaporized at a low temperature),[88] as well as to the ion-molecule reaction (and therefore is concentration dependent).[89-90]

Thus, while the (M-16)$^+$ peak is a useful indication of the presence of the *N*-oxide group, unimolecular fragmentation of the oxygen atom is not a favored pathway from the radical

TABLE 4
Melting Point of *N*-Oxides, °C

N-oxide	Anhydrous	Hydrate	Picrate
Pyridine	66		179.5
Quinoline	62	54(2H$_2$0)	141
Isoquinoline	105—106	103.5—6.5($^1/_2$ H$_2$0)	165
Phenazine	226.5		

TABLE 5
Bond Distances and Angles in the Crystalline State

Substrate	Bond N-O	Distances N-C$_\alpha$	Å N-X$_\alpha$	Bond angle, °C C$_\alpha$-N-X$_\alpha$	Ref.
Pyridine 1-oxide[a]	1.34	1.34		122	48
Pyridine 1-oxide.HCl	1.37	1.33		127	49
(Pyridine 1-oxide)$_4$Cu(C10$_4$)$_2$	1.34	1.34		123.7	50
2-Hydroxymethylpyridine 1-oxide	1.32	1.36(C$_2$)	1.35(C$_6$)	121.5	52
4-Nitropyridine 1-oxide	1.26	1.38		115.4	53
8-Hydroxyquinoline 1-oxide	1.33	1.35(C$_2$)	1.48(C$_{8a}$)	120.4	51
Quinoline-8-diazonium 1-oxide fluoborate[b]	1.31	1.35(C$_2$)	1.38(C$_{8a}$)	119.6	54
Phenazine 5-oxide	1.24	1.35		119	55
1,6-Dihydroxyphenazine 5-oxide (iodinin)	1.30	1.38	1.36	120.9	56
2-Phenylisatogen	1.26	1.37(C$_2$)	1.46(C$_{7a}$)	110.2	57
4-Methyl-3(*p*-bromophenyl) furoxan	1.24	1.31(C$_3$)	1.43(O)	108.3	58
5-Chlorobenzofuroxan	1.11	1.46(C$_{7a}$)	1.40(O)	103	59a
Benzotrifuroxan	1.23	1.33(C)	1.48(O)	107	60
3-Cyano-4,4,5,5-tetramethyl-4,5-dihydroisoxazole 2-oxide	1.24	1.30(C$_3$)	1.39(O)	111.6	61

[a]   Two discrete forms of the molecule are present in the crystal packing. Average parameters are given.

[b]   The oxygen atom interacts with the α-nitrogen of the diazonium group ("incipient nucleophilic substitution").

cation, which mainly rearranges. Indeed, several of the processes that are usually observed under electron impact, e.g., the loss of CO from the molecular ion, demand previous skeletal rearrangement.[85,89,91-93] It has been observed that in most cases the fragmentation of each *N*-oxide resembles that of the respective photoproducts, and it is reasonable that the molecular ion rearranges along the same pathways as the excited state (e.g., to yield lactams, oxazepines, acylpyrroles; compare with Chapter 4).

Substituents in α greatly affect the mass spectrum, e.g., loss of OH (and McLafferty-type rearrangement when apppropriate) is characteristic of α-alkyl[84,85,94-96] and loss of H of α-aryl derivatives.[85]

As for five-membered rings, the loss of oxygen is often observed, but other fragmentations are typical of each system (e.g., loss of NO for pyrazole dioxides,[97] loss of two NO fragments, and, to a minor extent, retro-"Beirut" cleavage for benzofuroxan);[84,98a] in particular cases, different fragmentations take over completely, e.g., N$_2$ elimination from 1,2,4-triazole 4-oxide.[98b] Negative ion spectra, characterized by the loss of H, O, and OH, and ring fragmentation, have been also reported[99] (Table 10).

TABLE 6

^{1}H Chemical Shifts ($\delta \geqslant N \rightarrow O$) Of N-Oxides in CDC1$_3$, and Increment Observed in the Protonated Form ($\Delta\delta \geqslant N^+$-OH) and in the Corresponding Base ($\Delta\delta \geqslant N$)

N-oxide		1	2	3	4	5	6	7	8
							Position		
Pyridine	a, $\delta \geqslant N \rightarrow O$		8.23	7.30	7.30				
	b, $\Delta\delta \geqslant N^+$-OH		+0.37	+0.42	+0.62				
	c, $\Delta\delta \geqslant N$		+0.36	−0.05	+0.34				
Quinoline	a		8.53	7.27	7.72	7.85	7.62	7.74	8.76
	b		+0.49	+0.37	+0.65	+0.21	+0.18	+0.23	−0.09
	c		+0.38	+0.06	+0.37	−0.07	−0.16	−0.01	−0.63
Isoquinoline	a	8.76		8.13	7.66				
	b	+0.60		+0.27					
	c	+0.47		+0.40	−0.03				
Acridine	a	7.79	7.41	7.65	8.79				8.04[b]
	b	+0.35	+0.29	+0.38	−0.08				+0.98[b]
	c	+0.01	−0.02	+0.03	−0.58				+0.49[b]
1-Azaphenanthrene	a		8.68	7.39					10.85[a]
	c		+0.31	+0.11					1.56[a]

[a] Proton in position 10.
[b] Proton in position 9.

Adapted from Hamm, P. et al. *Helv. Chem. Acta*, 54, 2363, 1971.

TABLE 7

^{13}C Chemical Shifts ($\delta \geqslant N \rightarrow O$) of N-Oxides in CDC1$_3$ and Increment Observed in the Corresponding Bases ($\Delta\delta \geqslant N$)

N-oxide		1	2	3	4	5	6	7	8	4a	8a
						Positions					
Pyridine	a, $\delta \geqslant N \rightarrow O$		139.1	126.4	126.4						
	b, $\Delta\delta \geqslant N$		+9.9	−2.2	+9.8						
Quinoline	a		135.2	121.1	125.3	128.1	128.6	130.0	119.2	130.3	141.2
	b		+14.9	−0.4	+10.2	−0.5		−0.9	+10.2	−2.2	+7.0
Isoquinoline	a	135.9		136.6	124.2	126.6	129.4	128.8	124.7	129.4	128.5
	b	+16.7		+6.5	−3.9				+2.8	+7.2	−0.7
Quinoxaline	a		130.1	146.1		130.2	130.2	131.7	118.9	146.0	137.5
	b		+14.9	+1.1		−0.7		−2.0	+10.6	−2.9	+5.5

Adapted from Günther, H., et al., *Heterocycles*, 11, 337, 1978.

E. VIBRATIONAL AND ROTATIONAL SPECTRA

An intensive (reported molar absorptivities, $\varepsilon \approx 100$ or more) band in the region 1200—1350 cm^{-1} is observed in the infrared spectra of heterocyclic N-oxides, unless it is obscured by bands attributable to substituents, and is ascribed to stretching of the N-O bond.[6,100-106] The frequency and intensity of the band depend the presence of substituents (e.g., electron-withdrawing groups in position 2 and 4 tend to increase the frequency in that the double-bond character is increased; see canonical formula **3**) and on the nature of the solvent (in particular, formation of a hydrogen bond in protonating solvents is apparent); the position of the band has been used to characterize metal complexes of the N-oxides.[107] Linear correlation analysis of the N-O stretching frequency with substituent parameters,[6] as well as the effect of the N-

TABLE 8
^{15}N and ^{17}O Chemical Shifts of Azine N-Oxides

N-oxide	$\delta(^{15}N)^a$	$\delta(^{15}N)^b$	$\delta(^{17}O)$	$\delta(^{17}O)^e$
Pyridine	80.6	129.5	349[c]	274
4-Methoxypyridine	100.2	155.3	312[c]	253
4-Acetylpyridine			389[c]	
4-Nitropyridine	67.2	106.7	414[c]	329
Quinoline	95.3	144.1	343[d]	

a In DMSO, vs. external HNO_3.[75]
b In trifluoroacetic acid, vs. external HNO_3, the protonated form is present.[75]
c In acetonitrile, vs. external H_2O.[81]
d In acetonitrile, vs. external H_2O.[80]
e In water.[81]

TABLE 9
Linear Correlations with ^{14,15}N and ^{17}O Chemical Shifts of Azine N-Oxides

Chemical shift	N-oxide	Correlation	Ref.
^{15}N	Pyridine	Substituent (in 3 and 4) chemical shifts (SCSs) correlates with ^{13}C SCSs of benzenes	18
^{15}N	Pyridine	SCSs (in 3 and 4) correlate with $\sigma_i + \sigma^+_\pi + \sigma^-_\pi$	18
^{15}N	Various	Correlate with ^{13}C of the α-carbon	12
^{15}N	Various	Correlate with IR N-O absorption frequencies	12
^{15}N	Pyridine	SCSs (in 3 and 4) do not correlate with the respective ^{17}O SCSs	18
^{14}N	Various	Correlate with calculated charge (on the basis of INDO/S MO calculations)	76
^{17}O	Pyridine	SCSs (in 3 and 4) correlate with $\sigma_i + \sigma^+_\pi + \sigma^-_\pi$	18,77,78
^{17}O	Pyridine	SCSs (in 4) correlate with O SCSs of anisoles	81
^{17}O	Pyridine	SCSs (in 4) correlate with ionization potentials	79,81
^{17}O	Pyridine	SCSs (in 3) do not correlate with ΔpKa	78

TABLE 10
Relative Abundance (%) of the Peaks in the Mass Spectra of N-Oxides at 70 eV

N-oxide	M^+	$(M-16)^+$	$(M-17)^+$	$(M-27)^+$	$(M-28)^+$	Ref.
Pyridine	75	29	17			84
4-Methylpyridine	100	16	8			84
2-Methylpyridine	69	28	100			84
Quinoline[a]	100	3	4	18	8	85
Isoquinoline[a]	100	10	6	32	5	85
Phenanthridine	100	17	4	7	57	85
Quinoxaline	100	16	8		14	96
Benzofurazan	100	20				98a

a With the ion source at 70°C; with the ion source at 180°C, the $(M-16)^+$ peak is much stronger, 8% or 37% of M^+ for quinoline and 17% or 36% for isoquinoline N-oxides in two different spectrometers.

oxide group on other aromatic vibrations,[108,109] have been reported (e.g., the $\sigma_R°$ value for the substitution of the $\geqslant N\rightarrow O$ group for a CH in the benzene ring has been determined from the integrated intensities of the absorptions at ca. 1600 cm^{-1}, the former group resulting in a donor of considerable strength).[109]

Complete vibrational assignment of the infrared spectra of simple N-oxides has been obtained,[110,111] and Raman,[112,113] as well as microwave, spectra[1,2,4,15,114] have likewise been analyzed.

F. ELECTRONIC SPECTRA

The ultraviolet absorption of pyridine 1-oxide is characterized by a strong (log ϵ > 4) absorption at 255—285 nm (blue shifted in polar solvent), and further in apolar solvents by a weaker (log ϵ ≈ 2) band extending to 350 nm.[103,115,116] In polar solvents, the latter band is strongly blue shifted and almost disappears under the previously mentioned strong absorption, and thus was at first identified as a $n\pi^*$ transition; however, it was later shown that both calculations[22-27,117-120] and experimental analysis[121,122] require that this is a $\pi\pi^*$ band with a strong internal charge-transfer character (a transition from a π orbital that is highly localized on the oxygen atom to one delocalized on the ring). Similar characteristics are observed in the spectra of the other azine *N*-oxides: the first absorption band is considerably to the red with respect to the corresponding azine, but undergoes a strong blue shift in polar solvents, and the spectrum in acids is very near to that of the azinium ion.[116]

Monocyclic and bicyclic *N*-oxides usually fluoresce weakly or not at all (and correspondingly react from the singlet excited state with a high quantum yield, see Chapter 4), but higher members of the series (e.g., phenazine 5-oxide,[123] alloxazine *N*-oxides[124]) fluoresce strongly. Phosphorescence in glass at low temperature has been reported for several *N*-oxides, notably nitro derivatives.[125]

A $\pi\pi^*$ band with a strong internal charge-transfer character and a corresponding blue shift in polar solvents is also characteristic of cyclic nitrones[126] and five-membered heterocycles, e.g., benzofuroxans.[127]

The chiroptical properties of several amines,[128] amino acids,[129,130] and oxiranes[131] containing *N*-oxidoheteroaryl substituents have been studied.

G. ELECTROCHEMICAL MEASUREMENTS

Chemical reduction of pyridine 1-oxide is not as easy as that of aliphatic *N*-oxides (see Chapter 4). Correspondingly, cathodic reduction occurs at a more negative potential (e.g., in water, $E_{1/2}^{red}$ is –0.7 V for dimethylaniline *N*-oxide and –1.41 V for pyridine 1-oxide).[132-136] As expected, fused benzo rings and aza substitution make the reduction easier. *N*-oxides are reduced at a less negative potential than the corresponding azines (e.g., –1.7 for pyridine in water).[132b] In comparing the data, attention should be given to the medium effect, since with *N*-oxides the observed wave is strongly pH dependent, reduction being easier under acidic conditions[132] (this also holds for cyclic nitrones).[137] The radical anions of several simple heteroaromatic *N*-oxides prepared by cathodic reduction have been characterized by electron paramagnetic resonance.[133,134]

A major difference with azines is that heteocyclic *N*-oxides are also oxidized at accessible potentials, and, although the radical cations are less stable than the corresponding anions, in several cases, e.g., phenazine 5,10-dioxide, cyclic voltammetry shows reversible waves for both oxidation and reduction and both radical ions have been characterized.[135,136,138,139] The observed $E_{1/2}$ have been found to linearly correlate with the ionization potential determined by PES (see Section IB) in the gas phase[139] (Table 11).

In view of the importance of redox processes in their methabolism, electrochemical studies of *N*-oxides known as carcinogens have been carried out,[140,141] and a correlation between the biological activity of certain quinoxaline and phenazine *N*-oxides and their reduction potential has been proposed.[142] Polarographic and voltammetric methods have been proved useful for analytical purposes and have been developed, particularly for *N*-oxides used as drugs.[141,143,144]

H. QUALITATIVE AND QUANTITATIVE ANALYSIS

As appears from the foregoing, *N*-oxidation causes characteristic changes in the physical properties of nitrogen heterocycles, and practically all spectroscopic techniques mentioned above are used for the identification and quantitative analysis of *N*-oxides. Further popular techniques include liquid-phase chromatography (e.g., in TLC, low-molecular weight *N*-oxides are much more strongly adsorbed than the corresponding bases, and require more polar

TABLE 11
Half-Way Redox Potential of *N*-Oxides in Acetonitrile by Polarography[a]

N-oxide	Reduction $E_{1/2}$, V vs. SCE	Oxidation $E_{1/2}$, V vs. SCE
Pyridine	−2.297	1.802
Pyrazine	−1.837	2.312
Quinoline	−1.809	1.537
Isoquinoline	−1.946	1.600
Quinoxaline	−1.419	1.971
Acridine	−1.300	1.280
Phenazine	−0.972	1.745

[a] Similar values are obtained by cyclic voltammetry.[136]

Adapted from Miyaraki, H., et al. *Bull. Chem. Soc. Jpn.*, 45, 780, 1972.

— or protic — solvents for elution; the effect is much smaller with the *N*-oxides of polynuclear azines[145]) and redox titration.[147] The oxidizing power of *N*-oxides has been exploited in a color test based on the formation of crystal violet from *N, N*-dimethylaniline on gentle heating under acid conditions.[146]

Color test for N-oxides. Dimethylaniline (0.2 ml), concentrated hydrochloric acid (0.05 ml) and the material to be tested (0.1 g) are boiled in a test tube for 1 min. Ethanol (1 ml) is added to the cooled residue, and the development of an intense blue color indicates a positive test.[146]

III. EQUILIBRIA

A. DISSOCIATION CONSTANTS

Azine *N*-oxides are weak bases, the pKa for the dissociation of the cation being around 1 in water,[150-159] i.e., at a much lower value than observed both with parent azines (≈5) and with aliphatic *N*-oxides (4—5). The difference is essentially due to the stabilization gained by the unprotonated form through a solvation effect, since the gas-phase pKa of heteroaromatic *N*-oxides has been found to be very near to that of the corresponding bases.[160] The attending thermodynamic parameters and their (linear) correlation with pKa have been determined for several compounds, [153,155,156] and the problems with a free energy-enthalpy correlation for such weak bases have been discussed (see Table 12).[156] For the properties of *N*-oxide salts, see Chapter 4.

Various correlations with susbtituent parameters of the effect on the basicity of pyridine 1-oxides have been evaluated; LArSr treatment appears to be the best suited, since it allows the description of the dual resonance effect of the *N*-oxide function. Indeed, application of the three-parameter equation $\log k/k_0 = \rho \, (\sigma^\circ + r^+ \Delta \, \bar{\sigma}_R^+ + r^- \Delta \, \bar{\sigma}_R^-)$ to *meta* and *para* pyridine 1-oxide is satisfactory (coefficient 0.997) with the values $\log k/k_0 = 2.60 \, (\sigma^\circ + 0.59 \, \Delta \, \bar{\sigma}_R^+ + 0.22 \, \Delta \, \bar{\sigma}_R^-) - 0.06$ (the overall ρ value is identical to the one obtained for the *meta* substituents only). The effect is even more pronounced in the gas phase (solvation smooths down the substitutent effect in the same way that it diminishes the basicity, see above).[151,160] The contribution of canonical forms **2** and **3** is apparent in the marked increase *both* in r⁻ (by 0.7) *and* in r⁺ (by 0.39) with respect to the corresponding correlation for pyridines. For the treatment of the substituent effect on the basis of spectral characteristics, see Section II.

Other molecular sites may compete with the *N*-oxide function, e.g., in pyrazine and quinoxaline mono *N*-oxides, the first protonation occurs at the unoxidized nitrogen atom;[161,162] however, in aminopyridine 1-oxides, the oxygen atom is protonated first (pK$_a$ ≈ 2 to 4).[149,163]

TABLE 12
pKa Values of *N*-Oxide Conjugate Acids in Water at 25°C

N-oxide	pKa	Ref.
Pyridine[a]	0.79	155
4-Methoxypyridine	2.05	151
4-Nitropyridine	−1.51	151
Quinoline	0.86	158
4-Nitroquinoline	−1.39	158
Isoquinoline	1.01	148
Acridine	1.37	159
Pyrimidine	−0.5	157
Pyrazine	−0.15	161
Quinoxaline	0.25	161
Phenazine	−4.88	155

[a] pKa in nitromethane 8.75.[152]

The stability of the 1,10-phenanthroline 1-oxide cation has been imputed to the hydrogen bond shown.[164] The anion **7** is in pH-dependent equilibrium with the ring-closed cation **6**, and since the pK of the first equilibrium is lower than the latter one, the neutral species is always metastable[165] (a related example of direct anion-cation interconversion, again based on a reversible ring opening, is the reaction of *N*-alkoxypyridinum cations with bases; see Chapter 4).

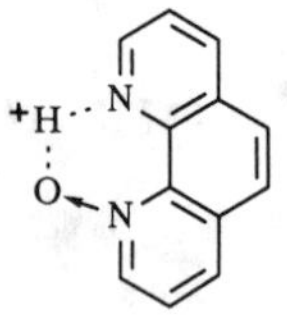

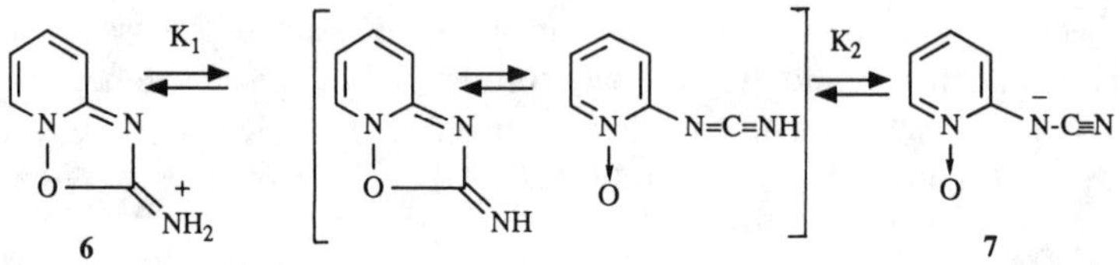

Linear correlation analyses have also been applied to the protolytic equilibria of substitutent groups, the electron-withdrawing effect of the *N*-oxide function resulting in an increased acidity,[166,167] e.g., of 3- and 4-pyridinecarboxy acid 1-oxides (not the 2-isomer, due to intramolecular hydrogen bonding; see below).

B. FORMATION OF COMPLEXES

1. Intramolecular Association

Acidic substituents in the α (or in *peri*) position form intramolecular hydrogen bonds with the *N*-oxide group. This is the case for 2-pyridinecarboxy acids,[168] carboxyamides,[169] and amidines;[170] for 8-hydroxyquinoline;[171] and for related compounds, including nonconjugated chromophores, as in some heteroaryl-acetic acid,[172] -methanol,[172] and -alkylsulfonic acid derivatives[173] (see, e.g., formulae **8—10,** and the following section for tautomeric hydroxy and thio derivatives). Association is demonstrated by the spectra, e.g., by a more intense (in nonconjugated species such as **8**[172] and in aliphatic derivatives such as **11**)[174] or less intense (in conjugated compounds such as **12**)[174] continuum in the infrared spectra of carboxylic acids, and shows the contribution of both structures ≥NO••H-O- and ≥N⁺-O-H••⁻O-. In turn, the polarity of the solvents influences the equilibrium between the two forms in a manner that is proportional to the polarizability of the proton (see the following section).[175] The hydrogen bond does not need to directly involve the ≥N→O group; for example, in products **13** and **14** the intramolecular bond appears to involve the π cloud.[176]

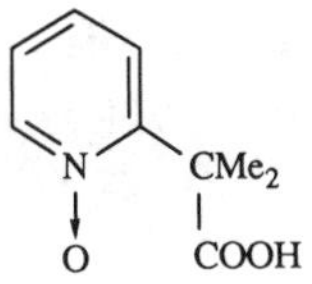

8 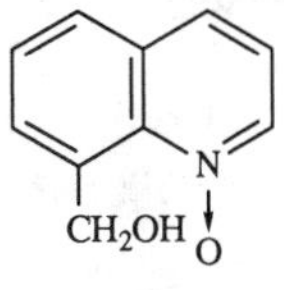   9 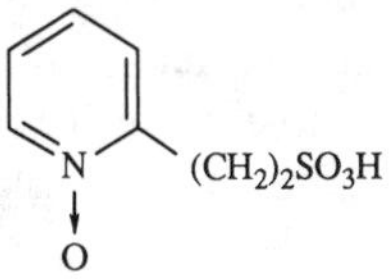   10

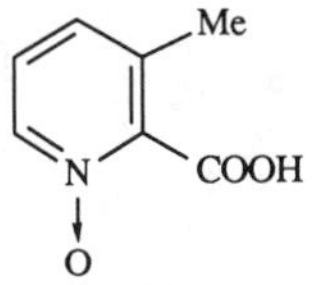

12 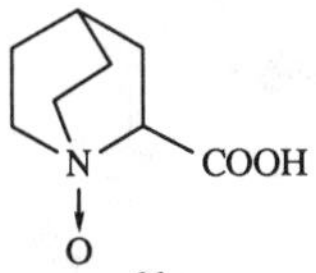   11

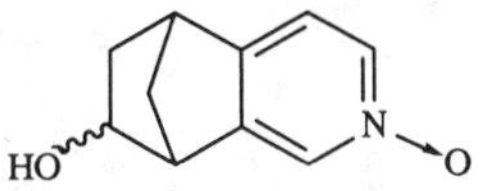

13 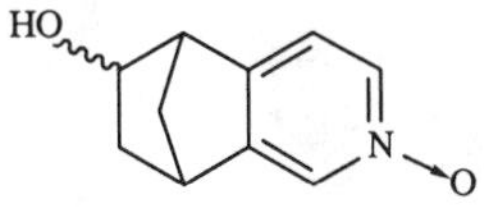   14

2. Intermolecular Association

Heterocyclic *N*-oxides are weakly associated in solution through a dipole-dipole interaction.[177] More important is intermolecular association through hydrogen bonding for compounds containing acidic functions, e.g., alcohols, phenols, and carboxylic acids (both inter- and intramolecular bonds in the appropriate cases, see above), or *N*-hydroxylated derivatives such as 1-hydroxyimidazole 3-oxide.[178] Some biindolenine 1,1'-dioxides show self-association.[179]

Hydroxylic acids (the most studied being haloacetic acids,[180-183] sulfonic acids,[184] phenols,[185-187] and alcohols[188]) strongly associate with *N*-oxides. These equilibria have been the subject of extensive spectroscopic work, the main aim (and the subject of some controversy) being to establish whether localized hydrogen bonding, characterized by a single-minimum well in the potential function, or delocalized, double-minimum (and fast-exchanging) hydrogen bonding is involved. The latter rationalization would explain the strong polarizability of the bond. UV measurements on the pyridine 1-oxide solution are useful for indirectly determining the association constants between alcohols and weak bases.[188b]

It is possible that association with phenols also involves a π contribution, in addition to H-bonding (see Section IIC for NMR evidence for weak association with benzene). Besides proton acids, other molecules that have been found to form complexes with heterocyclic *N*-oxides include halogens,[189,190] boron trifluoride,[191] and other Lewis acids.[192,193] The association between known carcinogens, particularly 4-nitroquinoline 1-oxide, and molecules of biological relevance, e.g., DNA, has been studied and related to the mechanism of action.[194-196] For stable metal complexes and the role of charge-transfer interaction in the chemistry of *N*-oxides, see Chapter 5.

C. TAUTOMERISM

The tautomerism of nitrogen heterocycles carrying a hydroxy-, amino, or thio group in the α or γ position is well known. The corresponding *N*-oxides are likewise involved in a tautomeric equilibrium with the *N*-hydroxy derivatives (formulae **15—18**). 2-Hydroxy pyridine 1-oxides are in the lactamic form (hydroxamic acids), and the same holds for the corresponding sulfurated derivatives. Notice that, since both forms are expected to be involved in a strong intramolecular hydrogen bond, the distinction may be not unambigous. In the corresponding 4-substituted compounds, both the oxo (or thione) tautomer and the hydroxy (or thiol) tautomer are present (in comparable amounts in aqueous solution).[149,197-199] The amino derivatives, on the contrary, exist as such.[149] The amino tautomers prevail more over the imino form in the case of the 1-oxides than do the corresponding pyridines. This has been attributed to the added stabilization of canonical form **19** (since here the negative charge is on the oxygen atom, rather than on the nitrogen, as in the pyridine) and the destabilization of form **20** (by the unfavorable inductive effect of the hydroxyl group). For the same reason, the oxo (and thione) tautomers are not as favored as they are in the pyridines, where they largely prevail.

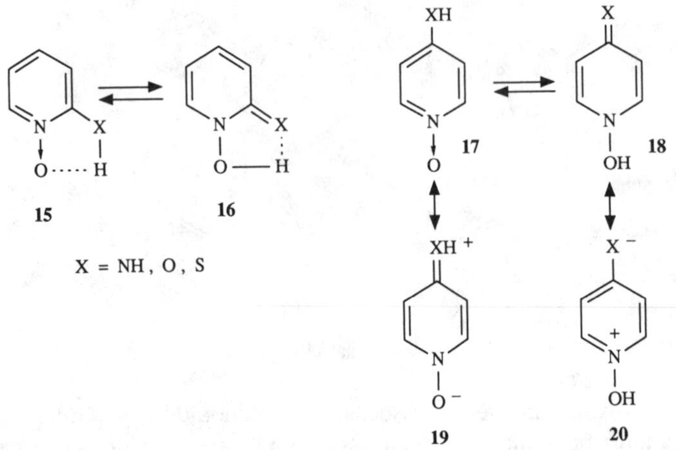

X = NH, O, S

Similar observations have been made for other azine *N*-oxides.[200] Under neutral conditions in guanine 3-oxide (**21**), the 3-hydroxy tautomer predominates, and the ionization sequence is 3, 9, 1.[201] The alloxazine 5,10-dioxide **22** exists as such, and not as a N(10)- or C(2)-hydroxy tautomer.[202] As for five-membered derivatives, it has been found that indolenine 1-oxide (**23**) is favored in a solvent capable of forming hydrogen bonds, while the *N*-hydroxy tautomer predominates in other cases.[203] In benzimidazole[204] and benzotriazole[205] derivatives, the 1-oxide again predominate in water, with the 1-hydroxy tautomer being important in other solvents, although substituents may affect the balance. Triphenylimidazole 3-oxide[206,206] and 1,2,4-triazole 4-oxide[207] exist predominantly as such.

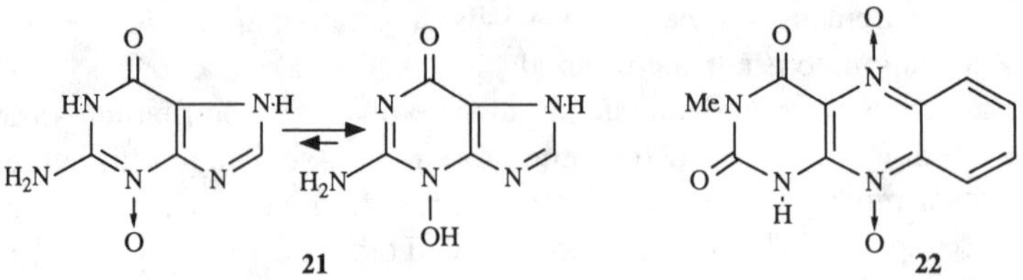

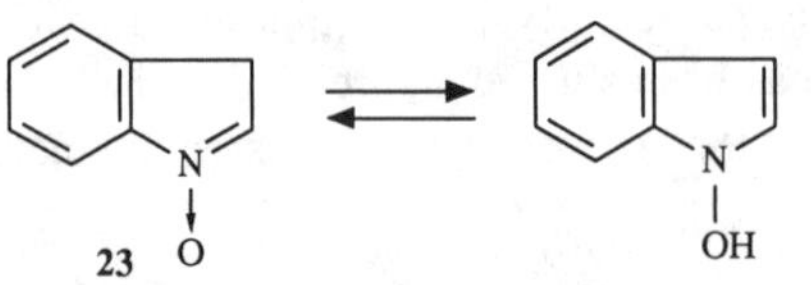

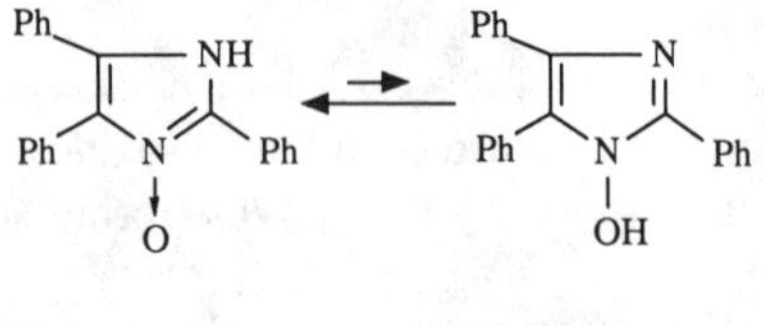

An important tautomerism involves furoxans and benzofuroxans. Indeed, the easy interconversion of substituted benzofuroxans at room temperature has long hindered the assignment of an unsymmetrical formula for the heterocyclic ring. Spectroscopic studies, in particular NMR experiments on the various nuclei, provided evidence for the tautomerism.[208-212] The ¹H NMR spectrum of benzofuroxan is unsymmetrical at low temperature, but collapses to an AA′ BB′ system at room temperature, and the equilibrium probably involves the transient formation of 1,2-dinitrosobenzene (experiments with ¹³C NMR lead to the same conclusion;[17] see Chapter 4). For most substituted benzofuroxans, the free-energy difference between tautomers is only 1—2 kJM⁻¹, and the activation ΔG is ≤50—60 kJM⁻¹.[210,211] In the **24**—**25** equilibrium, electron-withdrawing substituents favor the former structure and electron donors favor the latter one.[210]

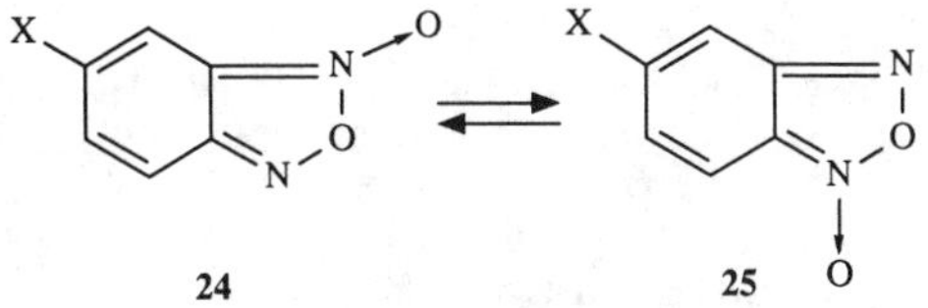

24   **25**

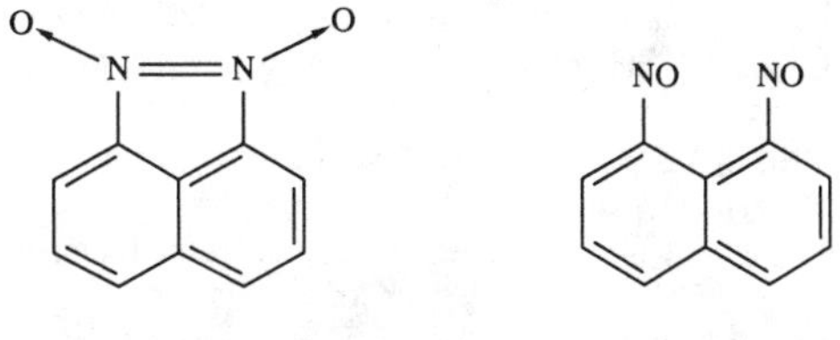

26

Heterocyclic derivatives containing the azo dioxide moiety are also potentially tautomeric with dinitroso derivatives. In the case of benz[c,d]indazole, **26**, the ring-closed structure, rather than dinitrosonaphthalene, is present in both the crystalline state and in solution (see also Section IC for photoelectron spectra).[213]

REFERENCES

1. **Snerling, O., Nilsen, C.J., Nygaard, L., Pedersen, E.J., and Sørensen, G.O.,** Microwave spectra of carbon-13 and nitrogen-15 labeled pyridine *N*-oxides and a preliminary ring structure, *J. Mol. Struct.,* 27, 205, 1975.
2. **Brown, R.D., Burden, F.R., and Garland, W.,** Microwave spectrum and dipole moment of pyridine *N*-oxide, *Chem. Phys. Lett.,* 7, 461, 1970.
3. **Chiang, J.F.,** Molecular structure of pyridine *N*-oxide, *J. Chem. Phys.,* 61, 1280, 1974.
4. **Mata, F., Quintana, M.J., and Sørensen, G.O.,** Microwave spectra of pyridine and monodeuteropyridines. Revised molecular structure of pyridine, *J. Mol. Struct.,* 42, 1, 1977.
5. **Vögtle, F. and Risler, H.,** Spatial requirements of pyridine *N*-oxide oxygen, *Angew. Chem. i.e.e.,* 11, 727, 1972.
6. **Shindo, H.,** Infrared spectra of subtituted pyridine l-oxide, *Chem. Pharm. Bull.,* 6, 117, 1958.
7. **Linton, E.P.,** The dipole moments of amine oxides, *J. Am. Chem. Soc.,* 62, 1945, 1940.
8. **Katritzky, A.R., Randall, E.W., and Sutton, L.E.,** The electric dipole moments of a series of 4-substituted pyridine and pyridine l-oxides, *J. Chem. Soc.,* 1769, 1957.
9. **Hamm, P. and Philpsborn, W.,** Proton resonances of aromatic *N*-oxides. Calculation of the chemical shifts caused by the field effect of the N-O group, *Helv. Chem. Acta,* 54, 2363, 1971.
10. **Anet, F.A.L. and Yavari, I.,** Carbon-13 nuclear magnetic resonance study of pyridine *N*-oxide, *J. Org. Chem.,* 41, 3589, 1976.
11. **Günther, H. and Gronenborn, A.,** ¹³C NMR spectra of aromatic *N*-oxides, *Heterocycles,* 11, 337, 1978.
12. **Paudler, W.W. and Jovanovic, M.V.,** Backdonation and interelationships between ¹⁵N, ¹³C chemical shifts and infrared absorption frequencies in heterocycle *N*-oxides, *Heterocycles,* 19, 93, 1982.

13. **Wayciesjes, P.M., Janes, N., Ganapathy, S., Hiyama, Y., Brown, T.L., and Oldfield, E.,** Nitrogen and oxygen nuclear quadrupole and nuclear magnetic resonance spectroscopic study of N-O bonding in pyridine 1-oxides, *Magn. Res. Chem.,* 23, 315, 1985.

14. **Cheng, C.P. and Brown, T.L.,** Oxygen-17 nuclear quadrupole double resonance spectroscopy. 3. Results for N-O, P-O and S-O bonds, *J. Am. Chem. Soc.,* 102, 6418, 1980.

15. **Chiang, J.F. and Song, J.J.,** Molecular structure of 4-nitro, 4-methyl and 4-chloro-pyridine-*N*-oxides, *J. Mol. Struct.,* 96, 151, 1982.

16. **Boykin, D.W., Balakrishnan, P., and Baumstark, A.L.,** Natural abundance ^{17}O NMR spectroscopy of heterocyclic *N*-oxides and di *N*-oxides. Structural effects, *J. Heterocycl. Chem.,* 22, 981, 1985.

17. **Anet, F.A.L. and Yavari, I.,** ^{13}C NMR spectra of benzofuroxan and related compounds, *Org. Magn. Reson.,* 8, 158, 1976.

18. **Sawada, M., Takai, Y., Yamano, S., Misumi, S., Hanafusa, T., and Tsuno, Y.,** Dual resonance functionality in pyridine 1-oxides. A double multinuclear NMR approach, *J. Org. Chem.,* 53, 191, 1988.

19. **Bird, C.W.,** The application of a new aromaticity index to six-membered ring heterocycles, *Tetrahedron,* 42, 89, 1986.

20. **Matsuoka, T., Shinada, M., Suematsu, F., Harano, K., and Hisano, T.,** Effect of aromaticity on 1,3-dipolar cycloaddition reactivity of substituted pyridine *N*-oxides and preparation of oxazolo[4,5-*b*]pyridine derivatives, *Chem. Pharm. Bull.,* 32, 2077, 1984.

21. **Devanneaux, J. and Labarre, J.F.,** Effect of the formation of donor-acceptor bond on the aromaticity of a planar hexagonal heterocyclic. Pyridine and its *N*-oxide, *J. Chim. Phys. Physicochim. Biol.,* 66, 1174, 1969.

22. **Sholz, M.,** The electronic structure of pyridine *N*-oxide, *J. Prakt. Chem.,* 323, 571, 1981.

23. **Ochiai, E.,** *Aromatic Amine Oxides,* Elsevier, Amsterdam, 1967, chap. 4.

24. **Ha, T.K.,** Ab initio SCF and CI study of the electronic spectrum of pyridine *N*-oxide, *Theor. Chim. Acta,* 43, 337, 1977.

25. **Leibovici, C. and Streith, J.,** A CNDO-CI analysis of ground and excited state properties of pyridine-*N*-oxide, *Tetrahedron Lett.,* 387, 1971.

26. **Evleth, E.M.,** The electronic structure of pyridine *N*-oxide and related compounds, *Theor. Chim. Acta,* 11, 145, 1968.

27. **Kobinata, S. and Nakagura, S.,** Electronic structures of pyridine *N*-oxide and 4-nitrosopyridine *N*-oxide, *Theor Chim. Acta,* 14, 415, 1969.

28. **Hinchliffe, A.,** Comparison of the electronic structure and properties of pyridine, pyridinium cation, and pyridine *N*-oxide, *J. Mol. Struct.,* 41, 159, 1977.

29. **Chadha, R.,** Molecular orbital study of some aromatic *N*-oxide systems, *Indian J. Chem.,* 25A, 7, 1986.

30. **Kulkarni, G.V., Ray, A., and Patel, C.C.,** Molecular orbital studies of some 4-substituted pyridine *N*-oxides and 4-nitroquinoline *N*-oxide, *J. Mol. Struct.,* 71, 253, 1981.

31. **Yamakawa, M., Kubota, T., and Akazawa, H.,** Electronic structure of aromatic amine *N*-oxides, *Theor. Chim. Acta,* 15, 244, 1969.

32. **Scholz, M., Götze, R., Kluge, G., Klasinc, L., and Novak, I.,** An interpretation of the photo-electron spectra of substituted pyridine *N*-oxide, *Z. Phys. Chem.(Leipzig),* 262, 897, 1981.

33. **Maier, J.P. and Muller, J.F.,** Ionisation energies of pyridine *N*-oxides determined by photoelectron spectroscopy, *J. Chem. Soc., Faraday Trans. 2,* 70, 1991, 1974.

34. **Weiner, M.A. and Lattman, M.,** Photoelectron spectra of 4-substituted pyridine *N*-oxides, *Tetrahedron Lett.,* 1709, 1974.

35. **Maier, J.P., Muller, J. F., and Kubota, T.,** Ionisation energies and the electronic structures of the *N*-oxides of diazabenzenes, *Helv. Chim. Acta,* 58, 1634, 1975.

36. **Maier, J.P., Muller, J.F., Kubota, T., and Yamakawa, M.,** Ionisation energies and the electronic structure of the *N*-oxides of azanaphthalenes and azaanthracenes, *Helv. Chim. Acta,* 58, 1641, 1975.

37. **Albini, A. and Mark, F.,** Photoelectron spectra of phenazine *N*-oxide and some of its derivatives, *J. Chem. Soc., Faraday Trans. 2,* 72, 463, 1976.

38. **Abramovitch, R.A. and Smith, E.M.,** Pyridine-1-oxides, in *Pyridine and its Derivatives,* Abramovitch, R.A, Ed., Wiley, New York, 1974, 46.

39. **Klopman, G.,** Chemical reactivity and the concept of charge- and frontier-controlled reactions, *J. Am. Chem. Soc.,* 90, 223, 1968.

40. **Houk, K.N., Caramella, P., Munchausen, L.L., Chang, Y.M., Battaglia, A., Sims, J., and Kaufman, D.C.,** Photoelectron spectra of nitrones and nitrile oxides, *J. Electron Spectros. Relat. Phenom.,* 10, 441, 1977.

41. **Distefano, G., Spunta, G., Colonna, F.P., and Pignataro, S.,** Transmission of electronic effects in substituted pyridine-*N*-oxides studied by ESCA. *Z. Naturforsch,* 31A, 856, 1976.

42. a. **Batich, C.D. and Donald, D.S.,** X-ray photoelectron spectroscopy of nitroso compounds: relative ionicity of the closed and open forms, *J. Am. Chem. Soc.,* 106, 2758, 1984; b. **Sjoegren, B., Freund, H.J., Salaneck, W.R., and Bigolow, R.W.,** Core ionization of nitrosobenzene-dimer compounds: phenazon di-*N*-oxide, *Chem. Phys.,* 118, 101, 1987; c. **Bus, J.,** X-ray photoelectron spectrum of benzotris[*c*]furazan 2-oxide (hexanitrosobenzene), *Recl. Trav. Chim. Pays Bas,* 91, 552, 1972.

43. **Sharpe, A.N. and Walker, S.,** Electric dipole moments of substituted pyridines, pyridine 1-oxides and nitrobenzenes, *J. Chem. Soc.,* 4522, 1961.

44. **Bax, C.M., Katritzky, A.R., and Sutton, L.E.,** The electric dipole moments of pyridine- and trimethylamine-boron trihydride and trihalide complexes, *J. Chem. Soc.,* 1258, 1958.

45. **Schaofeng, L. and Pilcher, G.,** Enthalpy of formation of pyridine-*N*-oxide: the dissociation enthalpy of the nitrogen-oxygen bond, *J. Chem. Thermodyn.,* 20, 463, 1988.

46. **Kirchner, J.J., Acree, W.E., Pilcher, G., and Shaofeng, L.,** Enthalpies of combustion of four *N*-phenylmethylene benzenamine *N*-oxide derivatives, of *N*-phenylmethylene benzenamine, and of trans-diphenyldiazene *N*-oxide: the dissociation ethalpy of the (N-O) bonds, *J. Chem. Thermodyn.,* 18, 793, 1986.

47. a. **Mirosnichenko, E.A. and Lebedev, Y.A.,** N → O bond energy in 3-nitroisoxazoline 2-oxide, *Khim. Geterotsikl. Soedin.,* 963, 1969; b. **Matyushin, Y.N., Pepekin, V.I., Golova, S.P., Godovikova, T.I., and Khmeltskii, L.I.,** Enthalpies of formation of dimethylfuroxan and dimethylfurazan, *Izv. Akad. Nauk. SSSR, Ser. Khim.,* 181, 1971; c. **Pepekin, V.I., Matyushin, Y.N., Feshchenko, A.G., Smirnov, S.P., and Apin, A.Y.,** Dissociation energy of the N-O bond in benzofurazan 1-oxide, *Dokl Akad. Nauk SSSR,* 202, 91, 1972; d. **Solomonov, B.N., Antipin, I.S., Novikov, V.B., and Konovalov, A.I.,** Enthalpies of dissolution of *N*-pyridine oxide, *Dokl. Akad. Nauk SSSR,* 239, 607, 1978.

48. **Ülkü, D., Huddle, B.P., and Morrow, J.C.,** The crystal structure of pyridine 1-oxide, *Acta Cryst.,* B27, 432, 1971.

49. **Tsoucaris, G.,** Structure of pyridine oxide hydrochloride, *Acta Cryst.,* 14, 914, 1961.

50. **Lee, J.D., Brown, D.S., and Melson, B.C.A.,** The crystal structure of tetra(pyridine 1-oxide) copper(II) perchlorate, *Acta Cryst.,* B25, 1378, 1969.

51. **Desiderato, R., Terry, J.C., Freeman, G.R., and Levy, H.A.,** The molecular and crystal structure of 8-hydroxyquinoline-*N*-oxide, *Acta Cryst.,* B27, 2443, 1971.

52. **Desiderato, R. and Terry, J.C.,** The crystal structure of 2-hydroxymethylpyridine *N*-oxide, *J. Heterocycl. Chem.,* 8, 617, 1971.

53. **Eichhorn, E.L.,** On the structure of 4-nitropyridine-*N*-oxide, *Acta Cryst.,* 9, 787, 1956.

54. **Wallis, J.D. and Dunitz, J.D.,** Incipient nucleophilic attack on a nitrogen-nitrogen triple bond: crystal structure of quinoline-8-diazonium-1-oxide tetrafluoborate at 95K, *J. Chem. Soc., Chem. Commun.,* 671, 1984.

55. **Curti, R., Riganti, V., and Locchi, S.,** The crystal structure of *N*-oxyphenazine, *Acta Cryst.,* 14, 133, 1961.

56. **Hanson, A.W. and Huml, K.,** The crystal structure of iodinin, *Acta Cryst.,* B25, 1766, 1969.

57. **Adams, D.B., Hooper, M., Swain, C.J., Raper, E.S., and Stoddart, B.,** Isatogens: crystal structure, electron density calculations, and ^{13}C nuclear megnetic resonance spectra, *J. Chem. Soc., Perkin Trans.,* 1, 1005, 1986.

58. **Calleri, M., Ferraris, G., and Viterbo, D.,** The crystal and molecular structure of the 5-oxide of 4-methyl-3-(*p*-bromophenyl)-1,2,5oxadiazole, *Acta Cryst.,* B25, 1133, 1969.

59. a. **Britton, D. and Noland, W.E.,** The crystal and molecular structure of 5-chlorobenzofurazan 1-oxide and 5-bromobenzofurazan 1-oxide, *J. Org. Chem.,* 27, 3218, 1962; b. **Bricton, D. and Olson, J.M.,** Benzofurazan 1-oxide, *Acta Cryst.,* B35, 3076, 1979.

60. **Cady, H.H., Larson, A.C., and Cromer, D.T.,** The crystal structure of benzotrifuroxan, *Acta Cryst.,* 20, 336, 1966.

61. **Clifford, G.J. and Gilardi, R.,** Structure of a substituted isoxazoline, *Acta Cryst., Cryst. Struct. Commun.,* C43, 362, 1987.

62. **Casteel, J.F. and Sears, P.G.,** Dielectric constants, viscosities, and related physical properties of four liquid pyridine-*N*-oxides at several temperatures, *J. Chem. Eng. Data,* 19, 303, 1974.

63. **Abramovitch, R.A. and Davis, J.B.,** Proton magnetic resonance spectra of pyridine 1-oxides and their conjugate acids, *J. Chem. Soc. (B),* 1137, 1966.

64. **Wamsler, T., Nielsen, J.T., Pedersen, E.J., and Schaumburg, K.,** NMR studies of pyridine-*N*-oxide. Determination of spectroscopic constants from [^{15}N]-, [4-^{2}H]-, and the parent species, *J. Magn. Res.,* 31, 177, 1978.

65. **Ohta, A., Akita, Y., and Takagai, C.,** Proton magnetic resonance spectra of some alkylpyridine and alkylpyrazine *N*-oxides, *Heterocycles,* 6, 1881, 1977.

66. a. **Parham, W.E., Olson, P.E., and Reddy, K.R.,** 1,3-Bridged aromatic systems. X. Stereospecific reductions with lithium aluminum deuteride, *J. Org. Chem.,* 39, 2432, 1974; b. **Parham, W.E., Sloan, K.B., Reddy, K.R., and Olson, P.E.,** Stereochemical configuration of substituents in the 1-position of 12,13-benzo-16-chloro [10](2,4)pyridinophanes, *J. Org. Chem.,* 38, 927, 1973.

67. **Paudler, W.W. and Humphrey, S.A.,** Solvent effects in nuclear magnetic resonance spectra of some pyrazines, pyrimidines, and their *N*-oxides, *Org. Magn. Res.,* 3, 217, 1971.

68. **Horrocks, W.D. and Sipe, J.P.,** Lanthanide shift reagents. Survey, *J. Am. Chem. Soc.,* 93, 6800, 1971.

69. **Fletton, R.A., Green, G.F.H., and Page, J.E.,** Lanthanide-induced displacements in the NMR spectra of simple heterocycle *N*-oxides, *Chem. Ind.,* 167, 1972.

70. **Rackam, D.M.,** Equilibrium binding constants for europium shift reagent with *N*-oxide (NO), sulfoxide (SO), and phosphate (PO) functions, *Spectrosc. Lett.,* 13, 513, 1980.

71. a. **Johnson, B.F.G., Lewis, J., McArdle, P., and Norton, J.R.,** Applications of lanthanoid shift reagents to carbon-13 and proton nuclear magnetic resonance spectroscopy, *J. Chem. Soc., Dalton Trans.,* 1253, 1974; b. **Tori, K., Yoshimura, Y., Kainosho, M., and Ajisaka, K.,** Evidence for the presence of contact term contribution to lanthanide induced shifts of ^{1}H and ^{13}C NMR spectra of pyridine N-oxides, *Tetrahedron Lett.,* 1573, 1973.

72. **Barchiesi, E., Bradamante, S., Carfagna, C., Ferraccioli, R., and Pagani, G.A.,** Evaluation of the polar-inductive and mesomeric effects exerted on contigous functionalities by N-oxidopyridinium group, *J. Chem. Soc., Perkin Trans.,* 2, 1009, 1987.

73. **Sojka, S.A., Dinan, F.J., and Kolarczyk, R.,** Carbon-13 nuclear magnetic resonance spectra of substituted pyridine N-oxides, *J. Org. Chem.,* 44, 307, 1979.

74. **Singh, S.P., Parmar, S.S., Farnum, S.A., and Sternberg, V.I.,** Fourier transform ^{13}C NMR analysis of benzodiazepines, *J. Heterocycl. Chem.,* 15, 1083, 1978.

75. **Yavari, I. and Roberts, J.D.,** Nitrogen-15 nuclear magnetic resonance spectroscopy. Pyridine N-oxides and quinoline N-oxides, *Org. Magn. Reson.,* 12, 87, 1979.

76. a. **Witanowski, M., Stefaniak, L., Kamienski, B., and Webb, G.A.,** Localization of N-oxide groups by means of nitrogen chemical shifts, *Org. Magn. Res.,* 14, 305, 1980; b. **Cowden, W.B. and Waring, P.,** Can nitrogen-15 NMR be used to determine the site of N-oxidation of pyrimidine-2,4-diamine?, *Aust. J. Chem.,* 34, 1539, 1981.

77. **Sawada, M., Takai, Y., Kimura, S., Misumi, S., and Tsuno, Y.,** A ^{17}O NMR study. Substituent chemical shifts of 4-substituted pyridine 1-oxides in DMSO: importance of dual enhanced resonance contributions with pi-donor and pi-acceptor substituents, *Tetrahedron Lett.,* 3013, 1986.

78. **Sawada, M., Takai, Y., Kimura, S., Yamano, S., Misumi, S., Hanafusa, T., and Tsuno, Y.,** Oxygen-17 NMR spectroscopy; σ_I, dependence for the substituent chemical shifts of 3-substituted pyridine 1-oxides, *Tetrahedron Lett.,* 5649, 1986.

79. **Joergensen, K.A.,** On the correlation of ionization potentials and ^{17}O chemical shifts in 4-substituted pyridine-N-oxides and 4-substituted N-(benzylidene)phenylamine-N-oxides, *Chem. Phys.,* 114, 443, 1987.

80. **Boykin, D.W., Balakrishnan, P., and Baumstark, A.L.,** ^{17}O NMR spectroscopy of heterocycles. Steric effects for N-oxides, *Magn. Res. Chem.,* 23, 695, 1985.

81. **Boykin, D.W., Barmstark, P.G., and Balakrishnan, P.,** ^{17}O NMR spectroscopy of 4-substituted pyridine N-oxides: substituent and solvent effects, *Magn. Res. Chem.,* 23, 276, 1985.

82. **Stefaniak, L.,** Nitrogen-14 nuclear magnetic resonance of azine-N-oxides, *Spectrochim. Acta,* 32A, 345, 1976.

83. **Tortorella, V., Bettoni, G., and Sciacovelli, O.,** NMR investigation of the conformational properties of N-[2-pyridyl-N-oxide]-amino derivatives, *Tetrahedron,* 29, 1345, 1973.

84. **Bild, N. and Hesse, M.,** Mass spectra of N-oxides, *Helv. Chim. Acta,* 50, 1885, 1967.

85. **Buchardt, O., Duffield, A.M., and Shapiro, R.H.,** Mass spectra of quinoline and isoquinoline N-oxides, *Tetrahedron,* 24, 3139, 1968.

86. **Schueller, D. and Harke, H.P.,** Mass spectra of nicotine N-oxides and related tetrahydroxazines, *Org. Mass. Spectrom,* 7, 839, 1973.

87. **Grigg, R. and Odell, B.G.,** Mass spectra of N-oxides, *J. Chem. Soc. (B),* 218, 1966.

88. **Duffield, A.M. and Buchardt, O.,** Thermal fragmentation of quinoline and isoquinoline N-oxides in the ion source of a mass spectrometer, *Acta Chem. Scand.,* 26, 2423, 1972.

89. **Simonotti, L., Facchetti, S., Bettinetti, G.F., and Albini, A.,** Phenazine 5-oxide, molecular ion fragmentation and isomerization under electron impact, *Gazz. Chim. Ital.,* 106, 49, 1976.

90. **Uchiba, M.,** Addition reaction with 2-methyl-2-nitrosopropane and pressure dependance of ionized picoline N-oxides in the gas phase, *J. Chem. Soc., Perkin Trans.,* 2, 1349, 1982.

91. **Bowie, J.H., Cooks, R.G., Jamieson, N.C., and Lewis, G.E.,** Skeletal rearrangement fragments in the mass spectra of aromatic N-oxides, *Aust. J. Chem.,* 20, 2545, 1967.

92. **Kubo, A., Sakai, S., Yamada, S., Yokoe, I., Kaneko, C., Tatematsu, A., Yoshizumi, H., Hayashi, E., and Nakata, H.,** Novel skeletal rearrangement of aromatic N-oxides upon electron impact, *Chem. Pharm. Bull.,* 15, 1079, 1967.

93. **Kubo, A., Sakai, S., Yamada, S., Yokoe, I., and Kaneko, C.,** Mass spectra of azanaphthalene N-oxides and their photochemical reaction products, *Chem. Pharm. Bull.,* 16, 1533, 1968.

94. **Tatematsu, A., Yoshizumi, H., Hayashi, B., and Nakata, H.,** M-16 and M-17 ions from aromatic N-oxides on electron impact, *Tetrahedron Lett.,* 2985, 1967.

95. **Lightner, D.A., Nicoletti, R., Quistad, G.B., and Irwin, E.,** Mass spectral fragmentations of alkylpyridine N-oxides, *Org. Mass. Spectrom.,* 4, 571, 1970.

96. **Bartoszek, M., Salzwedel, D., Stumm, G., and Nilas, H.J.,** OH· loss and molecular rearrangements of quinoxaline N-oxides studied by interpretation of mass spectra and application of linear discriminant analysis, *Org. Mass. Spectrom.,* 22, 259, 1987.

97. **Kotali, A., Papageorgiu, V.P., and Tsoungas, P.G.,** Electron-impact mass spectra of pyrazole and pyrazoline 1,2-dioxides. A comparative study with related systems, *Org. Mass. Spectrom,* 21, 435, 1986.

98. a. **Dyall, L.K.,** Mass spectral fragmentation of benzofurazan l-oxide, *Org. Mass Spectrom.,* 22, 519, 1987; b. **Becker, H.G.O., Beyer, D., and Timpe, H.J.,** Mass spectra of 1,2,4-triazole 4-*N*-oxides, *J. Prakt. Chem.,* 312, 869, 1971.

99. a. **Fal'ko, V.S., Dzhemilev, U.M., Khvostenko, V.I., and Tolstikov, G.A.,** Mass spectrometry of negative ions of picoline *N*-oxides, *Khim. Geterotsikl. Soedin.,* 661, 1974; b. **Paudler, W.W. and Humphrey, S.A.,** Negative-ion mass spectra of some pyridines, pyrazines, and their *N*-oxides, *Org. Mass. Spectrom.,* 4, 513, 1970.

100. **Sartori, G., Costa, G., and Blasina, P.,** Pyridine *N*-oxide. The N-O bond in the infrared spectrum, *Gazz. Chim. Ital.,* 85, 1085, 1955.

101. **Shindo, H.,** Infrared spectra of alkylpyridine l-oxides, *Pharm. Bull.,* 4, 460, 1956.

102. a. **Katritzky, A.R. and Hands, A.R.,** Infrared spectra of 2-substituted pyridine l-oxides, *J. Chem. Soc.,* 2195, 1958; b. **Katritzky, A.R. and Gardner, J.N.,** Infrared spectra of 4-substituted pyridine l-oxides, *J. Chem. Soc.,* 2192, 1958; c. **Katritzky, A.R., Beard, J.A.T., and Coats, N.A.,** Infrared spectra of 3-substituted pyridine l-oxides, *J. Chem. Soc.,* 3680, 1959.

103. **Ghersetti, S., Maccagnani, G., Mangini, A., and Montanari, F.,** Infrared and near ultraviolet absorption spectra of pyridine *N*-oxide derivatives, *J. Heterocycl. Chem.,* 6, 859, 1969.

104. **Shindo, H.,** Infrared spectra of substituted pyrazines and their *N*-oxides, *Chem. Pharm. Bull.,* 8, 33, 1960.

105. **Ghersetti, S., Giorgianni, S., Minari, M., and Spunta, G.,** Infrared spectral studies of quinoline *N*-oxides and isoquinoline *N*-oxides, *Spectrosc. Lett.,* 6, 167, 1973.

106. **Graja, A.,** Infrared spectroscopic investigations of picoline *N*-oxide and its complexes, *Adv. Mol. Relaxation Processes,* 5, 149, 1973.

107. **Nathan, L.C. and Ragsdale, R.O.,** Trends in the nitrogen-oxygen stretching frequency of 4-substituted pyridine *N*-oxide coordination compounds, *Inorg. Chim. Acta,* 10, 177, 1974.

108. **Katritzky, A.R., Palmer, C.R., Swinbourne, F.J., Tidwell, T.T., and Topsom, R.D.,** Infrared intensities as a quantitative measure of intramolecular interactions. VI. Pyridine, pyridine l-oxide, and monosubstituted derivatives. v_{16} band near 1600 cm^{-1}, *J. Am. Chem. Soc.,* 91, 636, 1969.

109. **Tupitsyn, I.F., Zatsepina, N.N., and Kolodina, N.S.,** Electronic interactions and infrared intensities of aromatic heterocyclic molecules. II. CH stretching vibrations of the aromatics CH bonds in six-membered nitrogen heterocycles and their *N*-oxides, *Reakts. Sposobnost Org. Soedin,* 6, 11, 1969.

110. **Gambi, A. and Ghersetti, S.,** Normal coordinate analysis of pyridine *N*-oxide, *Spectrosc. Lett.,* 10, 627, 1977.

111. **Varsanyi, G., Szoke, S., Keresztury, G., and Gelleri, A.,** Pyridine *N*-oxide: vibrational assignments and thermodynamic quantities, *Acta Chim. (Budapest),* 65, 73, 1970.

112. **Pujari, P.R. and Bist, H.D.,** Lattice Raman spectrum of pyridine *N*-oxide, *J. Raman Spectrosc.,* 1, 255, 1973.

113. **Joyeux. M. and Nguyen, Q.D.,** Vibrational spectroscopy study of 4-nitropyridine *N*-oxide. II. Internal vibration, *J. Raman Spectrosc.,* 19, 441, 1988.

114. **Endo, K., Saito, Y., Aota, M., and Furuhashi, A.,** Microwave spectra of 2-methylpyridine l-oxide and 2-methyl-d$_3$-pyridine l-oxide, *Nippon Kagaku Kaishi,* 253, 1987.

115. **Ito, M. and Hata, N.,** The ultra-violet absorption spectrum of pyridine *N*-oxide, *Bull. Chem. Soc. Jpn.,* 28, 260, 1955.

116. **Kubota, T.,** Electronic spectra and electronic structure of some basic heterocyclic *N*-oxides, *Bull. Chem. Soc. Jpn.,* 35, 946, 1962.

117. **Seibold, K., Wagniere, G., and Labhart, H.,** Interpretation of the UV spectra of pyridine l-oxide, *Helv. Chim. Acta,* 52, 789, 1969.

118. **Hochstrasser, R.M. and Wiersma, D.A.,** Structure and dipole moment of the first electronically excited state of pyridine *N*-oxide, *J. Chem. Phys.,* 55, 5339, 1971.

119. **Scholz, M.,** M.O. calculations of the electronic spectrum of pyridine *N*-oxide: a critical analysis, *J. Prakt. Chem.,* 324, 85, 1982.

120. **Gondo, Y., Gondo, Y., and Kanda, Y.,** Electronic structure and spectra of pyridazine *N*-oxide and pyridine *N*-oxide, *Mem. Fac. Sci., Kyushu Univ., Ser. C,* 13, 249, 1982.

121. **Brand, J.C.D. and Tang, K.T.,** Rotational analysis of the 342 nm band of pyridine *N*-oxide, *J. Mol. Spectrosc.,* 39, 171, 1971.

122. **Bist, H.D., Parihar, J.S., and Brand, J.C.D.,** The 341 nm band system of pyridine *N*-oxide. Analysis of the in-plane vibrational structure, *J. Mol. Spectrosc.,* 59, 435, 1976.

123. **Kraessig, R., Bergmann, D., Fliegen, N., Kummer, F., Seiffert, W., and Zimmermann, H.,** Polarization of the electronic bands of aromatic compounds. 8. Phenazine, phenazine *N*-oxide, phenazine *N,N'*-dioxide, *Ber. Bunsenges. Phys. Chem.,* 74, 617, 1970.

124. **Berezovskii, V.M., Aksel'rod, Z.I., Grigor'eva, N.D., Mel'nikov, V.Z., and Kirilova, N.I.,** Synthesis, reactions, and structure of alloxazine 9,10-dioxides, *Dokl. Akad. Nauk. SSSR,* 198, 829, 1971.

125. **Yamakawa, M., Kubota, T., Ezumi, K., and Mizuno, Y.,** Absorption and phosphorescence spectra of 4-nitropyridine *N*-oxides and 4- and 3-nitroquinoline *N*-oxides, *Spectrochim. Acta,* 30A, 2103, 1974.

126. **Kaminsky, L.S. and Lamchen, M.,** Ultraviolet absorption of the l-pyrroline l-oxides, *J. Chem. Soc. (B),* 1085, 1968.

127. **Sosa, R. and Paoloni, L.,** Spectra and the structure of the *N*-oxides of benzofurazan (benzofuroxans), *Tetrahedron,* 25, 4197, 1969.

128. **Bettoni, G., Catsiotis, S., Perrone, R., and Tortorella, V.,** The chiroptical properties of some substituted *N*-(2-pyridyl) derivatives of amines, *Gazz. Chim. Ital.,* 107, 111, 1977.

129. **El-Abadelah, M.M., Sabri, S.S., Nazer, M.Z., and Za'ater, M.F.,** Synthesis and chiroptical properties of some N-(3-methyl-2-quinoxaloyl) L-aminoacids and their dioxides, *Tetrahedron,* 32, 2931, 1976.

130. **El-Abadelah, M.M., Anani, A.A., Khan, Z.H., and Hassan, A.M.,** Chiroptical properties of some *N*-5(6)benzofuroxanoyl-L-α-amino acids and esters, *J. Heterocycl. Chem.,* 17, 213, 1980.

131. **Gottarelli, G. and Samori, B.,** Circular dichroism of (-)-(S)-trans-1,2-di(4-pyridine l-oxide) oxirane. Pyridine l-oxide 280 and 215 nm transitions, *Spectrochim. Acta,* 30A, 417, 1974.

132. a. **Kubota, T., and Miyazaki, H.,** Polarography of pyridine *N*-oxide and its alkyl derivatives, *Bull. Chem. Soc. Jpn.,* 35, 1549, 1962; b. **Emerson, T.K. and Rees, C.W.,** The deoxygenation of heterocyclic *N*-oxides. Part II. Polarographic reduction, *J. Chem. Soc.,* 1923, 1962.

133. **Kubota, T., Nishikida, K., Miyazaki, H., Iwatani, K., and Oishi, Y.,** Electronic spin resonance and polarographic studies of the anion radicals of heterocyclic amine *N*-oxides, *J. Am. Chem. Soc.,* 90, 5080, 1968.

134. **Kubota, T., Oishi, Y., Nishikida, K., and Miyazaki, H.,** Hydrogen-bonding effects on the electron spin resonance spectra of the anion radicals of several aromatic amine *N*-oxides, *Bull. Chem. Soc. Jpn.,* 43, 1622, 1970.

135. **Miyazaki, H., Kubota, T., and Yamakawa, M.,** Characterization of the electronic spectra of heterocyclic amine *N*-oxides by means of the nonaqueous oxidation and reduction potentials and the substituent effects on them, *Bull. Chem. Soc. Jpn.,* 45, 780, 1972.

136. **Miyazaki, H., Matsuhisa, Y., and Kubota, T.,** Cyclic voltammetry of aromatic amine *N*-oxides in nonaqueous solvents and the stability of the free radicals produced, *Bull. Chem. Soc. Jpn.,* 54, 3850, 1981.

137. **Leibzon, V.N., Mairanovski, S.G., and Konnik, E.I.,** Polarographic reduction of isoxazoline *N*-oxides, *Izv. Akad. Nauk. SSSR, Ser. Khim.,* 1429, 1971.

138. **Stuewe, A., Weber-Schaefer, M., and Baumgaertel, H.,** Electrochemical oxidation of aromatic *N*-oxides, *Tetrahedron Lett.,* 427, 1972.

139. **Loutfy, R.O.,** Solution and gas phase ionization energies and electron affinities of substituted pyridine *N*-oxides, *J. Chem. Phys.,* 66, 4781, 1977.

140. **Kano, K., Uno, B., Kaida, N., Zhang, Z.X., Kubota, T., Takahashi, K., and Kawazoe, Y.,** Voltammetric and spectroscopic studies of the carcinogen 4-(hydroxylamino)quinoline *N*-oxide and its analogs, *Chem. Pharm. Bull.,* 35, 1702, 1987.

141. **Burmicz, J.S. and Smyth, W.F.,** Differential-pulse polarographic analysis of 4-nitroquinoline *N*-oxide and its principal metabolites, *Anal. Proced. (London),* 17, 284, 1980.

142. **Crawford, P.W., Scamehorn, R.G., Hollstein, U., Ryan, M.D., and Kovacic, P.,** Cyclic voltammetry of phenazines and quinoxalines including mono and di-*N*-oxides, *Chem-Biol. Interact.,* 60, 67, 1986.

143. **Ma, T.S., Hackman, M.R., and Brooks, M.A.,** Determination of the *N*-oxide function by differential pulse polarography, *Mikrochim. Acta,* 2, 617, 1975.

144. **Janssen, R.W. and Discher., C.A.,** Controlled potential coulometric analysis of amine oxides of pharmaceutical interest, *J. Pharm. Sci.,* 60, 798, 1971.

145. **Bieganowska, M. and Wawrzynowicz, T.,** Comparison of chromatographic parameters of heterocyclic bases and their *N*-oxides in thin layers of silica gel, *Chem. Anal. (Warsaw),* 21, 211, 1976.

146. **Coats, N.A. and Katritzky, A.R.,** Some reactions of l-methoxypyridinium salts and a color test for *N*-oxides, *J. Org. Chem.,* 24, 1836, 1959.

147. **Brooks, R.T. and Sternglanz, P.D.,** Titanometric determination of the *N*-oxide group in pyridine *N*-oxide and related compounds, *Anal. Chem.,* 31, 561, 1959.

148. **Jaffé, H.H. and Doak, G.O.,** The basicities of substituted pyridines and their-oxides, *J. Am. Chem. Soc.,* 77, 4441, 1955.

149. **Gardner, J.N. and Katritzky, A.R.,** The tautomerism of 2- and 4-amino- and -hydroxy-pyridine l-oxide, *J. Chem. Soc.,* 4375, 1957.

150. **Johnson, C.D., Katritzky, A.R., Ridgewell, B.J., Shakir, N., and White, A.M.,** The applicability of Hammet acidity functions to substituted pyridine and pyridine l-oxide, *Tetrahedron,* 21, 1055, 1965.

151. **Sawada, M., Yukawa, Y., Hanafusa, T., and Tsuno, Y.,** Importance of dual resonance susceptibilities for pi-donor and pi-acceptor substituents regarding a quantitative description of substituent effects. The case of basicity of pyridine *N*-oxides, *Tetrahedron Lett.,* 4013, 1980.

152. **Korolev, B.A., Osmolovoskaya, L.A., and Dyumaev, K.M.,** Basicity of subsited pyridines and pyridine *N*-oxides in nitromethane, *Zh. Obshch. Khim.,* 49, 898, 1979.

153. **Klofutar, C., Paljk, S., and Kremser, D.,** Thermodynamic of protonation of pyridine *N*-oxide and its methyl substituted derivatives in aqueous media, *Spectrochim. Acta,* 29A, 139, 1973.

154. **Johnson, C.D., Katritzky, A.R., and Shakir, N.,** Amide acidity function. Its extension and application to *N*-oxides, *J. Chem. Soc. (B),* 1235, 1967.

155. **Cook, M.J., Dassanyake, N.L., Johnson, C.D., Katritzky, A.R., and Toone, T.W.,** Free energy-enthalpy correlation for protonation of pyridine bases and azine N-oxides and temperature variation in the H_O and H_A acidity function, *J. Chem. Soc., Perkin Trans.,* 2, 1069, 1974.

156. **Cox, R.A. and Yates, K.,** Thermodynamics of protonation of weak bases in sulfuric acid-water media, determined using the excess acidity method, *Can. J. Chem.,* 62, 2155, 1984.

157. **Lezina, V.P., Kozlova, M.M., Gashev, S.B., Gol'tsova, L.V., and Stepanyants, A.U.,** Investigation of acid-base equilibria of N-containing heterocycles by NMR methods. 2. Substituted pyridine and pyrimidine l-oxides, *Khim. Geterotsikl. Soedin,* 1369, 1987.

158. **Whyman, R., Copley, D.B., and Hatfield, W.E.,** Magnetic and spectral properties of complexes with substituted quinoline N-oxides, *J. Am. Chem. Soc.,* 89, 3135, 1967.

159. **Kubota, T. and Miyazaki, H.,** The effect of pH on the absorption and fluorescence spectra of acridine N-oxide and phenazine mono- and di-N-oxides, *Nippon Kagaku Zasshi,* 79, 916 and 924, 1958.

160. **Mishima, M., Terasaki, T., Fujio, M., Tsuno, Y., Takai, Y., and Sawada, M.,** Substituent effect on the gas phase basicity of pyridine N-oxide, *Mem. Fac. Sci., Kyushu Univ., Ser. C,* 16, 77, 1987.

161. **Dvoryantseva, G.G., Kaganskii, M.M., Musatova, I.S., and Elina, A.S.,** Electronic spectra and structure of protonated N-oxides of pyrazines and quinoxalines, *Khim. Geterotsikl. Soedin,* 1554, 1974.

162. **Kaganskii, M.M., Dvoryansteva, G.G., and Elina, A.S.,** Structure of cations of pyrazine and aminopyrazine N-oxides studied by a UV-spectroscopic method, *Dokl. Akad. Nauk. SSSR,* 197, 832, 1971.

163. **Forsythe, P., Frampton, R., Johnson, C.D., and Katritzky, A.R.,** Protonation of dimethylaminopyridines and their N-oxides, *J. Chem. Soc., Perkin Trans.,* 2, 671, 1972.

164. **Corey, E.J., Borror, A.L., and Foglia, T.,** Transformation in the 1,10-phenanthroline series, *J. Org. Chem.,* 30, 288, 1965.

165. **Clarkson, R., Dowell, R.I., and Taylor, P.J.,** The reversible cation-anion isomerization of 2-imino-2*H*-pyrido[1,2-*b*] [l,2,4]thia(oxa)diazole hydrobromide, *Tetrahedron Lett.,* 485, 1982.

166. **Dimitrijevic, D.M., Tadic, Z.D., Misic-Vukovic, M.M., and Muskatirovic, M.,** Investigation of eletronic effects in the reaction of diazodiphenylmethane with pyridine and pyridine N-oxide carboxylic acids, *J. Chem. Soc., Perkin Trans.,* 2, 1051, 1974.

167. **Lezina, V.P., Stepanyants, A.U., Golovkina, N.I., and Smirnov, L.D.,** NMR study of acid-base reactions of 3-hydroxypyridine, 3-hydroxypyridine N-oxide, and β–hydroxypyridinecarboxylic acids, *Izv. Akad. Nauk. SSSR, Ser Khim.,* 98, 1980.

168. **Szafran, M. and Brzezinski, B.,** Infrared spectra of some pyridine carboxylic acid N-oxides, *Bull. Acad. Pol. Sci., Ser. Sci. Chim.,* 18, 247, 1970.

169. **Dziembowska, T. and Szafran, M.,** Infrared spectra of 4-substituted quinoline 2-carboxyamide N-oxides, *Rocz. Chem.,* 48, 611, 1974.

170. **Rzeszowska, K., and Zommer-Urbanska, S.,** Investigation of the protolytic equilibria and spectroscopic analysis of amidoxime of picolinic acid, *Rocz. Chem.,* 51, 1609, 1977.

171. **Ghuge, K.D., Umapathy, P., and Sen, D.N.,** Effects of substituents on the intramolecular hydrogen bond in 8-quinolinol-N-oxides, *J. Indian Chem. Soc.,* 57, 967, 1980.

172. **Brzezinski, B. and Zundel, G.,** Proton polarizability of intramolecular hydrogen bonds with molecules nonconjugated and conjugated between donor and acceptor groups, *Chem. Phys. Lett.,* 75, 500, 1980.

173. **Brzezinski, B. and Zundel, G.,** A noncharged, easily polarizable, intramolecular, hydrogen bond in 2-(α-pyridyl N-oxide)ethanesulfonic acid, *J. Mol. Struct.,* 68, 315, 1980.

174. **Brzezinski, B. and Zundel, G.,** Electronic structure of molecules and infrared continuums caused by intramolecular hydrogen bond with great proton polarizability, *J. Phys. Chem.,* 86, 5133, 1982.

175. **Brzezinski, B. and Zundel, G.,** Influence of solvents on intramolecular hydrogen bonds with large proton polarizability, *J. Magn. Reson.,* 48, 361, 1982.

176. **Takasuka, M., Irie, T., and Tanida, H.,** Intramolecular OH...π hydrogen bonding in 6- and 7-hydroxy-5,6,7,8-tetrahydro-5,8-methano isoquinolines and their N-oxides, *J. Chem. Soc., Perkin Trans.,* 2, 1673, 1985.

177. **Grundwald, M., Szafran, M., and Kraglewski, M.,** Dipolar association of pyridine N-oxide in benzene at 25°C, *Adv. Mol. Relaxation Interact. Processes,* 18, 53, 1980.

178. **Burchard, W. and Zimmermann, H.,** Association products of l-hydroxy-2,4,5-trimethylimidazole 3-oxide. III. Model calculations for determination of equilibrium constants, *Makrom. Chem.,* 126, 288, 1969.

179. **Tosi, G., Stipa, P., and Bocelli, G.,** 2,2'-Diphenyl-$\Delta^{3,3'}$-bi-3*H*-indole-1,1'-dioxide: molecular interactions and crystal structure, *Monatsh. Chem.,* 119, 487, 1988.

180. a. **Kreevoy, M.M. and Chang, K.C.,** Ultraviolet spectra and structure of complexes of pyridine 1-oxide and oxygen acids, *J. Phys. Chem.,* 80, 259, 1976; b. **Boehner, U. and Zundel, G.,** Broad single-minimum proton potential and proton polarizability of the hydrogen bonds in trichloroacetic acid-pyridine N-oxide systems as a function of donor and acceptor properties and environment. Infrared studies, *J. Chem. Soc., Faraday Trans. l,* 81, 1425, 1985.

181. **Hadzi, D. and Smerkoli, R.,** Proton chemical shifts and thermodynamic data on the association of chloroacetic acids with some oxygen bases, *J. Chem. Soc., Faraday Trans. 1,* 72, 1188, 1976.

182. **Brycki, B. and Szafran, M.,** Infrared and proton nuclear magnetic resonance studies of hydrogen bonds in some pyridine *N*-oxide trifluoroacetates and their deuterated analogs in dichloromethane, *J. Chem. Soc., Perkin Trans. 2,* 1333, 1982.

183. **Nogaj, B.,** Hydrogen-bond theories and models based on nuclear quadrupole resonance spectroscopy studies, *J. Phys. Chem.,* 91, 5863, 1987.

184. **Boehner, U. and Zundel, G.,** Sulfonic acid-oxygen base systems as a function of the Δ pKa, *J. Phys. Chem.,* 89, 1408, 1985.

185. **Kulevsky, N. and Lewis, L.,** Thermodynamic of hydrogen bond formation between phenol and some diazine *N*-oxides, *J. Phys. Chem.,* 76, 3502, 1972.

186. **Bueno, W.A. and Lucisano, Y.M.,** Sensitivity of hydrogen bond formation to substituent parameters, *Spectrochim. Acta,* 35A, 381, 1979.

187. **Nelson, J.H., Nathan, L.C., and Ragsdale, R.O.,** Hydrogen-bonding interaction of aromatic amine oxides with phenols, *J. Am. Chem. Soc.,* 90, 5754, 1968.

188. a. **Kubota, T. and Miyazaki, H.,** Ultraviolet absorption spectra of derivatives of heterocyclic *N*-oxides. General properties of the solvent effect and the substituting effect, *Chem. Pharm. Bull.,* 9, 948, 1961; b. **Benamou, C. and Bellon, L.,** Electronic spectra of pyridine *N*-oxide as intermediates for the measure of the association constants between proton donors and weak bases, *Bull. Soc. Chim. Fr.,* I-321, 1984.

189. **Beggiato, G., Aloisi, G.G., and Mazzucato, U.,** Complexes between pyridine l-oxides and halogens, *J. Chem. Soc., Faraday Trans. 1,* 70, 628, 1974.

190. **Giera, J., Sobeczyk, L., Lux, F., and Paetzold, R.,** Correlation between the dipole moments and thermodynamical data of iodine complexes with organic oxides, sulfides, and selenides, *J. Phys. Chem.,* 84, 2602, 1980.

191. **Tanida, H.,** Nitration of quinoline homolog *N*-oxides, *Yakugaku Zasshi,* 78, 1079, 1958.

192. **Popp, C.J., Nathan, L.C., McKean, T.E., and Ragsdale, R.O.,** Donor properties of acridine *N*-oxide, *J. Chem. Soc. (A),* 2394, 1970.

193. **Popp, C.J., Nelson, J.H., and Ragsdale, R.O.,** Thermodynamic and infrared studies of tertiary amino oxides with bis(2,4-pentanedionato) oxovanadium (IV), *J. Am. Chem. Soc.,* 91, 610, 1969.

194. **Okano, T., Isobe, A., and Matsumoto, H.,** Charge transfer in the molecular interaction of carcinogenic 4-nitroquinoline l-oxides and DNA, with special reference to analysis at the nucleoside level, *Gann,* 63, 427, 1972.

195. **Okano, T., Takenaka, S., and Sato, Y.,** Interaction of the carcinogen 4-nitroquinoline l-oxide with protein and aromatic amino acids, *Chem. Pharm. Bull.,* 16, 556, 1968.

196. **Okano, T. and Uekama, K.,** Stoichiometric interactions between 4-nitroquinoline l-oxide and deoxyribonucleosides with special reference to the mode of binding of the carcinogen to DNA, *Chem. Pharm. Bull.,* 15, 1812, 1967.

197. **Jones, R.A.Y., Katritzky, A.R., and Lagowsky, J.M.,** Investigation of tautomeric equilibria by proton magnetic resonance spectroscopy, *Chem. Ind.,* 870, 1960.

198. **Jones, R.A. and Katritzky, A.R.,** Tautomerism of mercapto- and acylamino-pyridine l-oxides, *J. Chem. Soc.,* 2937, 1960.

199. **Katritzky, A.R. and Jones, R.A.,** Infrared absorption of heteroaromatic and benzenoid six-membered monocyclic nuclei. Part X. Pyridones and pyridthiones, *J. Chem. Soc.,* 2947, 1960

200. a. **Ionescu, M., Katritzky, A.R., and Ternai, B.,** The tautomerism of 9-hydroxy- and 9-mercapto-acridine l0-oxide, *Tetrahedron 22,* 3227, 1966; b. **Kaneko, C.,** Infrared spectra of 2- and 4-hydroxyquinoline l-oxide and their derivatives, *Yakugaku Zasshi,* 79, 428, 1959.

201. **Parham, J.C., Winn, T.G., and Brown, G.B.,** Tautomeric structures of the 3-*N*-oxides of xanthine and guanine, *J. Org. Chem.,* 36, 2639, 1971.

202. **Perlman, K.L., Pfeiderer, W.E., and Gottlieb, M.,** Synthesis and structure of alloxazine 5,10-dioxides, *J. Org. Chem.,* 42, 2203, 1977.

203. **Mousseron-Canet, M. and Boca, J.P.,** Indolic nitrones and their tautomeric *N*-hydroxyindoles, *Bull. Soc. Chim. Fr.,* 1296, 1967.

204. **Chua, S.O., Cook, M.J., and Katritzky, A.R.,** Imidazole 3-oxide vs. 3-hydroxy-3*H*-imidazole equilibrium, *J. Chem. Soc. (B),* 2350, 1971.

205. **Boyle, F.T. and Jones, R.A.Y.,** Tautomerism of benzotriazole l-oxide and its 4- and 6-nitroderivatives with the corresponding l-hydroxybenzotriazoles, *J. Chem. Soc., Perkin Trans. 2,* 160, 1973.

206. **Maquestiau, A., Van Haverbeke, Y., Flammang, R., Chua, S.O., Cook, M.J., and Katritzky, A.R.,** Thermal degradation of imidazole *N*-oxides in the source of a mass spectrometer, *Bull. Soc. Chim. Belg.,* 83, 105, 1974.

207. **Becker, H.G.O., Goermar, G., Haufe, H., and Timpe H.J.,** Structure of 4-*N*-hydroxy-1,2,4-triazoles, *J. Prakt. Chem.,* 314, 101, 1972.

208. **Englert, G.,** Nuclear resonance of furoxans, *Z. Electrochem.,* 65, 854, 1961.

209. **Harris, R.K., Katritzky, A.R., Øksne S., Bailey, A.S., and Paterson, W.G.,** Proton resonance spectra and the structure of benzofuroxan and its nitro derivatives, *J. Chem. Soc.,* 197, 1963.

210. **Boulton, A.J., Katritzky, A.R., Sewell, M.J., and Wallis, B.,** Nuclear magnetic resonance spectra and tautomerism of some substituted benzofuroxans, *J. Chem. Soc. (B),* 914, 1967.

211. **Boulton, A.J., Halls, P.J., and Katritzky, A.R.,** Effect of methyl and aza-substituents on the tautomeric equilibrium in benzofuroxan, *J. Chem. Soc. (B),* 636, 1970.

212. a. **Calvino, R., Gasco, A., Menziani, E., and Serafino, A.,** Chloromethylfuroxans, *J. Heterocycl. Chem.,* 20, 783, 1983; b. **Calvino, R., Ferrarotti, B., Gasco, A., Serafino, A., and Pelizzetti, E.,** Studies of 3-phenylfuroxan, *Gazz. Chim. Ital.,* 113, 811, 1983.

213. **Alder, R.W., Niazi, G.A., and Whiting, M.C.,** Benz[*c,d*]inidazole l-oxide, benz[*c,d*]indazole 1,2-dioxide(1,8-dinitrosonaphthalene) and related compounds, *J. Chem. Soc. (C),* 1693, 1970.

Chapter 3

SYNTHESIS

The methods for obtaining heterocycles N-oxides can be classified into three main groups, i.e., (1) preparation from a preformed heterocycle, either by adding the oxygen atom (oxidation of a sp² hybridized nitrogen atom) or by oxidizing a cyclic hydroxylamino derivative; (2) synthesis by cyclization, usually with simultaneous formation of the ring and of the N-oxide function; and (3) synthesis from a preformed N-oxide by introducing or modifying a substituent, and this includes the formation of a new fused ring, or alternatively, modification of the ring structure itself, e.g., ring expansion of quinazoline to benzo-1,4-diazepine oxides.

The first two groups of methods are discussed in the two sections of this chapter, and the last one in Chapter 4, as this is considered to be a reaction of the appropriate substrate. It is merely referred to in the present chapter as one of the methods modifying the ring.

I. PREPARATION FROM PREFORMED HETEROCYCLES

The reaction of nitrogen heterocycles with peroxidic derivatives offers easy access to the corresponding N-oxides. As opposed to the case of open-chain and cyclic amines,[1a] the sp² hybridized nitrogen atom of imines, both when this function is isolated and when it is a part of an aromatic heterocyclic ring, is not oxidized by hydrogen peroxide (see below for exceptions). However, these two classes of compounds react smoothly with peracids. Under this condition, isolated imines gives either nitrones or oxaziridines ("isonitrones"),[1b] and the latter compounds can be readily rearranged to the former ones; however, oxidation of imines is not an important method for preparing nitrones. Heteroaromatic N-oxides are formed from aza heterocycles, and the ready availability of such substrates makes the reaction a major method for the preparation of N-oxides; indeed, the structure of these compounds was first unambiguously recognized from their formation by N-oxidation of heterocycles (see Chapter 1).

The attack by peracids is an electrophilic process. The reaction between pyridine and perbenzoic acid has been shown to involve the two neutral species and is quenched at low pH when the pyridinium cation is present. The observed second-order rate constant correlates with the substituent's σ constants and with the pK values, except for the 2,6-disubstituted derivatives, due to steric hindering.[2,3] A study with aza-naphthalenes and -phenanthrenes showed that N-oxidation is less sensitive to steric effects than quaternarization with methyl iodide.[4]

A. N-OXIDATION BY PERCARBOXYLIC ACIDS

Reaction of heterocycles with anhydrous percarboxylic acids, often at room temperature or slightly above, smoothly produces the corresponding N-oxides. Reagents used include both aliphatic derivatives, such as peracetic[5] and perlauric acids,[6] and more usually, aromatic compounds, such as perbenzoic acid[7-10] and its derivatives, many of which are more stable than the parent compound (e.g., the 3-chloro,[11-12] 4-carboxymethyl,[13] nitro,[14] and pentafluoro[15] substituted perbenzoic acids), as well as monoperphthalic acid.[16] Peracids may be also obtained *in situ*, e.g., by adding maleic[17] or phthalic anhydride[18] in portions to hydrogen peroxide.

Alternatively, a mixture of hydrogen peroxide and a carboxylic acid (usually acetic[19-23] or formic,[24,25] but others, e.g., benzoic, phthalic, or succinic, have been reported)[7,19] is used, and the oxidizing agent is the peracid present in equilibrium. In this case, higher temperatures are required and there is a greater risk of side reactions (see Section IC2).

A mixture of glacial acetic acid and 30—40% hydrogen peroxide has been largely used; the reaction is usually carried out at 60—80°C for several hours with a 20—70% excess of hydrogen peroxide over the theoretical amount (often added in portions while the reaction proceeds). The solution is then concentrated under reduced pressure, and the residue is basified and extracted (concentration is required, in particular, with low molecular weight *N*-oxides, in view of the free solubility in water in these compounds). The pure *N*-oxide is obtained by distillation or recrystallization. It is advisable to decompose the excess peroxide before concentration (e.g., by adding manganese dioxide or palladium on carbon or formaldehyde), since the occurrence of severe explosions at this step has been reported.[26,27] Basification with 50% aqueous NaOH and extraction has been recommended as an alternative to concentration.[23]

The most popular oxidant is now 3-chloroperbenzoic acid, a crystalline solid that is stable to storage in the refrigerator, which is commercially available with a ca. 55% titre (the remainder being ca. 10% 3-chlorobenzoic acid and 35% water). When the reaction is carried out at room temperature in chloroform, the 3-chlorobenzoic acid formed precipitates out and is filtered, simplifying the work-up. No excess is required; however, it must also be remembered that this reagent, like the other peracids mentioned, is shock sensitive. A safe and equally effective reagent is magnesium monoperphthalate.[28]

Peracid oxidation is also effective with five-membered heterocycles,[30,32,33] though side reaction might be important, e.g., with some benzothiazole, ring cleavage to aminobenzenesulfonic acid is observed,[34a] and benzofuroxan does not survive treatment with peracids, being converted to *o*-dinitrobenzene[34b] (for cleavage under oxidative conditions, see also Chapter 4).

As mentioned above, peracid oxidation is slow with less basic (electron-withdrawing substituted or aza-substituted) azines. In this case it is advantageous to either enhance the reactivity of the peracid by protonating it with mineral acids[35] or to employ stronger percarboxylic derivatives, such as performic,[25] pertrifluoroacetic,[36-39] or perdichloromaleic acids[40] (Table 1).

Isoquinoline 2-oxide. *A mixture of freshly distilled isoquinoline (64.5 g), glacial acetic acid (150 ml), and 30% hydrogen peroxide (50 ml) is heated at 60—70°C for 12 h, with the addition of another 40 ml H$_2$O$_2$ after 3 h. Concentration under vacuum on the water bath, repeated after the addition of 50 ml water, gives a red liquid. This is treated with excess solid potassium carbonate and 300 ml chloroform. Separation, washing of the residue with chloroform, and evaporation of the combined extracts gives a liquid, which when poured into 40 ml of ethyl acetate gives 48.5 g of the product as the hemihydrate.[21]*

2-Phenylquinoline 1-oxide. *2-Phenylquinoline (41 g, 17.9 mmol) and ca. 80% m-chloroperbenzoic acid (50 g, 23 mmol) are dissolved in chloroform (0.7 l). After 2 d at room temperature in the dark, the solution is shaken with an excess 1 N NaOH, washed with NaOH, and then aqueous NaCl, dried, and evaporated to yield the product (42 g, 95%).[12]*

4-Dimethylaminopyridine 1-oxide. *30% Hydrogen peroxide (2.8 g, 25 mmol) is added dropwise to a stirred mixture of 4-dimethylaminopyridine (1.83 g, 15 mmol), benzonitrile (2.3 g, 22 mmol), and sodium hydrogen carbonate (0.75 g, 9 mmol) at room temperature. After 12 h the solvent is evaporated, the residue treated with water (50 ml), cooled at 0°C, filtered (from benzamide), and evaporated. Refluxing 2-3 h with 20% sulfuric acid and CH$_2$Cl$_2$ extraction gives 1.43 g (69%) of the product.[100]*

Explosion while using peracetic acid. *A mixture of 2,5-dimethylpyrazine (55 g), glacial acetic acid (200 ml), and 30% hydrogen peroxide is heated at 56°C for 16 h, and then let stand at room temperature for 24 h. A trace of MnO$_2$ is added to decompose excess peroxide, but no test to confirm complete decomposition is done. The solution is concentrated under reduced pressure to yield an oil (150 ml) that detonates on standing.[26]*

A mixture of nitriles and hydrogen peroxide has also been found to be effective, probably due to the formation of iminoperoxy acids,[41] and the formation of similar intermediates explains the formation of carboxyamidopyridine 1-oxides from the 2- and 3-pyridinecarbonitriles by direct reaction with hydrogen peroxide in the absence of additives.[42] Percarboxylic acids are also obtained from isocyanates and H$_2$O$_2$.[43]

TABLE 1
N-Oxidation of Heterocycles with Percarboxylic Acids

Substituted	Reagent, conditions	Work-up	Yield,%	Ref.
Pyridine	AcOH, H_2O, 70—80°C, 12 h	Conc., bas., ext. $CHCl_3$, dist.	96	20
	PhCN, H_2O_2, NaOH, MeOH, exoth. react., 25—30°C, 3 h	Conc., ext. H_2O, $CHCl_3$, dist.	79	41
	Mg monoperphthalate, H_2O, 45°C, 1.5 h		92	28
	$PhCONHCO_3H$, C_6H_6, exoth. reac.	Evap., ext. C_6H_6	Quant.	43
	$C_6F_5CO_3H$, CH_2Cl_2, 25°C, 10 min		94	15
3,5-Lutidine	AcOH, H_2O, 90°C, 21 h	Bas., ext. $CHCl_3$	30	23
2-Bromopyridine	40%, AcO_2H, 45—50°C, 24 h	Conc., bas., ext. $CHCl_3$, ext. HCl	70	8
	$PhCO_3H$, $CHCl_3$, r.t., 4 d	Ext. HCl	60	8
Pentachloropyridine	AcOH, H_2SO_4, H_2O_2 at 0°, then 20°C, 48 h	Shake with base, evap.	85	35
1,6-Di-4-pyridinylhexane	AcO_2H, $CHCl_3$, r.t. 5h		20a	5
Nicotine	Perlauric ac., petr. eth., r.t. 40 h	Evap., ext. H_2O	ca. 90a	6
Quinoline	AcOH, H_2O_2, 67—70°C, 9 h	Conc. bas.	92	20
2-Phenylquinoline	H_2O_2, phthal. anh., 2 h at ice t.; r.t., 15 h	Bas., ext. $CHCl_3$	70	18
8-Methoxyquinoline	m-$ClC_6H_4CO_3H$, $CHCl_3$, r.t., 48 h	Shake NaOH, evap.	95	12
	$PhCO_3H$, $CHCl_3$, r.t., overnight		36	9
Isoquinoline	AcOH, H_2O_2, 60—70°C, 12 h	Conc., bas., ext. $CHCl_3$	63	21
Phenanthridine	AcOH, H_2O_2, 80—85°C, 5 h	Conc. bas.	46	22a
Acridine	Monoperphthalic ac., $CHCl_3$-Et_2O, 5°C 5 d	Filter, ground with NH_3, recryst.	87	16
Benzofuro[2,3-g] isoquinoline	$PhCO_3H$, $CHCl_3$, r.t., 4h		50	10
	p-$NO_2C_6H_4CO_3H$, CH_2ClCH_2Cl, refl., 24 h	Conc. ground with NaOH aq	89	14a
2,6-Dimethyl-4-chloro pyrimidine	Maleic anh., H_2O_2, $CHCl_3$, 5°C,	Wash base, evap.	34b	17b
3,6-Dichloropyridazine	Dichloromaleic anh., H_2O_2, $CHCl_3$, 7°C, 96 h	Filter, shake with base, evap. recr.	86	39
2-Phenylquinoxaline	HCOOH, H_2O_2, 50°C, overnight	Conc. pour ice	95c	25
Guanine	CF_3CO_3H, H_2O_2, exoth. reac., r.t. overnight	Conc., ext. HCl	62d	37-38
2,2-Dimethyl-4,5-diphenylimidazole	AcOH, H_2O_2, 25°C, 2 h	Dil. H_2O, recryst.	88e	30a

TABLE 1 (continued)
N-Oxidation of Heterocycles with Percarboxylic Acids

Substituted	Reagent, conditions	Work-up	Yield,%	Ref.
1-(Pyrid-2-yl)-benzotriazole	40% AcO_2H, 100°C, 1 h	Conc., ctg.	d	33a
4-Phenyl,1,2,3-thiadiazole	AcOH, H_2O_2, 40—50°C, 2 d	Dil. H_2O, ext Et_2O, ctg.	36	32
5,10,15,20-Tetramesityl-porphyrinato iron(III)	m-$ClC_6H_4CO_3H$, PhMe, 0°C			29
Adenosine	AcOH, H_2O_2, 30°C		b	31

Note: Abbreviations. Conc(entrate), bas(ify), ext(ract), dist(il), exoth(ermal), react(ion), phthal(ic) anh(ydride), r(oom) t(emperature), evap(orate), petr(oleum) eth(er), recr(ystallize), c(hroma)t(o)g(raphy), dil(ute).

[a] 1-3 Oxide, overnight with excess peracid the 1,1'-dioxide.
[b] 1-Oxide.
[c] 4-Oxide; a small amount of 1,4-dioxide also formed.
[d] 3-Oxide.
[e] 1,3 Dioxide.

TABLE 2
Various Reagents for *N*-Oxidation

Substrate	Conditions, work-up	% Yield	Ref.
Pyrazine	35% H_2O_2, Na_2WO_4, 40°C, 3.5 h. dilute, conc., ext. C_6H_6	47	49
Pyrazine	30% H_2O_2, Na_2WO_4, 70°C, 6 h, dilute, ctg.	17[a]	49
Tetrachloropyrazine	60% H_2O_2, 96% H_2SO_4, 22°C, 20 h, dilute, filter, ext. Me_2CO	70[b]	45
4-Phenylpyrimidine	H_2NOSO_3H, MeOH, 70°C, 4 h, bas., ext. CH_2Cl_2, ctg.	29	57
Benzo[*c*]cinnoline	hv, MeOH, O_2	48	61

Note: Abbreviations: Conc(entrate), ext(ract), c(hroma)t(o)g(raphy), bas(ify).

[a] 1,4-Dioxide.
[b] 1,4-Dioxide; a 30% yield of the 1-oxide also obtained.

B. OTHER REAGENTS FOR *N*-OXIDATION

The activity in *N*-oxidation is not limited to percarboxylic acids, and inorganic peroxy acids, or a mixture of hydrogen peroxide and a mineral (polyphosphoric,[44] sulfuric[45]) acid may also be effective. Oxidation by H_2O_2-polyphosphoric acid has been found to be advantageous with azines of low basicity. With peroxymonosulfuric acid (Caro's acid) and its salts, catalysis by ketones has been observed, and dioxiranes formed *in situ* might be involved in this case.[46,47] A mixture of H_2O_2 and Na_2WO_4[48,49] and related reagents [e.g., alkyl peroxide — $MoCl_5$;[50] ozonides — $Mo(CO)_6$[51]] are convenient and avoid some of the side reactions often observed with peracids (see Section IC2). Hydrogen peroxide alone is, as mentioned above, usually not sufficient for the oxidation of azines, though successful examples have been reported, e.g., with aminoacridines;[56a] pyridine 1-oxides are obtained by heating at 115—120°C in a closed vessel.[56b]

1,6-Naphthyridine 6-oxide. 1,6-Naphthyridine (1.62 g) in 10% H_2O_2 (13 ml) containing 0.1 g Na_2WO_4 is heated at 40°C for 6 h, then cooled, diluted with 13 ml water, treated with Na_2CO_3, concentrated, and treated in this way again until a KI test showed that no H_2O_2 was present, and finally evaporated; recrystallization of the residue gave the product (0.45 g, 26%).[49]

Isolated examples of other reagents include hypofluoric acid,[52] hypochlorite under basic conditions,[53] and Pb(IV) derivatives.[54] Tris(2,2′-bipyridyl)iron(III) complexes were dispro-portionate in basic solutions to the *N*-oxide iron(II) complexes.[55]

Hydroxylamine *O*-sulfonic acid may cause both *N*-amination and *N*-oxidation,[57,58] and the latter process takes over with several pyrimidines; oxidation of phenazine by hydroxylamine has also been observed.[59] Formation of *N*-oxides by radiolysis[60] or photolysis[61] of heterocycles has been reported; a limitation of such methods of course is that *N*-oxides are themselves quite photolabile (Table 2).

C. SELECTIVITY AND SIDE PROCESSES IN *N*-OXIDATION

1. Regioselectivity in the *N*-Oxidation of Polyazines

Oxidation of polyazines usually can be limited at the mono *N*-oxide stage, more drastic conditions being required for a second attack. When both electronic and steric effect cooperate, e.g., in pyridazines, formation of dioxides requires drastic conditions (prolonged reflux with AcOH/H_2O_2),[62] or changing the reagent (e.g., 3-amino-1,2,4-triazine gives only the 2-oxide with percarboxylic acids and the 2,4-dioxide with H_2O_2-polyphosphoric acid[63]). H_2O_2-Na_2WO_4 appears to be a suitable reagent for the stepwise oxidation of diazines.[49] In some cases, the reaction does not proceed to the di-*N*-oxide stage, e.g., in 1,8-naphthiridine derivatives[64] and 1,10-phenanthroline.[65] *N*-oxidation is a reversible process, and benzo[*a*]phenazine 12 oxide yields both the 5,12-dioxide and the less-hindered 5-oxide when treated with peracetic acid.[66]

In asymmetric polyazines, regioselective mono *N*-oxidation is observed in several cases.

TABLE 3
Selective *N*-Oxidation of Some Azines[a]

Substrate	Oxidant	Position of oxidation (% yield)	Ref.
3-Acetamidopyridazine	AcOH, H_2O_2	2 (50)	80
3-Aminopyridazine	H_2O_2, polyphosphoric acid	1 (65)[b]	44
4-Methylpyrimidine	*m*-$ClC_6H_4CO_3H$, $CHCl_3$	1 (21), 3 (23)	69,75
4-Methoxypyrimidine	AcOH, H_2O_2	1 (29)	75
4-Aminopyrimidine	*m*-$ClC_6H_4CO_3H$, $CHCl_3$	1 (36)	75
2-Aminopyrazine	*m*-$ClC_6H_4CO_3H$, Me_2CO	1 (68)	72
2-Chloropyrazine	AcOH, H_2O_2	4 (61)	78,79
	$H_2S_2O_5$	1 (22)	78
1,6-Naphthyridine	$PhCO_3H$, $CHCl_3$	1 (12), 6 (38)	81
1,7-Naphthyridine	$PhCO_3H$, $CHCl_3$	1 (3), 7 (45)	81
1-Methylphthalazine	Monoperphthalic acid	2 (17), 3 (32)	71
1-Benzylphthalazine	Monoperphthalic acid	2 (61), 3 (20)	71
4-Methyl-1,2,3-triazine	*m*-$ClC_6H_4CO_3H$, CH_2Cl_2	2 (30), 3 (11)	82
5-Methyl-1,2,4-triazine	$PhCO_3H$, $CHCl_3$	1 (27)	73a
2-Amino-5-phenyl-1,2,4-triazine	AcOH, H_2O_2	2 (80)	73b
3-Aminobenzo-1,2,4-triazine	AcOH, H_2O_2	2 (46)	74
6-Methylpurine	*m*-$ClC_6H_4CO_3H$, Et_2O	1 (32), 3 (19)	85
6-Aminopurine	AcOH, H_2O_2	1 (84)	77
6-Methoxypurine	AcOH, H_2O_2	3 (72)	76
4-Methyl-5,7-diphenyl-3*H*-1,2-diazepine	*m*-$ClC_6H_4CO_3H$, CH_2Cl_2	2 (77)	84
1,2-Benzodiazepine	*m*-$ClC_6H_4CO_3H$, CH_2Cl_2	1 (20), 2 (63)	83

[a] See further examples in Table 1.
[b] The 3-nitro derivative is formed.

Both steric and electronic effects have been considered as a possible cause, e.g., *N*-oxidation of 1,6-naphthiridine takes place at the most basic position,[67] while, on the other hand, the preferential oxidation at positions 7, and, respectively, 8, of 1,7- and 1,8-phenanthroline has been rationalized as a steric effect, the difference in electron density being negligible.[68]

In a study with 2-substituted quinoxalines, it was found that a methyl group slightly facilitates the *N*-oxidation of the neighboring nitrogen, other primary alkyl groups have no effect; secondary alkyl groups and the phenyl somewhat disfavor attack at the vicinal position; and chloro, alkoxy, and carboxy groups virtually hinder it. However, the result with 2,3-disubstituted quinoxaline often cannot be predicted on the basis of the effect of a single substituent.[22b] A substituent in *peri* (5) position also hinders oxidation of the vicinal nitrogen. Similar considerations (including the caveat) are useful for the rationalization of the substituent effect in other polyazines (see some examples in Table 3). Amino groups usually have a strong directing effect, e.g., 2-aminopyrazine yields the 1-oxide,[72] and 3-amino-1,2,4-triazine[73] and -benzotriazine[74] the 2-oxide; with 4-aminopyrimidine, the reaction is directed in *para* to the group.[75] 1-Aminopurine forms the 1-oxide, while with other substituents the oxidation is generally in 3.[76,77] On the other hand, the choice of the reagent may be determining, e.g., peracetic and peroxysulfuric acid give the opposite selectivity with 2-chloropyrazines[78-79] (Table 3).

When both a five- and a six-membered rings are present, the latter one is oxidized, as judged from the reaction of 1-(2-pyridyl)-benzotriazole.[33] This preference is maintained in fused systems, e.g., in purines,[38,76,77,85] and in some imidazo-[86,87] and triazolo-pyrazines.[88]

2. Competitive Oxidation Processes

a. At a Nitrogen Atom

In 2-(2-pyridyl)-4(3H)quinazolinone, both an imine function and a pyridine ring are present; with AcOH/H_2O_2 only the latter moiety is oxidized, while the stronger pertrifluoroacetic acid affects both sites.[89a] In the bifunctional derivative **1**, only double attack has been reported, yielding both an oxirano N-oxide and a di-N-oxide.[89b]

The azo group is somewhat less readily oxidized than a ring nitrogen, and both azo- and azoxy-N-oxides are obtained from phenylazopyridines.[90-92] Although the cyano group remains unaltered in many cases (or is hydrolyzed, see Section IC3), 1-isoquinolinecarbonitrile yields the nitriloxide 2-oxide.[93]

As for the amino nitrogen, primary and secondary amino groups directly substituted on the ring are usually not affected under conditions causing oxidation at the ring, at least with aromatic peracids (and in polyazines dictate the oxidation position),[72,94,95] while it is best to protect them by acylation when peracetic acid is used. With strong acids (e.g., trifluoroacetic[39b] or polyphosphoric acid and H_2O_2), the NH_2 group is converted to NO_2. With tertiary amino group, the situation is more complex, the substituent being preferentially oxidized in 2-dimethylaminopyridine, quinoline, and pyrazines, and the ring nitrogen in the 4-substituted isomers.[94,96-99] A convenient reagent for ring N-oxidation is benzonitrile-H_2O_2-base, since then the ever-present protonation of the amino group when peracids are used is avoided.[100]

Of the three functions present in compound **2**, the dimethylamino group is oxidized first (by perbenzoic acid) and the pyridine nitrogen as the latter one.[101] Oxidation at the external nitrogen causes elimination of the group in 1-dimethylamino purine derivatives.[102]

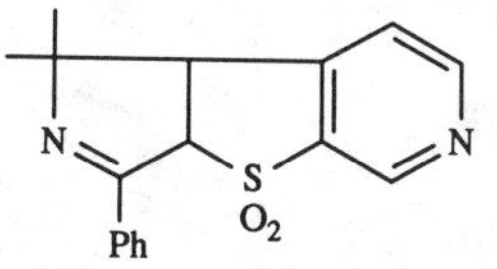

1 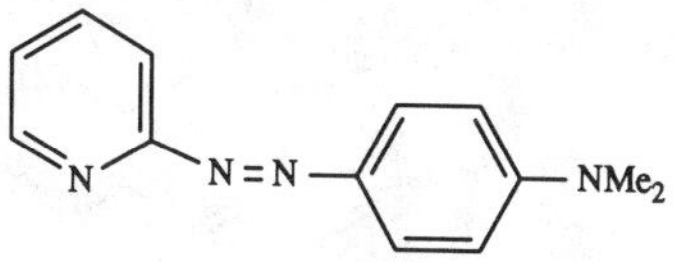   **2**

Selective oxidation of an aliphatic amino group present in the molecule can be readily accomplished, as mentioned before, by using H_2O_2;[1a,6,103a] for example, this is the way in which nicotine 1'-oxide is obtained, while the 1,1'-dioxide is formed with peracids.[6]

b. At a Sulfur Atom

Oxidation of a heterocyclic sulfur atom is usually slower than that of a ring nitrogen, both when both heteroatoms are part of the same ring (e.g., in 1,2,3-thiadiazole[32,103b] and 1,2,3,4-thiatriazole[104]) and when they are in fused rings (e.g., thieno-pyridine or -pyrazine),[105,106] and treatment with peracids of these compounds yields (one of the possible) N-oxides. Alternatively, a N,S,S-trioxide is obtained, as observed with some thiadiazoles[103b] and benzothieno–isoquinolines[14a] and cinnolines.[107] The product obtained from 10-(2-pyridyl)-phenothiazine is oxidized both at the pyridine ring and at the sulfur atom.[108]

Alkylthio substituents, as in 2-alkylthio-[109] and 3-alkylthio-2-chloro-pyridines,[14c] as well as alkythioalkyl groups, as in the pyridinophane **3**,[110] are oxidized preferentially by peracids and yield the corresponding sulfones or sulfoxides, with no reaction at the nitrogen; however, further oxidation of the nitrogen atom is usually possible, e.g., using stronger acids (pertrifluoroacetic,[111] 3,5-dinitroperbenzoic;[14c] m-chloroperbenzoic acid is sufficient with compound **3**).[110]

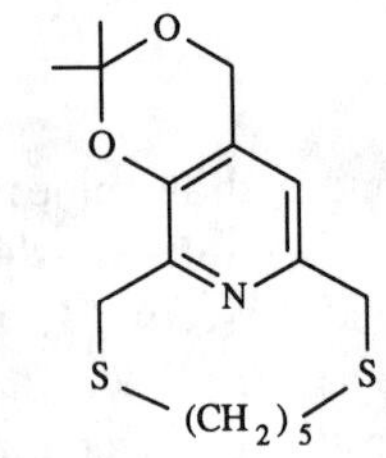

3

c. At a Carbon Atom

1. In the Side Chain

Under the condition required for *N*-oxidation, attack at the α-position of the side chain, when this is activated, as in the groups CH_2OH,[112] $CHRCN$,[113] $CHRCOOR'$, and $CHRCOR'$,[114,115] easily occurs and leads to acyl or hydroxy derivatives. Azafluorenes yield the *N*-oxides, but with diazafluorenes the corresponding fluorenones are formed in the first step, and then the reaction pursues with the *N*-oxidation.[116]

Epoxidation of a carbon-carbon double bond is competitive with attack at the nitrogen atom. Treatment of the acridine derivative **4** with one mole of peracid yields a mixture of the *N*-oxide (40%) and of the epoxide (23%), and double oxidation is obtained with excess reagent. In contrast, isomeric **5** is epoxidized but not *N*-oxidized,[117] and the related derivative **6** is only *N*-oxidized.[118] Various epoxide *N*-oxides have been obtained from vinylpyriridines.[119-121] In 1,4,5-triphenyl-3*H*-benzo[*c*]azepine, oxidation with *m*-chloroperbenzoic acid at room temperature is limited to the nitrogen, while epoxidation requires reflux.[121]

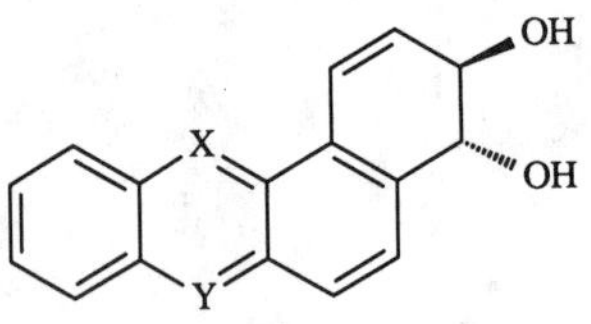

4   X = CH,   Y = N
5   X = N,   Y = CH

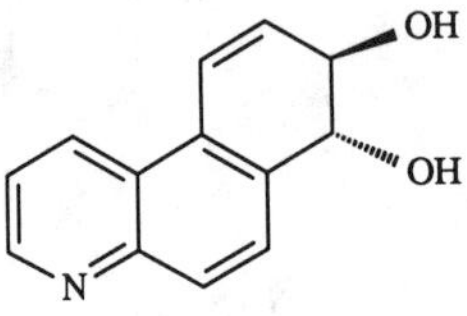

6

A formyl derivative is oxidized to the *N*-oxide of the carboxylic acid (but oxime, semicarbazone, or phenylhydrazones undergo ring *N*-oxidation without a change in the chain).[122,123] With the acylpyridine **7**, both Baeyer-Villiger oxidation and *N*-oxidation take place competitively.[124]

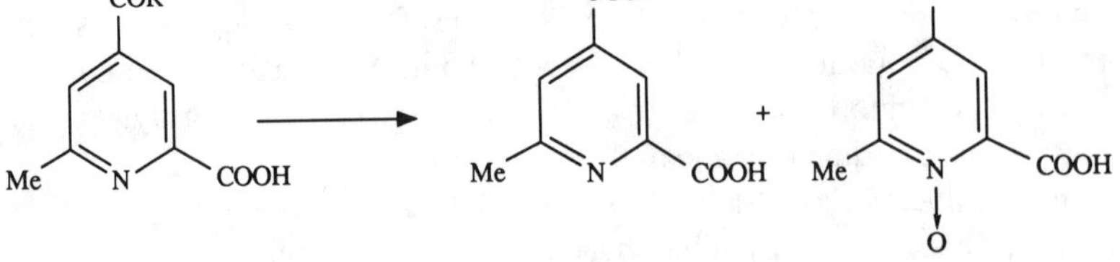

7   47%   31%

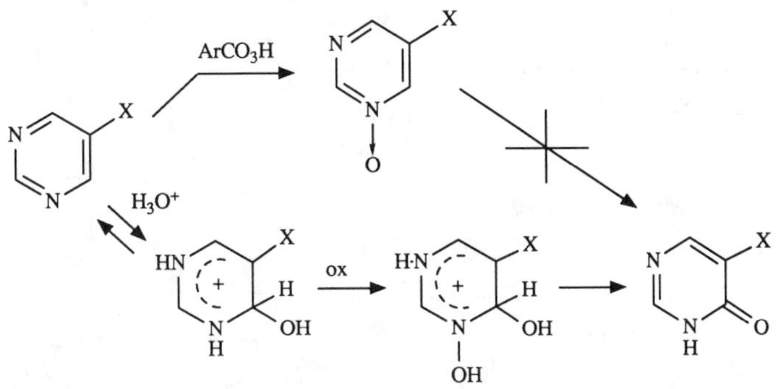

Scheme 1.

2. In the Ring

A side process in the N-oxidation of azines, usually accounting for at most a few percent of the products, is attack at a α or β ring carbon to yield lactams or β-hydroxyazines, respectively (e.g., from the reaction of 2-methylquinoline with peracetic acid a 4% yield of 3-hydroxy-2-methylquinoline is obtained[125]). In some cases, however, and in particular with electron-poor or hindered azines, the proportion changes, e.g., β-hydroxylation is the main process from 6-nitroquinoline and several 8-substituted quinolines.[126] 4,10-Diazapyrene yields only the 5,9-dioxo derivative,[127] and oxidation at a carbon atom predominates with several cinnolines[128,129] and quinazolines,[130] as well as with pyrimidines, unless substituents are present in positions 4 and 6.[70,75,131] For some of the last derivatives, it has been shown that formation of the N-oxides and of the pyrimidinone occur through independent pathways, and that in the latter case a covalent hydrate is the intermediate (Scheme 1, this path is favored under strongly acidic conditions).[131] Quinazolines also tend to give 4-quinazolinones[132] (Scheme 1).

Related pathways lead to ring cleavage or contraction, e.g., addition of peracetic acid at position 6 in pyrimidines explains the observed ring contraction to imidazoles[133] (see Scheme 2; a similar process has been reported for quinazolines).[134] 8-nitroquinoline yields 7-nitrooxindole via the 3-hydroxyquinoline (see above);[126b] cleavage of a fused 3-hydroxypyridine ring under oxidative conditions has also been reported for an anthridine[135] (Scheme 2).

3. Other Side Processes

(Partial) hydrolysis of labile groups (acyloxy[136] and, to a lesser degree, alkoxycarbonyl,[137] carboxyamido,[138] and in some cases, also carbonitrile)[139] occurs when peracetic acid is used for N-oxidation, and can be avoided by using aromatic peracids in nonpolar solvents (e.g., 3-benzoyloxypyridine gives the 1-oxide under this condition;[140] however, 4-acyloxy-pyridine and quinoline undergo hydrolysis even with perbenzoic acid).[136] Halogen atoms are sometimes affected, as has been observed for 1-haloisoquinolines[141] and pentabromopyridine.[142] 2,6-Disubstituted-4-chloropyrimidines yield 4-pyrimidones with peracetic and 1-oxides with permaleic acid.[143]

Other side reactions include the formation of an azoxy derivative from 4-azido-2-methylpyridine[144] and extrusion of SO$_2$ from the quinoxaline **8**.[145]

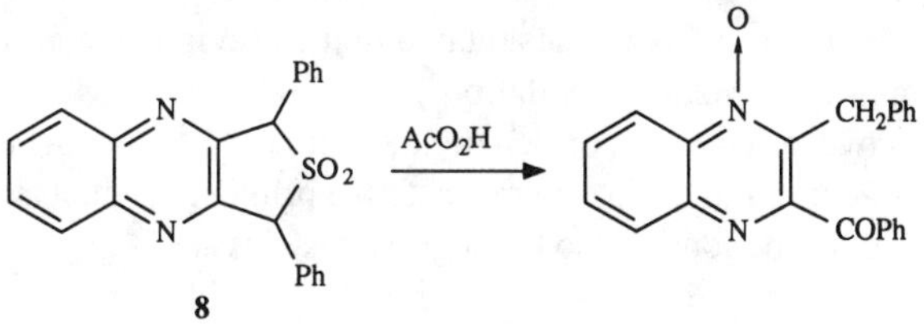

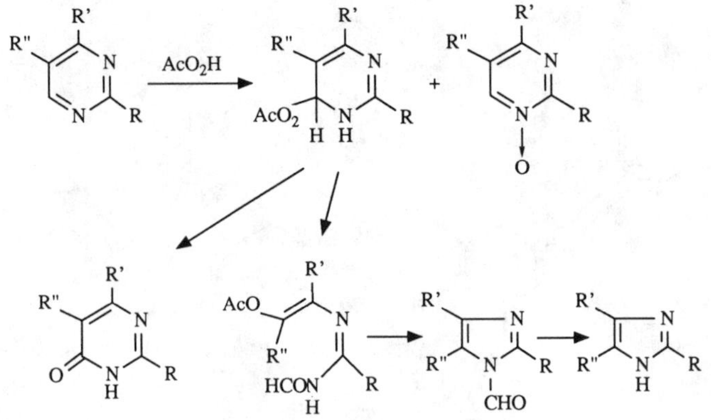

Scheme 2.

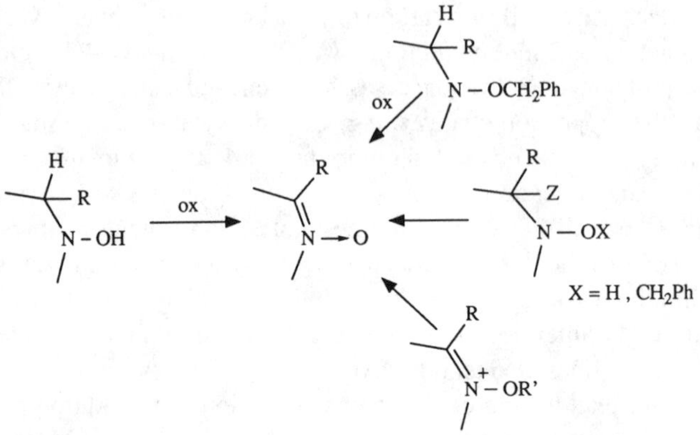

Scheme 3.

D. OTHER PREPARATIONS FROM PREFORMED HETEROCYCLES

Another approach starts from heterocycles that already contain the exocyclic oxygen atom, in most cases from cyclic hydroxylamines and their derivatives. Oxidation of secondary hydroxylamines is, in fact, a main entry for nitrones, including the cyclic derivatives, and has been applied particularly to the synthesis of compounds containing the isolated nitrone chromophore, although polyunsatured or aromatic compounds have also been obtained by this way (Scheme 3).

Typical conditions are reaction with inorganic oxidants, such as HgO,[146,147] PbO$_2$,[148-150] MnO$_2$,[151,152] SeO$_2$,[153] and Cu(II) acetate,[156] but palladium-catalyzed dehydrogenation,[155] and photochemically induced electron-transfer have also been used.[154] Four-,[149] five-,[147,153,157] six-,[148,150,152] and seven-membered[146] heterocycles have been prepared; in several cases, e.g., tetrahydropyridine[156] or 1,4-dihydrooxazine N-oxides,[151b] the nitrones are not stable and easily dimerize or polymerize. Oxindole and other indole 1-oxides have been obtained from 1-hydroxyindoles by autooxidation,[158] oxidative dimerization,[159] and oxidative addition.[160-161]

There are sparse examples of syntheses related to the previously mentioned dehydrogenations of hydroxylamines, i.e., peracid oxidation of benzoyloxyamines[162] and ionic elimination of a group in α from hydroxylamines or benzoyloxyamines[163-164] (see Scheme 3). The electrocyclic ring opening of oxaziridines, in several cases the primary products from the oxidation of imines (see beginning of Section I), also belongs to this class.

A nonoxidative approach involves demasking of the N-oxide function from its derivatives, notably hydrolysis of the salts of *N*-alkoxy heterocycles, a nice example being the cleavage of the oxazolo[165] or the oxadiazolo[166,167] ring in polycyclic derivatives such as the perchlorate **9**.[166]

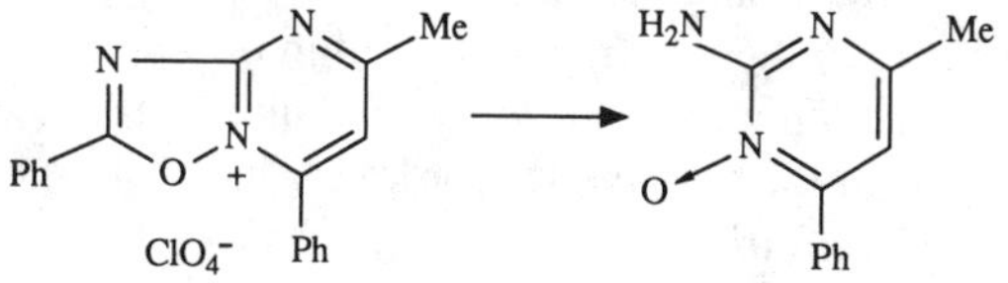

9

II. SYNTHESIS OF THE HETEROCYCLIC RING

The preparation of heterocyclic *N*-oxides by synthesis of the ring can be accomplished in three basic ways, that is (i) formation of the *N*-oxide chromophore by reaction of two functions present in the same molecule, (ii) formation of the heterocyclic ring by joining two (or more) molecules, one of the connections originating the *N*-oxide function,* or, (iii) intra- or intermolecular cyclization of open-chain compound(s) containing the *N*-oxide function (see Scheme 4).

The most common syntheses occur through paths i and ii, and will be discussed according to the function involved in the formation of the *N*-oxide function, independently from the intra- or intermolecular nature of the reaction and from the dimension of the ring formed. Classification is made on the basis of the function from which the *N*-oxide group comes, or is thought to come, and follows the degree of oxidation, thus subsequently including hydroxylamine derivatives, then nitroso and nitro derivatives. Syntheses from nitrile oxides (which formally might be considered to be examples of both paths ii and iii), and syntheses from preformed *N*-oxides are then discussed (Scheme 4).

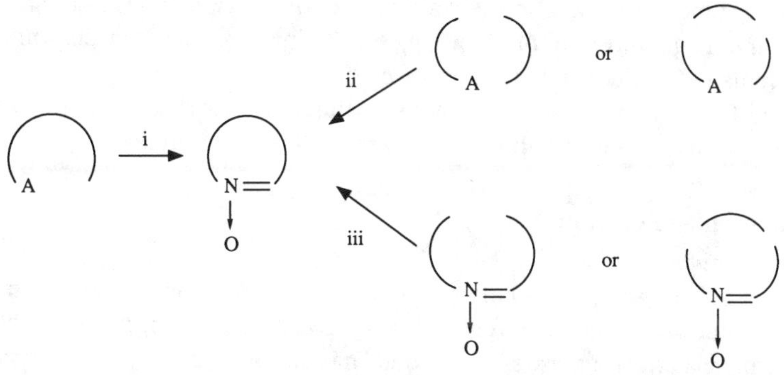

Scheme 4.

The classification is probably oversimplified and not unambiguous. At least in some cases, it appeared more useful to class the reactions according to the chemical function(s) present in the significant (though possibly hypothetical) intermediate, rather than in the starting material. Thus, the reaction of a diketone with hydroxylamine is classed as an intramolecular process from the mono-oxime of a ketone; likewise, reactions involving the preparation *in situ* of a hydroxylamino derivative from a nitro derivative are classed among the reactions of the former compounds, with only cross reference in the section devoted to the latter one. On the other hand, in other cases classification depends on the starting material, either because several mechanisms are possible or for convenience, e.g., because a small number of examples discourages one from

---

* What is meant here is that the heterocyclic ring is built by forming two (or more) bonds. This usually corresponds to the reactions of two molecules, but of course there are examples of the intramolecular version of this pathway, and this leads directly to a bicyclic *N*-oxide, see e.g., intramolecular synthesis of fused 1,2-oxazine 2-oxides in Section IIC2b2.

creating a specific section. In particular, syntheses probably involving an intramolecular redox process (as opposed to net reduction or oxidation of the substrate by an external reagent) are considered according to the original function.

The emphasis given here to the functions involved in the cyclization should give a better representation of the chemistry involved and should possibly suggest new applications. On the other hand, a recapitulation of the syntheses according to the nucleus formed (with examples) is offered at the end of the section.

The preparation of *N*-oxides by building the ring is not only an alternative to *N*-oxidation of preformed heterocycles, but may also be an original approach to that nucleus (see, in particular, the syntheses from nitro compounds presented in Section IIC), and indeed in several cases synthesis of the *N*-oxide and deoxygenation (see Chapter 4) is a convenient entry for the nonoxidized heterocycles.

It is difficult to draw a clear-cut distinction between reactions leading to the isolated nitrone function and those leading to aromatic *N*-oxides; therefore, the scope of this section is broader than in the other chapters, and includes both fully unsaturated and partially saturated heterocycles (not, however, aliphatic *N*-oxides), as well as tautomeric *N*-hydroxy derivatives.

A. FROM FUNCTIONAL DERIVATIVES OF HYDROXYLAMINE

A large variety of heterocyclic *N*-oxides have been obtained from derivatives of hydroxylamine, and most of the syntheses employed reproduce the well-known pathways for the synthesis of the corresponding nonoxidized heterocycles from ammonia derivatives.

Intramolecular and intermolecular reactions are successively considered. Within this classification, derivatives of hydroxylamine are dealt with according to their increasing degree of oxidation (hydroxylamines, oximes, hydroxamic acids, etc.). The general trend observed is the same in the three groups of compounds, and, other factors being identical, an increase in the oxidation of the function present in the starting material obviously leads to an increase of oxidation (or in conjugation) of the product, e.g., from a simple nitrone to an unsaturated nitrone (or a heteroaromatic *N*-oxide) and to an oxo-*N*-oxide.

Oximes and their potential tautomers, α,β-unsaturated hydroxylamines, are considered in either of the relevant sections, usually following the authors' formulation.

1. By Intramolecular Reaction

a. Of Hydroxylamines

The hydroxylamino group is usually introduced *in situ* by reduction of a nitro group or through a substitution reaction. Therefore, the easily available aromatic nitro derivatives are an attractive starting material for these syntheses, and most heterocyclic *N*-oxides prepared in this way contain a fused carbo- or hetero-cyclic ring.*

The choice of the reducing agent is critical, both in order to avoid overreduction of the nitro to an amino group (and thus cyclization to the unoxidized heterocycle) and reduction of the other group, which is meant to condense with the hydroxylamino, in particular, when this is a carbonyl group (Scheme 5).

1. With a Carbonyl Group

The condensation of a hydroxylamino with a carbonyl group (Scheme 5) is the most largely used method of this class. The former group often arises from the in-situ reduction of a nitro group, and the yield obtained depends on the reducing agent, with variations, even within a homogeneous series of substrates (e.g., 1-pyrroline 1-oxides are obtained from the reductive cyclization of γ-nitroalkylketones, the best results being obtained with Zn/NH$_4$Cl for the 5,5-dialkyl-2-aryl-, and with Fe and mildly acidic solutions for the 2-alkyl derivatives).[168] Typical

---

*   Reactions involving intramolecular redox processes of nitro compounds or in which the reduction to the hydroxylamine stage is not unambiguously defined are discussed in Section IIC.

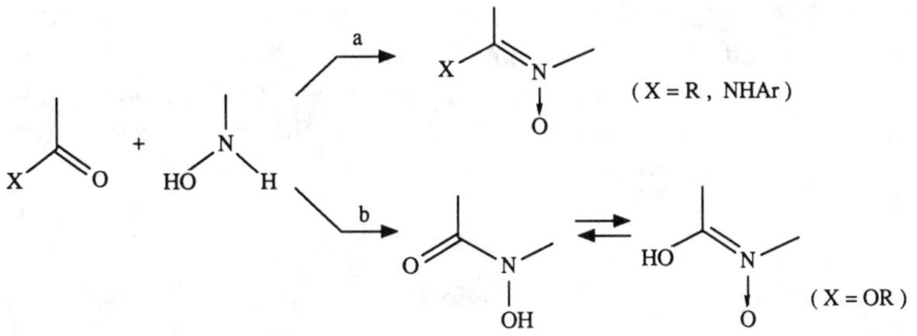

Scheme 5.

reducing agents are Zn/NH$_4$Cl,[168-172] Zn,[173] or Fe[168] and acids, TiCl$_3$/hν,[174] as well as electro-chemical reduction, a method that allows for the identification of the hydroxylamino intermediate.[175,176] Catalytic hydrogenation often gives clean results, both with aromatic[177-179] and with aliphatic nitro compounds.[180] Hydrogen transfer from ammonium formate in the presence of Pd/C is also effective.[181]

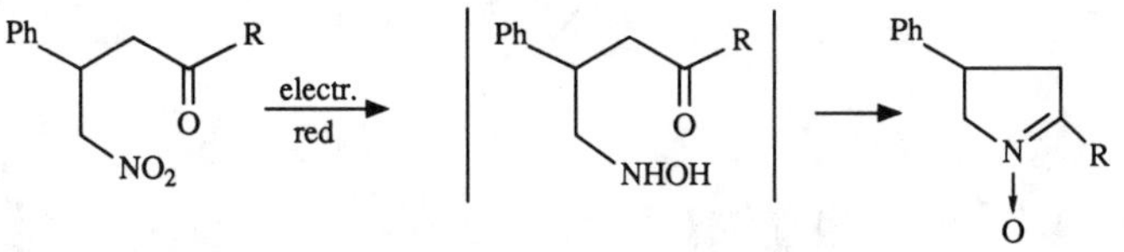

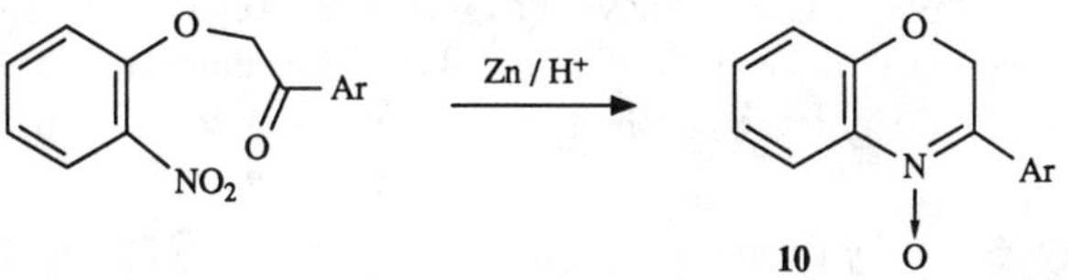

Apart from the above-mentioned synthesis of pyrroline 1-oxides from γ-nitroketones,[168,180,181] the most largely exploited path starts from aromatic nitro compounds. Cyclization involves a carbonyl group in a *ortho* side chain, and yields indolenine 1-oxides if the group is in the β position[170] (or, respectively, isatogens with α,β-diketones),[179b] or 3,4-dihydroquinoline 1-oxides if the group is in γ[171] (or, respectively, quinoline 1-oxides when there is a hydroxy group in the α or in the β position,[182] and 1-hydroxy-4-quinolones from α,γ-diketones).[183] 1,4-Benzoxazine N-oxides **10** are conveniently obtained from *o*-nitrophenoxyacetophenones.[169]

The hydroxylamino group can also be introduced by substitution of a halogen atom, rather than by reduction of a nitro group. Thus, imidazo[1,2-*a*]pyridine 1-oxides have been obtained from salts of 2-chloro-1-phenacylpyridinium (and probably the reaction involves substitution of the ring halogen to yield hydroxylamine **11**)[184] and 3,4-dihydroisoquinoline 2-oxide from *o*-(2-bromoethyl)benzaldehyde.[185]

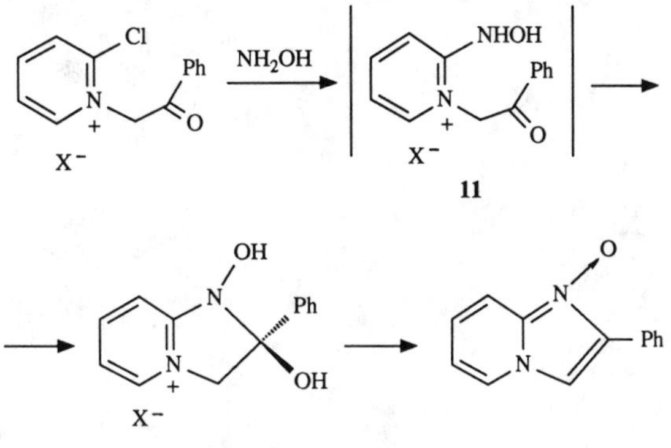

α,β-Unsatured hydroxylamines, or the tautomeric oximes, are intermediates in the synthesis of 4-hydroxylaminopyridine 1-oxides from γ-pyrones,[186a-186d] as well as in the synthesis of pyridine[186e] and isoquinoline *N*-oxides[186f] from, respectively, pyrilium and 2-benzopyrylium salts, a method that is advantageous over the *N*-oxidation of the azine, e.g., for polyarylpyridine 1-oxides.[186e]

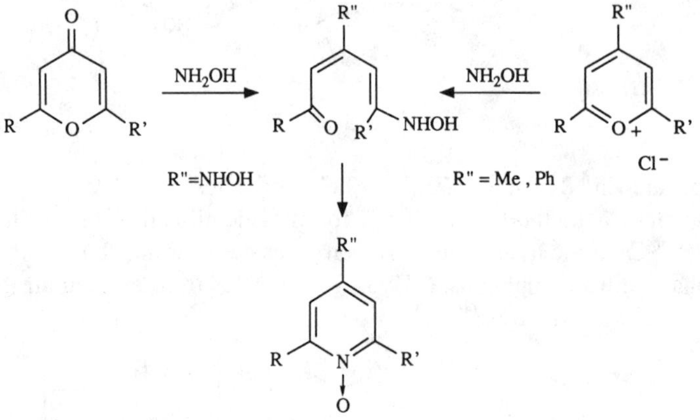

2. With the Carboxyl Group of Amides

This reaction proceeds in the same manner as when aldehydes and ketones are used (Scheme 5, path a). Reported examples involve *N*-acyl derivatives of aromatic amines, as typified by the synthesis of benzimidazole 1-oxides by reduction of *o*-nitroacetanilides.[187] Recent procedures mostly make use of catalytic hydrogenation in an acidic medium, often with excellent results.[188]

3. With the Carboxyl Group of Esters and Acids

The condensation of hydroxylamines with esters follows the general scheme of acylic substitution and leads to cyclic hydroxamic acids, which are tautomeric with α-hydroxy *N*-oxides (Scheme 5, path b). Six-membered heterocyclics with a fused benzo ring (quino-lines,[189,190] benzothiazines,[191] benzoxazines[192]) have been obtained with this method, usually from the aromatic nitro derivatives with various reducing agents; historically the first example is the synthesis of 1-hydroxy-2-quinolone by the reduction of ethyl *o*-nitrocinnamate with H_2S/NH_3.[189] Other reagents are zinc dust-ethanol,[194] and, particularly convenient, sodium borohy-dride and Pd/C;[190,191,193] under this condition *E* ethyl α-cyano-*o*-nitrocinnamate (**12**) yields 3-cyano-1-hydroxyquinolin-2(1*H*)-one due to fast *E/Z* isomerization,[193] while the other possible condensation involving the cyano group takes place by catalytic hydrogenation (see the following section). *o*-Nitrophenylpyruvate ester yields 3-amino-1-hydroxy-2-quinolones with $NaBH_4$, but not with H_2.[190,196]

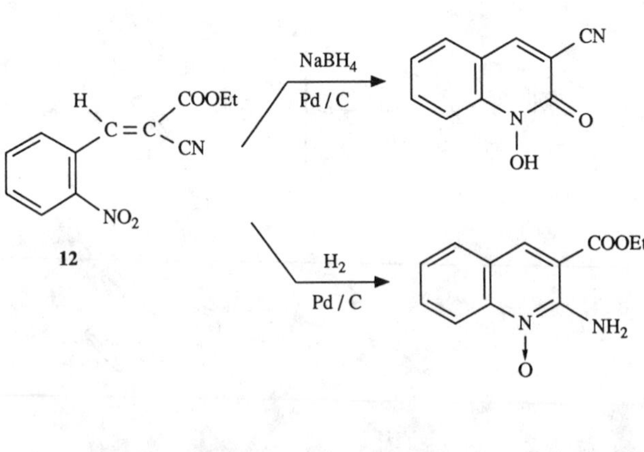

Nitro groups in the side chain analogously undergo reduction and cyclization, e.g., an isoquinoline derivative is obtained from 2-(2'-nitroacetyl)benzoic acid on treatment with $SnCl_2$.[195]

4. With a Cyano Group

The condensation with a cyano group leads to amino-substituted heterocyclic N-oxides, e.g., to 2-amino pyrroline 1-oxides from β-nitroalkylnitriles.[197] Benzo-fused five-membered rings are similarly obtained, e.g., 2-aminoindolenine 1-oxides from the reduction of o-nitrophenylacetonitrile derivatives with Sn/HCl[198] and 2-aminobenzothiazole 3-oxide from the cathodic reduction of o-nitrothiocyanobenzene.[199]

The results are analogous for six-membered heterocycles, e.g., 2-amino-3-phenyl- and 3-carboxyethyl-quinoline 1-oxides are obtained from the corresponding α-substituted o-nitrocinnamonitriles.[200a] Such products are likewise obtained by electrochemical reduction of Z 2-nitrocinnamonitrile[200b] and related carbonyl and carboxyl derivatives.[200c] If the side chain is saturated, i.e., starting from o-nitrophenylpropanenitriles, 2-amino-3,4-dihydroquinoline 1-oxides are obtained.[201] Similarly, o-nitrobenzylthiocyanates,[202] 1-(o-nitrophenyl) carbamoylnitrile,[203] and o-nitrobenzylcyanamide[203] yield 2-amino-benzo-1,3-thiazine, -quinoxaline and -quinazoline N-oxides, respectively. Seven-membered rings have been built analogously.[204b]

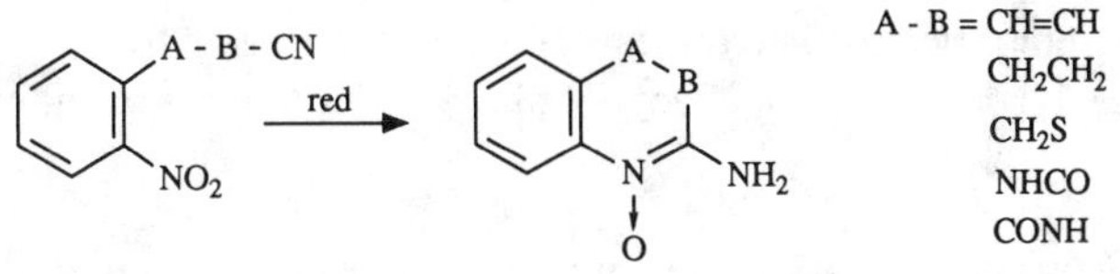

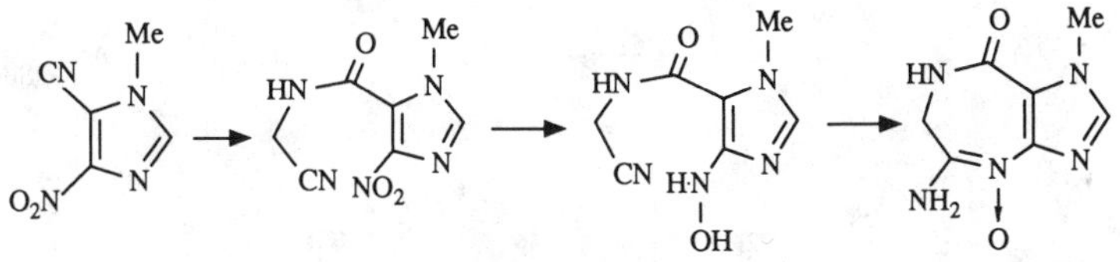

As in the previous cases, unsaturated hydroxylamines participate in this condensation, and this is the case in the reaction of enaminonitrile **13** with NH_2OH to yield a 2-aminopyridine 1-oxide;[204a] alternatively, such unsaturated hydroxylamines are obtained by base-catalyzed addition of ylidenemalonitriles ($R'' = CN$) to nitrile oxides (in this case, the cyclization to aminopyridine 1-oxides is accompanied by the formation of isoxazoles).[205]

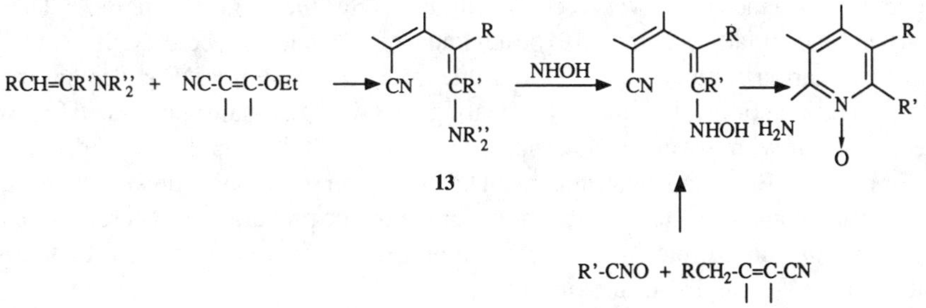

b. Of Oximes

1. With a Carbonyl Group

This condensation may be formulated either as involving the oxime function as such or, provided that there is a hydrogen atom in the β position, as taking place via the tautomeric α, β-unsaturated hydroxylamine; in this sense, several relevant examples are reported in the previous section. Here again, the oxime group is usually prepared *in situ*, though in this case, contrary to what happens with hydroxylamines, the oxime can often be isolated as such, and the following cyclization requires catalysis by strong acids.[206]

There are sparse examples of five-membered *N*-oxides synthesized in this way, e.g., 3*H*-3-oxopyrrole 1-oxides from the oxime of 1,3,4-triphenylbutane-1,2,4-trione,[207] and derivatives of naphto[2,3-*a*]pyrrole 1-oxide from 2-methyl-2-{2-(1,4-naphthoquinonyl)}propenal.[208] However, most of the reported examples involve the synthesis of azine *N*-oxides by reaction between 1,5-dicarbonyl derivatives and hydroxylamine. In this way, pyridine 1-oxide is obtained from glutaconaldehyde,[209] and 4,4′-azoxy-2,6-lutidine 1,1′-dioxide is obtained from diacetylacetone.[210a] 2-Azaquinolizinium oxides can be obtained analogously.[206] A limitation of the method is the poor availability of the required dicarbonyls. Alternative access to 3-alkoxypyridine 1-oxides is from 6*H*-6-alkoxy-3-acyl-4,5-dihydro-1,2-oxazines, via a δ-acyloxime.[210b]

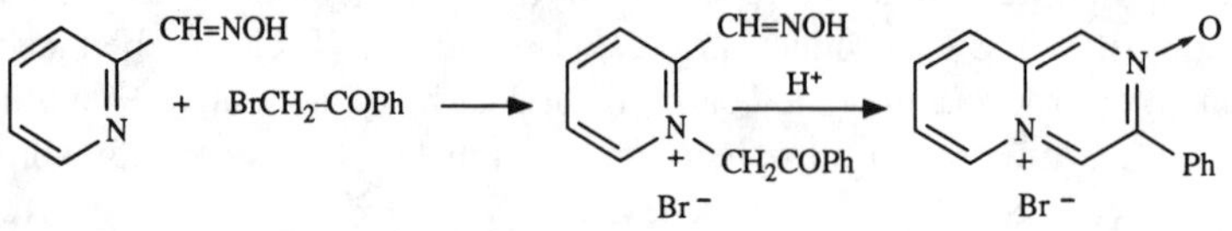

2. With the Carboxyl Group of Amides

The reaction between the amide and the oxime functions leads to the *N*-oxides of 1,3-diaza heterocycles, exemplified by the synthesis of the imidazole ring from the oxime (or rather, the corresponding silyl derivative) of *N*-acyl-*N*-phenyl-2-amino-2-propanone.[211]

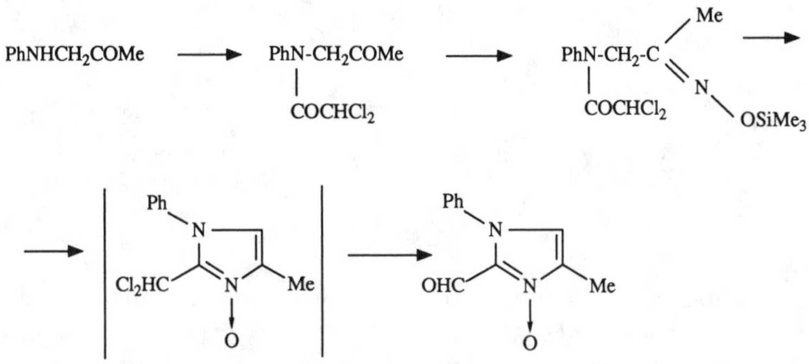

However, most of the reported reactions lead to quinazoline 3-oxides starting from the *o*-acylamino derivatives of aromatic aldo- and ketoximes. The *N*-oxide structure of the products obtained in the condensation was recognized much after the original reports.[212] These compounds had been initially considered to be acylindazoles[213] and, later, benzoxadiazepines.[214] Applications to various quinazoline[211,215] and thieno[2,3-*d*]pyrimidine *N*-oxides have been reported.[216] The reaction takes place by heating in HCl (gas)-saturated AcOH,[212] or more conveniently but with lower yield, concentrated sulfuric acid or polyphosphoric acid; with POCl$_3$ or PCl$_5$, the Beckmann rearrangement takes over; in some cases, however, cyclization to the *N*-oxide is spontaneous under conditions for the oxime preparation.[215] A related reaction is the synthesis of quinazoline 3-oxides from quinazolines by treatment with hydroxilamine, probably via 2-(*N*-formylbenzaldoxime).[132]

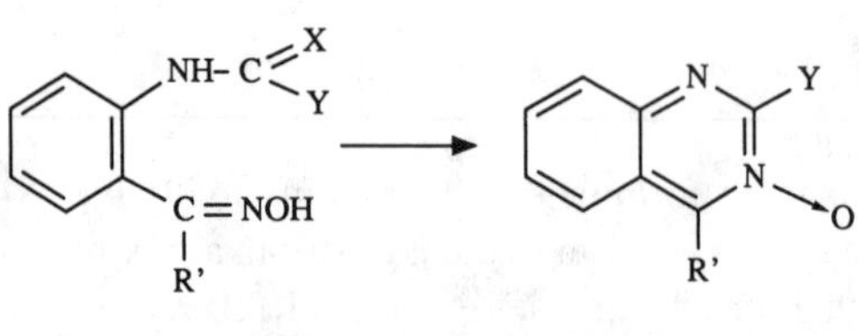

X = O , S   Y = R , NHR , SR

3. By Reaction with the Carboxyl Group of Acids and Esters, and with the (Thio)Carboxyl Group of Urethanes, Thiourethanes, and Related Derivatives

Again, as with hydroxylamines, the process taking place with esters and acids is substitution at the carboxyl group and leads, in this case, to cyclic hydroxamic acids. With (thio)urethans and related derivatives, one obtains either acylic substitution, as above, or condensation on the C=O (or carbon-sulfur) double bond, as with amides (see previous section).

Beginning from the last process, quinazoline 3-oxides are obtained via thiourea derivatives, e.g., by heating *o*-amino aromatic ketoximes with methylisothiocyanate,[217] or alternatively by treating *o*-amino ketones first with isothiocyanate and then with hydroxylamine.[218] The corresponding urethans require more drastic conditions (refluxing with $TiCl_4$ in benzene).

Related reactions lead to naphtho-1,2,4-triazine oxides[219-221] from the two monooximes of 1,2-napthoquinone (tautomeric with nitrosonaphthols) via the guanidylhydrazones[220,221] or methyl dithiocarbazones;[219] however, the semicarbazone of the 1-oxime yields a 2*H*-triazin-3-one 4-oxide.[221]

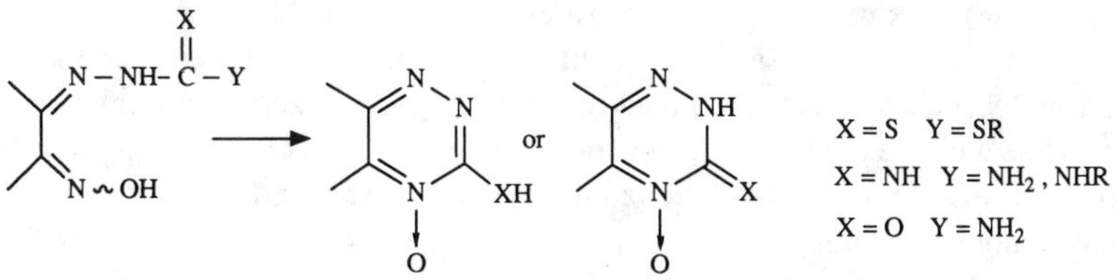

$$X = S \quad Y = SR$$
$$X = NH \quad Y = NH_2, NHR$$
$$X = O \quad Y = NH_2$$

Formation of oxo-*N*-oxides, as in the last case, is also observed from carbamic acids such as **14**.[222-223] On the other hand, *N*-ethoxyalkylidenhydrazones, such as **15**, cyclize to 1,2,4-triazine 4-oxides upon refluxing in high-boiling solvents.[224]

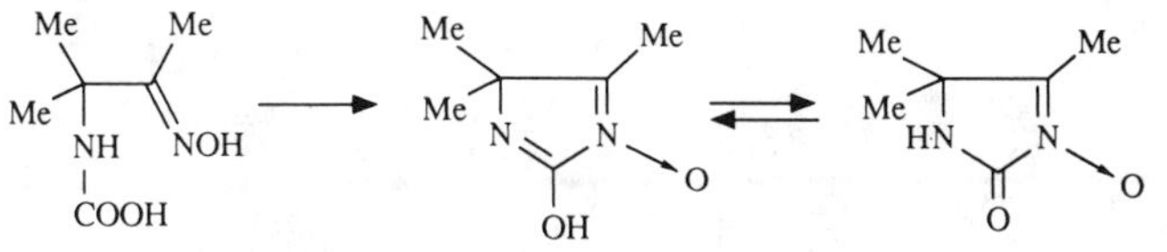

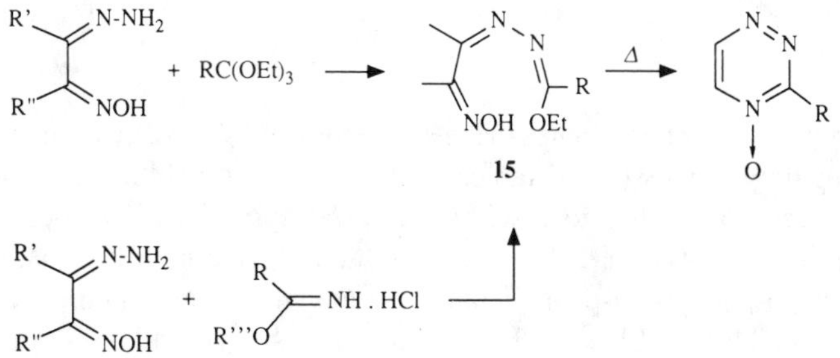

The oxime of *o*-carboxyphenylacetaldehyde yields *N*-hydroxycarbostyril when heated at 130°C.[225] 3-Hydroxyisoquinoline 2-oxides are obtained from 2-bromobenzaldoxime by condensation with malonic esters, β-keto esters, and ethyl cyanoacetate in the presence of sodium hydride and copper (I) bromide.[226a] Apparently only the ester function is involved in the cyclization, with no competition by the keto or the nitrile functions. Another method for preparing δ-carboxyl- (or cyano-) oximes is the attack of active methylene compounds to the isoxazole ring in isoxazole[3,4-*d*]pyrimidine derivatives, which likewise leads to the formation of a pyridine ring.[226b]

4. By Reaction with a Cyano Group

2-Amino *N*-oxides are formed in this case (compare Section IIA1a4). As an example, δ-ketonitriles yield 2-amino-3,4,5,6-tetrahydropyridine 1-oxides by treatment with hydroxylamine.[227] (Of course, depending on which group is attacked first, the cyclization might be formulated either as via δ-cyanooxime or via δ-ketoamidoxime, compare Section IIA1c1.[228]) Many of the reported reactions involve thiocyanates and lead to sulfurated heterocycles. 2-Aminothiazole 3-oxides[229,230] and 2-amino-1,3-thiazine 3-oxides[231] are obtained by treating either α-halo substituted oximes with barium thiocyanate[229] or α- or β-thiocyano aldehydes or ketones with hydroxylamine.[230,231]

5. By Displacement of a Leaving Group

In molecules containing both the oxime function and a saturated center open to nucleophilic attack, reaction under basic conditions usually take place according to both possible pathways, e.g., either the oxygen or the nitrogen atoms of the oxime group may be involved in the attack. Thus, γ- and δ-chloroketones with hydroxylamine cyclize both to pyrroline (or, respectively, tetrahydropyridine) 1-oxides and to 1,2-oxazines (or, respectively, 1,2-oxazepines),[232,233] the second path being favored by bases.[233] The configuration of the oxime is important, e.g., the *E* oximes of *o*-haloacetamidobenzophenones yield benzo-1,4-diazepine 4-oxides under basic conditions, while with the *Z* isomers the reaction is complex, with *O* alkylation predominating; with 2 mol of base, a good yield of a benzodiazepinone, resulting from further reaction of a benzoxadiazocinone, is obtained.[234,235]

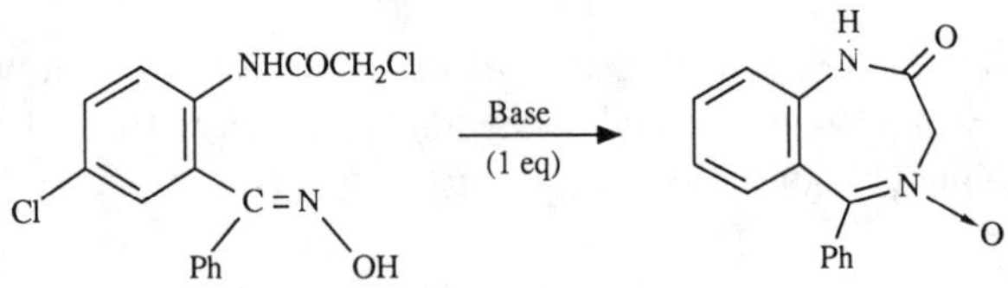

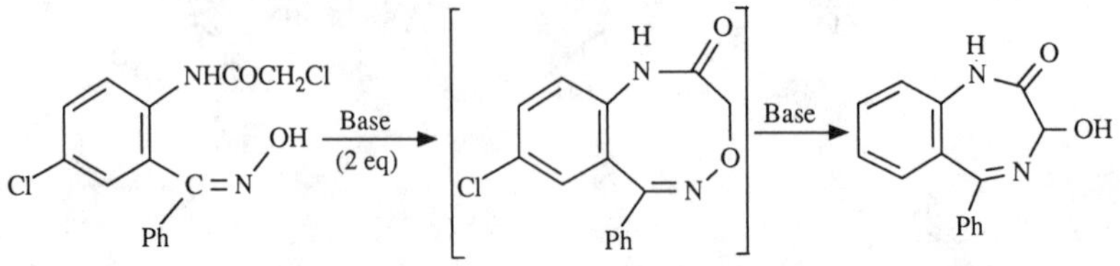

The same *E* oximes are intermediates in the conversion of 2-haloalkylquinazoline 3-oxide to benzo-1,4-diazepine 4-oxides (see Chapter 4).[235] Again, *E* α-(chloroacetamido)-ketoximes undergo base-induced cyclization to 3,6-dihydro-2(1*H*)-pyrazinone 4-oxides.[236]

Ring closure via elimination may also involve unsaturated centers and heteroatoms. Thus when 2,3,4,5-tetrachloro-5-phenyl-2,4-pentadienal is treated with hydroxylamine, cyclization with elimination of hydrochloric acid yields 3,4,5-trichloro-2-phenylpyridine 1-oxide.[237a] The intermediate unsaturated oximes can be obtained also by ring cleavage from *N*-arylpyridinium salts.[237b] The *E,E* dioximes of 1,3-diketones yield pyrazole 1-oxides when treated with SOCl$_2$; here -O-SO$_2$Cl is the leaving group from a nitrogen atom.[238] 2-Selenocyanatobenzophenone yields 2-phenyl-1,2-benzoisoselenazole 2-oxide when treated with hydroxylamine in pyridine, with CN as the leaving group (with the corresponding 3-methylacetophenone, however, the attack is at the C rather than the Se atom, and a 2-imino-1,3-benzoselenazine 3-oxide is formed; compare the reactivity of the sulfur analogs in the previous section.[239]

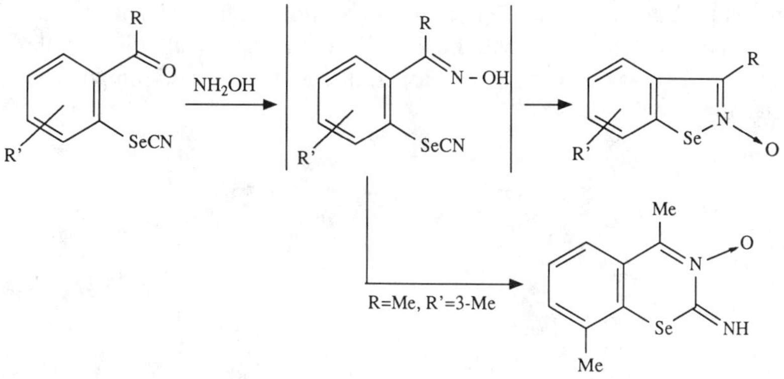

R=Me, R'=3-Me

6. With the Diazonium Group

The configuration of the oxime has an important role in this case also; thus, diazotization of the *E* oxime of *o*-aminobenzophenone yields benzo-1,2,3-triazine 3-oxide.[240] 2-Azapurine 1-oxides have been obtained in an analogous way.[241]

An interesting variation of this process takes place with α- or γ-methyl derivatives of aminoazines, since here nitrous acid attacks the methyl group (activated by the ring nitrogen) and the thus-formed oxime function reacts directly with the diazonium group. The reaction has been applied to 3-amino-4-methylquinolines[242] and 5-amino-4-methylpyrimidines.[243-244]

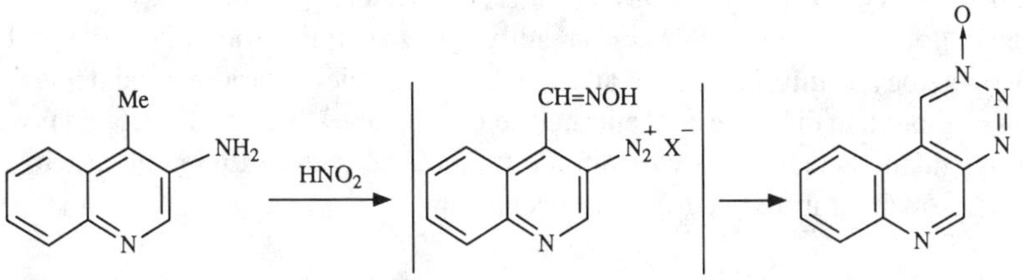

7. Other Reactions

There are isolated but interesting examples of cyclization involving carbon-carbon multiple bonds. Thus, the oxime function enters in a Michael addition onto α,β-unsaturated esters in compound **16**, yielding a tetrahydropyridine 1-oxide.[245] Isoquinoline 2-oxides have been obtained from *o*-ethynylbenzaldehydes, which in turn are easily available from *o*-bromobenzaldehydes.[246]

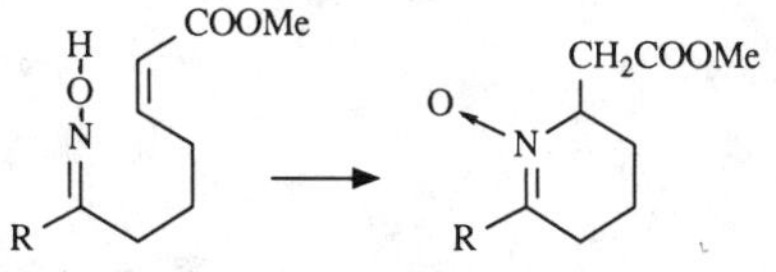

16

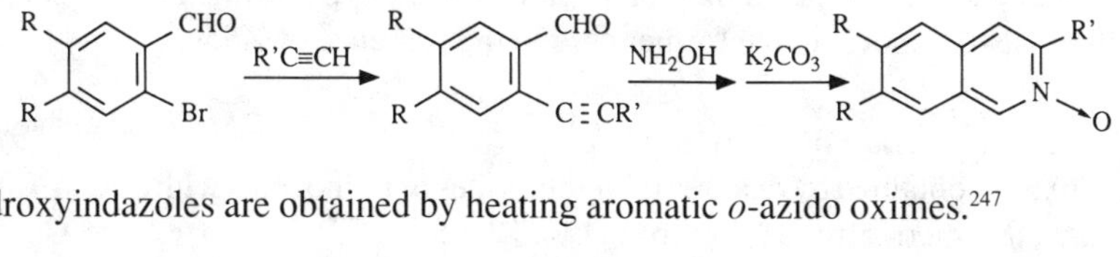

2-Hydroxyindazoles are obtained by heating aromatic *o*-azido oximes.[247]

c. Of Hydroxamic Acids, Amidoximes, and Related Compounds

1. With the Carbonyl Group of Aldehydes and Ketones

The reaction takes a course that is analogous to that observed with hydroxylamines. Thus, the hydroximinodithiocarbonic esters **17** cyclize with sulfuric acid to thiazole 3-oxides.[248] The δ-ketohydroxamic acids **18** cyclize with methanolic ammonia to yield directly, though in low yield, 1-hydroxy-2-oxopyrazines (racemic aspergillic acid and related compounds), rather than the expected dihydro derivatives.[249,250]

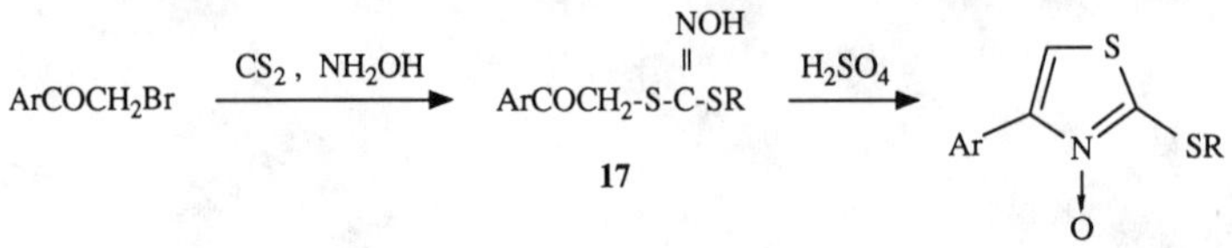

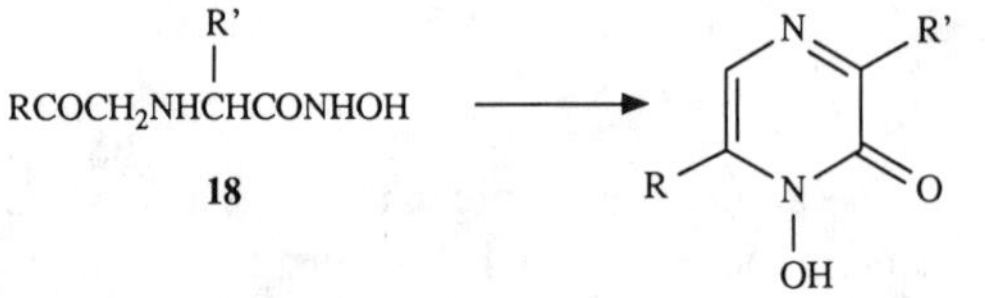

$$ArCOCH_2Br \xrightarrow{CS_2, NH_2OH} ArCOCH_2\text{-S-C-SR} \xrightarrow{H_2SO_4}$$

17

$$RCOCH_2NHCHCONHOH \longrightarrow$$

18

A similar pathway is involved in the conversion of 2-pyrones into 1-hydroxy-2-oxopyridine by reaction with hydroxylamine[251,252] (from 2-thiopyrone, 1-hydroxy-2-oximinopyridine is obtained).[252]

2. With the Carboxyl Group of Amides, Ureides, and Related Compounds

o-Acetamidobenzenhydroxamic acids readily cyclize to quinazoline 3-oxides.[253] The corresponding pyrazine and thiophene derivatives analogously yield pteridine[254] and thienopyrimidine oxides,[216] though in this case concentrated acid is required. A formally related process leads to 1-hydroxyguanine via the benzyloxyamide-thiourea derivative **19**, in this case by the action of MeI, and probably through a different mechanism.[255]

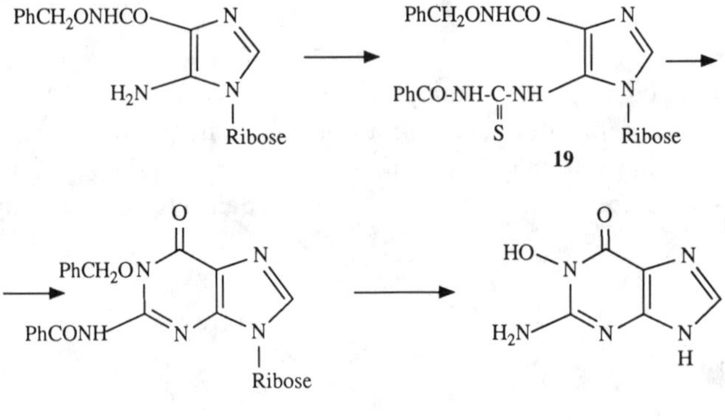

19

As for five-membered rings, 1-hydroxy-5-imidazolones[256] and 4-hydroxy-1,2,4-triazoles[257] have been obtained by thermal cyclization of β-acylaminohydroxamic and, respectively, N^3-hydroxy-N^1-acylhydrazones, in the former case in the presence of acids.

3. With the Carboxyl Group of Acids and Esters

The amidoximes obtained by reaction of the chlorides of iminoacids with esters of anthranilic acid readily cyclize thermally to 3-hydroxy-4-quinazolones.[258]

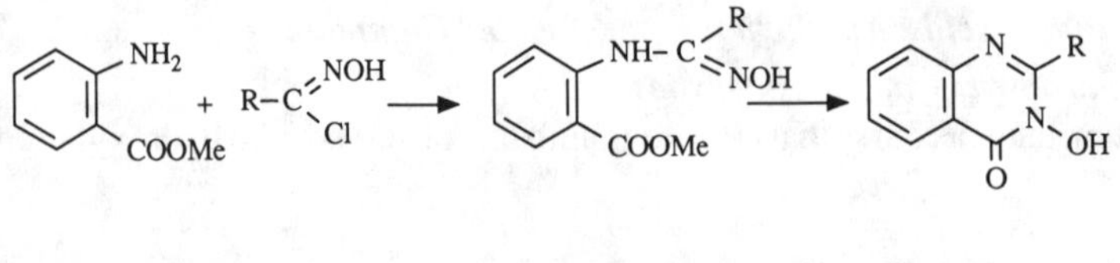

The same scheme for ring closure is found in the preparation of *N*-benzyloxyimides by the reaction of bicarboxylic (glutaconic, homophthalic) acids or their anhydrides with *O*-benzylhydroxylamine; deprotection by catalytic hydrogenation then gives *N*-hydroxy-pyridones, pyrimidones, or the analogous isoquinoline derivatives.[259,260] Both 1- and 3-hydroxyuracil have been prepared in a similar way, i.e., by condensation of a benzyloxyamide onto an urethan in the latter case, and by cyclization of the benzyloxyamide-acylvinyl ether **20** (vinylologous of an ester) in the former one.[261]

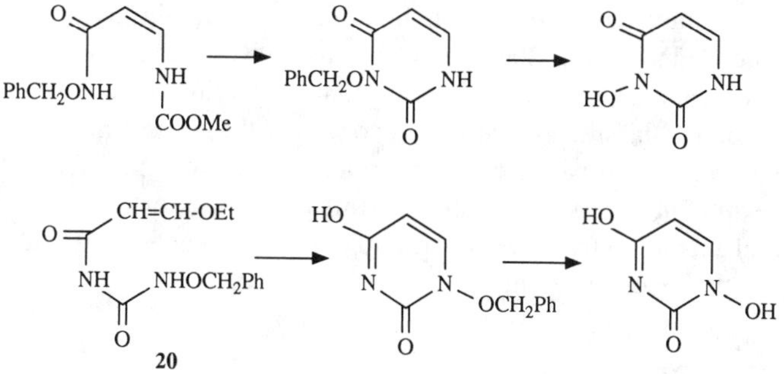

4. With a Cyano Group

Several syntheses of heterocyclic *N*-oxides by the reaction of bis-nitriles with hydroxylamine have been reported, and reasonably involve the intermediacy of amidoximes, though these have not been isolated.[262] Thus, triaminopyrimidine 1-oxides are formed when cyanoiminopropionitriles **21** are treated with hydroxylamine, and 2,6-diamino-1,3,5-triazine 1-oxides are obtained from the dicyanoimidate salts **22**.[263] With related substances, the reaction proceeds somewhat differently, (e.g., 2-amino-6-hydroxyaminopyrazine 1-oxide is obtained from iminodiacetonitrile)[264] or give no *N*-oxide, as in the case of glutaronitrile, which yields 2,6-dioximinopiperidine instead.[265]

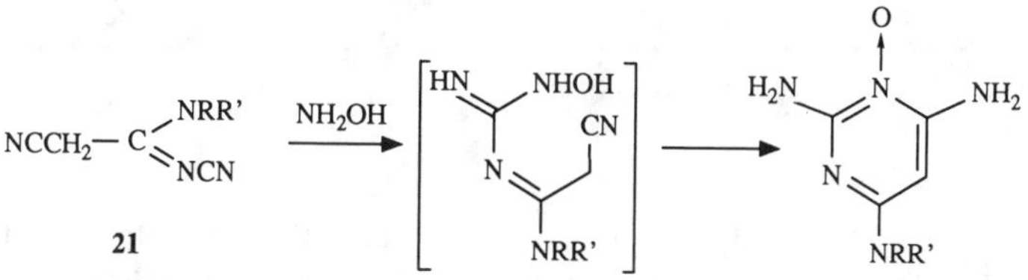

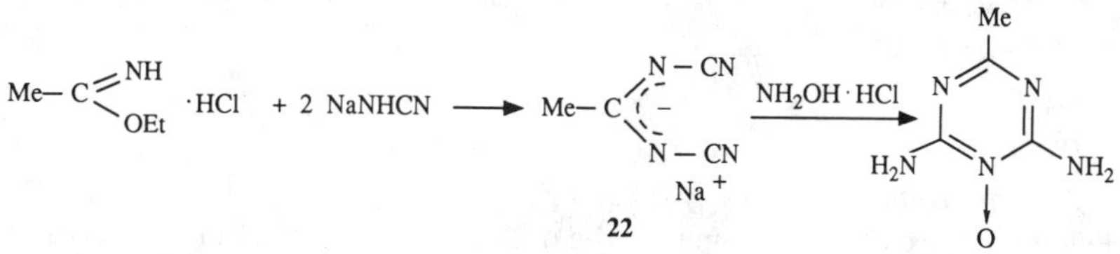

5. With a Diazonium Group

Benzo-1,2,3-triazine 3-oxides[253] and 2-azapurine 1-oxides[241] have been prepared from amidoxime precursors, similarly to the case of oximes.

6. By Elimination of a Leaving Group

2-Bromocinnamylideneglycine hydroxamic acid, but not the nonbrominated analog, cyclizes under basic conditions to 1-hydroxy-6-benzyl-2-pyrazinone.[266] The dichloroazine **23** gives a 4-hydroxy-1,2,4-triazole when treated with hydroxylamine.[267]

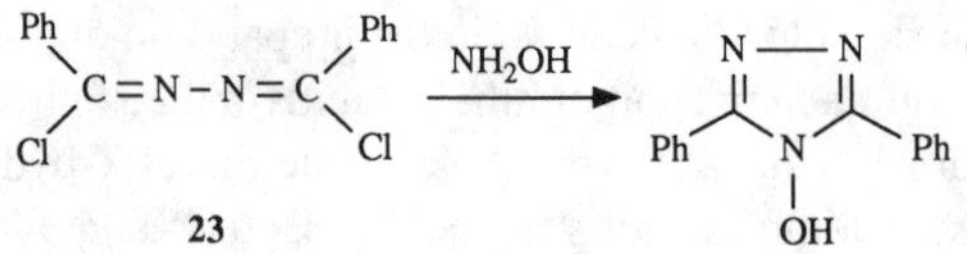

23

2. By Intermolecular Reaction

In the syntheses discussed in this section, the heterocyclic ring is built by forming at least two new bonds (path ii in Scheme 4), one of which simultaneously creates the *N*-oxide function. Thus, one of the compounds involved bears a substituent A containing a nitrogen and an oxygen atom, which is intended to enter in the *N*-oxide group, and a second substituent for the second bond; and the other reagent(s) also contain two suitable functions (each), or one function able to form both new bonds. Classification is done according to the group A in the first reagent (again hydroxylamine, oxime, hydroxamic acid, and related compounds in this order), and, when warranted by the amount of reported examples (this is the case for oximes), the reactions are further subdivided according to the second group in the first reagent and, finally, to the group reacting with A in the second reagent.

a. From Hydroxylamines

Organic hydroxylamines containing a hydroxy or amino group in α or β undergo condensation with orthoesters or amide acetals to yield cyclic nitrones, e.g., oxazoline 3-oxides from 2-hydroxylaminopropanols.[268] Position 2 in the indole ring enters in a similar reaction in the synthesis of carboline **24** from *N*-hydroxytrypthophan.[269]

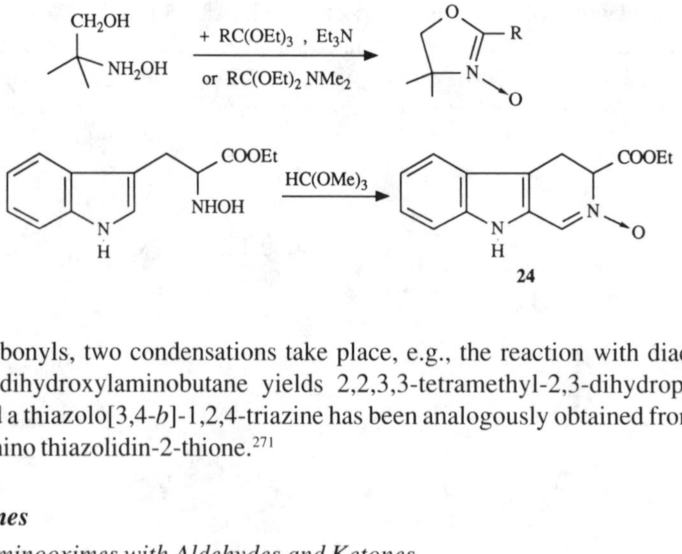

With dicarbonyls, two condensations take place, e.g., the reaction with diacetyl of 2,3,-dimethyl-2,3-dihydroxylaminobutane yields 2,2,3,3-tetramethyl-2,3-dihydropyrazine 1,4-dioxide,[270] and a thiazolo[3,4-*b*]-1,2,4-triazine has been analogously obtained from a 3-amino-2-hydroxylamino thiazolidin-2-thione.[271]

b. From Oximes

1. α- and β-Aminooximes with Aldehydes and Ketones

Condensation of a carbonyl group with the oxime and the amino function leads to five- (or six-, according to the position of the groups) -membered 1,3-diaza heterocycles. The first attack reasonably involves the more nucleophilic amino group, and the reaction has been envisaged as either involving an intermediate imine (**25**, and thus the mechanism is the same as in the intramolecular condensation between oxime and ketones or derivatives, see Section IIA1b1),[272a] or as proceeding via a zwitterion (**26**).[272b]

The stereochemistry of the oxime determines the course of the reaction. Indeed, in the first report on this subject concerning the reaction of benzaldehyde with the oximes of phenacylamines, it was noted that the Z-oximes yield imidazoline 3-oxides and the *E* isomers yield oxadiazines, and thus in the first case the nitrogen atom is involved in the nucleophilic attack, and the oxygen is involved in the latter one.[273] Later investigations supported this observation, at least when the reaction is carried out in ethanol; in acetic acid imidazoline oxides are obtained in both cases.[272b,274] In addition to the case of imidazolines, the reaction takes place analogously (from the suitable oxime isomer) in the synthesis of 1,2,5,6-tetrahydropyrimidine[275] and 1,2-dihydroquinazoline 3-oxides.[272a,276]

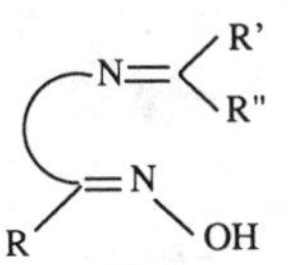

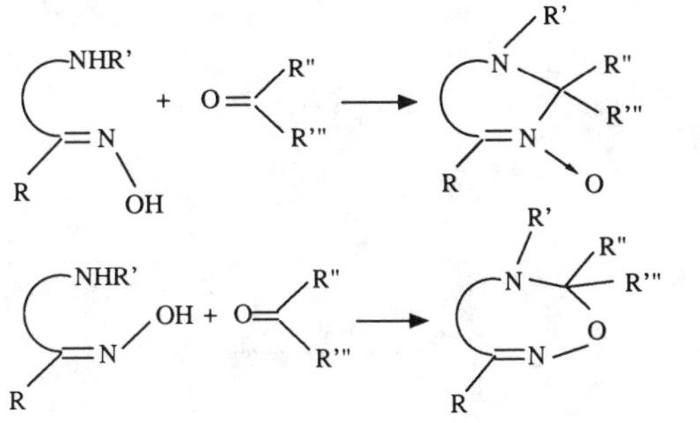

In a similar way, 2,3,5,6,7,8-hexahydrobenzo-1,2,4-triazine 4-oxide has been obtained via the hydrazone of 1,2-cyclohexandione mono-oxime, prepared *in situ* from an enamino-oxime.[277]

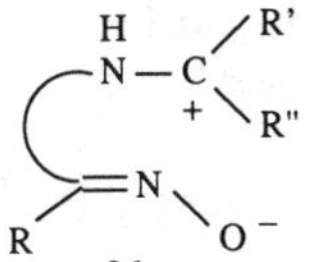

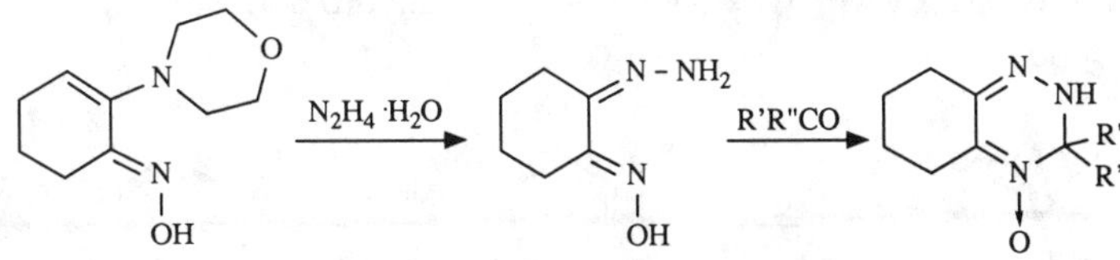

As in the previous section, reaction with dicarbonyl derivatives takes place with double condensation, e.g., reaction with 1,2-dicarbonyls yields pyrazine 1-oxides from α-aminoox-imes,[278] and 1,2,4-triazine 4-oxides from the hydrazone of 1,2-dicarbonyls mono-oximes.[224b]

2. α- and β-Aminooximes and Acid Derivatives

Already in 1891 von Auwers reported the formation of an heterocycle by reaction of *o*-aminoacetophenone oxime with acetic anhydride, though the structure of the product was first formulated as acetylindazole,[213] later as methylbenzoxadiazepine,[279-280] and was finally recognized as 2-methylquinazoline 3-oxide with the support of a crystallographic determination.[281]

The method has been largely applied, and, in addition to anhydrides,[213,279-281] formic acid,[234b,282] orthoesters,[283] *N,N*-dimethylformamide dimethylacetal,[282] acyl chlorides,[284] functional derivatives of carbonic acid such as phosgene, thiophosgene, and ethyl chlorocarbonate,[273,285-287] as well as phenylboronic acid[288] have been used for the synthesis of five-membered[273,282,285,286] and six-membered[283,284,287] diaza heterocyclic *N*-oxides, or their fused benzo or hetero derivatives (oxo-substituted *N*-oxides with phosgene and analogs, a dihydrobenzo-1,3,2-diazaborine 3-oxide from 2-aminobenzaldoxime and phenylboronic acid).[289] It is likely that an amide is formed in the first step, and therefore this reaction is strictly related to that discussed in Section IIA1b2. Related ring-opening/ring-closure processes have been reported, e.g., 2-aminopyrimidine 1-oxide from 2-chloropyrimidine.[282]

The oxime stereochemistry is also determining in this case, with competition between N and O attack. A thorough study with 2-amino-5-methoxybenzaldoxime showed that the *E* isomer yields 5-methoxyquinazoline 3-oxide with ethyl orthoformate, while the *Z* isomer gives 7-methoxybenzo-3,1,4-diazepine.[289] Previous reports of *N*-oxide formation from unseparated *E/Z* mixtures of isomeric oximes has been attributed to preliminary geometric isomerization or, more simply, to the reaction of the *E* isomer alone. *E*-α-aminooximes yield imidazoline 3-oxides with orthoesters, phosgene, thiophosgene, and ethyl chlorocarbonate,[223,285,286,290] while their *Z* isomers give 1,2,5-oxadiazines.[290] 1,2,4-triazine 4-oxide are obtained from α-hydrazone-oximes with orthoesters.[224b]

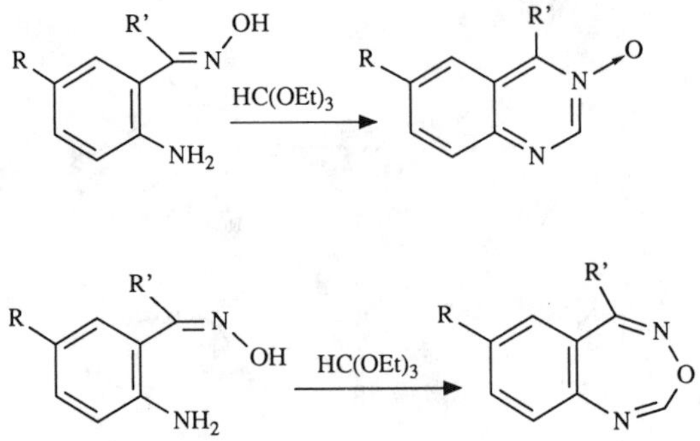

3. Hydroxylamino-Oximes with Aldehydes and Ketones

α- and β-hydroxylamino-oximes react with aldehydes and ketones and give 1-hydroxyimi-dazoline 3-oxides[291] and, respectively, 1-hydroxy-1,2,5,6-tetrahydropyrimidine 3-oxides.[152,292,293] The reaction produces a good yield and is a valuable source of the otherwise difficult to obtain pyrimidine 1,3-dioxides (by MnO_2 oxidation of the tetrahydro derivatives).[152]

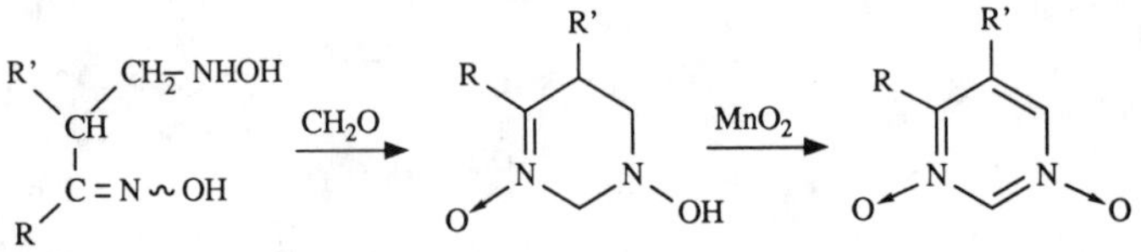

With α-dicarbonyls, the reaction involves either one or both of the groups, leading, respectively, to 2-acyl-1-hydroxy-3-imidazoline 3-oxides or to pyrazine 1,4-dioxides.[294,295] The former process is reversible.[295]

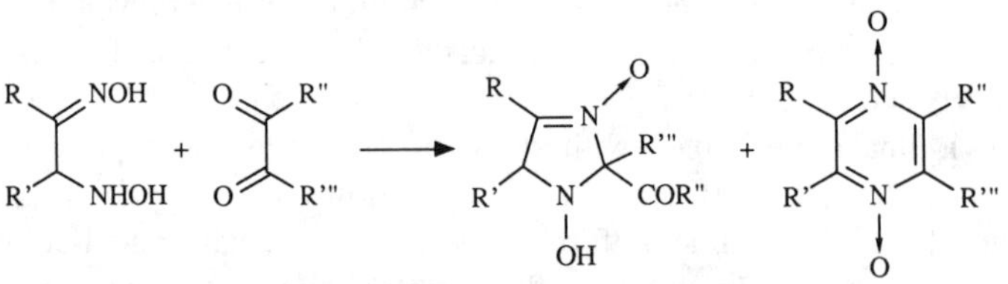

4. Oxo-Oximes with Aldehydes, Ketones, and Their Derivatives

Treating the mono-oximes of dicarbonyl derivatives with aldehydes in the presence of strong acids yields oxazole 3-oxides.[296-298] In some cases oxazoles, rather than the corresponding 3-oxides, are obtained, possibly because of mono-oxime dismutation to dioxime, and reduction of the latter to diimine by the aldehyde.[299]

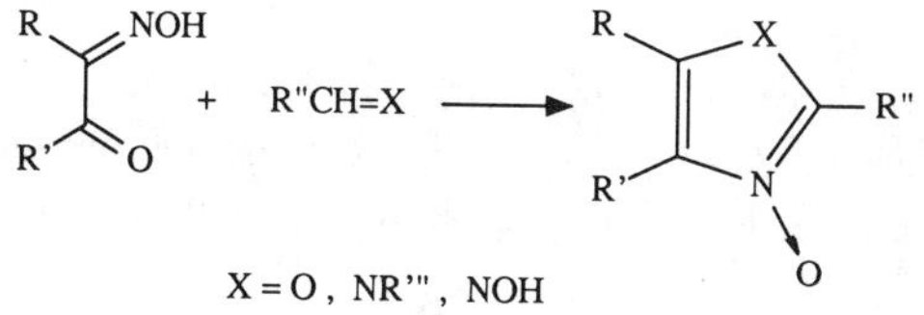

$$X = O, \ NR''', \ NOH$$

When an imine or an oxime is used in the place of the aldehyde, imidazole 3-oxides and, respectively, 1-hydroxyimidazole 3-oxides are obtained. As for the first reaction, both pre-formed imines[300,301] and a mixture of aldehyde and ammonia[302,303] can be used. The formation of the imidazole **27** when 3-oximinobutanone is treated with ammonium carbonate has been rationalized as a reaction of the imino group of 3-iminobutanone formed *in situ* with the starting material.[304] The cyclization of 6-amino-5-nitrosouracil, or its tautomeric oxime, with formal-deyde can be considered in this section, and leads to 7-hydroxyxanthine in low yield.[305]

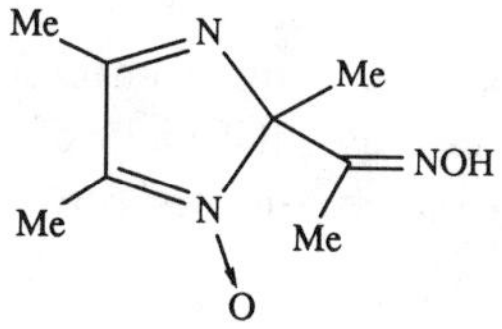

27

The reaction with oximes, originally reported to give 1,2,5-oxadiazines,[306] gives moderate to good yields of 1-hydroxyimidazole 3-oxides.[302,307,308] The formation of a tetrahydrobenzoderi-vative of this heterocycle from 2-hydroxy-2-methoxycyclohexanonoxime (obtained from the the nitrosation of cyclohexanone) in the presence of bases has been similarly rationalized, the same molecule playing both roles.[309] 1-Hydroxyimidazole 3-oxides can be conveniently obtained in a one-pot reaction from diacetyl, aldehydes, and hydroxylamine. The process might involve either the diacetyl mono-oxime and the aldoxime, or the dioxime and the unreacted aldehyde (compare Section IIA2b6).[310]

Although imidazole derivatives arising by reaction with a single carbonyl function are always formed in the syntheses mentioned above, even where the reagent has two potential sites for condensation, examples of the opposite behavior are known. Thus, acid hydrolysis of the dimethylacetal of phenylglyoxal 2-oxime yields a dihydroxydihydropyrazine dioxide,[311] and the reaction of the dimethylacetal of pyruvaldehyde 2-oxime with an α-acylaminoaldehyde offers a convenient approach to some pyrazine oxides.[312]

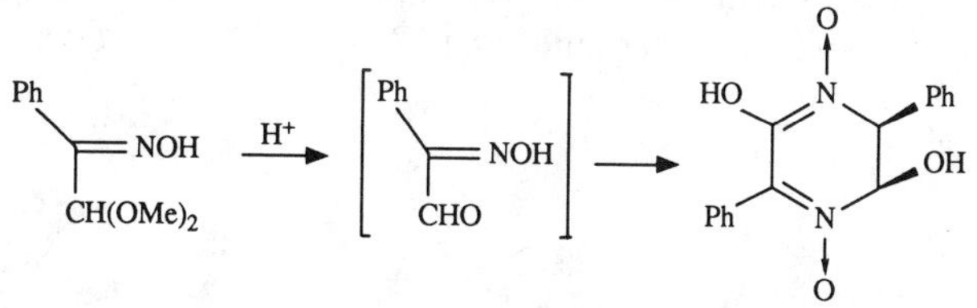

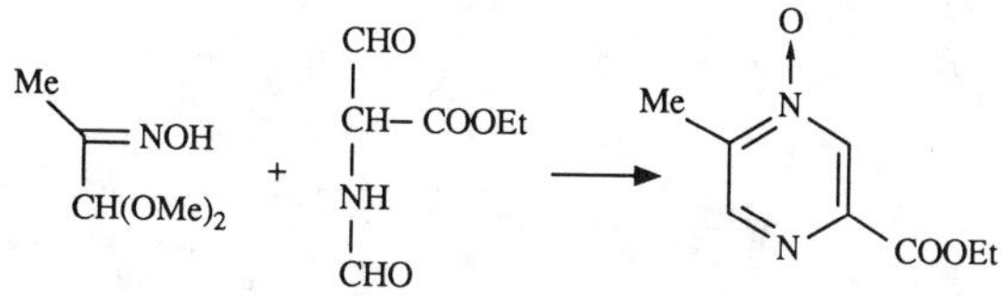

5. α-Oximinoketones with Aminonitriles

The reaction of pyruvaldehyde 1-oxime with α-aminopropionitrile affords 2-amino-3,5-dimethylpyrazine 1-oxide in moderate yield.[313] This synthesis has proven valuable for preparing 4 substituted pyrazine mono-oxides, which are useful as sulfamidic[314] or as intermediates for the synthesis of pteridines.[315-317]

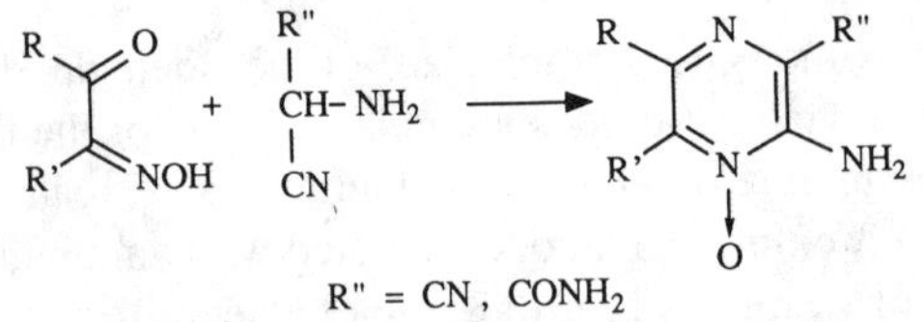

$$R'' = CN, CONH_2$$

The latter line of research has been extensively developed and put to use for the synthesis of natural compounds.[317-322] Noteworthy, *inter alia,* are the syntheses of L-*erytro*-biopterin (**28**) from the aldoxime obtained from 5-deoxy-L-arabinose,[317] as well as of DL-biopterin derivatives through a totally synthetic approach starting from crotonic acid,[320] and the synthesis of *Cypridina* etioluciferamine.[318] In one case, the yield of aminopyrazine oxide is drastically increased (from 3 to 51%) when the reaction is carried out in the presence of titanium tetrachloride.[318]

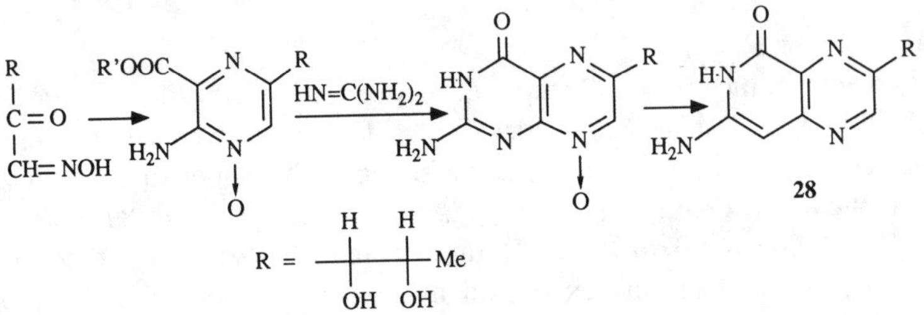

6. Dioximes with Aldehydes, Ketones, and Oximes

α-Dioximes react with aldehydes, ketones, or their oximes in the presence of strong acids and yield imidazole derivatives. When oximes are the reagent, one of the three oximino groups is probably hydrolyzed prior to condensation.[323] Thus, these reaction are related to those discussed in Section IIA2b4.

Thus, 1-hydroxyimidazole 3-oxides are obtained from the condensation of glyoxal dioxime with formaldehyde[324] and aromatic aldehydes,[325] and likewise of diacetyl dioxime.[310] The corresponding reaction with ketones affords 2*H*-imidazole 1,3-dioxides.[326] Treating glyoxal with hydroxylamine hydrochloride the imidazole **29** is obtained from two molecules of the substrate.[327]

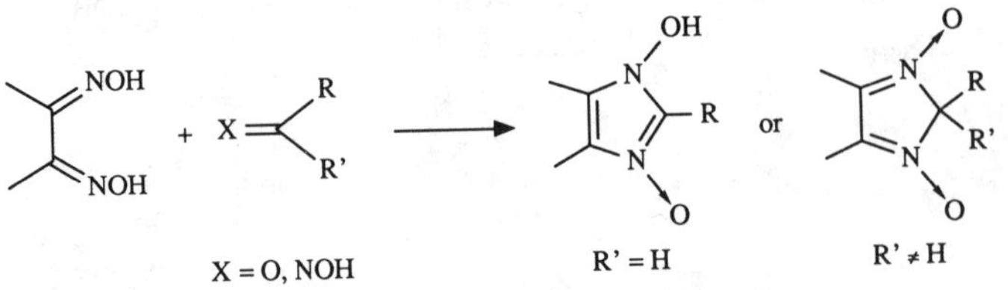

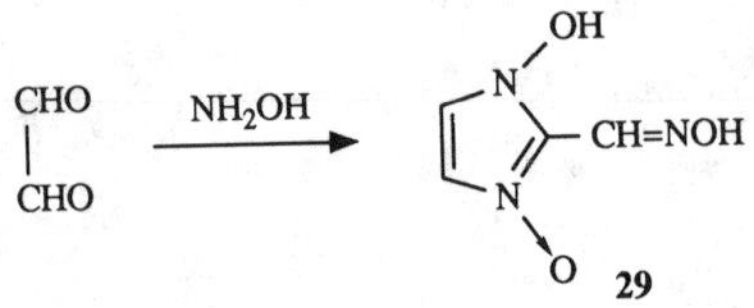

With phenanthrenequinone dioxime, it has been noticed that the reaction at room temperature affords only the imidazole, whereas at 60°C a 2,1,4-oxadiazine 4-oxide is also formed. [303,325] In view of the previously discussed effect of the oxime stereoisomerism on O vs. N attack in similar condensations, this is probably related to a shift of the isomeric equilibrium at higher temperature.

c. From Hydroxamic Acids, Amidoximes, and Related Compounds
1. With Aldehydes and Ketones
1-Hydroxy-2-pyrazinones are obtained by condensation either of α-aminohydroxamic acids with α-dicarbonyls,[224b,266,328a] or alternatively of α-oxohydroxamic acids with α-aminoketones.[328b] Similarly, hydrazinooximes react with diacetyl or benzil in the presence of gaseous hydrochloric acid and yield 1,2,4-triazine 4-oxide.[224b]

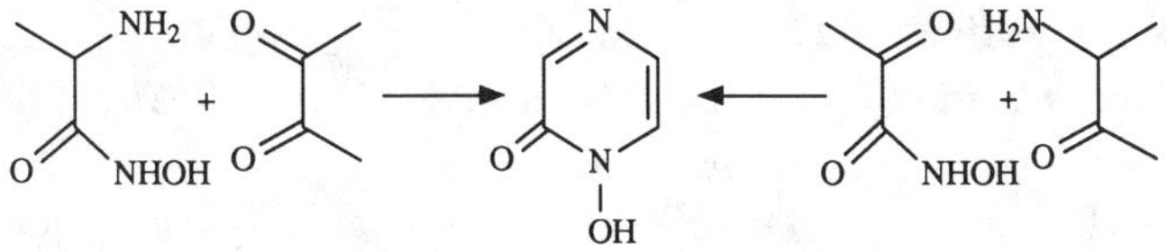

The reaction of o-aminobenzamidoxime with aldehydes leads both to 1,2-dihydro-4-aminoquinazoline 3-oxide (the oximino and the ring amino groups involved in the condensation) or to 1,2,3,4-tetrahydro-4-oximinoquinazoline (condensation involves both amino groups).[329]

Ethyl cyanoacetate and hydroxylamine react with acetylacetone in the presence of bases to yield a 1-hydroxy-2-pyridone,[330] and the yield is high when the preformed potassium salt of cyanoacetohydroxamic acid is used.[331] This is an application to N-oxides of the known Guareschi pyridine synthesis.

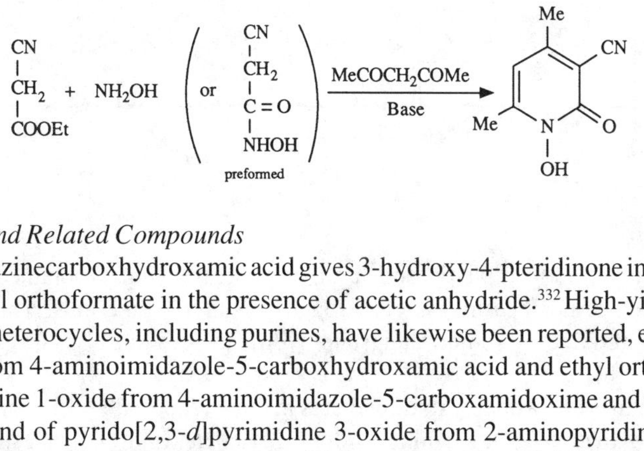

2. With Esters and Related Compounds
3-Aminopyrazinecarboxhydroxamic acid gives 3-hydroxy-4-pteridinone in high yield when heated with ethyl orthoformate in the presence of acetic anhydride.[332] High-yield syntheses of other binuclear heterocycles, including purines, have likewise been reported, e.g., of hypoxanthine 1-oxide from 4-aminoimidazole-5-carboxhydroxamic acid and ethyl orthoformate,[333] of 2-mercaptoadenine 1-oxide from 4-aminoimidazole-5-carboxamidoxime and carbon disulfide in pyridine,[334a] and of pyrido[2,3-d]pyrimidine 3-oxide from 2-aminopyridine-3-carboxamidoxime by refluxing with ethyl orthoformate.[334b] In a related process, N-hydroxyurea and ethyl acetoacetate condense in the presence of bases to give 1-hydroxy-6-methyluracil.[335]

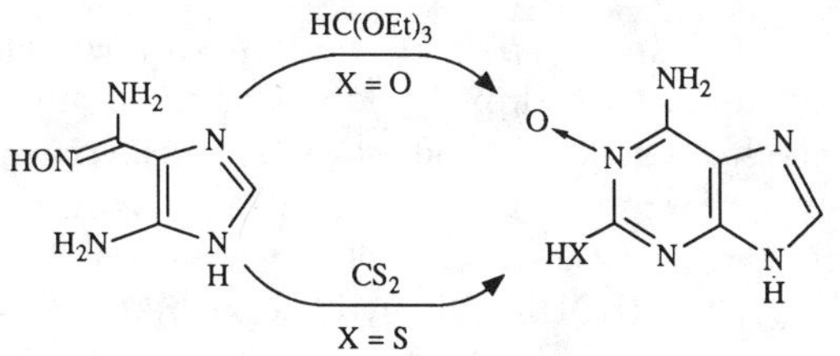

3. With Other Reagents

There are sparse examples of other synthetic pathways modeled on those already discussed for other hydroxylamino derivatives. Thus, amidoximes react with *N*-cyanoimidate and give 2-amino-1,3,5-triazine 1-oxides in moderate yield.[336]

B. FROM NITROSO DERIVATIVES

In this section syntheses of heterocyclic *N*-oxides starting from nitroso derivatives are described, as well as those involving a nitroso compound as a (hypothetical) intermediate. The material is subdivided in two parts, the first one considering reactions of either preformed nitroso derivatives and of their precursors, which are converted to the nitroso stage through redox processes, the latter one reviewing the syntheses of *N*-oxides initiated by direct nitrosation. Some of the reactions may be attributed to tautomeric oximes rather than the nitroso compounds.

1. From Nitroso Derivatives or Their Precursors

a. Intramolecular Cyclization from Nitroso Derivatives Formed In Situ

1. Azo Dioxides from Diamines and Dihydroxylamines

It is well known that nitroso derivatives (usually blue or green) exist in equilibrium with the colorless azo *N,N'*-dioxides. Aliphatic dinitroso derivatives in which the groups are at a convenient distance are easily converted to cyclic dioxides, and thus are a source of cyclic azoxy and azo derivatives.[337-339] As an example, the cyclic azo dioxides **30—32** are stable as such, and only in the case of the last one has formation of a blue color, indicative of the dinitroso form, been observed on heating.[337] This holds for aromatic derivatives as well,[340,341] and benzo[*cd*]indazole dioxide (**33**), potentially tautomeric with 1,8-dinitronaphthalene, is also in the heterocyclic form (in this case, strongly colored in red).[340]

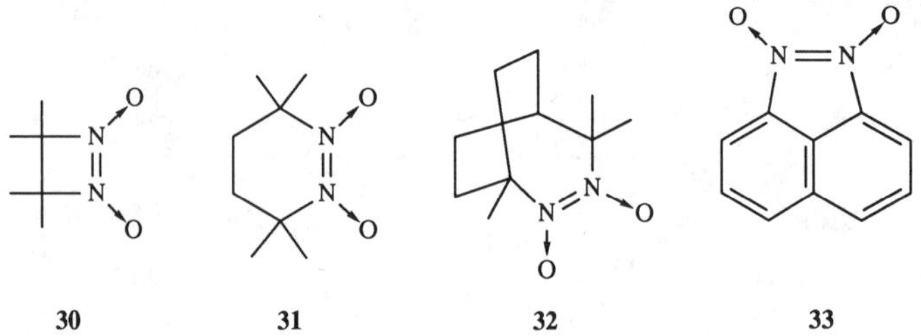

| 30 | 31 | 32 | 33 |

Such compounds, both aliphatic and aromatic, are frequently obtained through redox processes from substrates bearing other nitrogen-containing functions. Thus, diamines are oxidized to azo dioxide by sodium tungstate/hydrogen peroxide[337,341] or by aromatic peracids[339-341] (Scheme 6).

Dihydroxylamines are probably intermediates in these reactions and can be directly used as substrates, e.g., oxidation with bromine, sodium periodate, and manganese dioxide of 2,3-dimethyl-2,3-dihydroxylaminobutane yields compound **30**;[342] alternatively, dehydration at the hydroxylamino-oxime stage leads to cyclic azoxy derivatives.

2. Azo Dioxides, Furoxans, and Analogs from Dioximes and Hydroxylamino-Oximes

The hypobromite oxidation of α-hydroxylamino-oximes leads to 3-bromodiazetine 1,2-dioxides when there is no hydrogen in α (and thus intermediate formation of a dioxime is precluded), or otherwise to furoxans.[343a] At the other end of the oxidation degree, one can start at the nitro-oxime level, as is the case with carbohydrate furoxan **34**, which is prepared by treating a sulfonylribulose with $NH_2OH \cdot HCl$ in pyridine and refluxing with $NaNO_2$.[343b]

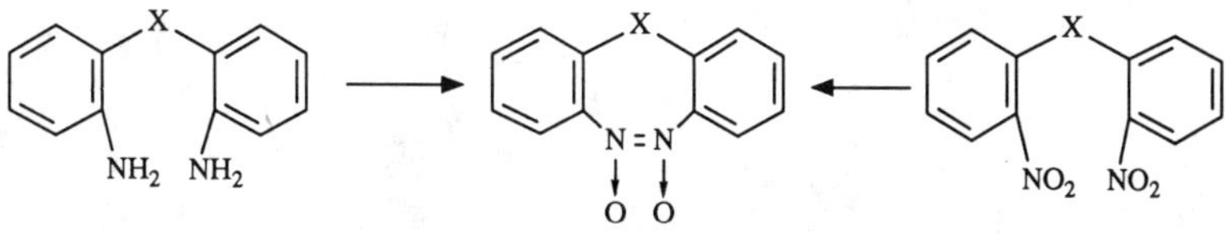

Scheme 6. Azo dioxides by oxidation of diamines, e.g., by H_2O_2/Na_2WO_4 ($X = CH_2CH_2$, CH_2OCH_2, CH_2NHCH_2), or by reduction of dinitro compounds, by Zn, Na_2S, H_2, or electrochemically ($X = CH_2$, CH_2CH_2, $CH=CH$, NH, S).

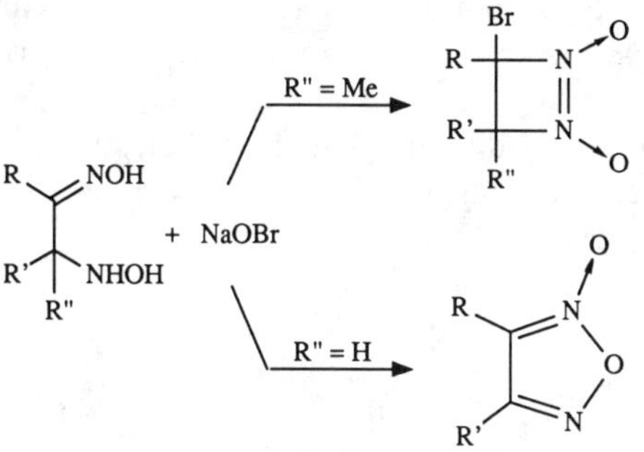

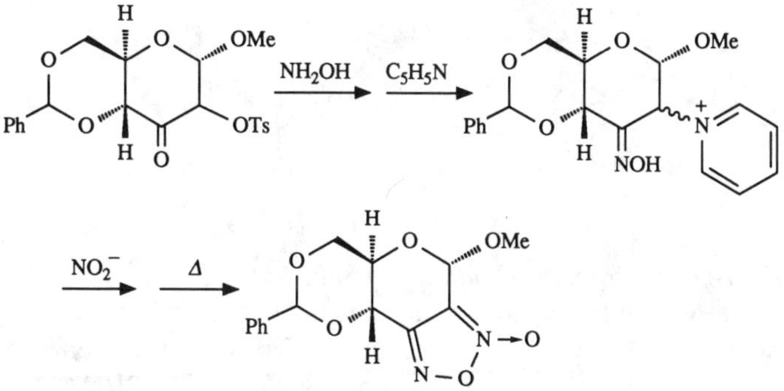

The general synthetic method for furoxans, however, is the dehydrogenation (formally occuring via dinitrosoalkenes; compare with Section D1) of α-dioximes. This is accomplished under a variety of conditions, such as treatment with halogens or alkali hypohalites,[343-347] lead tetraacetate,[348] potassium ferricyanide,[347,349-353] cerium(IV)ammonium nitrate,[347,354] nitrogen oxides,[354-356] N-iodosuccinimide,[357] phenyliodine ditrifluoroacetate,[358,359] as well as anodic oxidation.[360,361] Furoxans obtained in this way include terms with a fused strained ring,[344,346] and steroids,[344] as well as amino,[347,349-351] and aziridine[353] substituted derivatives.

Starting from unsymmetrical dioximes, two isomeric furoxans are obtained. Regioselectivity may be high, and depends both on the substituents present (e.g., p-methylphenylglyoxime yields more of the 4-substituted form and p-bromophenylglyoxime yields only the 5-substituted furoxan on anodic oxidation)[361] and the stereochemistry of the oxime functions (selective attack by oxygen on the nearest nitrogen in the *amphi* (i.e., *E,Z*, **35**) forms has been early proposed,[362] but in other cases no effect of configuration on product distribution has been detected),[356] or on the method chosen (e.g., 5-chloro-4-methylfuroxan predominates 95 to 5 when prepared from the corresponding dioxime by ceric ion oxidation, whereas a mixture of isomers is obtained with nitrogen oxides).[354] Different results may be accounted for by different oxidation mechanisms; however, in view of the possible change of the oximes configuration under the reaction conditions and of the often easy furoxan isomerization (see Chapter 2) it is not surprising that regioselectivity is not the rule.

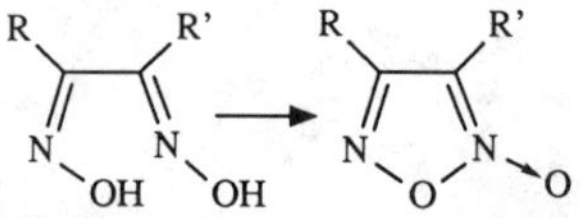

35

A related synthesis leads to 1,2,5-thia(selena)diazoles from the α-dioximes and S_2Cl_2[363] and Se_2Cl_2,[364] respectively.

β-Hydroxylamino-oximes give pyrazole dioxides by hypobromite oxidation,[365] and the same products are smoothly obtained from β-dioximes by treatment with lead tetraacetate (in this case, however, accompanied by 1,2,6-oxadiazine 1-oxides)[366,367] and phenyliodine bis(trifluoroacetate).[368]

Oxidation of 2-unsaturated 1,3-dioximes with $Pb(OAc)_4$ gives pyridazine dioxides,[369] while using $PhI(OCOCF_3)_2$ gives, rather, isoxazolo[4,5-*d*]isoxazoles[368,369]; saturated γ-dioximes do not cyclize with the latter reagent.[368]

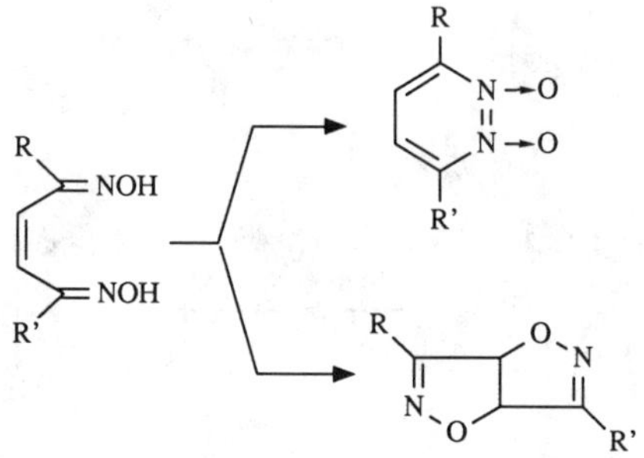

3. Heterocyclic N-Oxides from the Oxidation of Substituted Oximes

In the presence of a suitable substituent, oxidation of oximes leads to cyclization to a heterocyclic mono-*N*-oxide. Typical examples are the synthesis of 1,2-benzisoxazole 2-oxides by lead tetraacetate oxidation of *o*-hydroxyacetophenonoxime,[370] of 2*H*-1,2,3-triazole 1-oxides from anodic oxidation of α-hydrazino-oximes,[360] of 1,2,4-triazolo-pyridine and pyridazine 3-oxides form the chemical (bromine,[371a] lead tetraacetate[371b]) or electrochemical[360] oxidation of heterocyclic amidoximes.

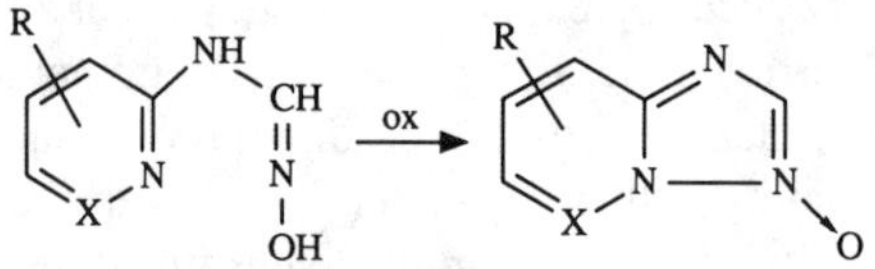

4. Azo N-Oxides and N,N'-Dioxides from the Reduction of Dinitro Derivatives

The reduction of aromatic nitro derivatives leads to azo, azoxy (possibly by dehydration of nitroso-hydroxylamines), and azo dioxide (either from dinitroso or from nitro-amines) derivatives. *N*-Oxides of benzo[*c*]cinnoline[372-375] and its benzo derivatives,[376] various polyaza-phenanthrenes,[377] and -pyrenes,[378-380] dibenzo-[*c,f*]1,2-diazepine,[381] -1,4,5-triazepine,[382] and -1,4,5-thiadiazepine,[383-385] dibenzo[*c,g*]1,2-diazocine,[386] dithienopyridazine[387-388] have been obtained from dinitro derivatives (see Scheme 6). Chemical (zinc in basic[372a,382] or neutral medium,[381] sulfides,[373,376,377,387] hydrogen and catalysts),[374,380,383] and electrochemical reduction[375,378,379,384,385] have been used; mixtures of products at different oxidation levels may be obtained, but selective processes have been reported; as an example, benzo[*c*]cinnoline 5,6-dioxide is conveniently prepared from 2,2'-dinitrobiphenyl by hydrogenation in the presence of

Raney nickel "of low activity"[374] or by deoxygenation with triphenylphosphine,[372b] while the 5-oxide is prepared by treatment with sodium sulfide[373] or by reoxidation of the bis-hydroxylamine obtained from cathodic reduction.[375] Photolysis of 2,2'-dinitrodiphenylmethane yields, among other products, dibenzo-1,2-diazepin-5-one oxide and dioxide.[389]

5. N-Oxides from Nitro- and Amino-Azo Derivatives

The reduction of 2-nitrophenylazoarenes leads to 2-arylbenzotriazole 1-oxides, or directly to the benzotriazoles. In view of the great industrial importance of the latter compounds as UV adsorbers, the procedure for their reduction has been extensively investigated. A good yield of the 1-oxides is obtained by reduction with ammonium sulfide,[390] hydrazine/NaOH with MnO_2 as a catalyst,[391,392] glucose in alkaline methanol at 40—45°C,[393] phosphorus, hypophosphorous acid and its salts,[394] thiourea S,S-dioxide,[395] as well as by catalytic hydrogenation in alkaline medium,[396,397] and by heating in alcohols in the presence of a hydrogen transfer catalyst (e.g., 2,3-dichloro-1,4-naphthoquinone).[398]

2-Arylbenzotriazole 1-oxides are likewise obtained by oxidation of 2-aminophenylazoarenes.[399] The irradiation of 1-(2'-nitrophenyl)pyrazoles in alcohols gives 2-vinylbenzotriazole 1-oxides, probably again via a nitroso intermediate.[400]

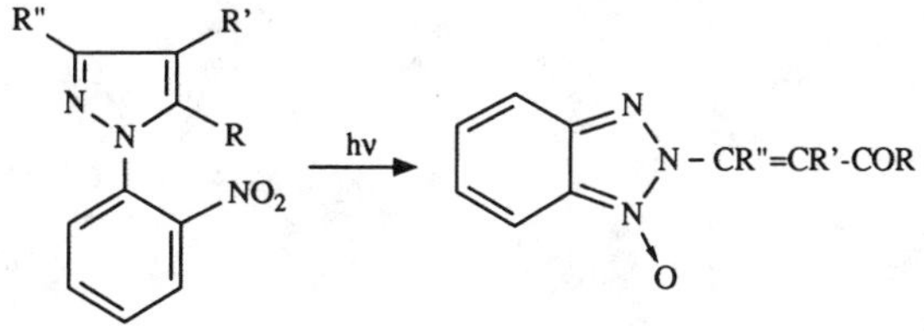

b. Reactions of Preformed Nitroso Derivatives

1. Cycloaddition of Nitroso-Alkenes, -Alkynes, and -Aromatics

Nitrosoalkenes undergo 1,4-cycloaddition with alkenes and yield 1,2-oxazines. However, with some electron-rich alkenes (2-methoxypropene,[401] indole[402]) low yields of pyrroline 1-oxides have been obtained through 1,3-cycloaddition. This second mode of reaction is also observed with nitrosoaromatics and nitrile oxides, and leads to 1-hydroxybenzimidazole 3-oxides.[403a] Base-induced dimerization of 1-nitroso-1-hexyne yields butyl-(1-hexynyl)furoxan,[403b] and, when aniline is present, anilinobutylfuroxan.[403c]

The "Beirut" synthesis of quinoxaline dioxides has been considered a 1,4-cycloaddition between o-dinitrosobenzene (valence tautomer of benzofuroxan) and enols (see Chapter 4 for discussion).

2. Reactions of α-Nitrosoamines and Related Compounds

Aromatic o-nitrosoamines (and phenols) undergo condensation with one- or two-carbon addends to form new rings. The reaction might involve the corresponding oximino-imine tautomer (see Section IIA2b4). Typical examples are the synthesis of benzo[h]quinoline and naphtho-1,4-oxazine N-oxides from 1-nitroso-2-naphthylamine and -naphthol with pyridinium betaines,[404] of pteridine 5-oxides from 4-amino-5-nitrosopyrimidines and again pyridinium betaines,[405] of fervenulin 4-oxide from 6-hydrazino-5-nitrosouracil and dimethylformamide/phosphorus oxychloride or triethyl orthoformate,[406] and of 7-hydroxyxanthine from the same reagent and formaldehyde.[305]

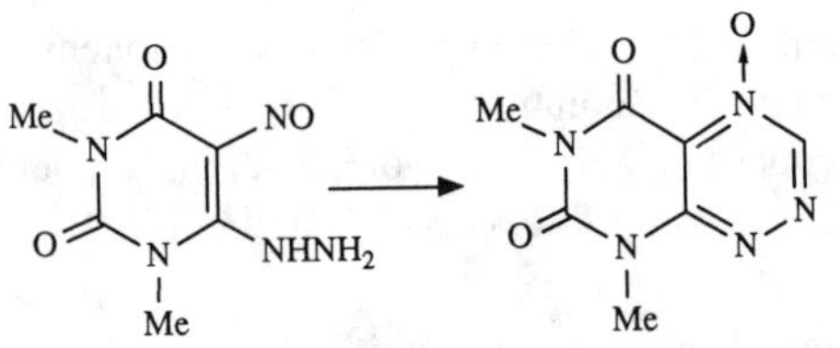

A related reaction from an aliphatic substrate is condensation between α-nitrosoamidines and hydrogen cyanide to give imidazole 1-oxides.[407]

3. Other Reactions

4-Substituted aromatic nitroso derivatives yield phenazine 5-oxides when treated with concentrated sulfuric acid; the reaction is limited to electron-donating substituents, and might involve *o*-nitroanilines as intermediates (compare Section IIC1a2).[408] The reaction of potassium ethylnitrosolate with benzyl halides does not stop at the stage of nitrosolic esters, but proceeds to 1,2,4-oxadiazole 4-oxides.[409a] 4-Alkylamino-5-nitrosouracils are oxidized by nitric acid, probably via the corresponding imine, and finally yield 7-hydroxyxanthine.[409b]

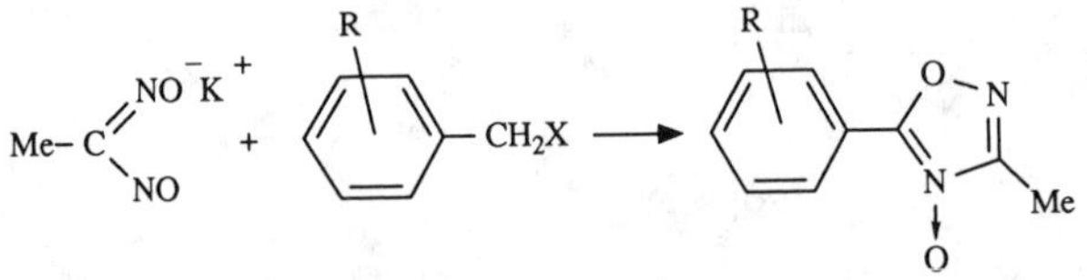

As for *N*-nitroso derivatives, cyclization of the *N*-nitrosoaniline **36** to give a benzopyrazole 2-oxide on treatment with sulfuric acid has been reported, and reasonably involves a benzylic carbocation.[410] α-Lithiated *N*-alkylnitrosoamines decompose at −73°C to yield tetrahydro-tetrazine oxides.[411]

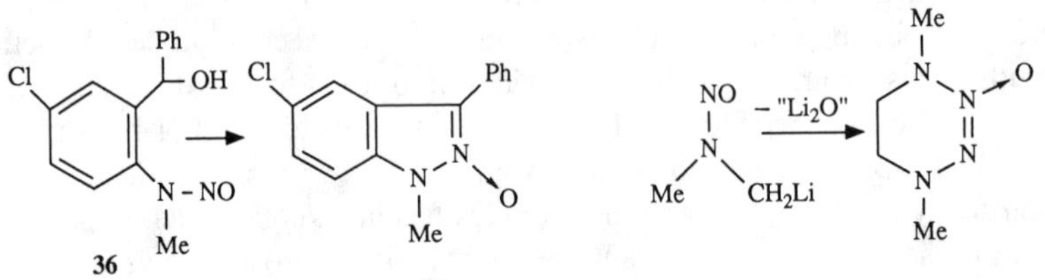

2. By Direct Nitrosation

Nitrosation of olefins, e.g., with dinitrogen trioxide, leads to α-nitroso-nitro derivatives ("pseudonitrosites", e.g., **37**), and these undergo rearrangement to nitro-oximes and acid-promoted dehydration to furoxans.[412,413] In some cases nitrofuroxans are formed, e.g., the 3-methyl-4-nitro derivative from propene.[414] The treatment of sorbic acid with sodium nitrite-sulfuric acid directly yields a furoxanylpropenoic acid.[415] The nitrosochlorination of some styrenes has been reported to yield 1,2,3-oxadiazoline 3-oxides;[416] however, the products formed by treatment with nitrous acid of related systems have been shown to be furoxans.[417]

The treatment of benzalacetone oxime and related compounds with butyl nitrite yields 1-hydroxypyrazole 2-oxides, and these can be isolated via copper salts.[418] With mesityl oxide oxime a 3H-pyrazole dioxide is obtained, possibly via a "pernitroso" [>C=N(O)-NO] derivative.[419]

The nitrosation of 1,3-dimethyl-6-hydroxylaminouracil leads to the 1,2,5-oxadiazolo[3,4-d]pyrimidine system,[420] and related syntheses of pyrimido[5,4-e]-1,2,4-triazine (fervenulin) 4-oxides from hydrazones of 6-hydrazinouracil,[421,422] and of isoalloxazine 5-oxides (including riboflavin) from 6-phenylaminouracils[423] have been reported. These "nitrosative cyclizations" involve a dehydrogenation, either on the part of nitrous acid itself or of an added oxidant, such as diethyl azodiformate.[422] Isoalloxazine 5-oxide has been also obtained from the 6-phenylamino-5-triphenylphosphoranylidenehydrazono derivative of uracil with nitrous acid.[424]

C. FROM NITRO COMPOUNDS

Nitro compounds, in particular the aromatic derivatives, are the starting materials for many of the syntheses via hydroxylamine or nitroso derivatives discussed in the previous sections, according to the relevant intermediate. A redox step might be involved in some of the reactions discussed in the present section as well, but in general it is an intramolecular process.

The syntheses starting from nitro derivatives include some useful and original (as opposed to preparation via hydroxylamine derivatives, usually a variation of the corresponding syntheses of the nonoxidized heterocycles via ammonia derivatives) approaches to various heterocyclic systems. The reactions are classed into two groups, the first one involving cyclization on the nitrogen and the latter cyclization on the oxygen atom of the nitro group, again differentiating intra- and intermolecular processes. Cyclization on the nitrogen or on the oxygen is referred to the end product and not to the mechanism; indeed some of the processes discussed in Section IIC1a involve preliminary oxygen attack, leading to intramolecular oxygen transfer.

1. Intramolecular Processes

a. Cyclization on the Nitrogen Atom

1. Reaction with a Saturated Carbon Atom

Most reactions of this section involve the interaction between the aromatic nitro group and a *ortho* side chain.[425] N-oxides of five- and six-membered (poly)aza heterocycles are prepared from o-nitroalkylaromatics, N-alkyl-o-nitroanilines, N-alkyl-o-nitrobenzamides, and related heterocyclic derivatives. The few examples involving aliphatic derivatives are discussed at the end of the section.

Reaction of nitrobenzenes with a unactivated o-alkyl chain is observed only as a photochemical process. Thus, 1,3,5-tris-*tert*-butylnitrobenzene yields a 3,3-dimethylindole 1-oxide as the primary product by irradiation, both in solution[426] and in the crystalline state.[427,428] The process probably involve initial hydrogen abstraction by the radicalic $n\pi^*$ state.

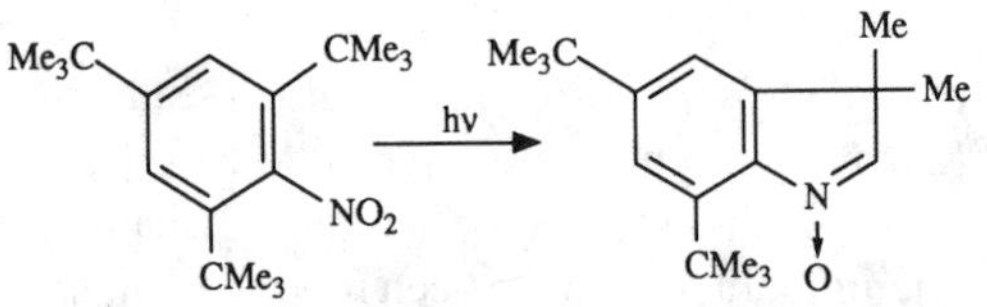

A much larger scope includes the reaction of nitro groups with "activated" (by an electron-withdrawing substituent) methylenes, the intramolecular version of the well-known synthesis of nitrones by condensation of carbanions with aromatic nitro compounds, and has been generally formulated as involving an "aldol-like" nucleophilic attack on the nitrogen atom (path a in Scheme 7). Another possibility is that an intramolecular redox reaction leads to an aromatic hydroxylamino derivative with a carbonyl group in the side chain. In this case, the actual cyclization step is the one discussed in Section IIA1a1,a4 (path b in Scheme 7, see below for some evidence[429]). When a methine, rather than a methylene, is involved, one of the activating groups is lost in the cyclization (Scheme 7).

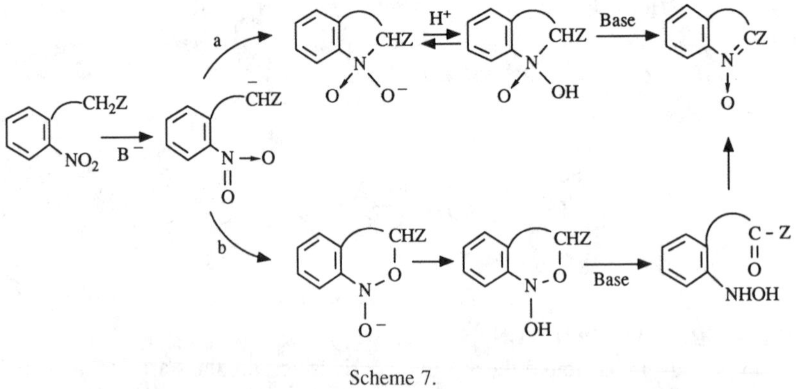

Scheme 7.

Thus, o-nitrobenzylidenesuccinic acids and esters undergo base-catalyzed cyclization to quinoline 1-oxides (under the reaction conditions, the carboxyl group in position 2 might be hydrolyzed and eliminated).[430,431] Heating 2-nitrobenzaldehydes with diethyl (diethoxyphosphinyl) succinate under alkaline conditions directly yields the heterocycle.[432a]

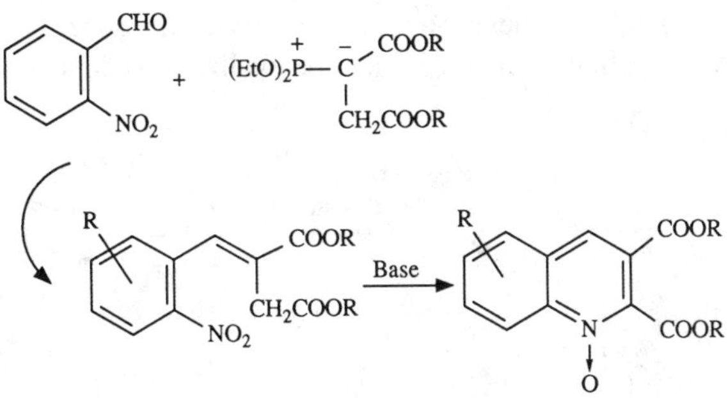

A route to kynurenic acid derivatives is offered by the spontaneous rearrangement at room temperature of the nitrophenylfurans **38**.[432b] Phenanthridine 5-oxide is obtained from the biphenyl derivatives **39** (X=COOR); with X=Br, treatment with cyanide yields directly 6-cyanophenantridine 5-oxide.[433]

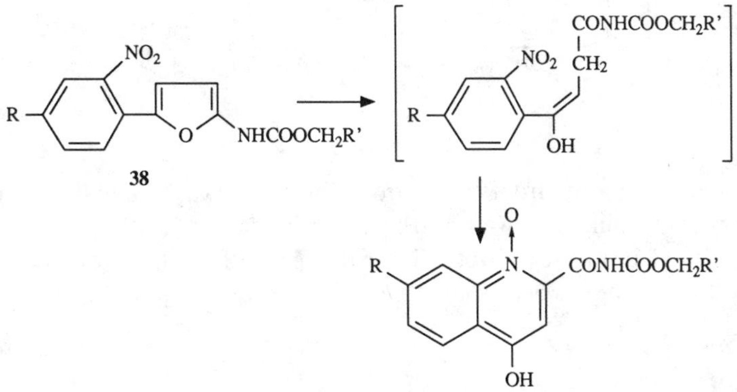

38

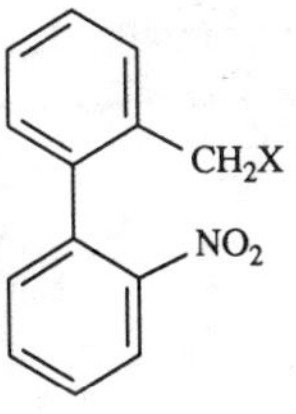

39

1,4-Dihydro-1-hydroxy-4-quinazolones are obtained from N-alkyl-o-nitrobenzamides[434] and 2-quinoxalinone 4-oxides from some o-nitroacetanilides.[435-439] The latter synthesis, a convenient entry to quinoxalinones obtained by deoxygenation (also in a one-pot procedure),[439a] is satisfactory, both with secondary and with tertiary amides. The activating groups (cyano, acyl, pyridinium) are easily substituted or eliminated; with the pyridinium group, however, ring opening occurs during the reaction forming an unsaturated side chain; reduction of this product affords 2-amino-3-hydroxyquinoxaline 1-oxides.[438]

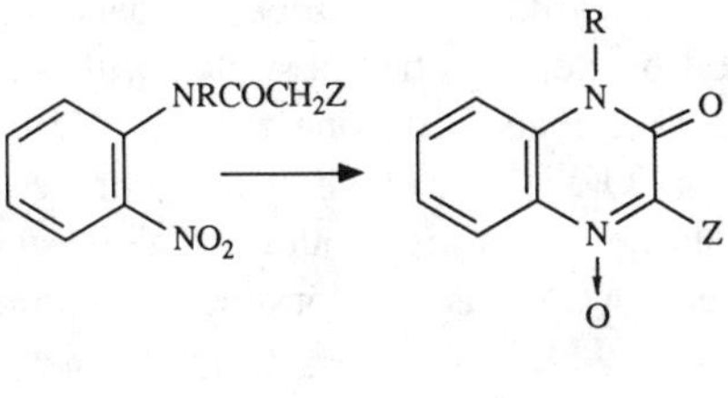

1-Hydroxyquinoxaline-2,3-diones are obtained from *N*-(*o*-nitrophenyl)sarcosine esters. The reaction begins through path b (Scheme 7), and the cyclization occurs on group Z.[429]

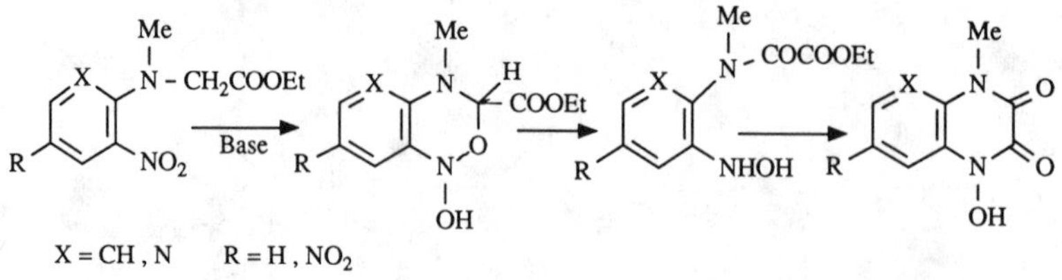

$$X = CH, N \quad R = H, NO_2$$

Pyrrolo- and pyrido-quinoxalines derivatives have been analogously prepared from the cyclic amines **40**,[440] and pteridines from 4-cyanacetamido-5-nitropyrimidines.[441]

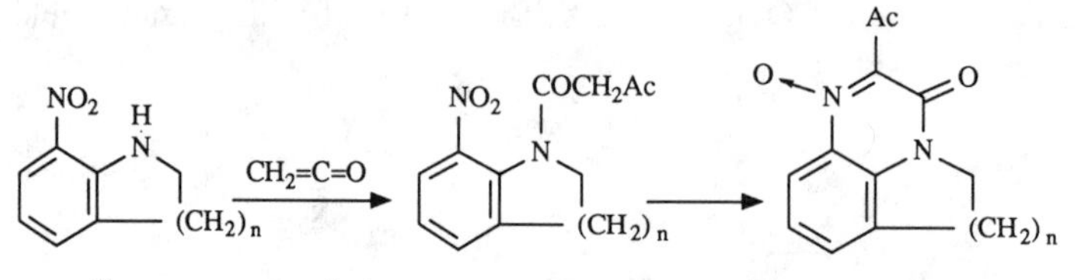

40   n = 1, 2

A related synthesis starting from *N*-(*o*-nitroaryl) (or heteroaryl) -amidines, e.g., **41**, offers another pathway to pteridines[442] as well as to benzo-,[443] imidazo-,[444] and pyrimido-quinoxaline[445] derivatives. A methylene in α to a nitrogen heterocycle is sufficiently acidic and participates in this kind of cyclization, as has been shown for arylmethylquinoxaline **42**.[446]

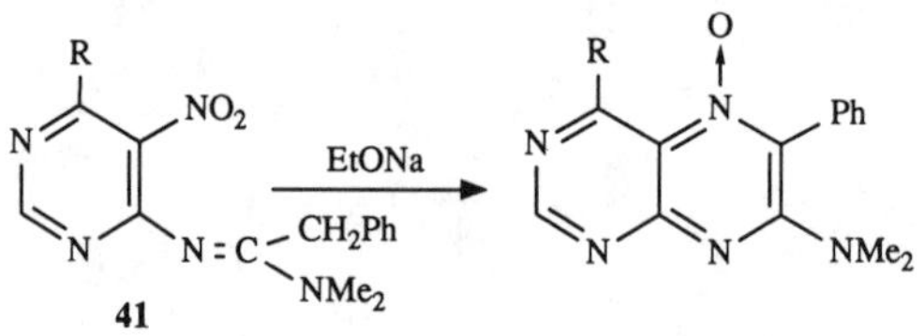

41

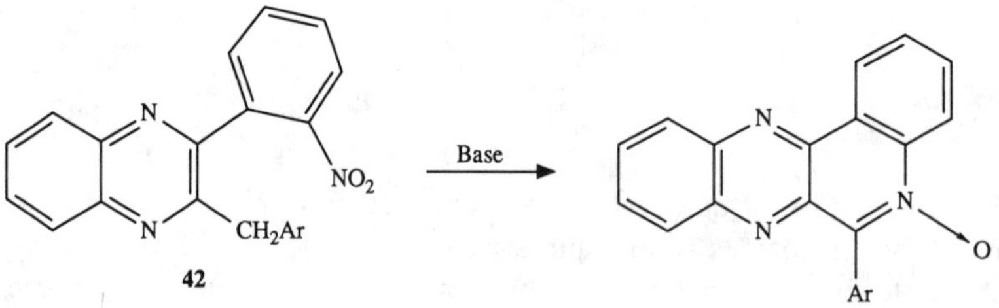

42

Five-membered heterocyclic *N*-oxides (or tautomeric *N*-hydroxy) analogously prepared are indole, benzimidazole, and benzothiazole derivatives. Thus, *o*-alkylnitrobenzenes with a β acidic proton (**43**) undergo base-catalyzed cyclization to 1-hydroxyindoles. The starting materials are available either by condensation of *o*-nitrobenzyl chloride with active methylene compounds[447] or by hydrogen cyanide addition to *o*-nitrobenzylidene derivatives[448] (this process and cyclization can be combined by heating the substrates with aqueous-ethanolic potassium cyanide).[448-450] The reaction takes place with one activating group (e.g., Y=*p*-nitrophenyl, Z=H)[448] or with two (e.g., Y,Z=COOR), one of which is eliminated (Scheme 8);[447] competitive cyclization can take place on the activating group, and yield quinoline 1-oxides (compare Section IIC1a3), but the use of weak bases favors the formation of 1-hydroxyindoles[448-450] (Scheme 8).

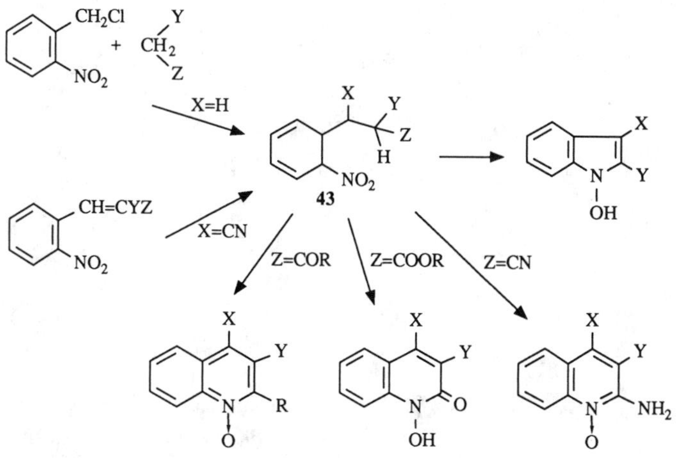

Scheme 8.

2-Nitrodeoxybenzoins yield 2-arylisatogens (3*H*-3-indolone 1-oxides);[451] the same type of products are obtained from the epoxides **44** by rearrangement with boron trifluoride followed by base-catalyzed cyclization with loss of a formyl group.[452]

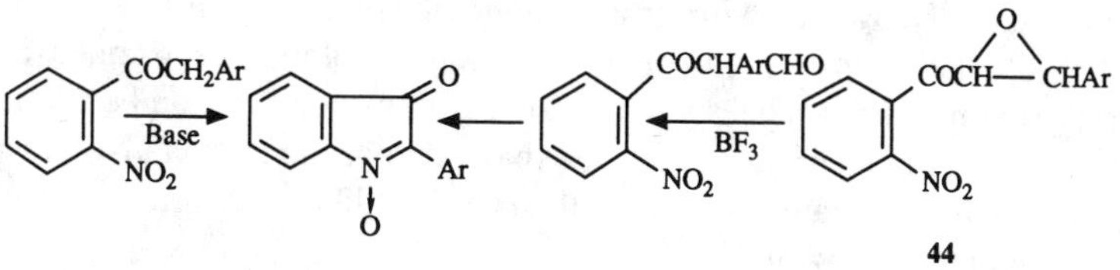

Isatogens are also obtained from the condensation of two molecules of either *o*-nitrobenzoyl-acetates or -acetone.[453] 1-Hydroxyisatins are prepared by treating *o*-nitrobenzoyldiazomethane and derivatives with acids, probably via intramolecular oxygen transfer from the nitro group to the carbocation formed by loss of nitrogen.[454-456]

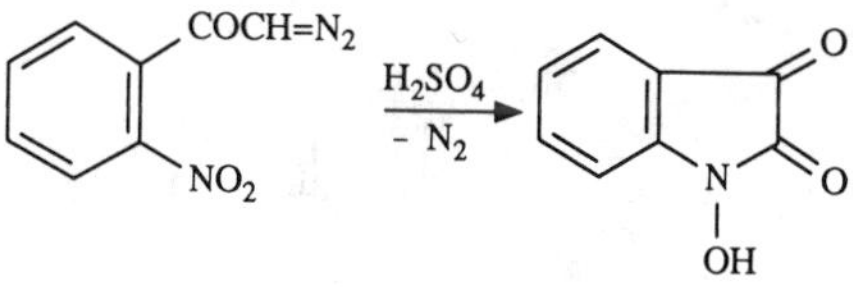

Benzimidazole 3-oxides are likewise easily accessible, and an activating electron-withdraw-ing group is not necessarily required. Thus, *N,N*-dialkyl-*o*-nitroanilines cyclize when heated with mineral acids, the process reasonably involving intramolecular hydrogen transfer.[457] Thermolysis at 200°C also affords benzimidazoles (probably via the 3-oxides)[458] and photolysis yields either the heterocycles or the 3-oxides, depending on the nature of *N*-alkyl groups and ring substituents, possibly through independent pathways (electron transfer and radical abstraction, respectively).[457] *o*-Nitrophenylazirines also yield benzimidazole oxides on irradiation.[459]

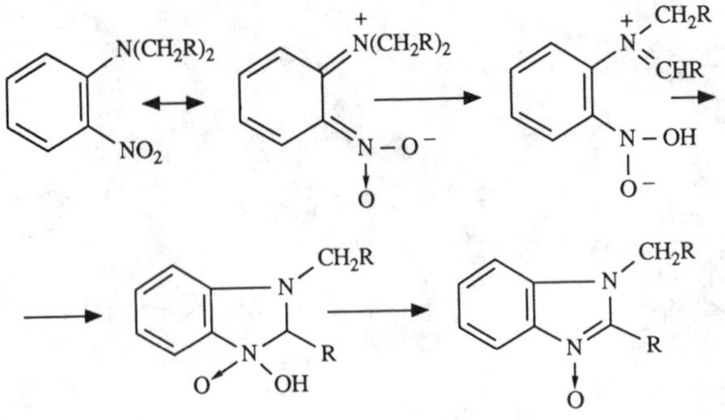

Base-catalyzed cyclization of activated substrates, such as 2-nitrophenylamino-substituted acetonitrile or alkyl acetates, is successful.[460-462] Peptides containing a *N*-terminal 2,4-dinitro-phenylglycine moiety undergo cyclization to 5-nitrobenzimidazole 3-oxide derivatives under mildly basic conditions (pH 8.3); 2,4-dinitrophenyl- or 2-nitrophenyl-glycine[463] (but not the esters)[462] are decarboxylated in the process (decarboxylation of amino acids derivatives is also observed when the benzimidazole synthesis is obtained by thermolysis[464] or photolysis[465]). Analogous to the case of 1-hydroxyindoles, the substrates can be obtained by HCN addition onto a double bond, the cyanide serving also as the cyclizing base. Thus, 2-phenylbenzimidazole 3-oxide is conveniently prepared by refluxing *N*-benzylidene-*o*-nitroaniline with potassium cyanide in methanol.[466] However, with tertiary amines the reaction is different, e.g., from *N*-(2-nitrophenyl)-*N*-alkylacetonitriles, 1-hydroxy-2-benzimidazolones are obtained (Scheme 9),[467] and the corresponding ethyl acetates yield quinoxalindiones rather than benzimidazoles (see page 66).[429] This has led to the suggestion these base-catalyzed reaction involves an intramolecular oxygen transfer in the anion **45**, rather than an aldol-like condensation (path b vs. path a in Scheme 7, compare Scheme 9).[429]

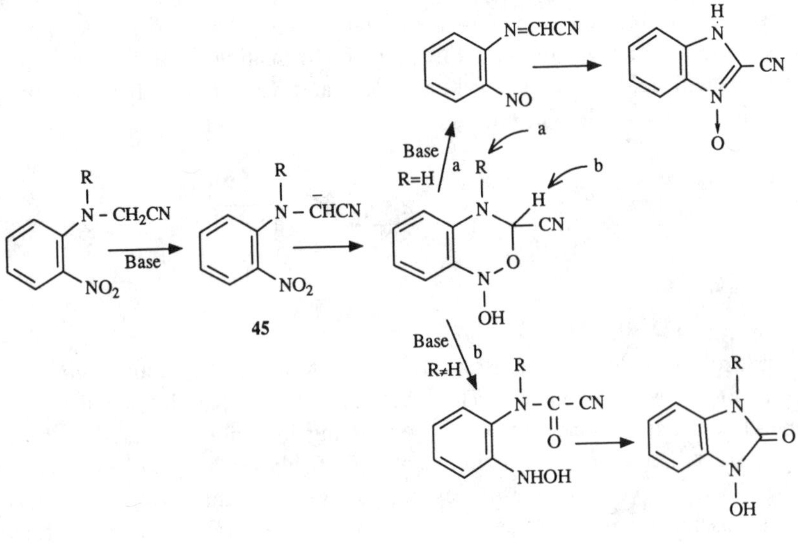

Scheme 9.

The same synthetic pathways have been exploited for various heterocycle-fused imidazole oxides, including imidazo[4,5-*b*]pyridines,[468] purine derivatives (guanine 7-oxides),[469,470] imidazo[4,5-*c*]pyrazoles,[471] thieno[2,3-*d*]imidazoles,[472] and pyrazino[1,2-*a*]benzimidazoles.[473]

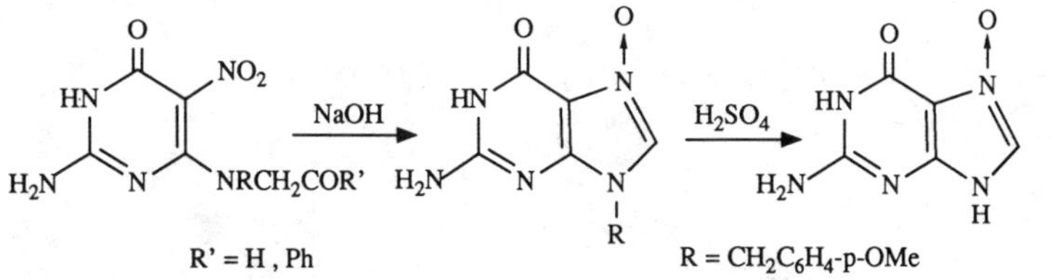

R' = H , Ph   R = CH$_2$C$_6$H$_4$-p-OMe

The corresponding reaction of 2-nitrophenyl sulfides for the synthesis of benzothiazole *N*-oxides is successful when further electron-withdrawing substituents are present in the ring, e.g., picryl chloride cyclize directly when treated with ethyl mercaptoacetate in the presence of triethylamine, and starting from 2,4-dinitrochlorobenzenes the sulfide is isolated and cyclized by bases;[474] similar synthesis of thiazoloquinazolines have been reported.[475]

Finally, some cyclization involving aliphatic nitro compounds are known and occur by elimination of a leaving group. Thus, by treating the morpholino salt of the 1,3,5-tricyano-1,3,5-trinitropentane dianion (or of its precursor, 2-nitroacrilonitrile) with sulfuric acid, 1,3,5-tricyanopyridine 1-oxide is obtained;[476] treatment of some α-bromo-γ-nitroketones with methanolic potassium hydroxide yields 3*H*-3-pyrrolone 1-oxides, obtained as the dimers **46**, again via a intramolecular redox pathway (compare Scheme 7).[477]

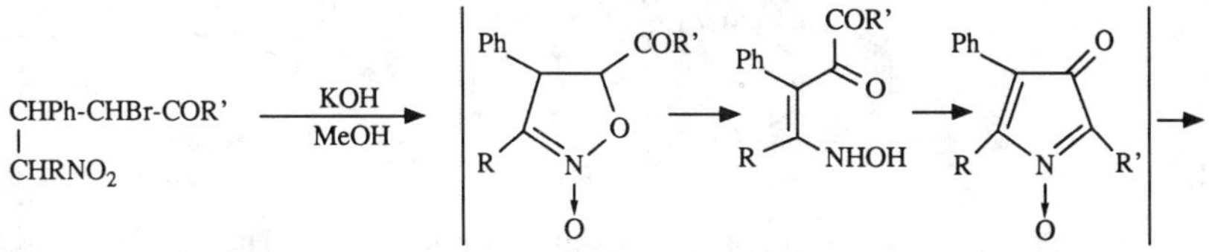

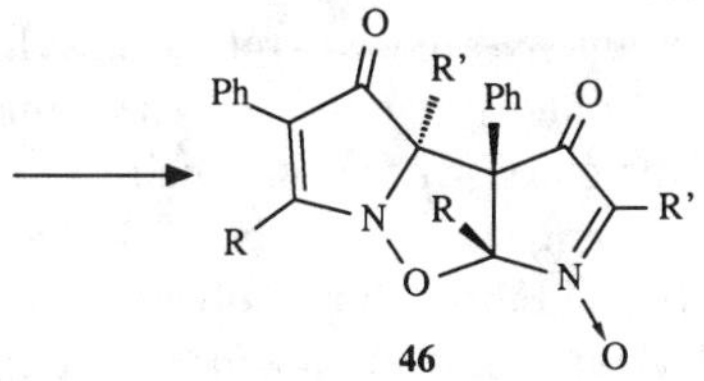

46

2. Reaction with the Carbon Atom of a C-C Multiple Bond

The cyanide-induced cyclizations of 2-nitrocinnammic derivatives[448-450] and *N*-benzylidene anilines[466] involve preliminary addition of HCN to the double bond, and thus have been considered among reactions on a saturated carbon (Section IIC1a1). Other reactions involving carbon-carbon multiple bonds include the synthesis of the 3-oxides of benzimidazole-2-carboxylic acid derivatives from 2-(2′-nitrophenylamino)acrylates and acrylamides (the process involves intramolecular oxygen transfer to the β position, and an aldehyde is eliminated),[478] the synthesis of 2-substituted quinoxaline 4-oxides from the enaminoketones **47**,[479] and the cyclization of *o*-nitrophenylacetylenes to isatogens, obtained both under basic (pyridine) and acid conditions,[480] as well as through the "catalytic" action of nitrosobenzene and by irradiation[481] (the same syntheses can be carried out by generating the triple bond from the pyridinium salts **48** by elimination with bases,[482,483] or by decomposition of 4-(2′nitrophenyl)-5-diazoniumisoxazole salts).[484] Furthermore, irradiation of *o*-nitrophenylpolyenes in the presence of iodine also gives 2-vinylisatogens.[485]

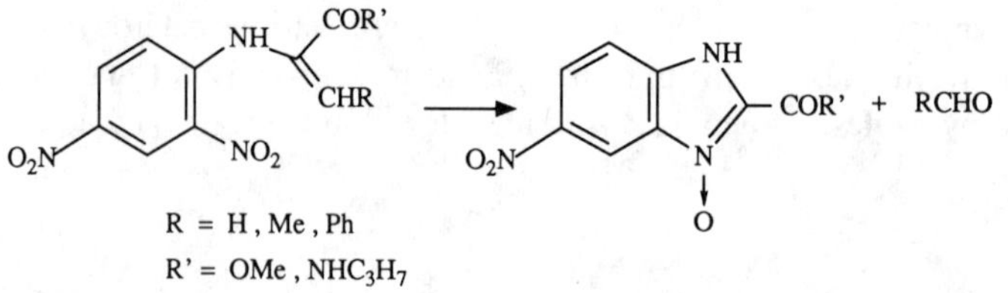

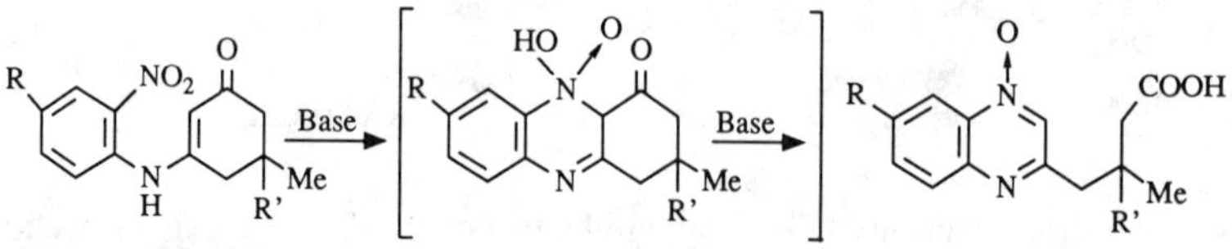

R = H , Me , Ph
R' = OMe , NHC3H7

47

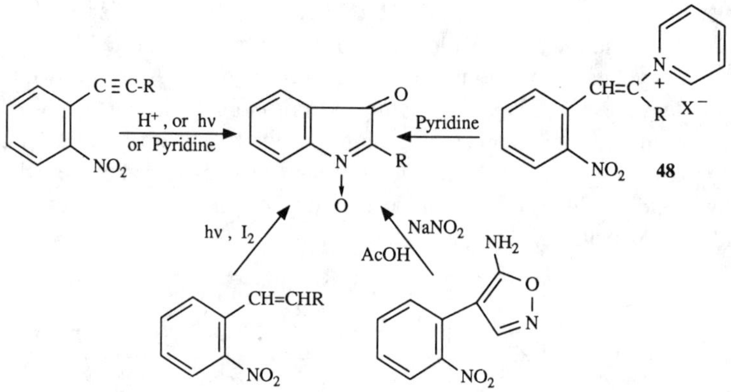

48

Several convenient syntheses involve attack at an aromatic ring. Thus, phenazines 5-oxides are obtained in good yield from 2-nitrodiphenylamines both under acid (oleum at 20—40°C) and under basic (powdered potassium hydroxide in refluxing xylene) conditions[486] (see Section IIC2a1 for the relation with the Wohl-Aue synthesis). The same reaction has been applied to the high yield synthesis of flavine oxides from 6-(N-methylanilino)-5-nitrouracil in sulfuric acid[487] (use of the Vilsmeier reagent leads directly to the deoxygenated heterocycles).[488] Phenazines 5-oxides are also obtained through photolysis of N-acyl- (but not of N-alkyl-) 2-nitrodiphenylamines.[489] A 10% yield of 2-methyl-7-methoxyquinoxaline 1-oxide is obtained from 1-(4-methoxyphenylamino)-2-nitropropene in H_2SO_4.[490] N-Hydroxyacridone is one of the products when o-nitrobenzhydrol is treated with $NaNO_2$ in H_2SO_4.[491]

3. Reaction with Carbonyl, Carboxyl, and Cyano Groups

o-Nitrobenzylmalonic esters and related compounds possess an acidic hydrogen and thus are a source of 1-hydroxyindoles, as has been previously mentioned (Section IIC1a1). Under these conditions, 4-cyano- (or 4-carboxamide-, due to the hydrolysis of the group under the reaction conditions), quinoline 1-oxides are also formed in moderate yield by the alternative condensation on one of the activating groups (a carbonyl, carboxyl, or cyano function; see scheme 8), and this pathway is favored when strong bases are used as catalysts.[448-450,492a] 1,3-dihydroxy-4-quinolones are obtained from trans-1-acyl-2-(2-nitrophenyl)-ethylene epoxide.[492b,492c]

These reactions are strictly related to the condensation of hydroxylamine derivatives, discussed in Section IIA1a1 (see, in particular, the electrochemical reduction of nitro derivatives, in which both 1-hydroxyindoles and quinoline 1-oxides are formed),[200b,200c] and mere consideration of the stoichiometry suggests a reduction step also in this case.

4. Reaction with a Nitrogen Atom

An important access to benzo-fused five- and six-membered 1,2-diaza heterocycles involves the intramolecular reaction of aromatic nitro derivatives containing an amino, hydroxylamino, hydrazino, or imino group to afford the *N*-oxides.

Thus, benzo[*c*]cinnoline 5-oxide is obtained from 2-amino-2′-nitrobiphenyl through base catalysis (sodium hydroxide[493,494] or, advantageously with less basic anilines, benzyltrimethyl-ammonium hydroxide[494]), and dibenzo[*c,h*]cinnoline oxides are similarly prepared.[495] Another useful way of preparing the amino nitro substrate is the decomposition of sulfonamides; thus, 3′-hydroxy-2-nitrobenzenesulfonanilide heated with alkali at 250°C gives 3-hydroxybenzo[*c*]cinnoline 6-oxide (easily deoxygenated under these conditions) via 2-amino-4-hydroxybiphenyl.[496]

4-Cyano-3-hydroxycinnoline 1-oxide has been obtained from 2-cyano-2-(2′-nitrophenyl)acetamide,[425a] but the synthesis can not been extended.[493] Various cinnoline 1-oxides can be prepared from a different process, involving the condensation between the nitro group and the imino ester function; the required intermediate is formed *in situ* by treating *N*-(*o*-nitrobenzylidene)anilines with potassium cyanide in methanol. The cinnoline oxides are the products whenever the aniline-derived ring bears an *ortho* substituent, while in the other case, indazoles are obtained (from the reduction of the corresponding *N*-oxides, in turn arising from the alternative cyclization onto the amino group; in the basic medium these product are either deoxygenated or rearranged to the cinnoline oxides[497]). It must been remembered that indazole 1-oxides are the isolated products when the reaction is carried out under very mild conditions (see page 72).

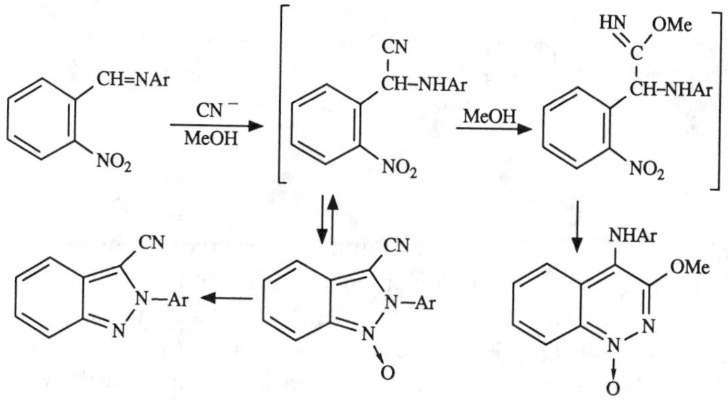

The base-catalyzed cyclization of *o*-nitrophenylureas and guanidine provides 3-substituted benzo-1,2,4-triazine 1-oxides, another example of the *N*-oxides offering an excellent entry to the corresponding heterocycles. Thus the guanidines **49**, conveniently prepared *in situ* by acid-catalyzed condensation of *o*-nitroanilines with cyanamide or its sodium salt,[498-500] or alternatively by reaction of *o*-chloronitrobenzenes with potassium guanidine,[501] or, in two steps, with sodium cyanamide and then with amines,[502] or again by Smiles rearrangement of the appropriate sulfonylguanidines,[503] are converted by bases into 3-aminobenzotriazine 1-oxides in high yield. A complication in these syntheses is the condensation of a second molecule of cyanamide with the aniline, thus obtaining a 3-guanidino as well as a 3-amino-benzotriazine 1-oxide, and the proclivity of the latter compounds to react further with bases to benzotriazoles (see Chapter 4); thus the base catalysis must be carefully adjusted.[504-506] Pyrido[2,3-*e*]-[505] and -[4,3-*e*]-1,2,4-triazine 1-oxides[506] are analogously prepared, as are benzo-1,2,4-triazin-3-ones starting from *o*-nitrophenylureas.[507]

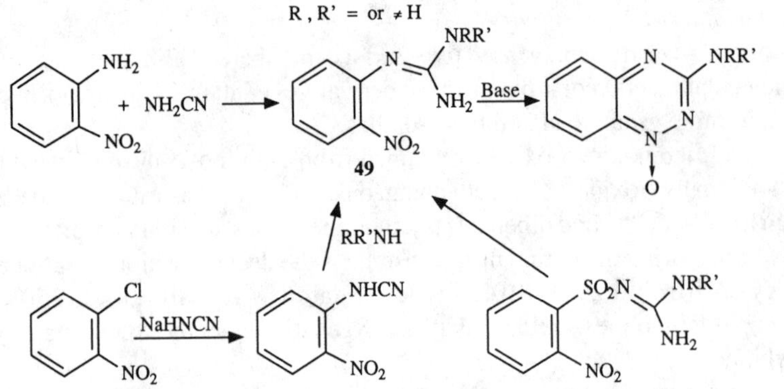

A related approach to the benzotriazine system, again forming *in situ* the amino group for the cyclization by amine addition to a cyano function, leads to pyrazolotriazines from the 2-nitrophenyl (or [4-(3-nitropyridyl)]) hydrazones of β-ketonitriles.[508]

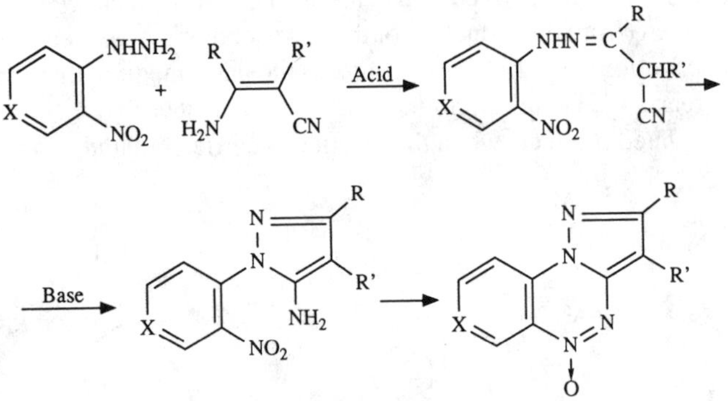

Five-membered heterocycles are also conveniently obtained with similar reactions. Thus 3-cyanoindazole 1-oxides[509-511] are formed from *o*-nitrobenzilidene anils upon treatment with cyanide, reasonably via previous HCN addition (compare with the 1-hydroxyindole synthesis in Section IIC1a1), as well as from *o*-nitromandelonitrile and amines, again through intermediate **50** (this has been independently prepared and cyclized).[512] Benzocinnoline oxides are formed competitively, the relative yield depending on the substituent in the *N*-aryl group and on the conditions (see page 71).[497] The anils are converted to 1-hydroxy-3-indazolones when refluxed with sodium carbonate in ethanol.[513]

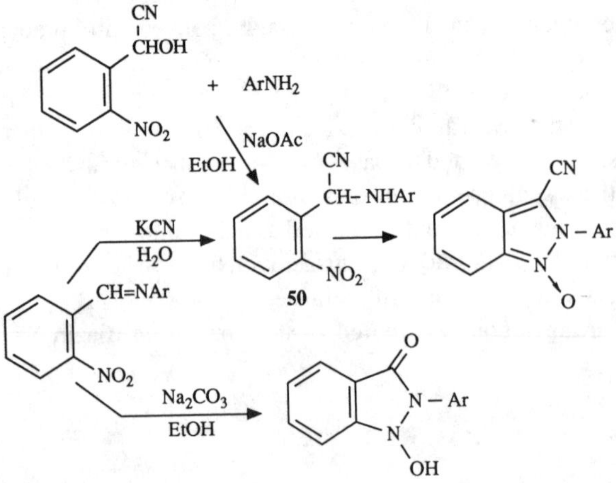

6-Bromomethyl-1,3-dimethyl-5-nitrouracil yields directly a 2-alkylpyrazole[4,3-d]pyrimidine 1-oxide when treated with primary aliphatic amines; the reaction stops at the bromine substitution when arylamine is used, but proceeds further with cyclization also in this case, provided that triethylamine is present.[514] A different approach to the pyrazole ring has been found in the pyrolysis of the N-phenyl-N′-(8-nitronaphthyl)carbodiimide, which gives benzo[c,d]indazole oxide.[515]

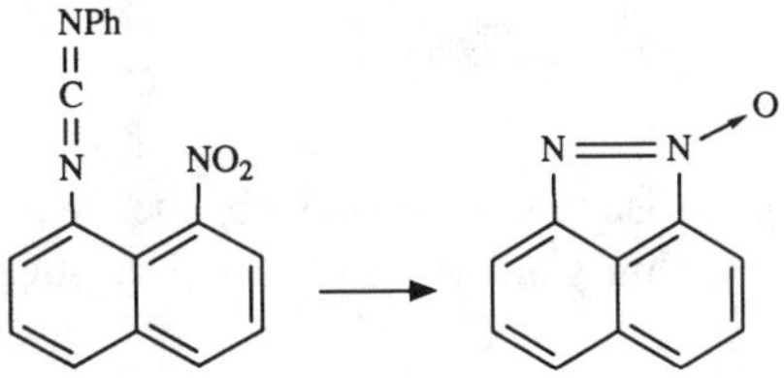

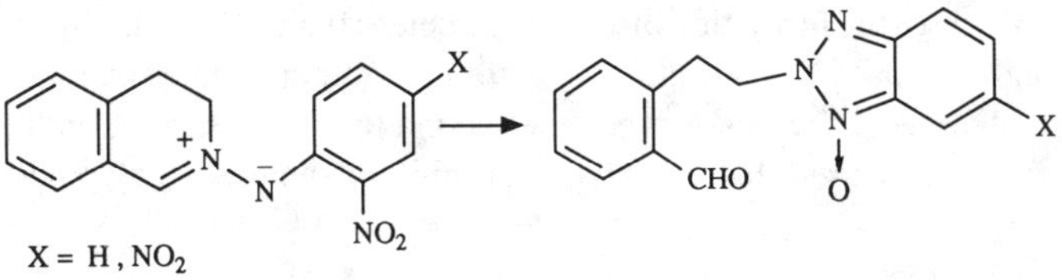

X = H , NO$_2$

54

1-Hydroxybenzotriazoles (**51**) tautomeric with N-oxides are prepared by base-catalyzed cyclization of o-nitrophenylhydrazines;[516] the substrates can be conveniently generated in situ from o-halogeno,[517,518] o-nitro,[519a] and o-methoxynitrobenzenes and hydrazine.[519b] Nontautomeric 2-substituted (**52**) and 3-substituted benzotriazole 1-oxides (**53**) are obtained from 2-substituted 1-(o-nitrophenyl)hydrazines, prepared form arylhydrazines and o-chloronitrobenzene,[520a] or, respectively, from 1-substituted-1-(o-nitrophenyl)hydrazines. Products of type **53** (R=CHRCH$_2$COR″), and not the expected pyrazolines, are obtained from 2,4-dinitrophenylhydrazine and some β-hydroxyketones).[520b,520c] Alkylation in 2 can also be accomplished in the same operation as cyclization, e.g., 2,4-dinitrophenylhydrazine yields 2-methyl-5-nitrobenzotriazole 3-oxide when treated with methyl iodide in DMSO at room temperature.[521]

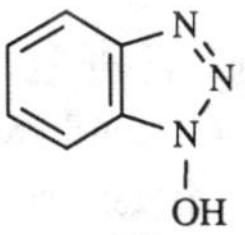

51 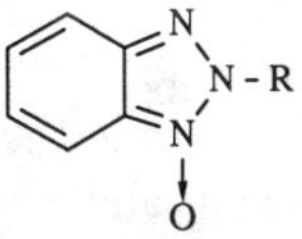   **52** 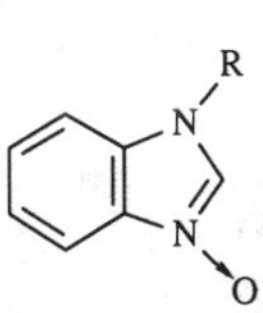   **53**

2-Nitrohydrazobenzenes behave similarly and give products of type **52**,[520a] with yields depending on the conditions. Thus, e.g., chloronitrobenzenes and phenylhydrazine yield benzotriazole 1-oxides when heated in AcOH, and the deoxygenated heterocycles in EtOH;[522] the condensation also takes place with acetic anhydride.[523] 2-Nitrohydrazobenzenes formed from the cathodic reduction of nitroazo- or azoxybenzenes have been found to give benzotriazole 1-oxides in ammonia buffer (a previous dehydration to a o-nitrosoazoderivative might be conceived, compare with Section IIB1a5).[524]

Related syntheses are the thermal (refluxing in toluene) conversion of 1-(2,4-dinitrophenyl)-2-methyl-3,3-pentamethylenediaziridine into 2-methyl-5-nitrobenzotriazole 3-oxide (the reaction does not occur via the hydrazine, since the latter cyclize only under acidic conditions)[521] and the rearrangement of the ylides **54**.[525] A photochemical synthesis starts from N^2-methyl-N^2-(*o*-nitrophenyl)benzaldehyde hydrazone or the analogous uracil derivative **55**.[526]

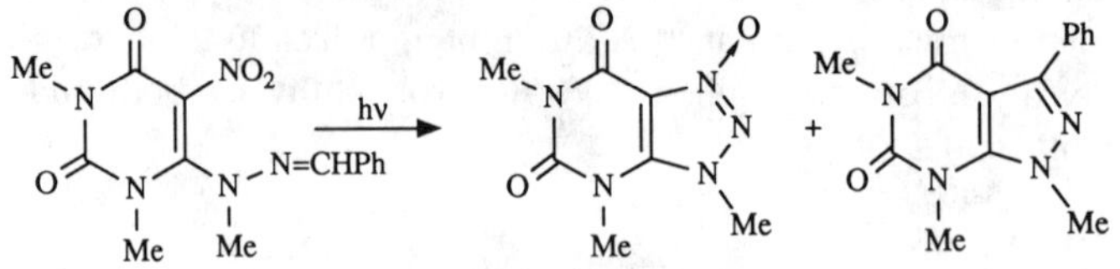

55

Finally, dehydrating 2,4-dinitrobenzensulfenamide (but not the 2- or the 4- mononitro analogs) with $(PhO)_2P(O)Cl$ in DMF-pyridine yields 5-nitrobenzo-1,2,3-thiadiazole 3-oxide.[527]

b. Cyclization on the Oxygen Atom

This section discusses the formation of 1,2-oxaza heterocycles *N*-oxides by reaction between a nitro group and a suitable function. The section is rather heterogeneous, since various mechanisms are involved. Obviously, when the second function reacting with the nitro group is oxygenated, it is not assured that the oxygen atom in the heterocycle comes from the nitro group.

1. Reaction with a Carbon Atom

2-Isoxazoline 2-oxides (cyclic nitrone esters) are obtained by treating 1-nitro-3-bromo-[528] or 1,3-dinitroalkanes[529] with bases (potassium acetate in the former case, sodium methylate in the latter one); the reaction actually involves a nitro group-assisted elimination of a leaving group, not previous formation of a double bond and addition.[530] *N*-nitroamides and sulfonamides form anions with strong bases, and these cyclize when there is a leaving group in the chain; with the β,γ-dibromo derivative **56**, formation of a five-membered ring is preferred, though a 1,2,3-oxadiazine 2-oxide is formed from an analogous γ-monobromo derivative.[531]

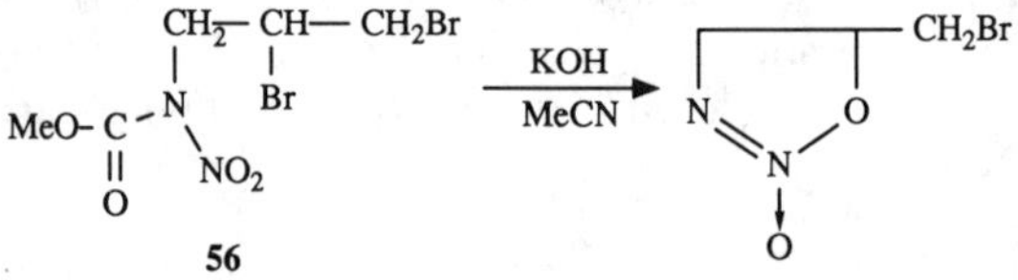

56

Addition of an aromatic nitro group onto nitrilimines yields 3-azo-2,1-benzisoxazole (anthranil) 1-oxides; in turn, the substrates are obtained by lead(IV)acetate oxidation of hydrazones (**57**)[532] or from the corresponding halides **58** in the presence of bases;[533] in the latter case, participation of the nitro group as an oxygen nucleophile has been kinetically evidenced.[534a] An anthranil 1-oxide is also obtained from 2-nitrophenyl-pentafluorophenylmethanol.[534b]

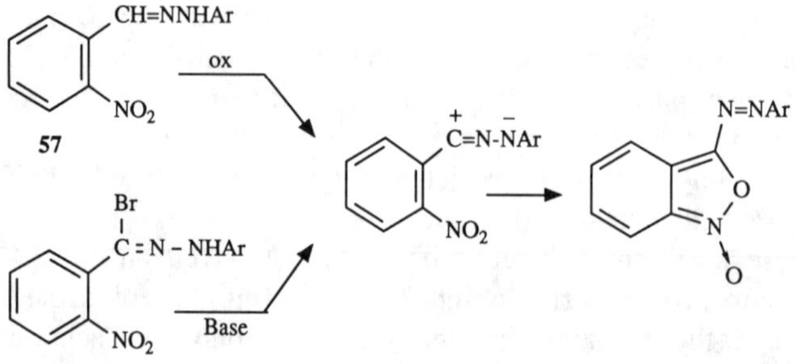

57

58

In some steroidal derivatives, a fused oxadiazole ring is formed from nitroimides (**59**, or from the tautomeric unsaturated nitroamines, **60**) with elimination of AcOH when such compounds are adsorbed on alumina; the mechanism is probably related to the previous ones.[535,536]

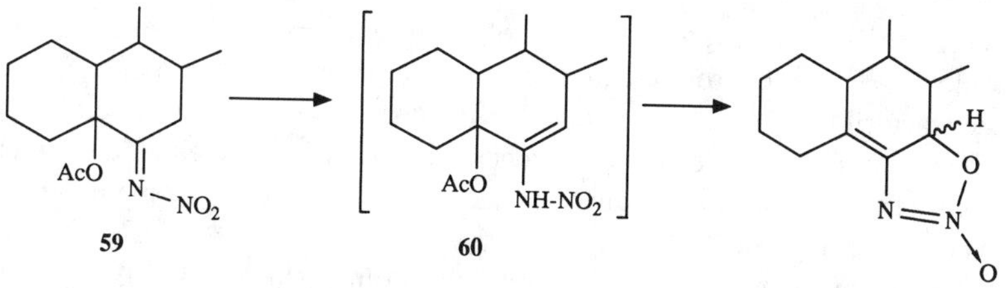

59   **60**

3-*Tert*-butyl-4,4-dimethyl-2-nitro-2-pentene undergoes spontaneous electrocyclic rearrangement to a 4*H*-1,2-oxazete 2-oxide, a process induced by the overcrowding around the double bond.[537] Other α, β-unsatured nitro compounds are formed by nitration of hindered allenes, and analogously cyclize.[538]

Some 2,1-benzisoxazolone *N*-oxides (cyclic nitronic anhydrides) are obtained by dicyclohexylcarbodiimide dehydration between a carboxyl and a nitro group.[539] Other analogs,[540] as well as nonfused isoxazolinone *N*-oxides,[541] have been proposed as intermediates, but have not been obtained in preparative useful reactions.

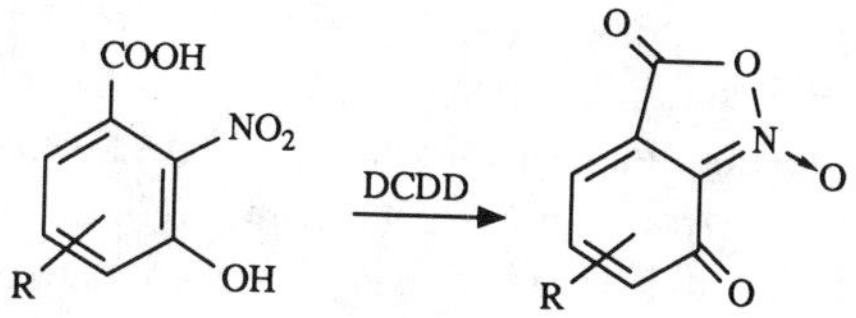

2. Reaction with a Nitrogen Atom

Some high-yield syntheses of (benzo)furoxans fall under this heading and formally involve cyclization of a nitro group unto a nitrene, arising either by fragmentation of azides or by oxidation of amines, though in most cases nitro-assisted elimination, rather than independent formation of a "free" nitrene function, is involved.

Furoxans are spontaneously formed at room temperature or below from 1,2-azidonitroalkenes, in turn obtained by the treatment of 1,2-dinitro-[542] or 1-halogeno-2-nitroalkenes[543] with the azide anion, or alternatively by nitration with NO_2BF_4 of vinyl azides.[544]

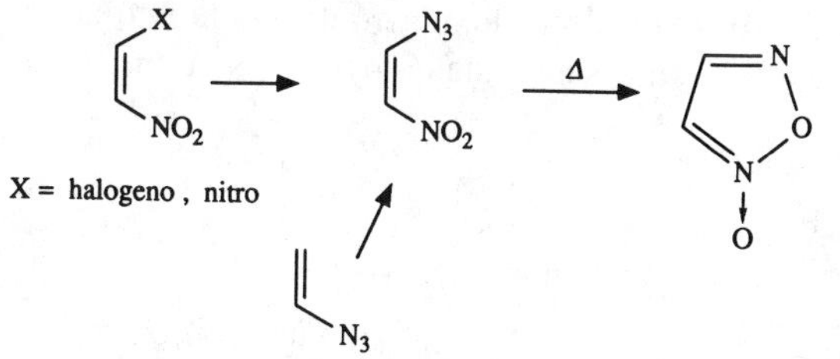

X = halogeno , nitro

The decomposition of *o*-nitrophenyl azides to yield benzofuroxans takes place at a lower temperature (80—120°C) than for the *meta* and *para* isomers, and is characterized by a low entropy of activation.[545,546] This suggests participation of the nitro group in the fragmentation of the azide function. This synthesis, usually by thermal decomposition,[547-550] though photochemical reactions give similar results,[548] has been largely applied for preparative purposes, including the syntheses of benzo-difuroxan and trifuroxan from polyazido-polynitrobenzenes,[549,550b] as well as of pyrido-,[549,551] quinolino-,[552] pyrimido-,[553-554] and thieno-furoxans.[554,555] Heating a *o*-

chloronitrobenzene with sodium azide is an alternative to the thermolysis of a preformed (e.g., from the hydrazine) azide;[550b,556] this reaction conveniently takes place at 60°C in CH_2Cl_2 in the presence of benzyltributylammonium bromide.[556] 4,5-Dinitroveratrole also gives a furoxan with NaN_3 in DMF.[557]

The other path to benzofuroxan is oxidation of *o*-nitroanilines,[558-563] generally accomplished with hypochlorite[558-559,561] (or from the preformed *N*-chloroaniline with bases),[559] or, in some cases less satisfactorily, with phenyliodonium bis(trifluoroacetate)[560-561] (or by electrochemical-generated iodonium ion).[563] Here again, the mechanism may involve either anchimeric assistance of the nitro group in the elimination of an anion (Cl⁻, PhI + AcO⁻) from the functionalized amine anion or a free nitrene.[559,561]

Another reaction of the nitro group with a nitrogen atom is illustrated by the formation of 4-formylbenzofuroxan from 7-nitroanthranil[564] and by the easy isomerization of 4- to 7-nitrobenzofurans.[565] These are examples of the general Boulton-Katritzky rearrangement of heterocycles (see also Chapter 2 for the benzofuroxan rearrangement).

2. Intermolecular Processes

a. Cyclization on the Nitrogen Atom

1. Various Attacks on a Carbon Atom

Several of the syntheses discussed in Section IIC1a1 can be realized by forming two new bonds to obtain the ring in a single operation, if not in a single step. Typical examples involve condensation between a nitro group and an activated methylene, with the second bond arising from reaction between an activated alkyl group and a carbonyl, as in the useful synthesis of quinoline 1-oxides from phosphoranes of diethyl succinate (see Section IIC1a1),[432a] and of pyrido[3,2-*d*]pyrimidine 5-oxide from 5-nitro-1,3,6-trimethyluracil and phenylacetaldehyde,[566] or from nucleophilic substitution, as in the synthesis of esters of benzothiazole-2-carboxylic acid by reaction of electron-withdrawing-substituted *o*-chloronitrobenzenes with thioacetates.[567]

Further syntheses involve attack at an aromatic position, as in the formation of *N*-hydroxy-acridones by treatment of *o*-nitrobenzaldehydes and aromatic hydrocarbons with cold, concentrated sulfuric acid,[568] and of acridine 10-oxide from the reaction of 2-nitrobenzyl chloride and benzene in the presence of $AlCl_3$.[569]

More important is the Wohl-Aue synthesis of phenazine 5-oxides by heating nitrobenzenes and arylamines with pulverized NaOH or KOH, a procedure that gives a moderate yield and several side products (azo and azoxy derivatives), but has the advantage of using readily available starting materials.[66,570-572] The reaction probably involves the intermediacy of 2-nitrodiphenylamines (compare with Section IIC1a2). A related process is the synthesis of pyrimido[4,5-*g*]pteridine 10-oxides by refluxing 5-nitro-6-chlorouracil and pyrimidinones, and again involves intermediated nitro-substituted dipyrimidyl amines.[573]

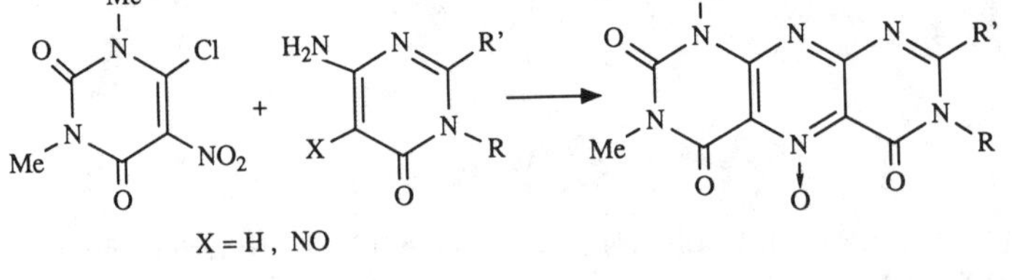

X = H, NO

2. Cycloaddition with Ynamines

1-Nitroalkenes undergo nucleophilic addition with ynamines to yield a zwitterion (**61**), and this evolves through formation either of a carbon-carbon bond (forming a nitrocyclobutene) or of a carbon-oxygen bond (forming a 4*H*-1,2-oxazine 2-oxide, **61′**). Unlike the corresponding 5,6-dihydro derivatives (see Section IIC2b2), compounds **61′** have only occasionally been

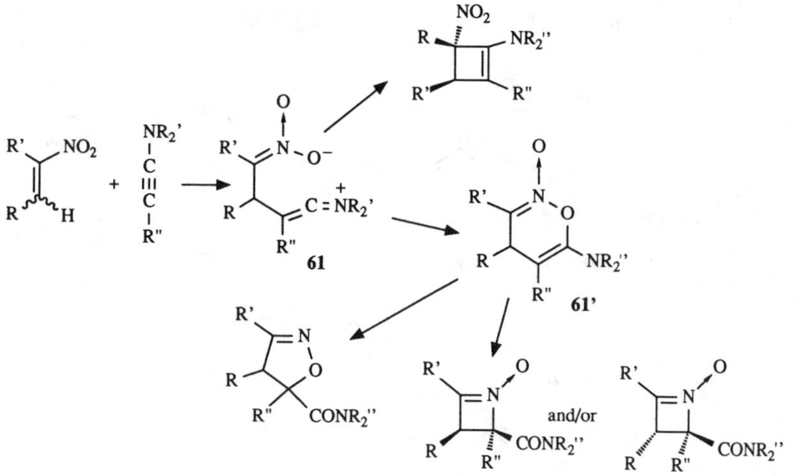

Scheme 10.

isolated (in the case of the reaction between 1-nitrocyclopentene and 1-phenyl-2-(1-pyrrolidinyl)acetylene), and generally rearrange to (*cis* or *trans*) substituted 2-carboxyamido-2,3-dihydroazete 1-oxides.[574-576] The latter is a stereospecific reaction, rationalized as either a symmetry-allowed ($_\sigma 2_s + _\pi 2_a$) or a symmetry-forbidden (but a heteroatom is involved) ($_\sigma 2_s + _\pi 2_s$) process, and its course can be related to the (calculated) preferred conformation of the primary adduct **61**. In turn, the azetine oxides undergo thermal electrocyclic rearrangement to open-chain azabutadiene N-oxides. This synthesis is discussed here because of the azetine structure of the final products, though from the mechanistic point of view it is rather related to the nitroalkene-enamine cycloaddition discussed in Section IIC2b2. The selectivity has been exploited and asymmetric induction in the azetine synthesis has been obtained using chiral ynamines[576] (Scheme 10).

The reaction has been extended to open-chain aliphatic- and aromatic-substituted nitroalkenes, as well as to nitrocycloalkenes, with both aliphatic and aromatic ynamines, the yield of azetine oxides typically ranges from 15% to 35%.[574-576] Side pathways (see Scheme 10) result from cleavage of the cyclic N-O bond in adduct **61** and with formation of a carboxamido isoxazoline (from 3-nitrobenzofuran and 1-phenyl-2-(1-pyrrolidinyl)acetylene, the same process goes further and the quinoline 1-oxide **62** is obtained by cleavage of the furan ring).[577a] The aromatic double bond in 5-nitropyrimidines participates into this reaction; when there is no substituent in positions 2 and 4, diazocines **63** are among the products and result from the addition of one mole of ynamine to the nitroalkene moiety (to form the azete ring) and of a second mole to a carbon-nitrogen double bond (to yield a fused nitroazetine, which undergoes electrocyclic ring opening).[577b]

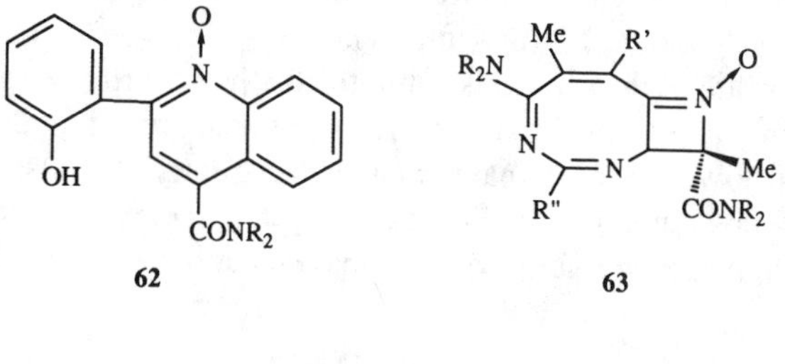

62

63

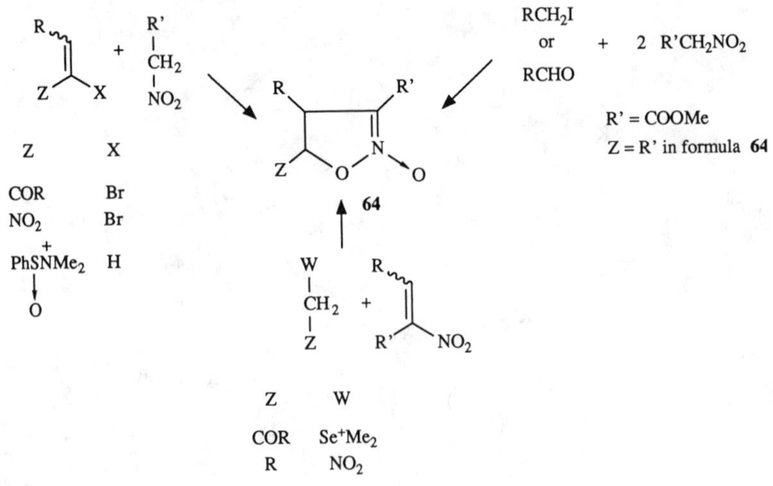

R' = COOMe

Z = R' in formula **64**

Scheme 11.

Z	X
COR	Br
NO$_2$	Br
PhSNMe$_2^+$ (with O below)	H

Z	W
COR	Se$^+$Me$_2$
R	NO$_2$

3. Attack on a Nitrogen Atom

The syntheses considered here lead to benzo-fused heterocycles and are simply variations of the processes discussed in Section IIC1a4. Here the *N*-oxide function arises from reaction of the nitro group with an amino group and the second bond from an aromatic nucleophilic substitution. Typical examples are the preparations of 3-aminopyrido[3,4-*e*]-1,2,4-triazine 1-oxide from 3-nitro-4-chloro-[578a] (or methoxy)[578b] pyridine and guanidine, and of triazoloquinoline oxides from 5,8-dimethoxy-6-nitroquinoline and alkylhydrazines.[578c]

b. Cyclization on the Oxygen Atom

1. Attack of the Nitro Group on a Carbon Atom, with Reactions Involving Carbanions or Carbenes

As it has been reported in Section IIC1a1, intramolecular attack through the oxygen atom of the nitro group in 1,3-bromonitro- or 1,3-dinitro derivatives leads to isoxazoline *N*-oxides. The precursors for such syntheses are often obtained from the reaction of the nitroalkane carbanion. Both carbon-carbon and carbon-oxygen bonds can be created in a single operation. Thus, isoxazoline 2-oxides are obtained through three main cyclization schemes, according to the fragments used, and the nitro group can be present either in the olefin undergoing the electrophilic attack or as the activating group in the carbanion. These are (1) a 3 + 2 scheme via Michael addition of nitroalkane anion onto α,β-unsaturated ketones, 1-nitroalkenes,[528-530,579,580] or other electron-poor olefins,[581] with accompanying C-O bond formation through elimination of a leaving group (bromo, nitro,[528-530,579,580] sulfinamide);[581] (2) a 1 + 2 + 2 scheme involving an aldehyde[582-584] (or its enamine)[585] or an alkyl iodide[586] and two molecules of a nitroacetic acid ester; and (3) a 4 + 1 scheme using a 1-nitroalkene and an active methylene carrying a leaving group (nitro,[580] alkylselennoium)[587] (Scheme 11).

The preparative value of the synthesis is limited by the occurrence of side processes (e.g., formation of cyclopropanes from nitroalkenes) and secondary reactions due to the lability of the isoxazolines oxides; however, this remains an interesting entry into this class of heterocyclics; among the many 4-substituted isoxazoline 2-oxides prepared, noteworthy are the glycosides **65**.[582] Furthermore, it has been found that two-phase reactions (alumina supported KF for the nitroalkene-nitroalkane system[588a] and molecular sieves for the aldehyde-nitroacetate system)[588b] are advantageous and lead to good yields of these N-oxides.

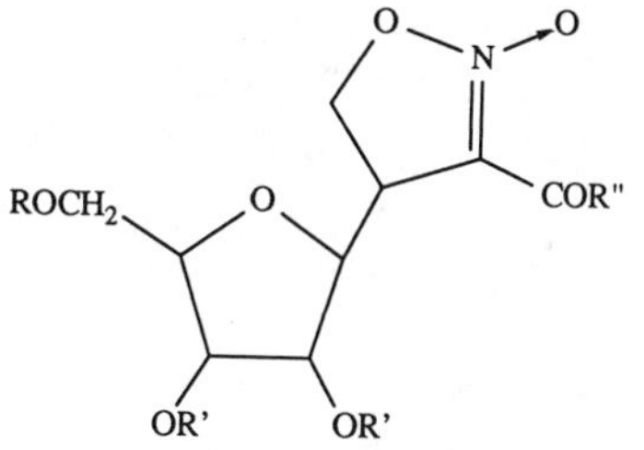

65

As far as the nitroalkane-nitroalkene reaction is concerned, an alternative to the not always easy availability of the latter substrates is the synthesis in moderate yield of 4,5-dihydro-3,4,5-triarylisoxazole 2-oxides from three moles of an arylnitromethane (or rather, the salts of the corresponding *aci* form) under oxidative conditions, a process that probably follows a similar mechanism.[589]

Another path for isoxazoline 2-oxides involves carbene precursors, either through a 4 + 1 scheme from 1-nitroalkenes[590] (or alternatively, trinitromethane derivatives)[591] and diazo compounds, or through a 3 + 2 scheme, obtained when the silver trinitromethane salt is silylated in the presence of cyclohexene.[592] A variation of this synthesis leads to the tricyclic compound **66** by the addition of three moles of diazomethane to 1,3,5-trinitrobenzene.[593]

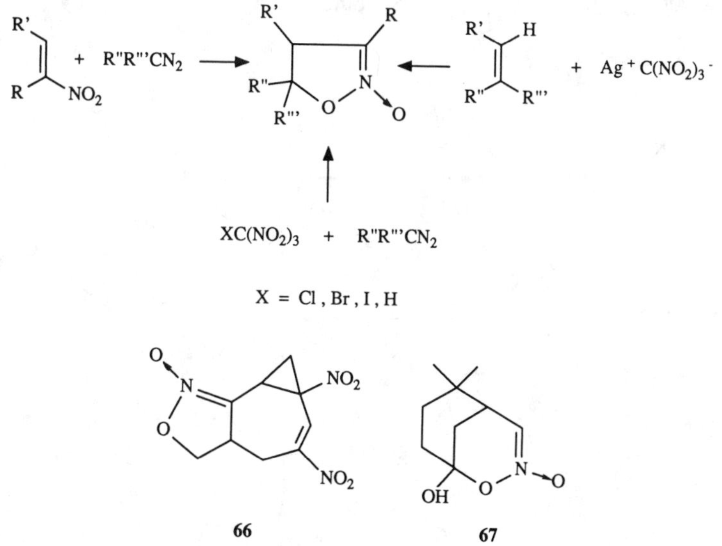

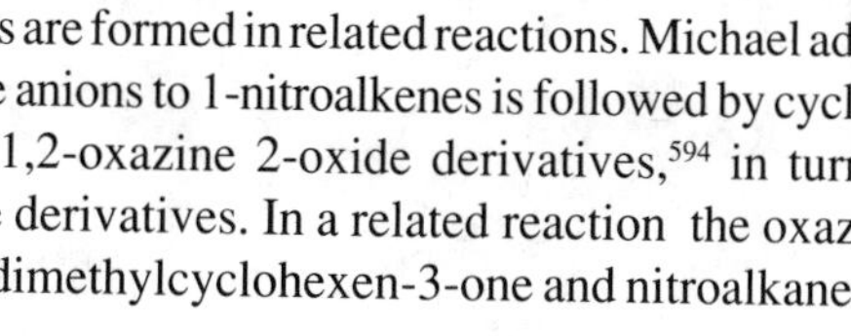

$$XC(NO_2)_3 \quad + \quad R"R"'CN_2$$

$$X = Cl, Br, I, H$$

66   67

Six-membered N-oxides are formed in related reactions. Michael addition of the acetoacetate or the 1,3-cyclohexandione anions to 1-nitroalkenes is followed by cyclization to monocyclic,[595b] or, respectively, bicyclic 1,2-oxazine 2-oxide derivatives,[594] in turn reactive intermediates further evolving to pyrrole derivatives. In a related reaction the oxazine **67** is obtained under basic conditions from 4,4-dimethylcyclohexen-3-one and nitroalkanes, along with open-chain nitro derivatives.[595, 595a]

2. Cycloaddition of Nitroalkenes to a Carbon-Carbon Double Bond

A valuable synthetic pathway leading to dihydro-1,2-oxazine 2-oxides involves the cycloaddition between a 1-nitroalkene and an alkene, formally a 1,4-dipolar cycloaddition. Thus, 1-nitrocyclohexene and cyclohexene react in the presence of $SnCl_4$ and give the tricyclic products **68—70** with, respectively, a 43%, 3%, and 7% yield.[596] Bicyclic derivatives are analogously obtained from the intramolecular cycloaddition of 1-nitro-1,7-dienes.[597]

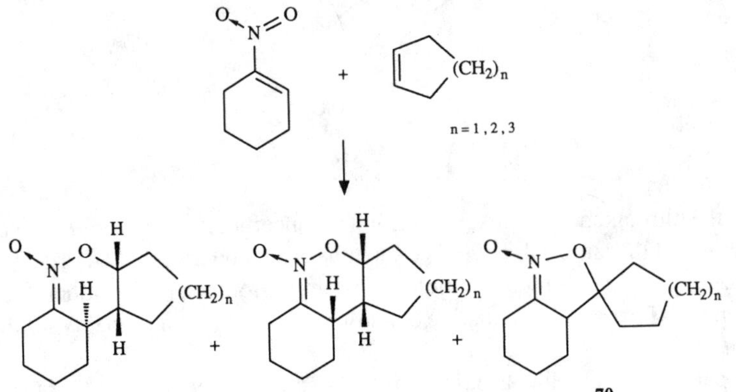

The cycloaddition requires no catalyst when enamines are used. Thus, 2-nitropropene and 1-(*N*-morpholino)cyclohexene yield 80% of the bicyclic oxazine oxide **71** in dry ether at $0°C$,[598] and related *N*-oxides are obtained from enamines and α-[599] and β-nitrostyrene,[600] as well as 1-phenyl-1-nitro-[601] and -2-nitro-propene[603] (Scheme 12).

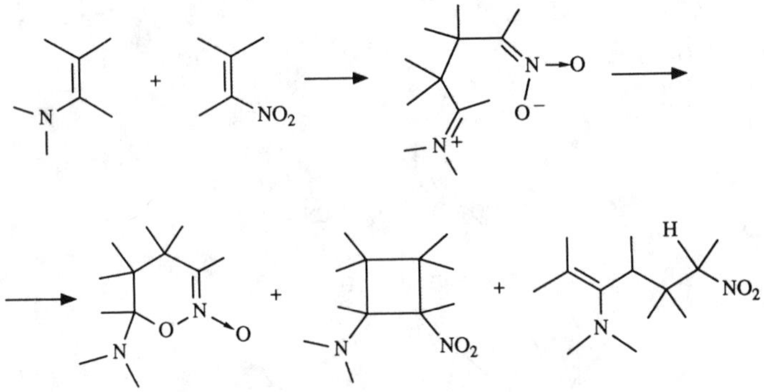

Scheme 12.

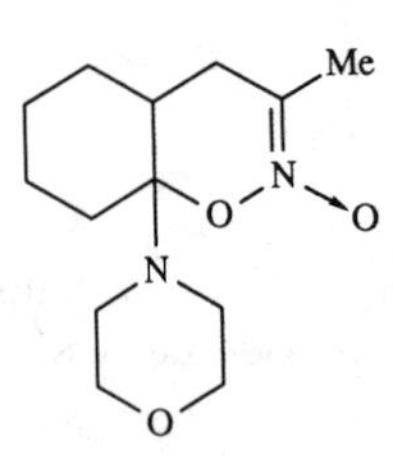

71

Other products, in particular nitrocyclobutanes and open-chain nitroalkyl derivatives, are formed besides oxazines, the ratio between them depending primarily on the substitution pattern of the nitroalkene; e.g., 1-nitropropene yields 85% of the open-chain adduct 1-(N-morpholino)-6-(2-methyl-2-nitroethyl)cyclohexene,[598] and in general oxazines are less readily isolated from the reaction of 1-unsubstituted 1-nitroalkenes (β-nitrostyrene does yield an oxazine oxide with 1-morpholino-1-cyclohexylethylene, although it is quite unstable, and with other enamines only open-chain derivatives are isolated,[600] whereas oxazines are readily obtained from β-substituted-β-nitrostyrenes).[603] Besides enamines, other electron-rich olefins undergo this cycloaddition, e.g., enols (see Section IIC2b1),[594] vinyl ethers,[604] ketene acetals (in the presence of zinc chloride),[605] and silylated enol ethers [in the presence of Ti(IV) derivatives].[606,607] Alkoxy-substituted oxazine oxides tend to be more stable than the above-mentioned amino derivatives.

The process is similar to the corresponding reaction with ynamines (compare with Scheme 10) and occurs via a zwitterion. Great attention has been given to the stereocontrol of the reaction, in particular as far as the new carbon-carbon bond (which is conserved when the oxazine opens to a nitroalkyl derivative, the product of a Michael addition) is concerned.[606-609] The selectivity may be high. Thus in the addition of cyclohexanone enamines and nitroalkenes, there is generally a strong preference (up to 99:1)[608] for the products resulting from an *endo* approach of the reagents (*like* using the Seebach-Prelog descriptors;[610] see formula **72**), but the stereochemistry can be reversed (in the case of β-nitrostyrene) to a predominantly (up to 4:1) *exo* (*unlike*, **73**) approach using silylated enol ether in the presence of dichloro(diisopropoxy)titanium (TiCl$_4$ gives a mixture).[606,607] With unsubstituted cycloalkenes the approach is *exo*.[597]

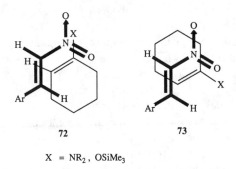

72   **73**

X = NR$_2$, OSiMe$_3$

3. Attack on a Heteroatom

An important process that would fall under this heading is the synthesis of furoxans from nitroalkanes, but since this involves the intermediacy of nitriloxides, a discussion is postponed to Section D1.

An interesting synthesis involves cyclization on a phosphorus atom and leads to 1,2,5-oxazaphospholine 2-oxides (tautomeric with the N-hydroxy forms) starting from nitroalkenes and P(III) derivatives [P(OR)$_3$,[611] PR'(OR)$_2$,[612] PR'$_2$OR].[613,614]

D. REACTIONS OF NITRILE OXIDES

1. Dimerization

Nitrile oxides are reactive species and often dimerize on standing to furoxans.[615] Thus, under a variety of conditions preparation of these derivatives leads directly to furoxans, unless particular care is taken or the reaction is diverted to other paths (rearrangement to isocyanate, trapping with dipolarophiles). This is a reversible reaction, and pyrolysis of furoxans is actually a method of generating nitrile oxides (see Scheme 13 and Chapter 4).

Only a hint is given of the typical reactions leading to nitrile oxides and hence to furoxans from the extended literature on the subject.

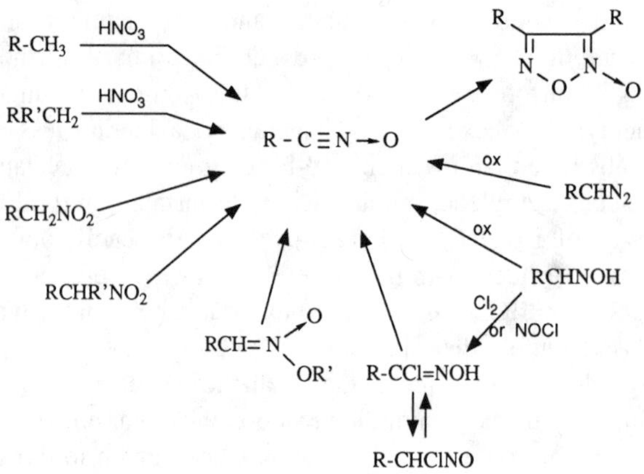

Scheme 13.

1. Nitration of "activated" (by electron-withdrawing substituents) alkyl groups, e.g., meth-ylketones to diacylfuroxans;[616-619] when a methylene, rather than a methyl group, is involved, one of the substituents is eliminated, e.g., diphenylfuroxan from benzyl bromide and sodium nitrite.[620] Dibenzoylfuroxan is obtained from dimethylphenacylsulfonium bromide and sodium nitrite.[621] Treatment of allyl acrylates with HNO_3/H_2SO_4 cleaves the double bond and gives bis(carboxyalkyl)furoxans.[622]

2. From preformed nitro derivatives, either through further nitration [and probably interme-diate formation of nitrolic acids, $R-C(=NOH)NO_2$, also possible, of course, in case 1 or through elimination paths; from the primary nitroalkanes the reaction is formally a dehydration of the *aci* form, while when there is an α substituent a group is eliminated, as in the corresponding case in 1. Observed reactions include spontaneous decomposition of *gem*-dinitro derivatives (alkyl dinitroacetates);[623] pyrolysis, e.g., of nitromalonates;[624] treatment of primary or secondary nitroalkanes with acids,[625] Lewis acids,[626] acyl chlorides, or anhydrides,[627-629] $POCl_3$ (e.g., in the synthesis of the thienofuroxan **74**, a key intermediate for biotin)[630] or $PhSO_2Cl$, all in the presence of triethylamine;[631] dehydroha-logenation of *gem*-bromonitro derivatives with PPh_3.[632] The nitroalkene **75** yields a bis(carboxyalkyl) furoxan upon treatment with BF_3.[633] The sulfonyl nitronic ester $PhSO_2CH=NO_2Me$ gives $PhSO_2CNO$, and the furoxan from it, in the presence of bases.[634] Nitroalkenes can be converted to hydroximyl chlorides, and these, in turn, can be dehydrohalogenated to nitrile oxides (see case 3).

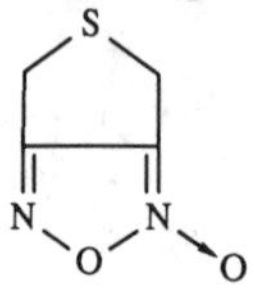

74 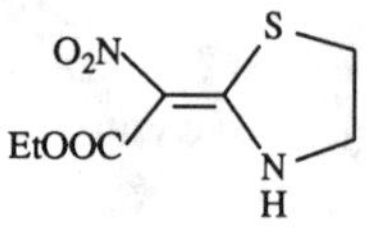   **75**

3. Oxidation of other nitrogen-containing substrates. The above-mentioned nitrolic acids $[R-C(=NOH)NO_2]$[619] can likewise be formed from oximes by treatment with nitrogen oxides.[636,637] *O*-aryl derivatives of oximes have been oxidized to furoxan with MnO_2.[638] Hydroximyl chlorides (α-chlorooximes) obtained by chlorination of oximes,[639-641] addi-

tion of the hydrogen halide of nitroalkenes,[635] nitrosation of sulfonium salts,[621,642] or other precursors (e.g., $CF_3CClNOH$ is conveniently obtained from the mixed anhydride $CF_3CClNOCOCF_3$, which in turn is prepared from N,O-bis(trifluoroacetyl) hydroxylamine)[643] are converted to nitrile oxides by treatment with bases (and in some cases spontaneously).[641] Bromohydrazones are likewise a source of nitrile oxides,[640] as are aliphatic diazo compounds carrying an electron-withdrawing substituent (acyl,[644-646] sulfonyl,[645,647] trifluoromethyl)[648] when treated by nitrous acid at 0°C[646] or by nitrogen oxides[644-645,648] or nitrosyl chloride,[647] and fulminic acid and its salts upon reaction with halides.[649]

Dimerization is a general process with the nitrile oxides (fulminic acid itself does not yield furoxan; however, among the products there is a tetramer with a furoxan structure, compound **76**),[650] and although steric hindering slows down the process,[651] even bulky radicals such as mesityl[652] and 1-adamantyl[653] do not preclude obtaining the furoxans under appropriate conditions. A mixture of two aromatic nitrile oxides gives both the symmetric and the mixed furoxans.[654] Kinetic studies are in accordance with a concerted pathway for the dimerization,[651,655] however, a stepwise mechanism is possible; thus the nitrile oxide can be regarded as a nitrosocarbene, and the carbon-carbon bond can be formed in the first step, the 1,2-dinitroalkene thus formed cyclizing at a later stage.[652,655,656]

At any rate, furoxans are not the only possible products. Thus, dimerization of aromatic nitrile oxides yields mainly 1,4,2,5-dioxadiazines, **77** (with minor amounts of 1,2,4-oxadiazole 4-oxides, **78**) under basic (pyridine,[657] triethylamine[658]) catalysis, and with BF_3 either compound **78** (ratio substrate/catalyst = 2) or compound **77** (excess catalyst) is formed[659a] (furthermore, nonoxidized 1,2,4-dioxadiazole is usually formed when nitrile oxides are generated from α-chlorooximes).[659b] A dioxadiazine is obtained also from trifluoroacetonitrile oxide unless the presence of the precursor chlorooxime is avoided.[643]

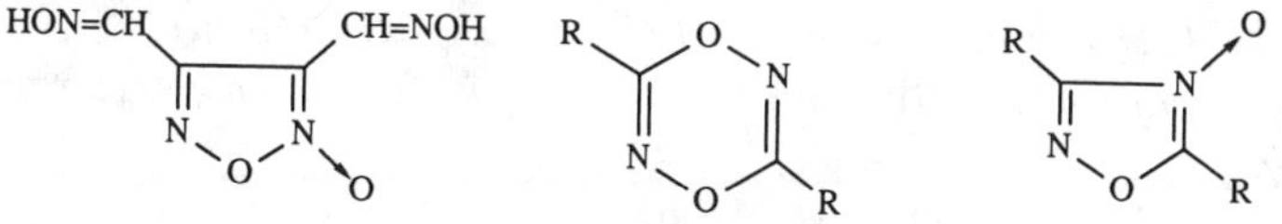

76	**77**	**78**

2. Other Reactions

Addition across the carbon-nitrogen bond of nitrile oxides also yields heterocyclic N-oxides. This is the case for the synthesis of 1-arylpyrazole[660] and 1-aryl-1,2,4-triazole 2-oxides[661] from sulfurated ylides, and of imidazole[662] and 1,2,4-triazole N-oxides[663] from isoxazol-5-ones and O-benzoylamidoximes, respectively.

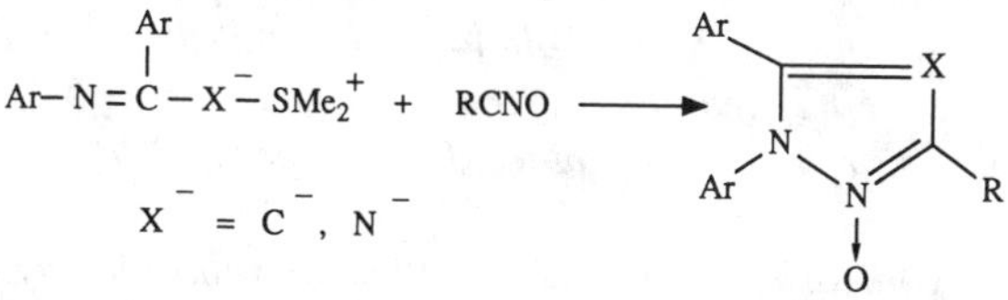

Cyclization may likewise involve a benzo ring, as in the synthesis of benzimidazole 1-oxides from N-aryl-S,S-dimethylsulfimides,[664] and of 1-hydroxybenzimidazole 3-oxides from nitrosobenzene.[403a]

E. CYCLIZATION OF COMPOUNDS CONTAINING THE PREFORMED N-OXIDE FUNCTION

In view of the many convenient ways of forming nitrone, azoxy, and related functions from the reaction of two groups, most syntheses of heterocyclic *N*-oxides occur through paths i and ii in Scheme 4. However, the access through path iii from open-chain compounds containing the preformed *N*-oxide function remains a possibility. Typical examples are the synthesis of 1*H*-2,3-dihydro-1,4-benzodiazepin-2-one 4-oxides by forming the lactam bond in *N*-(2-amino-α-phenylbenzylidene)glycine *N*-oxides,[665] and of cinnoline 2-oxide by condensation between the methyl and the carbonyl group in 2-methylazoxybenzophenone.[666] Particularly interesting is the electrocyclic ring closure (a photochemical reaction, reversible in the dark) of 2-azoxy-(1-naphthol) to a dinaphtopyridazine *N*-oxide.[667]

F. SURVEY OF THE SYNTHESES BY THE NUCLEUS FORMED

As was mentioned at the beginning of Section II, formation of both the ring and the *N*-oxide function is not only an alternative approach to *N*-oxides, but in several cases it is a convenient entry for the heterocycles themselves, which are obtained by following deoxygenation.

Reference to the main synthetic pathways is made in Table 4 according to the relevant section (Section IIA, from hydroxylamine derivatives; Section IIB, from nitroso, and Section IIC, from nitro compounds; Section IID from nitrile oxides).

A choice of experimental procedures is given below, following the order: azine, azole, and heterocycles with more than one nitrogen atom in different rings (Table 4).

1-Hydroxy-4,6-dimethyl-3-carbonitrile-2-pyridone. *A solution of potassium cyanohydroxamate (1.4 g, 10 mmol — prepared in 50% yield by mixing KOH, NH₂OH, and ethyl cyanoacetate in methanol) and acetylacetone (1 g, 10 mmol) in water (20 ml) containing piperidine (1 ml) is refluxed for 30 min, cooled, and acidified with acetic acid to give the product (1.4 g, 85%) (Section IIA2c1).*[331]

2,3,5,6-Tetraphenylpyridine 1-oxide. *2,3,5,6.Tetraphenylpyrylium bromide (2 g, 4.3 mmol), NH₂OH.HCl (1.5 g), and NaOAc.3 H₂O (10 g) in AcOH (40 ml) are refluxed for 5 min, and then poured into water (200 ml). Filtration and recrystallization gives the product (1.51 g, 88%) (Section IIA1a3).*[186e]

Ethyl 5-methyl-2-pyrazincarboxylate 4-oxide. *A mixture of the hydrochloride of glycine ethyl ester (5 g, 35.8 mmol), sodium ethoxide (5 g, 73.5 mmol), and ethyl formate (70 ml) is stirred for 15 h and allowed to stand overnight. Evaporation yields N,α-diformylglycine ethyl ester, which is satisfactory for the next step; 5 g (27.6 mmol) of this substance and 3.7 g (27.8 mmol) of pyruvaldoxime dimethylacetal are heated at 60°C for 6 h in acetone (125 ml) containing 35% HCl (5 ml). Evaporation, extraction with H₂O-CH₂Cl₂, and recrystallization from methanol give the product (2.8 g, 56%) (Section IIA2b4).*[312]

3-Methyl-9-N-morpholino-5,6,7,8,9,10-hexahydro-[1,2,4H]benzoxazine 2-oxide. *2-Nitropropene (2.8 g, 32 mmol, prepared by dehydration of 2-nitropropanol) in dry ether (5 ml) is added dropwise under stirring to a cooled solution of 1-N-morpholinocyclohexene (5.5 g, 32 mmol) in the same solvent (5 ml), maintaining the temperature under 5°C. The product (6.5 g, 80% yield) crystallizes out after a few minutes. It is stored in a dessicator in vacuo at 0°C (Section IICb2).*[598]

Diethyl quinoline-2,3-dicarboxylate 1-oxide. *(Diethyoxyphosphinyl)succinate (7.93 g, 27 mmol) is added to a cold solution of 2-nitrobenzaldehyde (3.8 g, 25 mmol) and sodium (0.62 g, 27 mmol) in ethanol (45 ml), and the mixture is stirred for 1.5 h in an ice bath. Evaporation, extraction with AcOEt-H₂O, and recrystallization yield the product (3.05 g, 40%). (Section IIC1a1).*[432a]

TABLE 4
Preparation of Heterocyclic *N*-Oxides by Synthesis of the Ring

Section[a]

a. Azine *N*-Oxides

Pyridine	A1a1, a4, b1, b5,b7, c1; A2c1
	C1a1, C2a1.
Pyridazine	B1a2
Pyrimidine	A1b6, c3, c4; A2b1, b3
Pyrazine	A1b5, c1, c4, c6
	A2a, b1, b3, b4, b5, c1
1,2,3-Triazine	A1c5
1,2,4-Triazine	A1b3; A2b1, b2, c1
1,3,5-Triazine	A1c4; A2c3
1,2-Oxazine	C2b1, b2
3-Thiazine	A1b4
Quinoline	A1a1, a3, a4, b3, b6, b2;
	C1a1, a3
Isoquinoline	A1a1, a3, b7
Cinnoline	C1a4
Quinazoline	A1a4, b2, b3, c3; A2b1, c1
Quinoxaline	A1a4; B1b1; C1a1, a2
1,2,3-Benzotriazine	A1b6
1,2,4-Benzotriazine	C1c4
1,4-Benzoxazine	A1a1, a3
1,3-Benzothriazine	A1a3, a4
Phenanthridine	C1a1
Acridine	C1a2
Benzo[*c*]cinnoline	C1a4
Phenazine	B1b3; C2a1, a2

b. Azole *N*-Oxides

Pyrrole	A1a1, b1, b5; B1b1
Pyrazole	A1b5; B1a2; D2
Imidazole	A1b2, c2, c3; A2b1, b3, b4, b6;
	B1b2; D2
1,2,3-Triazole	B1a3
1,2,4 Triazole	A1c2, c6; D2
Isoxazole	C1b1; C2b1
Oxazole	A2a, b4
Thiazole	A1b4, c1
1,2,4-Oxadiazole	B1b3
1,2,5-Oxadiazole	
(Furazan)	B1a2, b2; C1b2; D1
Indole	A1a1; C1a1, a2
Indazole	A1b7; C1a4
Benzimidazole	A1a2; C1a1, a2; D2
Benzotriazole	B1a5; C1a4
Benzisoxazole	B1a3; C1b1
Benzoxazole	A1a4
Benzofurazan	C1b2;
Benzothiazole	A1a4; C1a1, a2; D2

c. Other *N*-Oxides

Azete	C2a2
Diazete	B1a2
1,4-Benzodiazepine	A1b5

TABLE 4 (continued)
Preparation of Heterocyclic *N*-Oxides by Synthesis of the Ring

Section[a]

d. *N*-Oxides of Heterocycles with Nitrogen Atoms in Different Rings

Pteridine	A1c2; A2c2; B1b2, C1a1
Pyrimido[4,5-*b*]quinoxaline	B2
Pyrimido[5,4-*c*]-1,2,4-triazine	B2
Purine	A2b4, c2

[a]   Refer to the indicated subsections in Section II, see Table of Contents.

6-Chloro-quinoxalin-2-one 4-oxide. Diketene (17.7 ml) is added dropwise to a refluxing solution of 4-chloro-2-nitroaniline (34 g, 0.28 mol) in benzene containing triethylamine (1 ml) to yield N-(4-chloro-2-nitrophenyl)-2-oxo-butanoic amide (40 g, 70% yield). 50 g of this product are heated at 80°C in 6.5% aqueous NaOH to give the product (31 g, 65%) (Section IIC1a1).[439b]

Benzo[c]cinnoline 5,6-dioxide. A solution of 2,2'-dinitrobiphenyl (2.4 g, 10 mmol) in absolute ethanol (200 ml) containing 4% aqueous NaOH (4 ml) and Raney nickel W-6 is hydrogenated until 480 ml H_2 are adsorbed (ca. 1.5 h); then it is filtered and evaporated. Extraction with chloroform and precipitation with ether of the warm solution yields the product (1.05 g, 50% yield) (Section IIB1a4).[374]

7-Chloro-3-hydroxy-1,2,4-benzotriazine 1-oxide. A suspension of 4-chloro-2-nitrophenylurea (330 g, 1.53 mol) in 30% aqueous NaOH (9 l) is heated at 90—95°C while stirring for 30 min, and then it is acidified with AcOH. Filtration of the cooled solution, dissolution of the residue in 5% aqueous NaOH, and reprecipitation with concentrated HCl gives the product (266 g, 88% yield). (Section IIC1a4).[507]

Phenazine 5-oxide. A mixture of aniline (93 g, 1 mol), nitrobenzene(375 g, 3 mol), finely ground KOH (400 g) and benzene (0.8 l) is stirred and refluxed for 5 h and then allowed to stand overnight and steam distilled until some azobenzene passed over. The precipitate obtained from the cooled residue is washed with water and then with EtOH (250 ml), cooled, filtered, and finally briefly refluxed with 500 ml EtOH; the product (66 g, 34% yield) crystallized in a pure state on cooling (Section IIC2a1).[571a]

1-Hydroxy-2,4,5-trimethyl-imidazole 3-oxide. A solution of dimethylglyoxime (5.8 g, 50 mmol) and acetaldoxime (3 g, 50 mmol) in methanol (0.45 l) is treated with concentrated HCl (5 ml), allowed to stand for 2 d, and then refluxed for 1 h. Neutralization with Na_2CO_3 (7 g) and concentration gives the product (5.2 g, 73%) (Section IIA2b4).[310]

3,5-Bis(carboxymethyl)-4-phenyl-4,5-dihydroisoxazole 2-oxide. Benzaldehyde (1.06 g, 10 mmol), and then, dropwise, i-PrNH$_2$ (0.9 g, 15 mmol) are added to molecular sieves powder (4 A, 5 g) at 0—5°C; after 3 h at room temperature and evaporation of excess i-PrNH$_2$, methyl nitroacetate (2.6 g, 22 mmol) in 1:1 ether-methanol (8 ml) is added at 0°C, and then the mixture is refluxed for 6 h. Filtration through celite (with washing with 2 ×20 ml CH$_2$Cl$_2$), evaporation, and recrystallization yield the product (2.1 g, 71%) (Section IICb1).[588b]

5,7-Di-tert-butyl-3,3-dimethyl-3H-indole 1-oxide. Finely powdered 1,3,5-tris-tert-butyl-2-nitrobenzene (50 g) is spread on a 50 ×30 reflecting rectangular tin pan and irradiated with two 300-W sunlight lamps from a distance of 40 cm. After 45 min the crystals are collected, ground, and spread again, and the process is repeated once more after another 45 min. Dissolution of the slightly sticky crystals in 500 ml warm cyclohexane and chromatography on deactivated silica gel yield 42.8 g (86%) of the starting material and 2.5 g (5%) of the title product. Higher conversion is avoided in view of the photosensitivity of the N-oxide (Section IIC1a1).[427b]

2-(p-Tolyl)-1-hydroxybenzimidazole. *A solution of the o-nitroanil of p-tolualdehyde (1.2 g, 5 mmol) and KCN (0.65 g, 10 mmol) in MeOH (10 ml) is refluxed for 5 h, then cooled, refluxed, and acidified with concentrated HCl. Filtration and recrystallization gives the product (0.67 g, 61%) (Section IIC1a4).[497b]*

Riboflavin 5-oxide. *A mixture of 6-chlorouracil (1.47 g, 10 mmol) and N-D-ribityl-3,4-xylidine (2.55 g, 10 mmol) in DMF (0.8 ml) is heated at 155°C with stirring for 7 min. Dilution with 10 ml ethanol after cooling gave, after a few days, crystals of 6-(N-D-ribityl-3,4-xylidino)uracil (2.5 g, 68%). This material (3.7 g, 10 mmol) and sodium nitrite (3.5 g, 50 mmol) in AcOH (20 ml) are stirred at 40 C for 30 min and then allowed to stand for 2 h at room temperature and filtered. Dilution of the filtrate with ether (50 ml) gives the product overnight (3.3 g, 83%) (Section IIB2).[423]*

Hypoxanthine 1-oxide. *Methyl 4-nitroimidazole-5-carboxylate (7 g, 40 mmol) is added portionwise with stirring to a solution of hydroxylamine hydrochloride (4.2 g, 61 mmol) and NaOH (6 g, 150 mmol) in water (120 ml) After 2 h stirring and 3 d standing, the solution is acidified to pH 6 with diluted HCl, and 4-nitroimidazole-5-hydroxamic acid (6 g, 85%) is precipitated. 3 g of this compound in 40 ml DMF containing 0.3 g PtO$_2$ are hydrogenated at 60 psi (5 min). To the filtered solution, 20 ml ethyl orthoformate are quickly added, and the resulting mixture is refluxed at 155°C for 20 min, and then is poured into 100 ml of ice water. Evaporation and three recrystallizations give the product (1.8 g, 53% yield) (Section IIA2c2).[333]*

REFERENCES

1. a. **Bodendorf, K. and Binder, B.,** Properties of amine oxides, *Arch. Pharm.*, 287, 326, 1954; b. **Ogata, Y. and Sawaki, Y.,** Peracid oxidation of imines. Kinetics and mechanism of competitive formation of nitrones and oxaziranes from cyclic and acyclic imines, *J. Am. Chem. Soc.*, 95, 4692, 1973.
2. **Modena, G. and Todesco, P.E.,** On the N-oxidation of pyridine and some alkylpyridines, *Gazz. Chim. Ital.*, 90, 702, 1960.
3. **Dondoni, A., Modena, G., and Todesco, P.E.,** Polar and steric effects on the oxidation of alkyl- and halogeno-pyridines, *Gazz. Chim. Ital.*, 91, 613, 1961.
4. **Foucart, J., Nasielski, J., and Vander Donckt, E.,** Perbenzoic acid oxidation of monoaza-aromatic bases, *Bull. Soc. Chim. Belg.*, 75, 17, 1966.
5. **Ames, D.E. and Archibald, J.L.,** Some dipyridylalkanes, *J. Chem. Soc.*, 1475, 1962.
6. **Taylor, E.C. and Boyer, N.E.,** Nicotine-1-oxide, nicotine-1′-oxide, and nicotine-1,1′-dioxide, *J. Org. Chem.*, 24, 275, 1959.
7. **Ochiai, E., Katada, M., and Hayashi, E.,** A synthesis of pyridine 1-oxide, *Yakugaku Zasshi*, 67, 33, 1947.
8. **Shaw, E., Bernstein, J., Losee, K., and Lott, W.A.,** Substituted 2-bromopyridine N-oxides and their conversion to cyclic thiohydroxamic acid, *J. Am. Chem. Soc.*, 72, 4362, 1950.
9. **Shchukina, M.N. and Savitzkaya, N.V.,** N-oxides of 8-hydroxyquinoline and its ethers, *Zh. Obshch Khim.*, 22, 1224, 1952.
10. **Pushkareva, Z.V. and Varyukhina, L.V.,** Preparation and properties of N-oxides of some acridine derivatives, *Dokl. Akad. Nauk SSSR*, 103, 257, 1955.
11. **Markgraf, J.H. and Carson, C.G.,** The rate determining step for the reaction of acridine N-oxide with acetic anhydride, *J. Org. Chem.*, 29, 2806, 1964.
12. **Buchardt, O., Jensen, B., and Kjøller Larsen, I.,** The formation of benz[d]-1,3-oxazepines in the photolysis of quinoline N-oxides in solution, *Acta Chem. Scand.*, 21, 1841, 1967.
13. **Vostrikov, N.S., Dzhemilev, U.M., Bylina, G.S., Moiseenkov, A.M., Semenkovskii, A.V., and Tolstikov, G.A.,** Oxidation by p-carbomethoxyperbenzoic acid, *Izv. Akad. Nauk. SSSR, Ser. Khim.*, 2337, 1978.
14. a. **Fujiwara, A.N., Acton, E.M., and Goodman, L.,** N-Oxides and S-oxides of ellipticine analogs, *J. Heterocycl. Chem.*, 6, 389, 1969; b. **Rastetter, W.H., Richard, T.J., and Lewis, M.D.,** 3,5-Dinitrobenzoic acid. A crystalline, storable substitute for peroxytrifluoroacetic acid, *J. Org. Chem.*, 43, 3163, 1978; c. **Ponticello, G.S., Hartman, R.D., Lumma W.C., and Baldwin, J.J.,** Synthesis of novel 3-(alkylthio)-2-halopyridines and related derivatives, *J. Org. Chem.*, 44, 3080, 1978.

15. **Dzhemilev, U.M., Vostrikov, N.S., Moiseenkov, A.M., and Tolstikov, G.A.,** Pentafluoroperoxybenzoic acid as a new highly active and stable oxidizing reagent, *Izv. Akad. Nauk SSSR, Ser. Khim.,* 1320, 1981.
16. **Mamalis, P. and Petrow, V.,** Some heterocyclic *N*-oxides, *J. Chem. Soc.,* 703, 1950.
17. a. **Yamazaki, M., Honjo, N., Noda, K., Chono, Y., and Hamana, M.,** Syntheses and reactions of 2-chloro-, 2-bromoquinoline 1-oxides, and 2-bromolepidine 1-oxides, *Yakugaku Zasshi,* 86, 749, 1966.; b. **Kato, T., Yamanaka, H., and Hiranuma, H.,** *N*-oxidation of 4-chloropyrimide derivatives, *Chem. Pharm. Bull.,* 16, 1337, 1968.
18. **Colonna, M. and Risaliti, A.,** On aromatic *N*-oxides. Action of organo-magnesium derivatives, *Gazz. Chim. Ital.,* 83, 58, 1953.
19. **Ochiai, E., Ikehara, M., Kato, T., and Ikekawa, N.,** A new method of preparation of basic components of coal tar, *Yakugaku Zasshi,* 71, 1385, 1951.
20. **Ochiai, E.,** Recent Japanese work on the chemistry of pyridine 1-oxide and related compounds, *J. Org. Chem.,* 18, 534, 1953.
21. **Robinson, M.M. and Robinson, B.L.,** The rearrangement of isoquinoline-*N*-oxides, *J. Org. Chem.,* 21, 1337, 1956.
22. a. **Hayashi, E. and Hotta, Y.,** Phenanthridine 5-oxide, *Yakugaku Zasshi,* 89, 834, 1960; b. **Hayashi, E., Iijima, C., and Nagasawa, Y.,** *N*-oxidation of 2-substituted and 2,3-disubstituted quinoxalines, *Yakugaku Zasshi,* 84, 163, 1964.
23. **Hershenson, F.M. and Bauer, L.,** Steric and electronic effects influencing the deoxidative substitution of pyridine *N*-oxides by mercaptans in acetic anhydride, *J. Org. Chem.,* 34, 655, 1969.
24. **Yoneda, F. and Nitta, Y.,** Action of acetic anhydride on pyridazine *N*-oxides, *Chem. Pharm. Bull.,* 11, 269, 1963.
25. **Landquist, J.K. and Stacey, G.J.,** Oxides of *Py*-substituted quinoxalines, *J. Chem. Soc.,* 2822, 1953.
26. **Wommack, J.B.,** Peroxide explosion, *Chem. Eng. News,* 55(50), 5, 1977.
27. **Dholakia, S., Gillard, R.D., and Lancashire, R.J.,** 2,2′-Bipyridyl mono-*N*-oxide, *Chem. Ind.,* 963, 1977.
28. **Brougham, P., Cooper, M.S., Cummerson, D.A., Heaney, H., and Thompson, N.,** Oxidation reactions using magnesium monoperphtalate: a comparison with *m*-chloroperoxybenzoic acid, *Synthesis,* 1015, 1987.
29. **Groves, J.T. and Watanabe, Y.,** Preparation and characterization of an iron (III) porphyrin *N*-oxide, *J. Am. Chem. Soc.,* 108, 7836, 1986.
30. a. **Katritzky, A.R., Borja, S.B., Marquet, J., and Sammes, M.P.,** Quaternary salts of 2*H*-imidazoles, *J. Chem. Soc., Perkin Trans. 1,* 2065, 1983; b. **Hansen, G.R. and Boyd, R.L.,** Preparation and some reactions of 3-oxazoline *N*-oxides, *J. Heterocycl. Chem.,* 7, 911, 1970.
31. **Fuji, T., Saito, T., Itaya, T., and Yokoyama, K.,** Convenient methods for the synthesis of 2′,3′-*O*-isopropylideneadenosine 1-oxide, *Chem. Pharm. Bull.,* 21, 209, 1973.
32. **Braun, P., Zeller, K.P., Meier, H., and Mueller, E.,** 1,2,3-Thiadiazole 2-oxides, *Tetrahedron,* 28, 5655, 1972.
33. a. **Hubert, A.J. and Anthoine, G.,** Benzotriazole *N*-oxides, *Bull. Soc. Chim. Belg.,* 78, 553, 1969; b. **Rondeau, R.E., Steppel, R.N., Rosenberg, H.M., and Knaak, L.E.,** Structural differentiation of isomeric *N*-oxides, *J. Heterocycl. Chem.,* 10, 495, 1973.
34. a. **Hisano, T. and Koga, H.,** Reaction of benzothiazole derivatives with organic peracids, *Yakugaku Zasshi,* 91, 1013, 1971; b. **Boyer, J.H. and Ellzey, S.E.,** Oxidation of nitrosoaromatic compounds with peroxytrifluoroacetic acid, *J. Org. Chem.,* 24, 2038, 1959.
35. **Chivers, G.E. and Suschitzky, H.,** New method of preparing *N*-oxides from polyhalogenated *N*-heteroaromatic compounds, *J. Chem. Soc., Chem. Commun.,* 28, 1971.
36. **Kobayashi, Y. and Kumadaki, I.,** Preparation and reactions of *N*-oxides of trifluoromethylated pyridine, *Chem. Pharm. Bull.,* 17, 510, 1969.
37. **Delia, T.J. and Brown, G.B.,** The oxidation of guanine at position 7, *J. Org. Chem.,* 31, 178, 1966.
38. **Wölcke, U. and Brown, G.B.,** On the structures of 3-hydroxyxanthine and guanine 3-oxide, *J. Org. Chem.,* 34, 978, 1969.
39. a. **Liotta, R. and Hoff, W.S.,** Trifluoroperacetic acid. Oxidation of aromatic rings, *J. Org. Chem.,* 45, 2887, 1980; b. **Taylor, E.C. and Driscoll, J.S.,** 3-Nitropyridine 1-oxide, *J. Org. Chem.,* 25, 1716, 1960.
40. **Pollak, A., Zupan, M., and Sket, B.,** Peroxydichloromaleic acid. New reagent for *N*-oxidation of heterocyclic compounds, *Synthesis,* 495, 1973.
41. **Payne, G.B., Deming, P.H., and Williams, P.H.,** Alkali-catalyzed epoxidation and oxidation using a nitrile as co-reactant, *J. Org. Chem.,* 26, 659, 1961.
42. **Payne, G.B.,** Oxidation of cyanopyridines, *J. Org. Chem.,* 26, 668, 1961.
43. **Hoeffx, E. and Ganshow, S.,** Reaction of acyl- and arylsulfonyl- isocyanates with hydrogen peroxide, *J. Prakt. Chem.,* 314, 145, 1972.
44. **Pollak, A., Stanovink, B., and Tisler, M.,** Oxidation products of some simple and bicyclic pyridazines, *J. Org. Chem.,* 35, 2478, 1970.
45. **Kyriacou, D.,** Synthesis of tetrachloropyrazine bis-*N*-oxide, *J. Heterocycl. Chem.,* 8, 697, 1971.

46. **Gallopo, A.R. and Edwards, J.O.,** Kinetics and mechanism of the oxidation of pyridine by Caro's acid catalized by ketones, *J. Org. Chem.,* 46, 1684, 1981.

47. **Murray, R.W. and Jeyaraman, R.,** Dioxiranes: synthesis and reactions of methyldioxiranes, *J. Org. Chem.,* 50, 2847, 1985.

48. **Hamana, M., Nomura, S., and Kawakita, T.,** Preparation of aromatic N-oxides of the pyridine series by oxidation with hydrogen peroxide-sodium tungstate, *Yakugaku Zasshi,* 91, 134, 1971.

49. **Kobayashi, Y., Kumadaki, I., Sato, H., Sekine, Y., and Hara, T.,** N-oxidation of diazabenzene and diazanaphthalene, *Chem. Pharm. Bull.,* 22, 2097, 1974.

50. **Tolstikov, G.A., Emileev, U.M., Yur'ev, V.P., Gershanov, F.B., and Rafikov, S.R.,** Hydroperoxide oxidation of some nitrogen containing compounds catalyzed by metals, *Tetrahedron Lett.,* 2807, 1971.

51. **Smetana, R.D.,** Catalytic preparation of amine oxides, U.S. patent 3657251, 1972, *Chem. Abst.,* 77, 5355u, 1972.

52. **Andrews, L.E., Bonnet, R., Ridge, R.J., and Appelman, E.H.,** The preparation and reactions of porphyrin N-oxides, *J. Chem. Soc., Perkin Tran. 1,* 103, 1983.

53. **Franklin, J., Trouillet, P., and Delplanque-Janssens, F.,** Process for the preparation of heterocyclic aromatic amine N-oxides, French patent 2587705, 1987, *Chem. Abst.,* 109, 6415e, 1988.

54. **Volodarskii, L.B., Fust, L.A., and Kobrin, V.S.,** Preparation and covalent hydration of 4H-imidazole derivatives, *Khim. Geterotsikl. Soedin,* 1246, 1972.

55. **Nad, G., Pedersen, B.N., and Bjerbhakke, E.,** Dissociation and dioxygen formation in hydroxide solutions of tris (2,2'-dipyridyl)iron (III) and tris (1,10-phenanthroline)iron (III): rates and stoichiometry, *J. Am. Chem. Soc.,* 105, 1913, 1983.

56. a. **Mueller, R. and Boehland, H.,** Aminoacridine N-oxides, *Z. Chem.,* 19, 214, 1979; b. **Ryzhakov, A.V.,** Preparation of N-oxides of pyridines, USSR patent 1361144 A1, 1987, *Chem. Abst.,* 109, 190254a, 1988.

57. **Kasuga, K., Hirobe, M., and Okamoto, T.,** Syntheses of pyrazolodiazines by cyclization of N-aminodiazinium salts, *Chem. Pharm. Bull.,* 22, 1814, 1974.

58. **Kasuga, K., Hirobe, M., and Okamoto, T.,** Reaction of quinazolines with hydroxylamine-O-sulfonic acid, *Yakugaku Zasshi,* 94, 945, 1974.

59. **Hlavaty, J., Volke, J., and Bakos, V.,** Electrochemical reduction of 2,2'-dinitrodiphenyl ether and 2,2'-dinitrodiphenylamine at mercury cathodes, *Collect. Czech. Chem. Commun.,* 48, 379, 1983.

60. **Takamuku, S., Maeda, R., and Sakurai, H.,** Radiation-induced oxidation of heterocompounds in liquld carbon dioxide, *Radiat. Phys. Chem.,* 18, 1081, 1981.

61. **Shibata, K., Yogo, T., Hashimoto, M., Noda, H., and Matsui, M.,** Photo-azoxylation of aromatic azo compounds, *Chem. Express,* 2, 9, 1987.

62. **Suzuki, I., Nakadate, M., and Sueyoshi, S.,** Synthesis of pyridazine 1,2-dioxides, *Tetrahedron Lett.,* 1855, 1968.

63. **Jovanovic, M.V.,** Synthesis of the first monosubstituted 1,2,4-triazine di-N-oxide, *Heterocycles,* 24, 951, 1986.

64. **Van Dahm, R.A., Pokorny, D.J., and Paudler, W.W.,** Literature correction. Preparation of 2,4-dimethyl-7-ethoxy-1,8-naphthyridine 1-oxides, *J. Heterocycl. Chem.,* 9, 1001, 1972.

65. **Gillard, R.D.,** Nonexistence of 1,10-phenanthroline N,N'-dioxide, *Inorg. Chim. Acta,* 53, L 173, 1981.

66. **Patcher, I.J. and Kloetzel, M.,** Structure of benzo[a]phenazine oxides and syntheses of 1,6-dimethoxyphenazine and 1,6-dichlorophenazine, *J. Am. Chem. Soc.,* 73, 4958, 1951.

67. **Kobayashi, Y., Kumadaki, I., and Sato, H.,** Basicity of 1,6-naphthyridine and its N-oxides. Correlation to peracid oxidation, *Chem. Pharm. Bull.,* 22, 2812, 1974.

68. **Mlochowski, J. and Kloc, K.,** Reactivity of phenanthrolines. I. Synthesis of N-oxides, *Rocz. Chem.,* 47, 727, 1973.

69. **Ogata, M., Watanabe, H., Tori, K., and Kano, H.,** 4-Methylpyrimidine N-oxides, *Tetrahedron Lett.,* 19, 1964.

70. **Kato, T., Yamanaka, H., and Shibata, T.,** N-oxidation of pyrimidine derivatives, *Yakugaku Zasshi,* 87, 1096, 1967.

71. **Hayashi, E., Oishi, E., Tezuka, T., and Ema, K.,** N-oxidation of 1-methyl-and 1-benzylphthalazine, *Yakugaku Zasshi,* 88, 1333, 1968.

72. **Deady, L.W.,** Ring nitrogen oxidation of amino substituted nitrogen heterocycles with m-chloroperbenzoic acid, *Synth. Commun.,* 7, 509, 1977.

73. a. **Paudler, W.W. and Chen, T.K.,** Synthesis and characterization of 1,2,4-triazines N-oxides, *J. Org. Chem.,* 36, 787, 1971; b. **Sasaki, T. and Minamoto, K.,** Structural studies on the oxidation products of 3-amino-5-phenyl-as-triazine with organic peracids, *J. Org. Chem.,* 31, 3917, 1966.

74. **Mason, J.C. and Tennant, G.,** Synthesis and nuclear magnetic resonance spectra of 3-aminobenzo-1,2,4-triazines and their mono- and di-N-oxides, *J. Chem. Soc.,* (B), 911, 1970.

75. **Jovanovic, M.,** Syntheses of some pyrimidine N-oxides, *Can. J. Chem.,* 62, 1176, 1984.

76. **Kawashima, H. and Kumashiro, I.,** The synthesis of purine 3-N-oxides, *Bull. Chem. Soc. Jpn.,* 42, 750, 1969.

77. **Stevens, M.A., Margareth, D.I., Smith, H.W., and Brown, G.B.,** Monooxides of aminopurines, *J. Am. Chem. Soc.,* 80, 2755, 1958.

78. **Sato, B.,** Peracetic and peroxysulfuric acid *N*-oxidation of phenyl- and chlorophenylpyrazines, *J. Org. Chem.*, 43, 3367, 1978.
79. **Mixan, C.E. and Pews, R.G.,** Selective *N*-oxidations of chlorinated pyrazines and quinoxalines, *J. Org. Chem.*, 42, 1869, 1977.
80. **Horie, T. and Ueda, T.,** *N*-oxidation of 3-aminopyridazines, *Chem. Pharm. Bull.*, 11, 114, 1963.
81. **Paudler, W.W., Pokorning, D.J., and Cornich, S.J.,** Naphthyridine chemistry. XII. Syntheses of *N*-oxides, *J. Heterocycl. Chem.*, 7, 291, 1970.
82. **Ohsawa, A., Arai, H., Ohnishi, H., and Igeta, H.,** Synthesis, oxidation, and reduction of monocyclic 1,2,3-triazines, *J. Chem. Soc. Chem. Commun.*, 1182, 1980.
83. **Tsuchiya, T. and Kurita, J.,** Synthesis of 3-substituted 1*H*-1,2-benzodiazepines, *Chem. Pharm. Bull.*, 26, 1896, 1978.
84. **Argo, C.B., Robertson, I.R., and Sharp, J.T.,** The preparation of 3*H*-1,2-diazepine 2-oxides and their rearrangement to give 3-alkenyl-3*H*-pyrazole 2-oxides, *J. Chem. Soc., Perkin Trans.*, 1, 2611, 1984.
85. **Giner-Sorolla, A., Gryte, C., Cox, M.L., and Parham, J.C.,** Synthesis of purine 3-oxide, 6-methylpurine 3-oxide, and related derivatives, *J. Org. Chem.*, 36, 1228, 1971.
86. **Elina, A.S., Musatova, I.S., and Tsyrul'nikova, L.G.,** *N*-oxides of imidazo[4,5-*b*]quinoxalines and imidazo[4,5-*b*]pyrazines, *Khim. Geterotsikl. Soedin*, 1266, 1972.
87. **Abushanab, E., Bindra, A.P., Goodman, L., and Peterson, H.,** Imidazo[1,5-*a*]pyrazine system, *J. Org. Chem.*, 38, 2049, 1973.
88. **Hardy, C.R. and Parrick, J.,** Ring opening or rearrangement versus *N*-oxidation in the action of peracids upon pyrrolo[2,3-*b*]pyridines, pyrrolo[2,3-*b*]pyrazines, and triazolo[1,5-*a*]- and triazolo[4,3-*a*]pyrazine, *J. Chem. Soc., Perkin Trans. 1*, 506, 1980.
89. a. **Hisano, T., Ichikawa, M., Muraoka, K., Yabuta, Y., Kido, Y., and Shibata, M.,** Sulfurations and oxidations of 2,3-disubstituted 4 (3*H*)-quinazolinones, *Chem. Pharm. Bull.*, 24, 2244, 1976; b. **Fisher, U. and Schneider, F.,** 1,3-Dipolar addition of 2-benzonitrilio-2-propanide to 7-methylthieno[2,3-*c*]pyridine 1,1′-dioxide and subsequent reactions, *Helv. Chim. Acta*, 66, 971, 1983.
90. **Pentimalli, L.,** Peracid oxidation of 2-*p*-nitrophenilazopyridine, *Gazz. Chim. Ital.*, 93, 404, 1963.
91. **Colonna, M., Risaliti, A., and Pentinalli, L.,** Oxidation of arylazopyridines, *Gazz. Chim. Ital.*, 86, 1067, 1956.
92. **Buncel, E., Keum, S.R., Cygler, M., Varughese, K.I., and Birnbaum, G.I.,** Synthesis and structure determination of isomeric α and β-phenylazopyridines, *N*-oxides, and methiodides, *Can. J. Chem.*, 62, 1628, 1984.
93. **Szafran, M. and Siepak, J.,** Isoquinaldic acid *N*-oxide, *Rocz. Chem.*, 43, 473, 1969.
94. **Pentimalli, L.,** Oxidation of 2-aminopyridine with organic peracids, *Gazz. Chim. Ital.*, 94, 458, 1964.
95. **Pentimalli, L. and Milani, G.,** Peracid oxidation of (methylamino)quinolines, *Gazz. Chim. Ital.*, 113, 803, 1983.
96. **Wieczozek, J. S. and Plazek, E.,** On the *N*-oxides of 2-dimethylaminopyridine, *Recl. Trav. Chim. Pays Bas*, 83, 249, 1964.
97. **Pentimalli, L.,** Oxidation of 4-*p*-dimethylaminophenyl- and 4-dimethylamino-pyridine with organic peracids, *Gazz. Chim. Ital.*, 94, 902, 1964.
98. **Roberts, S.M. and Suschitzky, M.,** Oxidation of pentachloropyridine and its *N,N*-disubstituted aminoderivatives with peroxyacids, *J. Chem. Soc. (C)*, 1537, 1968.
99. **Lutz, W.B., Lazarus, S., Klutchko, S., and Meltzer, R.I.,** Amino derivatives of pyrazine *N*-oxides, *J. Org. Chem.*, 29, 1645, 1964.
100. **Katritzky, A.R., Rasala, D., and Brito-Palma, F.,** Regioselective *N*-oxidation of 4-dialkylaminopyridines with Payne's reagent, *J. Chem. Res. (S)*, 42, 1988.
101. **Pentimalli, L.,** Oxidation of 2-(*p*-dimethylamino)-phenylazopyridine, *Gazz. Chim. Ital.*, 89, 1843, 1959.
102. **Endo, T. and Zemlicka, J.,** An anomalous reaction course of oxidation of N^6,N^6-dialkyladenosines and related compounds with *m*-chloroperoxybenzoic acid, *J. Org. Chem.*, 53, 1887, 1988.
103. a. **Gray, A.P., Heitmeier, D.E., and Hortensine, J.T.,** Pyridylcyclobutanes, *J. Org. Chem.*, 36, 1449, 1971; b. **Braun, H.P. and Meier, H.,** Preparation and properties of cicloalkeno-1,2,3-thiadiazoles and their mono- and trioxides, *Tetrahedron*, 31, 63, 1975.
104. **Holm, A., Carlsen, L., Lawesson, S.O., and Kolind-Andersen, H.,** 5-Phenyl 1,2,3,4-thiatriazole-3-oxide. New class of heteroaromatic *N*-oxides, *Tetrahedron*, 31, 1783, 1975.
105. **Dressler, M.L. and Joullié, M.M.,** Synthesis and chemical reactivity of thieno[2,3-*c*]- and thieno[3,2-*c*]pyridines, *J. Heterocycl. Chem.*, 7, 1257, 1976.
106. **Klemm, L.H., Merrill, R.E., Lee, F.H.W., and Klopfenstein, C.E.,** Direct halogenation of thieno[2,3-*b*]pyridine, *J. Heterocycl. Chem.*, 11, 205, 1974.
107. **Chippendale, K.E., Iddon, B., and Suschitzky, H.,** Synthesis and reactions of [1]benzothieno[3,2-*c*]cinnolines, *J. Chem. Soc., Perkin Trans. 1*, 2030, 1972.
108. **Gilman, H. and Ranck, R.O.,** Some new *N*-substituted phenothiazine derivatives and their 5-oxides and 5,5-dioxides, *J. Org. Chem.*, 23, 1903, 1958.

109. **Walter, W., Koss, J., and Curts, J.,** Oxidation reactions at thiamidoesters and N-heteroaromatics with α- and γ-thiocarbonyl groups, *Liebigs Ann.Chem.,* 695, 77, 1966.

110. **Ando, M. and Kazuhara, H.,** Stepwise oxidation of the heteroatoms in O-protected pyridoxine like pyridinophane containing two groups, in the intramolecular bridge chain, *Bull. Chem. Soc. Jpn.,* 60, 818, 1987.

111. **Van Zwieten, P.A., Gerstenfeld, M., and Huisman, H.O.,** Synthesis of some heterocyclic-aromatic sulfides and sulfones, *Recl. Trav. Chim. Pays Bas,* 81, 604, 1962.

112. **Szafran, L. and Brzezinski, B.,** Oxidation of 2,6-1utidine N-oxide and its derivatives, *Rocz. Chem.,* 43, 653, 1969.

113. **Ochiai, E. and Takahashi, M.,** Synthesis of indole derivatives from corresponding hydrocarbostyril derivatives, *Chem. Pharm. Bull.,* 14, 1272, 1966; *Itsuu Kenkyusho Nempo,* 1, 1968.

114. **Skattebøl, L. and Boulette, B.,** Oxidation of pyridineacetic acid derivatives with peracids. An unusual α-hydroxylation, *J. Org. Chem.,* 34, 4150, 1969.

115. **Katritzky, A.R.,** The preparation of some substituted pyridine 1-oxides, *J. Chem. Soc.,* 2404, 1956.

116. **Kloc, K., Mlochovski, J., and Szulc, Z.,** Reactions at the nitrogen atom in azafluorene systems, *Can. J. Chem.,* 57, 1506, 1979.

117. **Engelhardt, U. and Schaefer-Ridder, M.,** N-oxidation versus epoxidation in polyclcic azarenes, *Tetrahedron Lett.,* 4687, 1981.

118. **Dubey, S.K. and Kumar, S.,** Synthesis of dihydrodiols and diol epoxides of benzo[f]quinoline, *J. Org. Chem.,* 51, 3407, 1986.

119. **Benderly, A., Fuller, G.B., Knaus, E.E., and Redda, K.,** Synthesis and reactions of 2-epoxyethylenepyridine 1-oxide with nitrogen, oxygen, and sulfur nucleophiles, *Can. J. Chem.,* 56, 2673, 1978.

120. a. **Avasthi, K. and Knaus, E.E.,** Synthesis and reactions of 1-[1-oxido-2-(3,4)-pyridinyl]-2-methyloxirans with nitrogen nucleophiles, *J. Heterocycl. Chem.,* 18, 375, 1981; b. **Malm, W.E., Lesiak, J.Z., and Muszkiet, B.,** Reactions of arylidene derivatives of cycloalkaneazaarenes with perbenzoic acid, *Rocz. Chem.,* 48, 177, 1974.

121. **Hassner, A. and Anderson, D.J.,** Cycloaddition of 1-azirines to 1,3-diphenylisobenzofuran and rearrangement of the adducts, *J. Org. Chem.,* 39, 2031, 1974.

122. **Katritzky, A.R. and Monro, A.M.,** Per-acid oxidation of some conjugated pyridines, *J. Chem. Soc.,* 150, 1958.

123. **Pentimalli, L. and Bruni, P.,** Reactivity of pyridinealdehydes phenylhydrazones, *Ann. Chim. (Rome),* 54, 180, 1964.

124. **Clarke, P.D., Fitton, A.O., Kosmirak, M., Suschitzky, H., and Suschitzky, J.L.,** Transformations of 3-formylchromones into pyrroles and pyridines, *J. Chem. Soc., Perkin Trans. 1,* 1747, 1985.

125. **Ochiai, E., Tanida, H., and Uyeda, S.,** Phenolic by-products in the preparation of quinaldine N-oxide, *Chem. Pharm. Bull.,* 5, 188, 1957.

126. a. **Ochiai, E., Kaneko, C., Shimada, I., Murata, Y., Kosuge, T., Miyashita, S., and Kawasaki, C.,** Formation of 3-hydroxyderivatives during the N-oxidation of quinoline derivatives with hydrogen peroxide in acetic acid solution, *Chem. Pharm. Bull.,* 8, 126, 1960; b. **Nakashima, T. and Suzuki, I.,** Ring contraction of 3-hydroxyquinolines to oxindoles with hydrogen peroxide in acetic acid, *Chem. Pharm. Bull.,* 17, 2293, 1969.

127. **Coffin, B. and Rabbins, R.F.,** 4,10-Diazapyrenes, *J. Chem. Soc.,* 3379, 1965.

128. **Suzuki, I., Nakashima, T., and Nagasawa, N.,** Oxidation of 5- and 8-nitrocinnoline, *Chem. Pharm. Bull.,* 13, 713, 1969.

129. **Palmer, M.H. and Russel, E.R.R.,** Cinnolines II. Reaction with hydrogen peroxide and acetic acid, *J. Chem. Soc. (C),* 2621, 1968.

130. **Yamanaka, H.,** Reaction of 4-alkoxyquinazolines with organic peracids, *Chem. Pharm. Bull.,* 7, 152, 1959.

131. **Kress, T.J.,** Synthesis of pyrimidine N-oxides and 4-pyrimidinones by reaction of 5-substituted pyrimidines with peracids. Evidence for covalent hydrates as reaction intermediates, *J. Org. Chem.,* 50, 3073, 1985.

132. **Adachi, K.,** A novel procedure of preparing quinazoline N-oxide from quinazoline and hydroxylamine, *Yakugaku Zasshi,* 77, 507, 1957.

133. **Yamanaka, H., Ogawa, S., and Sakamoto, T.,** Oxidation of 2,4-disubstituted pyrimidines with organic peracids, *Heterocycles,* 16, 573, 1981.

134. **Landquist, J.K. and Stacey, G.J.,** Oxides of py-substituted quinoxalines, *J. Chem. Soc.,* 2822, 1953.

135. **Carboni, S., Da Settimo, A., Segnini, D., and Tonetti, I.,** Anthrydines, *Gazz. Chim. Ital.,* 96, 1443, 1966.

136. **Ames, D.E., Franklin, C.S., and Grey, T.F.,** N-oxides of some hydroxy-and amino-quinolines, *J. Chem. Soc.,* 3079, 1956.

137. **Peterson, M.L.,** Nitrogen-substituted derivatives of 2,5-pyridinedicarboxylic acid, *J. Org. Chem.,* 25, 565, 1960.

138. **Gardner, T.S., Wenis, E., and Lee, J.,** Pyridine N-oxides with sulfur-containing groups, *J. Org. Chem.,* 22, 984, 1957.

139. **Furukawa, S.,** Mechanism of rearrangement of picoline 1-oxide derivatives with acetic anhydride, *Yakugaku Zasshi,* 59, 492, 1959.

140. **Cadogan, J.I.G.,** Reactions of pyridyl benzoates with some perbenzoic acids, *J. Chem. Soc.,* 2844, 1959.

141. a. **Robinson, M.M. and Robinson, B.L.,** Reaction of 1-chloroisoquinoline with peracetic acid, *J. Org. Chem.,* 1071, 1958; b. **Bellas, M. and Suschitzky, H.,** Fluoroisoquinoline *N*-oxides, *J. Chem. Soc.,* 4561, 1964.

142. **Collins, I. and Suschitzky, H.,** Nucleophilic substitution and peroxy-acid oxidation of pentabromopyridine and some of its *N,N*-dialkylamino- and bis(*N,N*-dialkylamino)-derivatives, *J. Chem. Soc. (C),* 1523, 1970.

143. **Kato, T., Yamanaka, H., and Hiranuma, H.,** *N*-oxidation of 4-chloropyrimidine derivatives, *Chem. Pharm. Bull.,* 16, 1337, 1968.

144. **Itai, T. and Kamiya, S.,** 4-Azidoquinoline and 4-azidopyridine derivatives, *Chem. Pharm. Bull.,* 9, 87, 1961.

145. **Moriconi, E.J., Misner, R.E., and Brady, T.E.,** Sulfur extrusion from 1,3-diphenyl-1,3-dihydrothieno[3,4-*b*]quinoxaline 2,2-dioxides. A new synthesis of 6-phenylbenzo[*b*]phenazines, *J. Org. Chem.,* 34, 1651, 1969

146. **Kreher, R. and Pawelczyk, H.,** Syntheses and reactions of 5*H*-dibenzo[*c,e*]azepine *N*-oxides, *Z. Naturforsch.,* 29B, 425, 1974.

147. **Tufariello, J.J., Milowsky, A.S., Al-Nuri, M., and Goldstein, S.,** The synthesis and cycloaddition reactions of 3-azabicyclo[3.1.0]hex–2-ene 3-oxide and 3-azabicyclo[3.2.0]hept-2-ene 3-oxide, *Tetrahedron Lett.,* 267, 1987.

148. **Volodarskii, L.B., Tikhonov, A.Y., and Fust, L.A.,** Condensation of syn-α-hydroxylamino oximes with acetone and preparation of 6*H*-1,2-oxadiazine 5-oxides, *Isv. Sib. Otd. Akad. Nauk SSSR, Ser. Khim. Nauk,* 91, 1971; *Chem. Abst.,* 76, 140737, 1972.

149. **Pennings, M.L.M. and Reinhoudt, D.N.,** Synthesis and oxidation of 1-hydroxyazetidines, *J. Org. Chem.,* 48, 4043, 1983.

150. **Sevast'yanova, T.K. and Volodarski, L.B.,** Conversion of *N*-(1-oxo-1-aryl-2-alkyl)nitrones to 4-hydroxy-2,3,4,5-tetrahydro-1,2,4-triazines, *Khim. Geterotsikl. Soedin.,* 134, 1973.

151. a. **Spence, T.W.M. and Tennant, G.,** Acid-catalyzed ring-opening reactions of substituted (*o*-nitrophenyl)ethylene oxides involving partecipation by the nitro group, *J. Chem. Soc. (C),* 3712, 1971; b. **Elsworth, J.F. and Lamchen, M.,** The synthesis and reactions of 2,3-dihydro-1,4-oxazine 4-oxide, a heterocyclic nitrone, *J. Chem. Soc. (C),* 2423, 1968.

152. **Tikhonov, A.Y. and Voladarskii, L.B.,** Synthesis of pyrimidine 1,3-dioxides and their transformation into isomeric pyrimidine mono-*N*-oxides, *Tetrahedron Lett.,* 2721, 1975.

153. **Bender, H. and Döpp, D.,** 2*H*-pyrrole-1-oxides by selenium dioxide dehydrogenation of pyrroline-1-oxides, *Tetrahedron Lett.,* 1833, 1980.

154. **Pandey, G., Kumaraswamy, G., and Krishna, A.,** Photosensitized single-electron-transfer oxidation of *N*-hydroxylamines: a convenient synthesis of cyclic nitrones, *Tetrahedron Lett.,* 2649, 1987.

155. **Murahasi, S., Mitsui, H., Watanabe, T., and Zenki, S.,** The reaction of *N*-mono- and *N,N*-disubstituted hydroxylamines with palladium catalyst, *Tetrahedron Lett.,* 1049, 1983.

156. **Thesing, J. and Mayer, H.,** Cyclic nitrones. I. Dimeric 2,3,4,5-tetrahydropyridine *N*-oxide, *Chem. Ber.,* 89, 2159, 1956.

157. **Kliegel, W., Enders, B., and Becker, H.,** Nitrones of acetone and isomeric *N*-hydroxyoxazolidines, *Liebigs Ann. Chem.,* 1712, 1982.

158. **Marchetti, L. and Passalacqua, V.,** Autoxidation of 1-hydroxy-2-phenylindole, *Ann.Chim. (Rome),* 57, 1251, 1967.

159. **Colonna, M. and Greci, L.,** Reactions of 2-phenyl-3-diazoindolenine with phenylhydroxylamine and indole ene-hydroxylamines[1-hydroxyindoles], *Gazz. Chim. Ital.,* 100, 757, 1970.

160. **Berti, C., Greci, L., Andreuzzi, R., and Trazza, A.,** Reaction of 2-phenylindole with primary aromatic amines. A chemical and electrochemical investigation, *J. Chem. Soc., Perkin Trans. 1,* 607, 1986.

161. **Colonna, M. and Bruni, P.,** Reactions of 1-hydroxy-2-phenylindole with nitrogen-heterocycle azo compounds (electron transfer processes), *Gazz. Chim. Ital.,* 97, 1584, 1967.

162. **Kreher, R. and Morgenstern, H.,** Preparation of cyclic nitrones from heterocyclic *N*-(benzyloxy)amines, *Chem. Ber.,* 115, 2679, 1982.

163. **Zheved, T.D. and Altukhov, K.V.,** Acid hydrolysis of derivatives of *N*-(2'-nitroethoxy)-3-cyano-3-nitroisoxazolidine, *Zh. Org. Khim.,* 12, 2028, 1976.

164. **Green, I.R. and Lamchen, M.,** Some reactions of 2-cyano-1-hydroxy-5,5-dimethylpyrrolidine with nucleophiles, *S. Afr. J. Chem.,* 30, 149, 1977.

165. **Zvilichovsky, G. and David, M.,** Synthesis, stability, and rearrangements of 2-imino-2*H*-isoxazolo 2,3-*a*]pyrimidines and 2-aminoisoxazolo[2,3-*a*]pyrimidinium salts, *J. Org. Chem.,* 48, 575, 1983.

166. **Buscemi, S., Macaluso, G., Frenna, V., and Vivona, N.,** Synthesis of 1,2,4-oxadiazolo[2,3-*a*]pyrimidinium systems and their ring opening into pyrimidine *N*-oxides, *J. Heterocycl. Chem.,* 23, 1175, 1986.

167. **Vivona, N., Buscemi, S., Frenna, V., and Ruccia, M.,** Rearrangement of *N*-(1,2,4-oxadiazol-3-yl)-β-enaminoketones into pyrimidine *N*-oxides, *J. Chem. Soc., Perkin Trans. 1,* 17, 1986.

168. **Bapat, J.B. and Black, D.S.C.,** Preparation of 1-pyrroline 1-oxides and 1-pyrrolines by reductive cyclization of γ-nitro carbonyl compounds, *Aust. J. Chem.,* 21, 2483, 1968.

169. **Battistoni, P., Bruni, P., and Fava, G.,** 3-Phenyl-2*H*-1,4-benzoxazine 4-oxides, *Tetrahedron,* 35, 1771, 1979.

170. **Mousseron-Canet, M. and Boca, J.P.,** Study of some indolic nitrones and of tautomeric N-hydroxyindoles, *Bull. Soc. Chim. France,* 1296, 1967.

171. **Bapat, J.B., Black, D.S.C., and Newland, G.,** Vinglogous hydroxamic acid structures of the reduction products from γ-nitro(o- and p-hydroxyphenyl)ketones, *Aust. J. Chem.,* 27, 1591, 1974.

172. **Paulsen, H. and Budzis, M.,** Synthesis of acyclic and cyclic carbohydrate nitrones, *Chem. Ber.,* 107, 1998, 1974.

173. **Pfoertner, K.H. and Foricher, J.,** New pathways for 1H- and 2H-pyrazoles, *Helv. Chim. Act.,* 63, 658, 1980.

174. **Goerlitzer, K. and Buss, D.,** 1-Acetyl-2,4,5-trimethyl-3,6-diazaphenanthrenes, *Arch. Pharm.,* 318, 21, 1985.

175. **Mugner, V. and Laviron, E.,** Synthesis of derivatives of benzo[c]cinnoline, 4,9-diazapyrene and a new heterocycle, 4,5-diazapyrene, *Bull. Soc. Chim. Fr.,* 2, 39, 1978.

176. **Cariom, M., Hazard, R., Jubault, M., and Tallec, A.,** Electrochemical reduction of γ-nitroketones. Experimental conditions for obtaining pyrroline and pyrrolidine derivatives, *Can. J. Chem.,* 61, 2359, 1983.

177. **Kurihara, T., Sano, H., and Hirano, H.,** Reductive cyclization of o-nitrobenzylideneacetylacetone analogs, *Chem. Phar. Bull.,* 23, 1155, 1975.

178. **Kurihara, T., Nasu, K., and Mihara, K.,** Synthesis of 1,4-dihydro-2-methyl-4-oxo-3-quinolineglyoxylic acid, *J. Heterocycl. Chem.,* 20, 289, 1983.

179. a. **Kurihara, T., Oshita, Y., and Sakamoto, Y.,** Reductive cyclization of 3-(o-nitrobenzylidene)-2,4-dioxopentanoic acid and its esters, *Heterocycles,* 6, 123, 1977; b. **Ruggli, P. and Hegedüs, B.,** The reduction of o-nitrobenzil and a further synthesis of 2-phenylisatogen, *Helv. Chim. Acta,* 22, 147, 1939.

180. **Turner, M.J., Luckenbach, L.A., and Turner, E.L.,** Synthesis of some substituted pyrroline 1-oxides by catalytic hydrogenation of aliphatic γ-nitrocarbonyl compounds, *Synth. Commun.,* 16, 1377, 1986.

181. **Zschiesche, R. and Reissig, H.U.,** An efficient synthesis of 5-membered cyclic nitrones from γ-nitroketones, *Tetrahedron Lett.,* 1685, 1988.

182. **Meisenheimer, J. and Stotz, E.,** On some quinaldine oxides, *Ber. Dtsch. Chem. Ges.,* 58, 2334, 1925.

183. **Gabriel, S. and Gerhard, W.,** Derivatives of some o-nitroketones, *Ber. Dtsch. Chem. Ges.,* 54, 1613, 1921.

184. **Hand, E.S. and Paudler, W.W.,** Imidazo[1,2-a]pyridine 1-oxide. Synthesis and chemistry of a novel type of N-oxide, *J. Org. Chem.,* 43, 658, 1978.

185. **Schmitz, E.,** 3,4-Dihydroisoquinoline N-oxide, *Chem. Ber.,* 91, 1488, 1958.

186. a. **Parisi, F., Bovina, P., and Quilico, A.,** On the action of hydroxylamine upon γ-pyrone, *Gazz. Chim. Ital.,* 90, 903, 1960; b. **Parisi, F., Bovina, P., and Quilico, A.,** On the aciton of hydroxylamine on 2,6-dimethylpyrone, *Gazz. Chim. Ital.,* 92, 1138, 1962; c. **Yates, P., Jorgenson, M.J., and Roy, S.K.,** Reactions of 4-pyrones with hydroxylamine, *Can. J. Chem.,* 40, 2146, 1962; d. **Crabbé, P., Haro, J., Rius, C., and Santos, E.,** Novel synthesis of pyridine N-oxides and isoxazolines, *J. Heterocycl. Chem.,* 9, 1189, 1972; e. **Pedersen, C.L., Harrit, H., and Buchardt, O.,** Pirylium salts and hydroxylamine in acidic medium, *Acta Chem. Scand.,* 24, 3435, 1970; f. **Dorofeenko, G.B. Sadekova, E.I., and Goncharova, V.M.,** 2-Benzopyrilium salts. X. Reaction with nitrogen bases and aromatic aldehydes, *Khim. Geterotsikl. Soedin.,* 1308, 1970.

187. **Kuhn, R. and Blau, W.,** The structure of the glycosine dye of Radziszewski and of the benzimidazole dye of van Niementowski, *Liebigs Ann. Chem.,* 615, 99, 1958.

188. **Kamel, M., Allam, M.A., and Abou-Zeid, N.Y.,** Novel studies on 5-amino-2-methylnaphth[1,2]imidazole and 5-amino-2-methyl-3-oxonaphth-[1,2]imidazole, *Tetrahedron,* 23, 1863, 1967.

189. a. **Friedländer, P. and Ostermaier, H.,** On carbostyril, *Ber. Dtsch. Chem. Ges.,* 14, 1916, 1881; b. **Friedländer, P.,** About cyclic hydroxylamine derivatives, *Ber. Dtsch. Chem. Ges.,* 47, 3369, 1914.

190. **Coutts, R.T., Mukherjee, G., Abramovitch, R.A., and Brewster, M.A.,** Reductive cyclization of some derivatives of methyl o-nitrophenylpyruvate and other ketoesters by catalyzed sodium borohydride, *J. Chem. Soc. (C),* 2207, 1969.

191. **Coutts, R.T. and Smith, E.M.,** The preparation of some benzothiazine hydroxamic acids, *Can. J. Chem.,* 45, 975, 1967.

192. **Honkanen, E. and Virtanen, A.I.,** The structure of the precursors of benzoxalinone in plants, *Acta Chem. Scand.,* 14, 1214, 1960.

193. **Coutts, R.T.,** Catalyzed sodium borohydride reduction of ortho-nitrocinnamates, *J. Chem. Soc. (C),* 713, 1969.

194. **Hansen, S.B. and Petrow, V.,** Some 10-substituted 1,3-dimethyl-2,9-diazaphenanthrene 9-oxides, *J. Chem. Soc.,* 350, 1953.

195. **Gabriel, S.,** Nitromethane and phthalic anhydride, *Ber. Dtsch. Chem. Ges.,* 36, 570, 1903.

196. **Baxter, I. and Swan, G.P.,** Catalytic hydrogenation of some derivatives of 2-nitrophenylpyruvic acid, *J. Chem. Soc. (C),* 2446, 1967.

197. **Buckley, G.D. and Elliot, T.J.,** Preparation of heterocyclic bases by reduction of 3-nitroalkyl cyanides, *J. Chem. Soc.,* 1508, 1947.

198. **Jawdosiuk, M. and Makosza, M.,** Synthesis of 2-aminoindolenine derivatives via reductive cyclization of 2-(o-nitrophenyl)-2-phenylalkanenitriles, *Rocz. Chem.,* 50, 857, 1976.

199. **Hlavaty, J., Volke, J., and Manousek, O.,** Formation of benzothiazole derivatives in the electrochemical reduction of o-nitrothiocyanobenzene, *Coll. Czech. Chem. Commun.,* 40, 3751, 1975.

200. a. **Bauer, K.H.,** On the catalytic reduction of *o*-nitrocinnamonitrile, *Ber. Dtcsch. Chem. Ges.,* 71, 2226, 1938;
b. **Hazard, R. and Tallec, A.,** Electrochemical preparation of *N*-hydroxyidoles. II. Oxidation at controlled
potential of some (*o*-hydroxylaminophenyl)alkenes, *Bull. Soc. Chim. Fr.,* 121, 1974; c. **Chibani, A., Hazard,
R., Jubault, M., and Tallec, A.,** Electrochemical synthesis of quinoline derivatives. II. Electrochemical
reduction of *o*-substituted nitrobenzylidenes, *Bull. Soc. Chim. Fr.,* 795, 1987.

201. **Makosza, M., Kumiotek-Skarzynska, I., and Jawdosiuk, M.,** Synthesis of substituted 3,4-dihydroquinolines
and their *N*-oxides via reductive cyclization of 2-phenyl-2-(*o*-nitrobenzyl)-alkanenitriles, *Synthesis,* 56, 1977.

202. **Hlavaty, J.,** Electrolytic reduction of *o*-nitrobenzylthiocyanate in buffered solutions on mercury, *Coll. Czech.
Chem. Commun.,* 50, 33, 1985.

203. **Taylor, E.C. and Jefford, C.W.,** New route to diazine-mono-*N*-oxides, *Chem. Ind.,* 1559, 1963.

204. a. **DeGraw, J.I. and Tagawa, H.,** An alternate synthesis of 6-substituted-5-deazapteridines, *J. Heterocycl.
Chem.,* 19, 1461, 1982; b. **Hosmane, R.S., Bhan, A., and Rauser, M.E.,** Synthesis of "fat" xanthine (fx), "fat"
guanine (fG), and "fat" hpoxanthine (fHx) analogs of the imidazo[4,5-*e*] [1,4]diazepine system, *Heterocycles,*
24, 2743, 1986.

205. **Gewald, K. and Hain, U.,** Reaction of nitrile oxides with ylidene malonitriles, *Z. Chem.,* 26, 434, 1986.

206. **Bradsher, C.K. and Telang, S.A.,** 2-Azaquinolizinium oxides, *J. Org. Chem.,* 31, 941, 1966.

207. **Kohler, E.P. and Addinall, C.R.,** Cyclic nitrones, *J. Am. Chem. Soc.,* 52, 1590, 1930.

208. **Petersen, S. and Heitzer, H.,** Reactions of naphthoquinonaldehyde, *Liebigs Ann. Chem.,* 740, 180, 1970.

209. **Baumgarten, P., Merländer, R., and Olshausen, J.,** On glutaconaldehyde: acetale, oxime and their ring
closure to pyridine 1-oxide, *Ber. Dtsch. Chem. Ges.,* 66, 1802, 1933.

210. a. **Wideman, L.G.,** Reaction of diacetylacetone with hydroxylamine, *J. Chem. Soc., Chem. Commun.,* 1309,
1970; b. **Gilchrist, T.L., Iskander, G.M., and Yagoub, A.K.,** Acid catalyzed rearrangement of 3- acyl-6-
alkoxy-5,6-dihydro-4*H*-1,2-oxazines: a route to 3-alkoxypyridine 1-oxides, *J. Chem. Soc., Perkin Trans. 1,*
2769, 1985.

211. **Mizuno, Y. and Inoue, Y.,** New synthesis of 2-formyl-4-methyl-1-phenylimidazole 3-oxide, *J. Chem. Soc.,
Chem. Commun.,* 124, 1978.

212. **Sternbach, L.H., Kaiser, S., and Reeder, E.,** Quinazoline 3-oxide structure of compounds previously
described in the literature as 3,1,4-benzoxadiazepines, *J. Am. Chem. Soc.,* 82, 475, 1960.

213. **Auwers, K. and Meyenburg, F.,** On a new synthesis of isoindazole derivatives, *Ber. Dtsch. Chem. Ger.,* 24,
2370, 1891.

214. **Bischler, A.,** About phenmiazine derivatives, *Ber. Dtsch Chem. Ges.,* 26, 1891, 1893.

215. **Kövendi, A. and Kircz, M.,** A new synthesis of quinazoline-*N³*-oxides and 1,2-dihydroquinazoline-*N³*-oxides,
Chem. Ber., 98, 1049, 1965.

216. **Fortea, J.,** Synthesis and *N*-oxide reactions of 4-phenyl- and 4-aminothieno[2,3-*d*]pyrimidine 3-oxides, *J.
Prakt. Chem.,* 317, 705, 1975.

217. **Metlesies, W., Silverman, G., and Sternbach, L.H.,** Synthesis and rearrangement reactions of a 3,1,4-
benzoxadiazepine, *Monatsh. Chem.,* 98, 633, 1967.

218. **Metallidis, A., Sotiriadis, A., and Theodoropoulos, D.,** A new synthesis of 2-amino derivatives of 6-chloro-
4-phenylquinazoline 3-oxide, *J. Heterocycl. Chem.,* 12, 359, 1975.

219. **Hajos, G., Messmer, A., Neszmelyi, A., and Parkanyi, L.,** Synthesis and structural study of azidonaphtho-*as*-
triazines: "annelation effect" in azide-tetrazole equilibria, *J. Org. Chem.,* 49, 3199, 1984.

220. **Fusco, R. and Bianchetti, G.,** On aminonaphthotriazines, II, *Gazz. Chim. Ital.,* 87, 446, 1957.

221. **Scott, F.L. and Lalor, F.J.,** Involvement of oxime groups with neighboring acyl functions, *Tetrahedron Lett.,*
641, 1964.

222. **Dornow, A. and Marquardt, H.H.,** The use of α-aminooximes for the synthesis of imidazole 3-oxides, *Chem.
Ber.,* 97, 2169, 1964.

223. **Gnichtel, H. and Schuster, K.E.,** On the synthesis of 2-oxo-3-imidazoline 3-oxides and their prototropism,
Chem. Ber., 111, 1171, 1978.

224. a. **Böhnisch, V., Burzer, G., and Neunhoffer, H.,** Syntheses of 1,2,4-triazine 4-oxides, *Liebigs Ann. Chem.,*
1713, 1977; b. **Neunhoeffer, H., Weischedel, F., and Böhnisch, V.,** Synthesis of 1,2,4-triazine 4-oxides,
Liebigs Ann. Chem., 750, 12, 1971.

225. **Robinson, M.M. and Robinson, B.L.,** Observations with *N*-hydroxycarbostyrils and other substituted
derivatives, *J. Am. Chem. Soc.,* 80, 3443, 1958.

226. a. **McKillop, A. and Rao, D.P.,** Copper-catalyzed condenstation of β-dicarbonyl compounds with 2-bro-
mobenzaldoximes: a simple, one-step synthesis of 3-hydroxyisoquinoline *N*-oxides, *Synthesis,* 760, 1977; b.
Prasad, A.S., Sandhu, J.S., and Baruah, J.N., Studies on pyrimidine analogs: reaction of isoxazolo[3,4-
d]pyrimidine with active methylene compounds, *Heterocycles,* 20, 787, 1983.

227. **Forrester, A.R., Irikawa, H., Thomson, R.H., Woo, S.O., and King, T.J.,** Iminils. Part 8. Intramolecular
addition to nitrile groups, *J. Chem. Soc., Perkin Trans. 1,* 1712, 1981.

228. **Konwar, D., Boruah, R.C., and Sandhu, J.S.,** A base catalysed reaction of arylmalonitrile and 2,1-
benzisoxazole, *Heterocycles,* 23, 2557, 1985.

229. **Dornow, A., Marquardt, H.H., and Paucksch, H.,** On the reactions of α-chlorooximes. II, *Chem. Ber.,* 97, 2165, 1964.

230. **Beyer, H. and Ruhlig, G.,** The synthesis of 3-substituted thiazolon-(2)-imides from α-rhodanketones, *Chem. Ber.,* 89, 107, 1956.

231. **Mühlstädt, M. and Schulze, B.,** On the reaction of β-thiocyanatovinylaldehyde with hydroxylamine, *Z. Chem.,* 21, 326, 1981.

232. **Schatzmiller, S. and Shalom, E.,** Synthesis and thermal cleavage of *O*-alkyl-*N*-vinylhydroxylamines, *Liebigs Ann. Chem.,* 897, 1983.

233. **Brandman, H.A. and Conley, R.T.,** Unambigous synthesis of a monocyclic 5,6-dihydro-1,2-oxazine, *J. Org. Chem.,* 38, 2236, 1973.

234. a. **Stempel, A., Douvan, I., Reeder, E., and Sternbach, L.H.,** 4,1,5-Benzoxadiazocin-2-ones, a novel ring system, *J. Org. Chem.,* 32, 2417, 1967; b. **Sternbach, L.H. and Reeder, E.,** Transformations of 7-chloro-2-methylamino-5-phenyl-3*H*-1,4-benzodiazepine 4-oxide, *J. Org. Chem.,* 26, 4936, 1961.

235. **Stempel, A., Reeder, E., and Sternbach, L.H.,** Mechanism of ring enlargement of quinazoline 3-oxides with alkali to 1,4-benzodiazepin-2-one 4-oxides, *J. Org. Chem.,* 30, 4267, 1965.

236. **Gnichtel, H., Schmitt, B., and Schunk, G.,** 3,6-Dihydro-2(1*H*)-pyrazinone 4-oxide, *Chem. Ber.,* 114, 2536, 1981.

237. a. **Roedig, A., Renk, H.A., Schaal, V., and Scheutzow, D.,** Stereoisomeric 2,3,4,5-tetrachloro-5-phenyl-2,4-pentadienals, *Chem. Ber.,* 107, 1136, 1974; b. **Tamura, Y., Tsujimoto, N., and Mano, M.,** Alternative preparation of pyridine *N*-oxide and *N*-aminopyridinium chloride, *Chem. Pharm. Bull.,* 19, 130, 1971.

238. **Gnichtel, H. and Boehringer, U.,** The Beckmann reaction of phenyl substituted 1,3-dioximes, *Chem. Ber.,* 113, 1507, 1980.

239. **Ames, D.E., Singh, A.G., and Smyth, W.F.,** Reaction of 2-acylphenylselenocyanate with hydroxylamine and phenylhydrazine, *Tetrahedron,* 39, 831, 1983.

240. **Meisenheiner, J., Senn, O., and Zimmermann, P.,** On the oximes of *o*-aminobenzo- and -acetophenone, *Ber. Dtsch. Chem. Ges.,* 60, 1736, 1927.

241. **Stevens, M.A., Smith, H.W., and Brown, G.B.,** *N*-oxides of azapurines, *J. Am. Chem. Soc.,* 82, 3189, 1960.

242. **Ockenden, D.W. and Schofield, K.,** Reactions of methazonic acid. Part III. A novel reaction of 3-amino lepidine, *J. Chem. Soc.,* 1915, 1953.

243. a. **Clark, J., Varvounis, G., and Bakavoli, M.,** Novel tricyclic compounds containing the pyrimido[5,4-*d*]-1,2,3-triazine system, *J. Chem. Soc. Perkin Trans. 1,* 711, 1986; b. **Clark, J. and Varvounis, G.,** Pyrimido[5,4-*d*]-1,2,3-triazines and some related tricyclic compounds, *J. Chem. Soc., Perkin Trans. 1,* 1475, 1984.

244. **Nemeryuk, M.P., Sedov, A.L., Safnova, T.S., Cerny, A., and Krepelka, J.,** Transformations of substituted 5-aminopyrimidines under conditions of the diazotization, *Coll. Czech. Chem. Commun.,* 51, 215, 1986.

245. **Armstrong, P., Grigg, R., Smendrakumar, S., and Warnock, W.J.,** Tandem intramolecular Michael addition and 1,3-dipolar cycloaddition reactions of oximes: versatile new carbon-carbon bond-forming methodology, *J. Chem. Soc., Chem. Commun.,* 1327, 1987.

246. **Sakamoto, T., Kondo, Y., Miura, N., Hayashi, K., and Yamanaka, H.,** A facile synthesis of isoquinoline *N*-oxides, *Heterocycles,* 24, 2311, 1986.

247. **Takada, K., Kan-Woon, T., and Boulton, A.J.,** 2-Hydroxyindazoles, *J. Org. Chem.,* 47, 4323, 1982.

248. **Pallas, M., Geissler, F., and Kalkofen, W.,** Synthesis of some 2-aralkylthio-substituted thiazole 3-oxides, *Z. Chem.,* 20, 257, 1980.

249. **Masaki, M., Chigira, Y., and Ohta, M.,** Total synthesis of racemic aspergillic acid and neoaspergillic acid, *J. Org. Chem.,* 31, 4143, 1966.

250. **Masaki, M. and Ohta, M.,** Synthesis of a homolog of aspergillic acid, *J. Org. Chem.,* 29, 3165, 1964.

251. **Wiley, R.H. and Slaymaker, S.C.,** 5-Aroyl-2-pyridines and 5-aroyl-2-pyridones, *J. Am. Chem. Soc.,* 78, 2393, 1956.

252. **El-Kholy, I.E.S., Rafla, F.K., and Soliman, G.,** 5-Aroyl-2-pyrones. The corresponding 2-thiopyrones, pyridones, 1-hydroxy- and 1-amino-2-pyridones, and related 4,5,6-triphenyl-2-pyridones, *J. Chem. Soc.,* 4490, 1961.

253. **Harrison, D. and Smith, A.C.B.,** The synthesis of some cyclic hydroxamic acid from *o*-aminocarboxylic acids, *J. Chem. Soc.,* 2157, 1960.

254. **Kocevar, M., Stanovnik, B., and Tisler, M.,** Ring transformation of some 4-aminopteridine 3-oxides and derivatives, *Tetrahedron,* 39, 823, 1983.

255. **Watson, A.A., Nesnow, S.C., and Brown, G.B.,** 1-Hydroxyguanine, *J. Org. Chem.,* 38, 3046, 1973.

256. **Shaw, E. and McDowell, J.,** Synthesis of cyclic hydroxamic acids with a five-membered ring, *J. Am. Chem. Soc.,* 71, 1691, 1949.

257. **Becker, H.G.O., Görmar, G., and Timpe, H.J.,** Preparation and reaction of 4-hydroxy-1,2,4-triazoles, *J. Prakt. Chem.,* 312, 610, 1970.

258. **Dornow, A. and Fisher, K.,** On some reactions of hydroxamic acid chlorides, *Chem. Ber.,* 99, 72, 1966.

259. a. **Ames, D.E. and Grey, T.F.,** N-Hydroxyimides. III. β-p-Methoxyphenylglutaconimides, *J. Chem. Soc.,* 2310, 1959; b. **Ames, D.E. and Grey, T.F.,** N-Hydroxyimide. II. Derivatives of homophthalic and phthalic acids, *J. Chem. Soc.,* 3518, 1955.

260. **Ames, D.E. and Grey, T.F.,** The synthesis of some N-hydroxyimides, *J. Chem. Soc.,* 631, 1955.

261. **Klötzer, W.,** On the two isomeric N-hydroxyuracils and their derivatives, *Monatsh. Chem.,* 95, 1729, 1964.

262. a. **McCall, J.M., TenBrink, R.E., and Unsprung, J.J.,** A new approach to triaminopyrimidine N-oxides, *J. Org. Chem.,* 40, 3304, 1975; b. **McCall, J.M. and TenBrink, R.E.,** 3-Amino-3-cyaniminopropanenitriles; useful precursors for 2-chloro-4,6-diamino- and 2,4,6-triamino-pyrimidine N-oxides, *Synthesis,* 673, 1978.

263. **Shaw, J.T.,** Mono-N-oxides of amino-substituted s-triazine derivatives, *J. Org. Chem.,* 27, 3890, 1962.

264. **Barot, N.R. and Elvidge, A.,** 2,6-Diaminopyrazine and its 1-oxide from iminodiacetonitrile, *J. Chem. Soc., Perkin Trans. 1,* 606, 1973.

265. **Elvidge, J.A., Linstead, R.P., and Salaman, A.M.,** Glutarimidine and the imidine form α-phenylglutaronitrile, *J. Chem. Soc.,* 208, 1959.

266. **Duun, G., Elvidge, J.A., Newbold, G.T., Ramsay, D.W.C., Spring, F.S., and Sweeney, W.,** Synthesis of cyclic hydroxamic acids related to aspergillic acid, *J. Chem. Soc.,* 2707, 1949.

267. **Stollé, R. and Thomä, K.,** On dibenzoylhydrazide chloride, *J. Prakt. Chem.,* 73, 288, 1906.

268. a. **Ashburn, S.P. and Coates, R.M.,** Preparation of oxazoline N-oxides and imidate N-oxides by amide acetal condensation and their [3+2]cycloaddition reaction, *J. Org. Chem.,* 50, 3076, 1985; b. **Ashburn, S.P. and Coates, R.M.,** Generation and [3+2]cycloaddition reactions of oxazoline N-oxides, *J. Org. Chem.,* 49, 3127, 1984.

269. **Plate, R., Hermkens, P.H.H., Smits, J.M.M., and Ottenheijim, H.C.J.,** Nitrone cycloaddition in the stereoselective synthesis of β-carbolines from N-hydroxytrytophan, *J. Org. Chem.,* 51, 309, 1986.

270. **Lamchen, M. and Mittag, T.W.,** Synthesis and properties of a monocyclic α-dinitrone, *J. Chem. Soc. (C),* 2300, 1966.

271. **Epshtein, S.P., Orlova, T.I., Rukasov, A.F., Taschi, V.P., and Putsykin, Y.G.,** The synthesis of 3-amino-4-(hydroxylamino)thiazolidine-2-thiones and 2,3-dimethyl-4a,5-dihydro-7-thioxothiazolo[3,4-b]-1,2,4-triazines, *Khim. Geterotsikl. Soedin.,* 554, 1987.

272. a. **Martin, V.V. and Volodarskii, L.B.,** Synthesis and some reactions of spatially hindered 3-imidazoline 3-oxides, *Khim. Geterotsikl. Soedin.,* 103, 1979; b. **Gnichtel, H.,** Imidazoline N-oxides from the treatment of anti-α-amino-oximes with aldehydes, *Chem. Ber.,* 103, 2411, 1970.

273. **Busch, M. and Strätz, F.,** On the isomerism of phenacylamino-oximes, *J. Prakt. Chem.,* 150, 1, 1937.

274. **Gnichtel, H. and Möller, B.,** Synthesis and reactions of 2-[α(E)-(hydroxyimino)benzyl]-3-imidazolin-3-oxides, *Chem. Ber.,* 114, 3170, 1981.

275. **Gnichtel, H. and Gau, B.P.,** Synthesis of 1,2,5,6-tetrahydropyrimidine 3-oxide derivatives, *Liebigs Ann. Chem.,* 2223, 1982.

276. **Walker, G.N., Engle, A.R., and Kempton, R.J.,** Novel synthesis of 1,4-benzadiazepines, isoindolo[2,1-d] [1,4]benzodiazepines, isoindolo[1,2-a][2]benzazepines, and indolo[2,3-d][2]benzazepines, based on the use of the Strecker reaction, *J. Org. Chem.,* 37, 3755, 1972.

277. **Fischer, R.H. and Weitz, H.M.,** Simple syntheses of substituted 2,3,5,6,7,8-hexahydro-1,2,4-benzotriazine 4-oxides, *Synthesis,* 794, 1975.

278. **Fujii, S. and Kobatake, H.,** 2(D-arabino-Tetrahydroxybutyl)pyrazine 4-N-oxide. A condensation product of 2-amino-2-deoxy-D-glucose oxime and glyoxal, *J. Org. Chem.,* 34, 3842, 1969.

279. **Auwers, K.,** The structure of stable and labile acylindazole, *Ber. Dtsch. Chem. Ges.,* 58, 2081, 1925.

280. **Meisenheimer, J. and Diedrich, A.,** On the isomeric acylindazoles of K.V. Auwers, *Ber. Dtsch. Chem. Ges.,* 57, 1715, 1924.

281. **Golic, L., Kaucic, V., Stanovnik, B., and Tisler, M.,** The structure of the product from the reaction of o-aminobenzaldoxime and acetic anhydride, *Tetrahedron Lett.,* 4301, 1975.

282. **Tomazic, A., Tisler, M., and Stanovnik, B.,** Syntheses and transformations of some heterocyclic hydroxylamines, *Tetrahedron,* 37, 1787, 1981.

283. **Neunhoeffer, H. and Böhnish, V.,** The synthesis of substituted 1,2,4-triazines, *Tetrahedron Lett.,* 1429, 1973.

284. **Sternbach, L.M., Reeder, E., Keller, O., and Metlesics, W.,** Substituted 2-amino-5-phenyl-3H-1,4-benzodiazepine 4-oxides, *J. Org. Chem.,* 26, 4488, 1961.

285. **Gnichtel, H., Griebenow, W., and Löwe, W.,** On the reaction of carboxylic acid derivatives with α-amino-oximes, *Chem. Ber.,* 105, 1865, 1972.

286. **Gnichtel, H., Walentowski, R., and Schuster, K.E.,** 2-Oxo-Δ³-imidazoline 3-oxides, *Chem. Ber.,* 105, 1701, 1972.

287. **Sulkowski, T.S. and Childress, S.J.,** The formation and subsequent rearrangement of 7 chloro-5-phenyl-3,1,4-benzoxadiazepin-2(1H)-one, *J. Org. Chem.,* 27, 4424, 1962.

288. **Yale, H.L.,** 2,3-Dihydro-1,3,5,2-oxadiazaboroles, 1,2-dihydro-1,3,2-benzodiazaborine 3-oxide, and 3,4-dihydro-2H-1,2,4,3-benzothia-diazaborine 1,1-dioxide, *J. Heterocycl. Chem.,* 8, 205, 1971.

289. **Brown, D.J. and England, B.T.**, The cyclization of *cis*- and *trans*-methoxybenzaldoximes to methoxy-3,1,4-benzoxadiazepines and methoxyquinazoline 3-oxide respectively, *Isr. J. Chem.*, 6, 569, 1968.

290. **Gnichtel, H., Exner, S., Bierbüsse, H., and Alterdinger, M.**, Reactions of *syn*- and *anti*-α-aminoketoxime with thiophosgene, *Chem. Ber.*, 104, 1512, 1971.

291. **Volodarsky, L.B., Lisack, A.N., and Koptyug, V.A.**, Synthesis of isomeric pairs of imidazole *N*-oxides, *Tetrahedron Lett.*, 1565, 1965.

292. **Volodarskii, L.B. and Tikhonov, A.Y.**, Synthesis of 1,3-di-*N*-oxides of pyrimidines, *Izv. Nauk SSSR, Ser. Khim.*, 1212, 1975.

293. **Tikhonov, A.Y. and Volodarskii, L.B.**, Reaction of 1,3-hydroxylamine oximes with formaldehyde, acetaldehyde, and acetone, *Khim. Geterotsikl. Soedin*, 252, 1977.

294. **Grigoreva, L.N., Amitina, S.A., and Volodarskii, L.B.**, Formation of derivatives of pyrazine 1,4-dioxide and 1-hydroxymidazole in the condensation of 1,2-hydroxylaminooxime acetates with 1-phenyl- and 1-(2-hetaryl)-1,2-dicarbonyl compounds, *Khim. Geterotsikl. Soedin.*, 1387, 1983.

295. a. **Volodarskii, L.B., Grigor'eva, L.N., and Tikhonov, A.Y.**, Synthesis and properties of 2,3- and 2,5-dihydropyrazine 1,4-dioxides based on 2-hydroxylamino-2-methylpropanal oxime, *Khim. Geterotsikl. Soedin.*, 1414, 1983; b. **Grigor'eva, L.N., Tikhonov, A.Y., Amitina, S.A., Volodarskii, L.B., and Korobeinicheva, I.K.**, Transformation of 2-acyl-1-hydroxy-3-imidazoline 3-oxides into pyrazine 1,4-dioxides, *Khim. Geterotsikl. Soedin.*, 331, 1986.

296. **Diels, O. and Riley, D.**, On the course of the reaction between aromatic aldehydes and diacetyl mono-oxime in the presence of strong acids, *Ber. Dtsch. Chem. Ges.*, 48, 897, 1915.

297. **Dilthey, W. and Friedrichsen, J.**, On oxido-oxazole, *J. Prakt. Chem.*, 127, 292, 1930.

298. **Cornforth, J.W. and Cornforth, R.H.**, A new synthesis of oxazoles and iminazoles including its application to the preparation of oxazole, *J. Chem. Soc.*, 96, 1947.

299. **Selwitz, C.M. and Kosak, A.I.**, The synthesis of imidazoles and oxazoles from α-diketone monooximes, *J. Am. Chem. Soc.*, 77, 5370, 1955.

300. **Ferguson, I.J., Schofield, K., Barnett, J.W., and Grimmet, M.R.**, Nitration of 1,4,5-trimethylimidazole 3-oxide and 1-methylpyrazole 2-oxide, and some reactions of the products, *J. Chem. Soc., Perkin Trans. 1*, 672, 1977.

301. **Ferguson, I.J. and Schofield, K.**, Syntheses and reactions of some imidazole 3-oxides, *J. Chem. Soc., Perkin Trans. 1*, 275, 1975.

302. **Ertel, H. and Heubach, G.**, 1-Hydroxyimidazoles and 1-hydroxyimidazole 3-oxides, *Justus Liebigs Ann. Chem.*, 1399, 1974.

303. **Volkamer, K., Baumgartel, H., and Zimmerman, H.**, *N*-oxides of imidazolyls, *Angew. Chem. Int. Ed. Engl.*, 6, 947, 1967.

304. **Bowness, W.G., Howe, R., and Rao, B.S.**, Application of the Bucherer hydantoin synthesis to diacetyl mono-oxime, *J. Chem. Soc., Perkin Trans. 1*, 2649, 1983.

305. **Zvilichovsky, G. and Brown, G.B.**, The cyclization of 6-amino-5-nitrosouracil with formaldehyde. Preparation and properties of 7-hydroxyxanthine, *J. Org. Chem.*, 37, 1871, 1972.

306. **Diels, O. and van der Leeden, R.**, On the condensation of isonitrosoketone with aldoximes: synthesis of oxadiazines, *Ber. Dtsch. Chem. Ges.*, 38, 3357, 1905.

307. **Wright, J.B.**, The preparation of 1-hydroxyimidazole 3-oxides, *J. Org. Chem.*, 29, 1620, 1964.

308. **Towliati, H.**, Note about the synthesis of imidazole derivatives, *Chem. Ber.*, 103, 3962, 1970.

309. **Rogic, M.M., Tetenbaum, M.T., and Swerdloff, M.D.**, 1-Hydroxy-2-(ω-methoxycarbonylbutyl)-4,5,6,7-tetrahydrobenzimidazole 3-oxide. An unusual product from the nitrosation of cyclohexanone, *J. Org. Chem.*, 42, 2748, 1977.

310. **Akagane, K. and Allan, G.G.**, One-step syntheses of 1-hydroxyimidazole 3-oxide, *Chem. Ind.*, 38, 1974.

311. **Karpetsky, T.P. and White, E.H.**, Formation of an unusual dihydropyrazine di-*N*-oxide during hydrolysis of an α-oximino acetal, *J. Org. Chem.*, 37, 339, 1972.

312. **Agnes, G., Felicioli, M.G., Ribaldone, G., and Santini, C.**, Synthesis of *N*-oxides of pyrazinecarboxylic acids and esters: an original and selective approach from aliphatic precursors, *Chim. Ind. (Milan)*, 70, 70, 1988.

313. **Sharp, W. and Spring, F.S.**, Synthesis of 3-aminopyrazine 1-oxides by the condensation of α-amino-nitriles with oximinomethyl ketones, *J. Chem. Soc.*, 932, 1951.

314. **Chillemi, F. and Palamidessi, G.**, Pyrazine derivatives, *Farmaco, Ed. Sc.*, 18, 566, 1963.

315. **Taylor, E.C. and Dumas, D.J.**, Preparation and chemistry of 2-amino-6-carbalkoxy-3-cyano-5-substituted pyrazine 1-oxides: synthesis of pterin-6-carboxaldehyde, *J. Org. Chem.*, 45, 2485, 1980.

316. **Taylor, E.C., Perlman, K.L., Sword, I.P., Séquin-Frey, M., and Jacobi, P.A.**, A new, general and unequivocal pterin synthesis, *J. Am. Chem. Soc.*, 95, 6407, 1973.

317. **Taylor, E.C. and Jacobi, P.A.**, A total synthesis of L-*erythro*-biopterin and some related 6-(polyhydroxyalkyl)pterins, *J. Am. Chem. Soc.*, 98, 2301, 1976.

318. **Karpetsky, T.P. and White, E.H.**, The synthesis of *Cypridina* etioluciferamine and the proof of structure of *Cypridina* luciferin, *Tetrahedron*, 29, 3761, 1973.

319. **Taylor, E.C., Abdulla, R.F., and Jacobi, P.A.,** Total synthesis of asperopterin B, *J. Org. Chem.,* 40, 2336, 1975.

320. **Pastor, S.D. and Nelson, A.L.,** Substituted 5-(DL-erythro-1′,2′-dihydroxypropyl)-pyrazines. Potential precursors for the synthesis of biopterin derivatives, *J. Heterocycl. Chem.,* 21, 657, 1984.

321. **Jacobi, P.A., Martinelli, M., and Taylor, E.C.,** Unequivocal synthesis of euglenapterin, *J. Org. Chem.,* 46, 5416, 1981.

322. **Taylor, E.C. and Abdulla, R.F.,** A new and unequivocal synthesis of isoxanthopterin-6-carboxylic acid (Cyprino-pourpre B), *Tetrahedron Lett.,* 2093, 1973.

323. **Franchetti, P. and Grifantini, M.,** Synthesis of imidazo[1,2-c]quinazoline and some of its methyl derivatives, *J. Heterocycl. Chem.,* 7, 1295, 1970.

324. **Franchetti, P., Grifanti, M., Lucarelli, C., and Stein, M.L.,** N-Hydroxyimidazole 3-oxides and their quaternary salts, *Farmaco Ed. Scient.,* 27,46, 1972.

325. **Volkamer, K. and Zimmerman, H.W.,** On aryl-substituted 1-hydroxyimidazoles and 1-hydroxyimidazole-N-oxides, *Chem. Ber.,* 102, 4177, 1969.

326. **Sansonov, V.A. and Volodarskii, L.B.,** Preparation and some properties of 2H-imidazole-1,3-dioxides, derivatives of alicyclic 1,2-dioximes, *Khim. Geterotsikl. Soedin.,* 808, 1980.

327. **Hayes, K.,** 1-Hydroxyimidazole 3-oxide and some 2-substituted derivatives, *J. Heterocycl. Chem.,* 11, 615, 1974.

328. a. **Safir, S.R. and Williams, J.H.,** Synthesis of cyclic hydroxamic acid derivatives of pyrazine, *J. Org. Chem.,* 17, 1298, 1952; b. **Ramsay, D.W.C. and Spring, F.J.,** Synthesis of homologs of aspergillic acid, *J. Chem. Soc.,* 3409, 1950.

329. **Gonçalves, H., Foulcher, C., and Mathis, F.,** Condensation reactions of ortho-aminobenzamidoxime with some aldehydes, *Bull. Soc. Chim. Fr.,* 2615, 1970.

330. **Rafla, F.K. and Khan, M.A.,** Synthesis and reactions of 1,2-dihydro-1-hydroxy-4,6-dimethyl-2-oxopyridine-3-carbonitrile, *J. Chem. Soc. (C),* 2044, 1971.

331. **Khan, M.A. and Rafla, F.K.,** Synthesis and reactions of potassium cyanoacetohydroxamate, *J. Chem. Soc., Perkin Trans. 1,* 327, 1974.

332. **Wright, W.B. and Smith, J.M.,** The preparation of 3-hydroxy-4-pteridinone, *J. Am. Chem. Soc.,* 77, 3927, 1955.

333. **Taylor, E.C., Cheng, C.C., and Vogl, O.,** Hypoxanthine-1-N-oxide, *J. Org. Chem.,* 24, 2019, 1959.

334. a. **Cresswell, R.M. and Brown, G.B.,** An activating effect on some displacement reactions, *J. Org. Chem.,* 28, 2560, 1963; b. **Vercek, B., Leban, I., Stanovnik, B., and Tisler, M.,** 1,2,4-Oxadiazollylpyrimidines and pyrido[2,3-d]-pyrimidine 3-oxides, *J. Org. Chem.,* 44, 1695, 1979.

335. **Cossey, A.L. and Phillips, J.N.,** A convenient route to N-3-hydroxyuracils, *Chem. Ind.,* 58, 1970.

336. **Huffman, K.R. and Schaefer, F.C.,** N-Cyanoimidates, *J. Org. Chem.,* 28, 1816, 1963.

337. **Greene, F.D. and Gilbert, K.E.,** Cyclic azo dioxides. Preparation, properties, and consideration of azo dioxide-nitrosolkane equilibria, *J. Org. Chem.,* 40, 1409, 1975.

338. **Lüttke, W. and Schabacker, V.,** Synthesis of 1,4-dichloro-2,3-diazabicyclo[2.2.2]-2-octene and its N^2 oxide, *Liebigs Ann. Chem.,* 687, 236, 1965.

339. **Morat, C., Rossat, A., and Rey, P.,** Preparation of Z- and E-1,8-dinitro-p-menthanes and determination of their stereochemistry, *Tetrahedron,* 31, 2927, 1975.

340. **Alder, R.W., Niazi, G.A., and Whiting, M.C.,** Benz[cd]indazole 1-oxide, benz[cd]indazole 1,2-dioxide (1,8-dinitrosonaphthalene), and related compounds, *J. Chem. Soc. (C),* 1693, 1970.

341. **Smith, M.A., Weinstein, B., and Greene, F.D.,** Cyclic azo dioxides. Synthesis and properties of bis(o-nitrosobenzyl) derivatives, *J. Org. Chem.,* 45, 4597, 1980.

342. **Ullman, E.F.,** Diazacyclobutanes, U.S. patent 4032519, 1977, *Chem. Abst.,* 87, 102306t, 1977.

343. a. **Volodarskii, L.B. and Tikhonova, L.A.,** Formation of furoxans and 1,2-diazetine 1,2-dioxides during oxidation of α-hydroxylaminoximes, *Khim. Geterostikl. Soedin.,* 748, 1975; b. **Wu, C.S., Szarek, W.A., and Jones, J.K.N.,** Synthesis of a carbohydratofuroxan derivatives, *J. Chem. Soc., Chem. Commun.,* 1117, 1972.

344. **Reimann, H. and Schneider, H.,** Steroid 16,17-furoxan derivatives, *Can. J. Chem.,* 46, 77, 1968.

345. **Von Dobeneck, H., Weil, E., Brunner, E., Deubel, H., and Wolkenstein, D.,** Annelated nitrogen-containing 5,5-ring systems from 3-pyrrolin-2-ones, *Liebigs Ann. Chem.,* 1424, 1978.

346. **Akrell, J., Altaf-ur-Rahman, M., Boulton,A.J., and Brown, R.C.,** Synthesis and reactions of some strained furazan N-oxides, *J. Chem. Soc., Perkin Trans. 1,* 1587, 1972.

347. **Gagneux, A.R. and Meier, R.,** Amino-furoxans. 1. Synthesis and structure, *Helv. Chim. Acta,* 53, 1883, 1970.

348. **Kropf, H. and Lambeck, R.,** Reactions with lead tetracetate. II. Aldoxime anhydride N-oxides from aromatic aldoximes, *Justus Liebig Ann.,* 700, 18, 1966.

349. **Andrianov, V.G. and Eremeev, A.V.,** Isomers of N-substituted 1-methyl-2-aminoglyoximes and their oxidation to furoxans, *Zh. Org. Khim.,* 20, 150, 1984.

350. **Walsotra, P., Trompen, W.P., and Hackmann, J.T.,** Amino-substituted glyoxines and furoxans, *Rec. Trav. Chim. Pays Bas,* 87, 452, 1968.

351. **Willer, R.L. and Moore, D.W.,** Synthesis and chemistry of some furazano- and furoxano[3,4-*b*]piperazines, *J. Org. Chem.,* 50, 5123, 1985.

352. **Akrell, J. and Boulton, A.J.,** Preparation and tautomerism of some acylfuroxans, *J. Chem. Soc., Perkin Trans. 1,* 351, 1973.

353. **Eremeev, A.V., Piskunova, I.P., Andrianov, V.G., and Liepins, E.,** Synthesis and study of aziridine dioximes, *Khim. Geterotsikl. Soedin,* 488, 1982.

354. **Calvino, R., Gasco, A., Menziani, E., and Serafino, A.,** Chloromethylfuroxans, *J. Heterocycl. Chem.,* 20, 783, 1983.

355. **Matyushin, Y.N., Pepekin, V.I., Nikolaeva, A.D., Lyapin, N.M., Nikolaeva, LV., Artyushin, A.V., and Apin, A.Y.,** Synthesis and thermochemical properties of dialkylfuroxans, *Izv. Akad. Nauk SSSR, Ser. Khim.,* 842, 1973.

356. **Burakevich, J.V., Lore, A.M., and Volpp, G.P.,** Phenylfurazan oxide. Structure, *J. Org. Chem.,* 36, 5, 1971.

357. **Talapatra, S.K., Chaudhuri, P., and Talapatra, B.,** *N*-iodosuccinimide, a convenient oxidative cyclizing agent in the synthesis of oxazole, isoxazole, benzofuran, furoxan, and 1,2,3-triazole-1-oxide derivatives, *Heterocycles,* 14, 1279, 1980.

358. **Spyroudis, S. and Varvoglis, A.,** Dehydrogenation with phenyliodine ditrifluoroacetate, *Synthesis,* 445, 1975.

359. **Nicolaides, D.N. and Gallos, J.K.,** A convenient synthesis of furoxano[3,4-*b*]quinoxalines and furazano[3,4-*b*]quinoxalines, *Synthesis,* 638, 1981.

360. **Tabakovic, I., Trkovnik, M., and Galijas, D.,** Anodic synthesis of some *N*-heterocycles, *J. Electroanal. Chem. Interfacial Electrochem.,* 86, 241, 1978.

361. **Jugelt, W., Tismer, M., and Rauh, M.,** Studies on the regioselectivity of the anodic cyclization of 1,2-dioximes to unsymmetrically substituted furoxans, *Z. Chem.,* 23, 29, 1983.

362. **Meisenheiner, J., Lange, H., and Lamparter, W.,** On the oximes of *p*-methoxybenzil, *Liebigs Ann. Chem.,* 444, 94, 1925.

363. **Pilgran, K.,** 3,4-Disubstituted and fused 1,2,5-thiadiazole *N*-oxides, *J. Org. Chem.,* 35, 1165, 1970.

364. **Pedersen, C.L.,** Disubstituted 1,2,5-selenadiazole *N*-oxides. Preparation and reactions, *Acta Chem. Scand.,* B30, 675, 1976.

365. **Volodarskii, LB. and Tikhonova, L.H.,** Formation of 1-pyrazoline 1,2-dioxides during oxidation of 1,3-hydroxylaminooximes, *Khim. Geterotsikl. Soedin,* 248, 1977.

366. **Stephanidou-Stephanatou, J.,** Oxidative cyclization of some 1,3-dioximes with lead tetracetate, *J. Heterocycl. Chem.,* 22, 293, 1985.

367. **Kotali, A. and Papageorgiu, V.P.,** Oxidation of the dioximes of 1,3-diketones with lead tetracetate, *J. Chem. Soc., Perkin Trans. 1,* 2083, 1985.

368. **Spyroudis, S. and Varvoglis, A.,** Oxidative cyclization of some 1,3- and 1,4-dioximes with phenyliodine (III) bis(trifluoracetate), *Chem. Chron.,* 11, 173, 1982.

369. **Ohsawa, A., Arai, H., and Igeta, H.,** Oxidative cyclization of 2-unsaturated 1,4-dioximes, *Heterocycles,* 9, 1367, 1978.

370. **Boulton, A.J. and Tsoungas, P.G.,** 1,2-Benzisoxazole *N*-oxides, *J. Chem. Soc., Chem. Commun.,* 421, 1980.

371. a. **Babic, K., Molan, S., Polanc, S., Stanovnik, B., Stres-Bratos, J., Tisler, M., and Vercek, B.,** Synthesis of some triazoloazine 3-oxides, *J. Heterocycl. Chem.,* 13, 487, 1976; b. **Gilchrist, T.L., Peek, M.E., and Rees, C.W.,** *N*-Aryl-*C*-nitroso-imines, *J. Chem. Soc. Chem. Commun.,* 913, 1975.

372. a. **Taüber, E.,** On diphenylenazone, a new, cyclic, nitrogen-containing, compound, *Ber. Dtsch. Chem. Ges.,* 24, 3081, 1891; b. **Bellaart, A.C.,** Reduction of aromatic nitro compounds with phosphine, *Tetrahedron,* 21, 3285, 1965.

373. **Ross, S.D., Kahan, G.J., and Leach, W.A.,** The chemical, catalytic and polarographic reduction of 2,2'-dinitrobiphenyl and the reduction products, *J. Am. Chem. Soc.,* 74, 4122, 1952.

374. **Kempter, F.E. and Castle, R.N.,** The synthesis of benzo[*c*]cinnoline 5,6-dioxides and related compounds, *J. Heterocycl. Chem.,* 6, 523, 1969.

375. **Laviron, E. and Lewandowska, T.,** Reduction of 2,2'-dinitrobiphenyl by linear sweep voltammetry, *Bull. Soc. Chim. Fr.,* 3177, 1970.

376. **Corbett, J.F. and Holt, P.R.,** Some unsymmetrical polycyclic cinnolines, *J. Chem. Soc.,* 3646, 1960.

377. **McBride, J.A.H.,** Synthesis of 2,7-diazabiphenylene by thermal extrusion of nitrogen from 2,7,9,10-tetraazaphenanthrene, *J. Chem. Soc., Chem. Commun.,* 359, 1974.

378. **Mugnier, Y. and Laviron, E.,** Synthesis of new heterocyclic compounds: 4,5-diazapyrene, 4,5-diazapyrene 4-oxide, and 4,9-diazapyrene 4,9-dioxide, *J. Heterocycl. Chem.,* 14, 351, 1977.

379. **Mugnier, Y. and Laviron, E.,** Preparation of 4,5,9,10-tetraazapyrene 4,5-dioxide, *Bull. Soc. Chim. Fr.,* 1496, 1976.

380. **Castle, R.N., Guither, W.D., Hilbert, P., Kempter, F.E., and Patel, N.R.,** Synthesis of 2,7-disubstituted-4,5,9,10-tetraazapyrenes, *J. Heterocycl. Chim.,* 6, 533, 1969.

381. **Catala, A. and Popp, F.D.,** 3,8-Dihalo-11*H*-dibenzo[*c,f*][1,2]diazepines, *J. Heterocycl. Chem.,* 1, 178, 1964.

382. **Grundon, M.F., Johnston, B.T., and Wasfi, A.S.,** Proximity effects in diaryl derivatives. Part I. The formation of seven-membered heterocyclic compounds, *J. Chem. Soc.,* 1436, 1963.

383. **Szmant, H.H. and Lapinski, R.L.,** Unsaturated seven-membered heterocyclic rings, *J. Am. Chem. Soc.,* 75, 6338, 1953.

384. **Mugnier, Y. and Laviron, E.,** Electrochemical study of 2,2′-dinitrodiphenylsulfide, *Electrochim. Acta,* 25, 1329, 1980.

385. **Hlavaty, J. and Volke, J.,** The electrochemical behavior of 2,2′-bis-(hydroxylamino)diphenyl sulfide, *Electrochim. Acta,* 29, 1399, 1984.

386. **Gansser, C.,** Derivatives of 5,6,11,12-tetrahydrodibenzo[*c,g*][1,2]-diazocine, *Eur. J. Med. Chem.-Chim. Ther.,* 10, 273, 1975.

387. **Shepherd, M.K.,** Cyclobuta[1,2-*c*:3,4-*c*′]dithiophene, *J. Chem. Soc. Chem. Commun.,* 880, 1985.

388. **Nonciaux, J.C., Guilard, R., and Laviron, E.,** Chemical and electrochemical reduction of dinitrobithienyls, *Bull. Soc. Chim. Fr.,* 41, 3318, 1973.

389. a. **Joshua, C.P. and Ramdas, P.K.,** Photolysis of 2,2′-dinitrodiphenylmethanes. New route to the dibenzo[*c,f*] [1,2]diazepine system, *Synthesis,* 873, 1974; b. **Christudhas, M. and Joshua, C.P.,** Irradiation of 5,5′-dimethyl-2,2′-dinitrodiphenylmethanes in neutral, acidic, and alkaline media, *Aust. J. Chem.,* 35, 2377, 1982.

390. **Elbs, K.,** On phentriazole, *J. Prakt. Chem.,* 108, 209, 1924.

391. **Wakisawa, Y. and Sumitani, M.,** 2-(2′-Hydroxyphenyl)benzotriazoles or their *N*-oxides, *Jpn. Kokai,* 78/63379, 1978; *Chem. Abst.,* 89, 180008c, 1978.

392. **Roberts, R.D. and Hardy, W.B.,** 2-Aryl-2*H*-benzotriazoles, U.S. patent 4224451, 1980; *Chem. Abst.,* 94, 121545f, 1981.

393. **Seino, S.,** 2-Phenyl-2*H*-benzotriazoles, *Jpn. Kokai,* 61/197570 A2, 1986; *Chem. Abst.,* 106, 50230c, 1987.

394. **Kaneoya, T., Maegawa, Y., and Okamuro, H.,** 2-Phenylbenzotriazoles, European patent 160246 A1, 1985, *Chem. Abst.,* 104, 129910s, 1986.

395. **Rosevear, J. and Wilshire, J.F.K.,** The reduction of some *o*-nitrophenylazo dyes with thiourea *S,S*-dioxide (formamidine sulfinic acid): a general synthesis of 2-aryl-2*H*-benzotriazoles and their 1-oxides, *Aust. J. Chem.,* 37, 2489, 1984.

396. **Ziegler, C.E. and Peterli, H.J.,** 2-Aryl-2*H*-benzotriazoles, Swiss patent 615165, 1980; *Chem. Abst.,* 93, 8181a, 1980.

397. **Jancis, E.H.,** Hydroxyarylbenzotriazoles and their *N*-oxides, Canadian patent 1155856 A1, 1983; *Chem. Abst.,* 100, 69327n, 1984.

398. **Jono, S.,** 2-Phenylbenzotriazole *N*-oxides, *Jpn. Kokai,* 59/172481 A2, 1984; *Chem. Abst.,* 102, 113511k, 1985.

399. **Charrier, G. and Crippa, G.B.,** Oxidation of *o*-aminoazo compounds by hydrogen peroxide in acetic acid, *Gazz. Chim. Ital.,* 53, 462, 1923; 56, 207, 1926.

400. **Bouchet, P., Coquelet, C., Elguero, J., and Jacquier, R.,** Photochemistry of nitrogen heterocycles. I. 1-*o*-nitrophenyl- and 1-(2′,4′-dinitrophenyl) pyrazoles: formation of benzotriazole 1-oxides, *Bull. Soc. Chem. Fr.,* 43, 184, 1976.

401. **Davies, D.E., Gilchrist, T.L., and Roberts, T.G.,** The cycloaddition of α-nitrostyrenes to olefins. Investigation of the scope and mechanism of the reaction, *J. Chem. Soc., Perkin Trans. 1,* 1275, 1983.

402. **Davies, D.E. and Gilchrist, T.L.,** Investigation of the retro-Diels-Alder reaction as a method for the generation of nitroso-olefins, *J. Chem. Soc., Perkin Trans. 1,* 1479, 1983.

403. a. **Minisci, F., Galli, R., and Quilico, A.,** Addition reaction of nitrile oxides on aromatic nitrosoderivatives. A novel synthesis of the benzimidazole ring, *Tetrahedron Lett.,* 785, 1963; b. **Robson, E., Tedder, J.M., and Woodcock, D.J.,** Nitrosoacethylenes, *J. Chem. Soc. (C),* 1324, 1968; c. **Tedder, J.M., and Woodcock, D.J.,** 4-Arylamino-3-butylfuroxans, unexpected products from 1-nitrosohex-1-yne, *J. Chem. Res. (S),* 356, 1978.

404. **Gerlach, K. and Kröhnke, F.,** The reaction of pyridinium betaines with α-nitroso-β-naphthol, *Chem. Ber.,* 95, 1124, 1962.

405. **Pachter, I.J., Nemeth, P.E., and Villani, A.J.,** Pteridines. III. Synthesis of some ketones, carbinols, and *N*-oxides, *J. Org. Chem.,* 28, 1197, 1963.

406. **Ichiba, M., Nishigaki, S., and Senga, K.,** Synthesis of fervenulin 4-oxides and its conversion to the antibiotics fervenulin and 2-methylfervenulone, *J. Org. Chem.,* 43, 469, 1978.

407. **Lillevik, H.A., Hossfeld, R.L., Lindstrom, H.V., Arnold, R.T., and Gortner, R.A.,** Technics in the synthesis of porphyrindin, *J. Org. Chem.,* 7, 164, 1942.

408. **Sawhney, S.N. and Boykin, D.W.,** A reinvestigation of the Bamberger-Ham phenazine synthesis: a limitation, *J. Heterocycl. Chem.,* 16, 397, 1979.

409. a. **Wiemer, D.F. and Leonard, N.J.,** Aralkylation of potassium ethylnitrosolate. Ring closure of nitrosolic acid esters, *J. Org. Chem,* 41, 2985, 1976; b. **Goldner, H., Dietz, G., and Carstens, E.,** Reactions with nitrosouracil derivatives. I. A new synthesis of xanthine, *Liebig Ann. Chem.,* 691, 143, 1966.

410. **Zenchoff, G.S., Walser, A., and Freyer, R.I.,** The synthesis of indazoles *via* 2,3-dihydroindazoles, *J. Heterocycl. Chem.,* 13, 33, 1976.

411. **Seebach, D., Dach, R., Enders, D., Renger, B., Jansen, M., and Brachtel, G.**, 1,4,5,6-Tetrahydro-*v*-tetrazine derivatives, *Helv. Chim. Acta*, 61, 1622, 1978.

412. **Barnes, J.F., Barrow, M.J., Harding, M.M., Paton, R.M., Sillitoe, A., Ashcroft, P.L., Bradbury, R., Crosby, J., Joyce, C.J., Holmes, D.R., and Milner, J.**, Synthesis and structure of some strained furazan *N*-oxides, *J. Chem. Soc., Perkin Trans. 1*, 293, 1983.

413. **Klamann, D., Koser, W., Weyerstahl, P., and Fligge, M.**, On pseudonitrosites, nitrooximes and furoxans, *Chem. Ber.*, 98, 1831, 1965.

414. **Fumasoni, S., Giacobbe, G., Martinelli, P., and Schippa, G.**, On the products from the reaction between propene and N_2O_4 and HNO_3, *Chim. Ind. (Milan)*, 1064, 1965.

415. **Osawa, T., Kito, Y., Namiki, M., and Tsuji, K.**, A new furoxan derivative and its precursors formed by the reaction of sorbic acid with sodium nitrite, *Tetrahedron Lett.*, 4399, 1979.

416. **Oglonlin, K.S., Girbasova, N.V., and Potekhin, A.A.**, Nitrosochlorination of 2-methyl-1-(*p*-ethoxyphenyl) propene, *Zh. Org. Khim.*, 9, 2236, 1973.

417. **Plücken, U., Winter, W., and Meier, H.**, Research on the structure of systems with an oxadiazole ring, *Justus Liebigs Ann. Chem.*, 1557, 1980.

418. **Hansen, J.F. and Vietti, D.E.**, Preparation and oxidation of 1-hydoxyprazoles and 1-hydroxypyrazole 2-oxides, *J. Org. Chem.*, 41, 2871, 1976.

419. **Freeman, J.P.**, The structure and reactions of pernitrosomesityl oxide, *J. Org. Chem.*, 27, 1309, 1962.

420. **Yoneda, F. and Sakuma, Y.**, Syntyeses and properties of 4,6-dimethyl[1,2,5]oxadiazolo [3,4-*d*] pyrimidine-5,7 (4*H*, 6*H*) dione 1-oxide, *J. Heterocyclic. Chem.*, 10, 993, 1973.

421. **Nishigaki, S., Kanazawa, H., Kanamori, Y., Ichiba, M., and Senga, K.**, A new synthesis of pyrimido [5,4-*e*]-*as*-triazine 4-oxides and their ring trasformation to pyrrolo[3,2-*d*]pyrimidines, *J. Heterocycl. Chem.*, 19, 1309, 1982.

422. **Yoneda, F., Nagamatsu, T., and Shinomura, K.**, A new synthesis of pyrimido [4,5-*e*]-*as*-triazines 4-oxides by nitrosative cyclization of aldehyde uracil-6-yl hydrazones in the presence of diethyl azodiformate, *J. Chem. Soc., Perkin Trans. 1*, 713, 1976.

423. **Yoneda, F., Sakuma, Y., Ichiba, M., and Shinomura, K.**, Synthesis of isoalloxazines and isoalloxazine 5-oxides. A new synthesis of riboflavin, *J. Am. Chem. Soc.*, 98, 830, 1976.

424. **Wessiak, A. and Bruice, T. C.**, Synthesis and study of 6-amino-5-oxo-3*H*,5*H*-uracil and derivatives. The structure of an intermediate proposed in mechanisms of flavine and pterin oxygenases, *J. Am. Chem. Soc.*, 105, 4809, 1983.

425. a. **Preston, P.N. and Tennant, G.**, Synthetic methods involving neighboring group interaction in *ortho*-substituted nitrobenzene derivatives, *Chem. Rev.*, 72, 627, 1972; b. **Loudon, J.D. and Tennant, G.**, Substituent interactions in *ortho*-substituted nitrobenzenes, *Quart. Rev.*, 18, 389, 1964.

426. **Döpp, D. and Sailer, K.H.**, Products of the photolysis of 1,3,5-tri-*tert*-butyl- 2-nitrobenzene, *Chem. Ber.*, 108, 301, 1975.

427. a. **Döpp, D. and Sailer, K.H.**, Photolysis of crystalline 2-nitro-1,3,5-tri-*tert*-butylbenzene, *Tetrahedron Lett.*, 2761, 1971; b. **Döpp, D.**, 5,7-Di-*tert*-butyl-3,3-dimethyl-3*H*-indole 1-oxide, *Org. Photochem. Synth.*, 2, 43, 1976.

428. **Padmanabhan, K., Venkatesan, K., Ramamurthy, V., Schimidt, R., and Döpp, D.**, Structure-reactivity correlation of photochemical reactions in organic crystals. Intramolecular hydrogen abstraction in an aromatic nitro compounds, *J. Chem. Soc., Perkin Trans.*, 2, 1153, 1987.

429. **McFarlane, M.D. and Smith, D.M.**, A new route to *N*-hydroxyquinoline-2,3-diones, *Tetrahedron Lett.*, 6363, 1987.

430. **Ahmad, Y. and Shamsi, S.A.**, Novel direct synthesis of quinoline derivatives, *Bull. Chem. Soc. Jpn.*, 39, 195, 1966.

431. **Ahmad, Y., Shamsi, S.A., and Qureshi, M. I.**, Further syntheses of quinoline *N*-oxide derivatives, *Rev. Roum. Chim.*, 25, 397, 1980.

432. a. **Kadin, S.B. and Lamphere, C.H.**, Reaction of 2-nitrobenzaldehydes with diethyl (diethoxyphosphinyl) succinate: a new synthesis of quinoline-2,3-dicarboxylic acid esters via their *N*-oxides, *J. Org. Chem.*, 49, 4999, 1984; b. **Yakushijin, K., Suzuki, R., Tohshima, T., and Furukawa, H.**, Transformation of 5-(2-nitropheny)-2-furylcarbamate into 4-hydroxy-2-quinoline-carboxamide 1-oxide, *Heterocycles*, 16, 751, 1981.

433. **Muth, C.W., Ellers, J.C., and Folmer, O.F.**, A novel ring closure involving a nitro group; preparation of phenanthridine 5-oxide, *J. Am. Chem. Soc.*, 79, 6500, 1957.

434. **Tennant, G. and Vaughan, K.**, The base-catalyzed cyclization of *N*-substitued *o*-nitrobenzamides to 1,4-dihydro-1-hydroxy-4-oxoquinazolines, *J. Chem. Soc. (C)*, 2287, 1966.

435. **Tennant, G.**, Extension of a quinoxaline *N*-oxide synthesis, *J. Chem. Soc.*, 2666, 1964.

436. **Tennant, G.**, A new synthesis of 2-hydroxyquinoxaline 4-oxides, *J. Chem. Soc.*, 2428, 1963.

437. **Fusco, R. and Rossi, S.**, On a new synthesis of quinoxalines. I, *Gazz. Chim. Ital.*, 94, 3, 1964.

438. **Fusco, R., Rossi, S., and Maiorana, S.**, On a new synthesis of quinoxalines. II, *Gazz. Chim. Ital.*, 95, 1237, 1965.

439. a. **Sakata, G., Makino, K., and Morimoto, K.,** The facile one pot synthesis of 6-substituted 2(1*H*)-quinoxalinones, *Heterocycles,* 23, 143, 1985; b. **Davis, R.F.,** Process for preparing 6-halo-2-chloroquinoxalines as herbicide intermediates, U.S. patent 4636562 A, 1987; *Chem. Abst.,* 106, 138474w, 1987.

440. **Makino, K., Sakata, G., Morimoto, K., and Ochiai, Y.,** A facile synthesis of novel tricyclic compounds, 2,3-dihydro-1*H*-pyrido[3,2,1-*i,j*]-quinoxalin-5-ones, *Heterocyles,* 23, 1729, 1985.

441. **Tennant, G. and Yacomeni, C.W.,** New synthetic route to 2-dialkylaminopteridin-7(8*H*)-ones and their 5-*N*-oxides, *J. Chem. Soc., Chem. Commun.,* 819, 1975.

442. **De Croix, B., Strauss, M.J., De Fusco, A., and Palmer, D.C.,** Pteridines from α-phenyl-*N,N*-dimethylacetamidine, *J. Org. Chem.,* 44, 1700, 1979.

443. **Strauss, M.J., Palmer, D.C., and Bard, R.R.,** Annulation of amidines on halonitroaromatics. A one-step route to quinoxaline and imidazoquinoxaline *N*-oxides and related structures, *J. Org. Chem.,* 43, 2041, 1978.

444. **Ellames, G.J. and Jaxa-Chamiec, A.A.,** Substituted dihydroimidazo[1.2-*a*]quinoxalines as anti-anaerobic agents, European patent, 214632 A2, 1987; *Chem. Abst.,* 107, 7221j, 1987.

445. **Ellames, G.J., Lawson, K.R., Jaxa-Chamiec, A.A., and Upton, R.M.,** Preparation and testing of substituted pyrimidoquinoxalines as bactericides, European patent 256545 A1, 1988; *Chem. Abst.,* 108, 204642u, 1988.

446. **Barnes, R.P., Graham, J.H., and Qureshi, M.A.S.,** The preparation of some benzyl-*o*-nitrophenylglyoxals and cyclization of their quinoxalines, *J. Org., Chem.,* 28, 2890, 1963.

447. **Reissert, A.,** Reactions of *o*-nitrobenzylmalonic esters. I. Treatment with alkali, syntheses of new indole derivatives, *Ber. Dtsch. Chem. Ges.,* 29, 639, 1896.

448. **Sword, I.P.,** Reactions of some benzylidene compounds with potassium cyanide, *J. Chem. Soc.(C),* 1916, 1970.

449. **Loudon, J.D. and Wellings, I.,** Substitutent interactions in *ortho*-substituted nitrobenzenes., Part I, *J. Chem. Soc.,* 3462, 1960.

450. **Loudon, J.D. and Tennant, G.,** Substituent interactions in ortho-substituted nitrobenzenes., Part II, *J. Chem. Soc.,* 3466, 1960.

451. **Kulkarni, S.N., Kamath, H.Y., and Bhamare, N.K.,** Synthesis of substituted phenylbenzyl ketones and 2-arylisatogens, *Ind. J. Chem.,* 278, 667, 1988.

452. **Kamath, H.V. and Kulkarni, S.N.,** A new synthesis of substituted 2-arylisatogens, *Synthesis,* 931, 1978.

453. a. **Coutts, R.T., Hooper, M., and Wibberley, D.G.,** Isatogens. Part I. Base-catalyzed condensation products of methyl and ethyl *o*-nitrobenzoylacetate and *o*-nitrobenzoylacetone, *J. Chem. Soc.,* 5205, 1961; b. **Hooper, M. and Wibberley, D.G.,** Isatogens. Part IV, *J. Chem. Soc.(C),* 1596, 1966.

454. **Arndt, F., Eistert, B., and Partale, W.,** Diazomethane and *o*-nitrocompounds. II. *N*-oxyisatine from o-nitrobenzoylchloride, *Ber. Dtsch. Chem. Ges.,* 60, 1364, 1927.

455. **Giovannini, E. and Portmann, P.,** On some oxindole and isatine derivatives. II. On the amino-, hydroxy-, and methoxy-substituted derivatives in position 5 and 6, *Helv. Chim. Acta,* 31, 1381, 1948.

456. **Moore, J. and Ahlstrom, D.H.,** Reactions of 1-diazo-3-(*o*-nitrophenyl)-acetone, *J. Org. Chem.,* 26, 5254, 1961.

457. **Fielden, R., Meth-Cohn, O., and Suschitzky, H.,** Action of acid on *NN*-disubstituted *o*-nitroanilines: benzimidazole *N*-oxide formation and nitro group rearrangements, *J. Chem. Soc., Perkin Trans. 1,* 696, 1973.

458. **Suschitzky, H. and Sutton, M.E.,** Thermal cyclization of aromatic nitro compounds, *Tetrahedron Lett.,* 3933, 1967.

459. **Heine, H.W., Blosick, G.J., and Lowrie, G.B.,** Photolysis of 1-(2,4,6-trinitrophenyl)-2,3-diphenylaziridine and 1-(2,4-dinitro-phenyl)-2-pheny-3-benzoylaziridine, *Tetrahedron Lett.,* 480, 1, 1968.

460. **Loudon, J.D. and Tennant, G.,** Substituent interactions in *ortho*-substituted nitrobenzenes. Part V., *J. Chem. Soc.,* 4268, 1963.

461. **Harvey, I.W., McFarlane, M.D., Moody, D.J., and Smith, D.M.,** 4- and 7-Amino-1*H*-benzimidazole 3-oxides, *J. Chem. Soc. Perkin Trans. 1,* 1939, 1988.

462. **McFarlane, M.D., Moody, D.J., and Smith, D.M.,** 5- and 6-Amino-1*H*-benzimidazole 3-oxides, *J. Chem. Soc., Perkin Trans. 1,* 691, 1988.

463. **Ljublinskaya, L.A. and Stepanov, V.M.,** Mild conversion of the 2,4-dinitrophenylglycyl moiety to a derivative of 6-nitrobenzimidazole 1-oxide, *Tetrahedron Lett.,* 4511, 1971.

464. **Goudie, R.S. and Preston, P.N.,** Thermolysis of *N-o*-nitrophenyl- and *N*-2,4-dinitrophenyl- α-amino-acids, *J. Chem. Soc.(C),* 1139, 1971.

465. **Neadle, D.J. and Pollit, R.J.,** The photolysis of *N*-2,4- dinitrophenylamino-acids to give 2-substituted 6-nitro benzimidazole 1-oxides, *J. Chem. Soc. (C),* 1764, 1967.

466. **Marshall, R. and Smith, D.M.,** Reactions of nucleophiles with *N*-benzylidene-*o*-nitroaniline, *J. Chem. Soc. (C),* 3510, 1971.

467. **Livingstone, D.B. and Tennant, G.,** Synthesis and reactivity of *N*-hydroxybenzimidazolones, *J. Chem. Soc. Chem. Commun.,* 96, 1973.

468. **Andrews, A.F., Smith, D.M., Hodson, H.F., and Thorogood, P.B.,** Synthetic approaches to some aza-analogs of benzimidazole *N*-oxides. I. The imidazo [4,5-*b*]pyridine series, *J. Chem. Soc., Perkin Trans. 1,* 2995, 1982.

469. **Brown, R., Joseph, M., Leigh, T., and Swain, M.L.,** Synthesis and reactions of 7,8-dihydro-8-methylpterin and 9-methylguanine 7-oxide, *J. Chem. Soc., Perkin Trans. 1,* 1003, 1977.

470. **Nobara, F., Nishii, M., Ogawa, K., Isano, K., Ubukata, M., Fujii, T., Itaya, T., and Saito, I.,** Synthesis of guanine 7-oxide, an antitumor antibiotic from *Streptomyces* species, *Tetrahedron Lett.,* 1287, 1987.

471. **Lange, M., Quell, R., Lettau, H., and Schubert, H.,** Imidazo[4,5-*c*]-pyrazoles from 4-nitro-5-benzylaminopyrazoles, *Z. Chem.,* 17, 94, 1977.

472. **Preston, P.N. and Sood, S.K.,** Approaches to the synthesis of 1*H*-[1]benzothieno[2,3-*d*]imidazoles and thieno [2,3-*d*] imidazoles, *J. Chem. Soc., Perkin Trans. 1,* 80, 1976.

473. **Tiwari, S.S. and Misra, S.B.,** Synthesis of pyrazino[1,2-*a*]benzimidazole 10-oxides as possible new anthelmintic agents, *Indian J. Chem.,* 148, 725, 1976.

474. **Knyazev, V.N., Drozd, V.N., and Mozhaeva, T.Y.,** Factors determining intramolecular nucleophilic attack on the nitrogen atom of an aromatic nitro group, *Izv. Timiryazevsk. S. Akad.,* 182, 1983; *Chem. Abst.,* 98, 143076p, 1983.

475. **Kanazawa, H., Senga, K., and Tamura, Z.,** Synthesis and chemical properties of thiazolo[4,5-*g*]quinazoline 3-oxides and thiazole[5,4-*f*]quinazoline 3-oxide, *Chem. Pharm. Bull.,* 33, 618, 1985.

476. **Garming, A., Redwan, D., Gelbke, P., Kern, D., and Dierkes, U.,** Synthesis and reactions of the salts of substituted 1,3-di-*aci*-nitro and 1,5-di-*aci*-nitro-3-nitrocompounds, *Liebigs Ann. Chem.,* 1744, 1975.

477. **Haddadin, M.J., Amalu, S.J., and Freeman, J.P.,** Self-condensation of 3*H*-pyrol-3-one 1-oxides, *J. Org. Chem.,* 49, 284, 1984.

478. **Luetzow, A.E. and Vercellotti, J.R.,** The synthesis of 2-carboxy-6-nitrobenzimidazole 1-oxides by intramolecular oxidation of α-(2,4-dinitrophenylamino)-α,β-unsaturated acyl derivatives, *J. Chem. Soc. (C),* 1750, 1967.

479. **Miyano, S., Abe, N., Takeda, K., and Sumoto, K.,** Applications of 3-(*o*-nitrtoanilino)-2-cyclohexen-1-ones in the synthesis of quinoxaline monoxides, *Synthesis,* 60, 1981.

480. **Pfeiffer, P.,** Photochemical synthesis of indole derivatives, *Liebigs Ann. Chem.,* 411, 72, 1916.

481. **Ruggli, P., Caspar, E., and Hegedüs, B.,** On the preparation of 2-phenylisatogen and 2-phenyl-6-ethoxycarbonylisatogen, *Helv. Chem. Acta,* 20, 250, 1937.

482. **Krönke, F. and Meyer-Delius, M.,** New syntheses of indole derivatives. I. A practical synthesis of isatogen, *Chem. Ber.,* 84, 932, 1951.

483. **Krönke, F. and Meyer-Delius, M.,** New syntheses of indole derivatives. II. Thermal cleavage of *N*-vinylpyridinium salts, *Chem. Ber.,* 84, 941, 1951.

484. **Beccalli, E.M., Manfredi, A., and Marchesini, A.,** Alkynes from 5-aminoisoxazoles, *J. Org. Chem.,* 50, 2372, 1985.

485. **Leznoff, C.C. and Hayward, R.J.,** Photocyclization reactions of aryl polyenes. IV. The syntheses of isatogens and isatogen-like compounds, *Can. J. Chem.,* 49, 3596, 1971.

486. **Cross, B., Williams, P.J., and Woodall, R.E.,** The preparation of phenazines by the cyclization of 2-nitrodiphenylamines, *J. Chem. Soc. (C),* 2085, 1971.

487. **Sakuma, Y., Nagamatsu, T., and Yoneda, F.,** New synthesis of flavins, *J. Chem. Soc., Chem. Commun.,* 977, 1975.

488. **Yoneda, F., Sakuma, Y., and Shinazuka, K.,** One step synthesis of 8-chloroflavins by the cyclization of 5-nitro-6- (*N*- substituted-anilino)uracils with the Vilsmeier reagent. Vilsmeier reagent as a reducing agent, *J. Chem. Soc. Chem. Commun.,* 681, 1977.

489. **Maki, Y., Suzuki, M., Hosokami, T., and Furuta, T.,** Photochemistry of *N*-acyl-2-nitrodiphenylamines. A novel photochemical synthesis of phenazine *N*-oxides, *J. Chem. Soc., Perkin Trans. 1,* 1354, 1974.

490. **Krowczynski, A. and Kozerski, L.,** Reaction of 2-nitroenamine in acids, *Heterocycles,* 24, 1209, 1986.

491. **Silberg, A., and Frenkel, Z.,** Mechanism of the Lehmstedt-Tanasescu reaction, *Rev. Roum. Chim.,* 10, 1035, 1965.

492. a. **Loudon, J.D. and Wellings, I.,** Substituent interaction in *ortho*-substituted nitrobenzenes, Part III. *J. Chem. Soc.,* 3470, 1960; b. **Sword, I.P.,** Reaction of 2,3-epoxy-3-(2-nitrophenyl)propiophenone (2-nitrochalcone epoxide) with hydrogen chloride, *J. Chem. Soc. (C),* 820, 1971; c. **Spence, T.W.M. and Tennant, G.,** Acid-catalysed ring-opening reactions of substituted *o*-nitrophenylethylene oxides involving participation by the nitro group, *J. Chem. Soc. (C),* 3712, 1971.

493. **Muth, C.W., Abraham, N., Linfield, M.L., Wotring, R.B., and Pacofsky, E.A.,** Preparations of phenanthridine 5-oxides and benzo[*c*]cinnoline 1-oxide, *J. Org. Chem.,* 25, 736, 1960.

494. **Barton, J.W. and Thomas, J.F.,** The synthesis and reactions of some benzo[*c*]cinnoline oxides, *J. Chem. Soc.,* 1265, 1964.

495. **Poesche, W.H.,** Dibenzo[*c,h*]cinnoline oxides, *J. Chem. Soc. (C),* 890, 1966.

496. **Waldau, E. and Pütter, R.,** Biaryls, stilbenes, benzo[*c*]cinnolines and dibenzo[*c,mn*]acridines from sulfonamides, *Angew. Chem. Int. Ed. Eng.,* 11, 826, 1972.

497. a. **Johnston, D., Smith, D.M., Shepherd, T., and Thompson, D.,** Formation of 4-arylamino-3-methoxycinnoline 1-oxides from *N*-*o*-nitro-benzylideneanilines, cyanide ion, and methanol: the intermediacy of 2-aryl-3-cyano-2*H*-indazole 1-oxides, *J. Chem. Soc., Perkin Trans. 1,* 495, 1987; b. **Johnston, D. and Smith, D.M.,** Cyanide induced cyclization of *o*-nitroanils, *J. Chem. Soc., Perkin Trans. 1,* 399, 1976.

498. **Wolf, F.J., Pfister, K., Wilson, R.M., and Robinson, C.A.,** Benzotriazines. I. Synthesis of compounds having antimalarial activity, *J. Am. Chem. Soc.,* 76, 3551, 1954.

499. **Jiu, J. and Mueller, G.P.,** Syntheses in the 1,2,4-benzotriazine series, *J. Org. Chem.,* 24, 813, 1959.

500. **Mason, J.C. and Tannant, G.,** Synthesis and nuclear magnetic resonance spectra of 2-aminobenzo-1,2,4-triazines and their mono- and di-*N*-oxides, *J. Chem. Soc. (B),* 911, 1970.

501. **Dolman, H., Peperkamp, H.A., and Moed, H.D.,** 3-Amino-6-chloro-7-sulfamoyl-1,2,4-benzotriazine 1-oxide, *Rec. Trav. Chim. Pays Bas,* 83, 1305, 1964.

502. **Hewlins, M.J.E. and Jones, H.O.,** Preparation of 3-dimethylamino-7-methyl-1,2,4-benzotriazine-1-oxide as an intermediate for azapropazone, World patent 87/4432 A1, 1987; *Chem. Abst.* 108, 21931t, 1988.

503. **Backer, H.J. and Moed, H.D.,** Intramolecular rearrangement of nitrophenylsulfonylguanidines, *Rec. Trav. Chim. Pays-Bas,* 66, 689, 1947.

504. **Carbon, J. A.,** The base catalyzed rearrangement of benzotriazine *N*-oxides to benzotriazoles, *J. Org. Chem.,* 27, 185, 1962.

505. **Carbon, J.A. and Tabata, S.H.,** Preparation and rearrangement of some pyrido[2,3-*e*]-*as*-triazine 1-oxides, *J. Org. Chem.,* 27, 2504, 1962.

506. **Lewis, A. and Shepherd, R.G.,** Cyclizations leading to pyrido[3,4-*e*]-and [4,3-*e*]-*as*-triazines. Ring interconversion of a pyrido-*as*-triazine 1-oxide to a pyridotriazole, *J. Heterocycl. Chem.,* 8, 47, 1971.

507. **Wolf, F.J., Wilson, R.M., Pfister, K., and Tishler, M.,** Synthesis of 3-amino-7-halo-1,2,4-benzotriazine 1-oxides, *J. Am. Chem. Soc.,* 76, 4611, 1954.

508. **Hauptmann, S., Blattman, G., and Schindler, W.,** Preparation and properties of pyrazolo[3,2-*c*]benzo-1,2,4-triazines and pyrazolo [3,2-*c*]-pyrido[4,3-*e*] [1,2,4]triazines, *J. Prakt. Chem.,* 318, 835, 1976.

509. **Heller, G. and Spielmeyer, G.,** Indazoles from *o*-nitromandelonitrile, *Ber. Dtsch. Chem. Ges.,* 58, 834, 1925.

510. **Behr, L.C.,** Position isomerism in the azoxybenzenes, *J. Am. Chem. Soc.,* 76, 3672, 1954.

511. **Behr, L.C., Alley, E.G., and Levand, O.,** *m*-Nitroazoxybenzenes, *J. Org. Chem.,* 27, 65, 1962.

512. **Reissert, A. and Lemmer, F.,** *o*-Nitrophenylanilinoacetonitrile and its reactions, *Ber. Dtsch. Chem. Ges.,* 59, 351, 1926.

513. **Secareanu, S. and Lupas, I.,** New research on Schiff bases of 2,4,6-trinitro- and 2,4-dinitrobenzaldehyde, *Bull. Soc. Chim. Fr.* [5], 2, 69, 1935.

514. **Hirota, K., Yamada, Y., Asao, T., and Senda, S.,** Reactions of 1,3,6-trimethyl-5-nitrouracil and its 6-bromomethyl analog with amines and hydrazines. Synthesis of pyrazolo[4,3-*d*]pyrimidine *N*-oxides and their ring expansion to pyrimido[5,4-*d*]pyrimidines, *J. Chem. Soc., Perkin Trans. 1,* 277, 1982.

515. **Houghton, P.G., Pipe, D.F., and Rees, C.W.,** Interamolecular reaction between nitro and carbodi-imide group; a new synthesis of 2-arylbenzotriazoles, *J. Chem. Soc. Perkin Trans. 1,* 1471, 1985.

516. **Nietzki, R. and Braunschweig, E.,** Action of alkali on *ortho*-nitrophenylhydrazine, *Ber. Dtsch. Chem. Ger.,* 27, 3381, 1894.

517. a. **Müller, E. and Zimmermann, G.,** On the action of hydrazine hydrate on some nitro- and chloro derivatives, *J. Prakt. Chem.,* 111, 277, 1925; b. **Leonard, N.J. and Golankiewicz, K.,** Thermolysis of substituted 1-acetoxybenzotriazoles. Carbon-to-oxygen migration of an alkyl group, *J. Org. Chem.,* 34, 359, 1969.

518. **König, W. and Geiger, R.,** A new method for the synthesis of peptides: activation of carboxyl group by dicyclohexylcarbodiimide in the presence of 1-hydroybenzotriazoles, *Chem. Ber.,* 103, 788, 1970.

519. a. **Vis, B.,** Some derivative of 1-hydroxy-1,2,3-benzotriazole, *Rec.Trav. Chim. Pays Bas,* 58, 847, 1939; b. **Brady, O.L. and Day, J.N.E.,** Some substituted hydroxybenzotriazoles and their methylation products, *J. Chem. Soc.,* 2258, 1923.

520. a. **Joshi, S.S. and Gupta, S.P.,** Some 2(*p*-tolyl)nitro-1,2,3-benzo triazoles and their 1-oxides, *J. Ind. Chem. Soc.,* 35, 681, 1958; b. **Shine, H.J., Fang, L.T., Mallory, H.E., Chamberlain, N.F., and Stehling, F.,** Reactions of vinyl ketones and β-alkoxy ketones: formation of substituted 6-nitrobenzotriazole oxides, *J. Org. Chem.,* 28, 2326, 1963; c. **Shine, H.J. and Tsai, J.Y.F.,** The formation of 1-(2,4-dinitrophenyl)pyrazolines from methylol ketones, *J. Org. Chem.,* 29, 443, 1964.

521. **Heine, H.W., Williard, P.G., and Hoye, T.R.,** The synthesis and reactions of some 1-(nitroaryl)diaziridines, *J. Org. Chem.,* 37, 2980, 1972.

522. **Kapil, R.S. and Joshi, S.S.,** Preparation of some polynitrophenylhydrazines, their hydrazones and some 2-arylbenzotriazoles, *J. Ind. Chem. Soc.,* 36, 417, 1959.

523. **Mangini, A.,** Research on the 1-chloro-3,4-dinitrobenzene series, *Gazz. Chim. Ital.,* 65, 1191, 1935.

524. **Hazard, R. and Tallec, A.,** Potential controlled electrochemical reduction of substituted azoxybenzenes. Study of nitro and dinitroazoxybenzenes, *Bull. Soc. Chim. Fr.,* 2917, 1971.

525. **Grashey, R.,** The nitro group as a 1,3-dipole in cycloadditions, *Angew. Chem. Int. Ed. Eng.,* 1, 158, 1962.

526. **Maki, Y., Izuta, K., and Suzuki, M.,** Novel photocyclizations of 1,3-dimethyl-5-nitro-6-benzylidenemethylhydrazinouracil, *Tetrahedron Lett.,* 1973, 1972.

527. **Koval, I.V., Tarasenko, A.I., Kremlev, M.M., and Naumenko, R.P.,** Arylsulfonylimination of bis(arylsulfenyl)imides, *Zh. Org. Khim.,* 22, 1178, 1986.

528. **Kohler, E.P. and Darling, S.F.,** A new type of cyclopropene derivative, *J. Am. Chem. Soc.,* 52, 1174, 1930.

529. **Kohler, E.P.,** Benzoyldiphenylisoxazoline oxide, *J. Am. Chem. Soc.,* 46, 1733, 1924.

530. **Nielsen, A.T. and Archibald, T.G.,** Mechanism of the Kohler synthesis of 2-isoxazoline 2-oxides from 1,3-dinitroakanes, *J. Org. Chem.,* 34, 984, 1969.

531. **Luk'yanov, O.A. and Ternikova, T.V.,** Reaction of *N*-(γ-haloalkyl)-*N*-nitramides with bases, *Izv. Akad. Nauk SSSR, Ser. Khim.,* 667, 1983.

532. **Gladstone, W.A.F., Aylward, J.B., and Norman, R.O.C.,** Oxidation of aldehyde hydrazones: a new method for the generation of nitrilimines, *J. Chem. Soc. (C),* 2587, 1969.

533. **Gibson, M.S.,** Intramolecular 1,3-dipolar additions involving nitro- and carbonyl groups, *Tetrahedron,* 8, 1377, 1962.

534. a. **Hegarty, A.F., Cashman, M., Aylward, J.B., and Scott, F.L.,** *ortho*-group participation in azocarbonium ion and 1,3-dipolar ion formation, *J. Chem. Soc. (B),* 1879, 1971; b. **Coe, P.L., Jukes, A.E., and Tatlow, J.C.,** The synthesis of some derivatives of 1,2,3,4-tetrafluoroacridine, *J.Chem. Soc. (C),* 2020, 1966.

535. **Narayanan, C.R., Ramaswamy, P.S., and Landge, A.B.,** A novel rearrangement of nitrimines, *Chem. Ind.,* 501, 1977.

536. **Onda, M., Yabuchi, R., Takeuchi, K., and Konda, Y.,** Alumina induced reaction of 3β,5α-diacetoxy-6-nitriminocholestane, *Chem. Pharm. Bull.,* 24, 1795, 1976.

537. **Berndt, A.,** Sterically induced intramolecular cyclization of an α,β-unsaturated nitro compound, *Angew. Chem. Int. Ed. Engl.,* 7, 637, 1968.

538. **Wieser, K. and Berndt, A.,** 4*H*-1,2-oxazete *N*-oxides from 1,1-di-*tert*-butylallenes, *Angew. Chem., Int. Ed. Engl.,* 14, 69, 1975.

539. **Garabadzhiu, A.V. and Glibin, E.N.,** 2-Nitro-3(5)-hydroxy-4-X-benzoic anhydrides, *Zh. Org. Khim.,* 18, 1243, 1982.

540. **Szmant, H.H. and Harmuth, C.M.,** Hydrazones of nitrophenyl and pyridyl aldehydes and ketones. New preparation of *o*-nitrobenzophenones and difference in behavior of *o*-nitrobenzophenone and *o*-nitrophenyl mesityl ketone, *J. Am. Chem. Soc.,* 81, 962, 1959.

541. **Corey, E.J., Seibel, W.L., and Kappos, J.C.,** Mechanism of the nitrous acid-induced dealkylation of trisubstituted (terminal isopropylidene)olefins to form acetylenes, *Tetrahedron Lett.,* 4921, 1987.

542. **Emmons, W.D. and Freeman, J.P.,** Reactions of dinitroolefins with nucleophilic reagents, *J. Org. Chem.,* 22, 456, 1957.

543. **Rappoport, Z. and Topol, A.,** Steroconvergence in nucleophilic vinylic substitution of an activated nitro olefin, *J. Am. Chem. Soc.,* 102, 406, 1980.

544. **Thakore, A.N., Buchshriber, J., and Oehlschlager, A.C.,** Vinyl azides as diazoenamines, *Can. J. Chem.,* 51, 2406, 1973.

545. **Fagley, T.F., Sutter, J.R., and Oglukian, R.L.,** Kinetics of thermal decomposition of *o*-nitrophenyl azide, *J. Am. Chem. Soc.,* 78, 5567, 1956.

546. **Birkimer, E.A., Norup, B., and Bak, T.A.,** An apparatus for following reaction evolving gas, *Acta Chem. Scand.,* 14, 1894, 1960.

547. **Smith, P.A.S. and Brown, B.B.,** The synthesis of heterocyclic compounds from aryl azides. I. Bromo and nitrocarbazoles, *J. Am. Chem. Soc.,* 73, 2435, 1951.

548. **Smith, P.A.S. and Boyer, J.H.,** Benzofurazan *N*-oxide, *Org. Syntheses,* 31, 14, 1951.

549. **Bailey, A.S. and Case, J.R.,** 4,6-Dinitrobenzofuroxan, nitrobenzodifuroxan, and benzotrifuroxan — a series of complex forming reagents for aromatic hydrocarbons, *Tetrahedron,* 3, 113, 1958.

550. a. **Boulton, A.J., Ghosh, P.B., and Katritzky, A.R.,** 5-Amino- and 5-hydroxybenzofuroxans, *J. Chem. Soc. (C),* 971, 1966; b. **Boulton, A.J., Gripper-Grey, A.C., and Katritzky, A.R.,** Furoxano- and furazano-benzofuroxan, *J. Chem. Soc.,* 5958, 1965.

551. a. **Dyall, L.K. and Wong, M.W.,** Identification of neighboring group effects in pyrolysis of azidopyridines and azidoquinolines, *Aust. J. Chem.,* 38, 1045, 1985; b. **Bailey, A.S., Heaton, M.W., and Murphy, J.I.,** Preparation of a nitropyrido[3,4-*c*]furoxan: 7-nitro[1,2,5]-oxadiazolo[3,4-*c*]pyridine 3-oxide, *J. Chem. Soc. (C),* 1211, 1971.

552. **Rahman, A., Boulton, A.J., Clifford, D.P., and Tiddy, G.J.T.,** The tautomerism of some naphtho- and quinolino-furoxans, *J. Chem. Soc. (B),* 1516, 1968.

553. **Nutiu, R. and Boulton, A.J.,** New furazano[3,4-*d*]pyrimidine *N*-oxides: preparation and structure, *J. Chem. Soc., Perkin Trans. 1,* 1327, 1976.

554. **Temple, C.J., Kussner, C.L., and Montgomery, J.A.,** Reaction of some 4-chloro-5-nitropyrimidines with sodium azide. Oxadiazolo[3,4-*d*]pyrimidine 1-oxides, *J. Org. Chem.,* 33, 2086, 1968.

555. **Boulton, A.J. and Middleton, D.,** Tropono[4,5,*c*-]thieno[2,3-*c*]-, and biphenyleno[2,3-*c*]furazan oxides, *J. Org. Chem.,* 39, 2956, 1974.

556. **Ayyangar, N.R., Kumar, S.M., and Srinivasan, K.V.,** Facile one-pot synthesis of 2,1,3-benzoxadiazole *N*-oxide (benzofuroxan) derivatives under phase transfer catalysis, *Synthesis,* 616, 1987.

557. **Eswaran, S.V. and Sajadiau, S.K.,** Synthesis of dimethoxy- and dioxano-annellated benzofuroxans from *o*-dinitroarenes, *J. Heterocycl. Chem.,* 25, 803, 1988.

558. **Green, A.G. and Rowe, F.M.,** Conversion of *o*-nitroamines into iso-oxadiazole oxides, *J. Chem. Soc.,* 103, 2023, 1913.

559. **Chapman, K.J., Dyall, L.K., and Frith, L.K.,** Mechanisms of cyclization of *N*-chloro-2-nitroanilines to benzofuroxan under alkaline condition, *Aust. J. Chem.,* 37, 341, 1984.

560. **Dyall, L.K. and Pausaker, K.H.,** The oxidation of 3-(or 6-) substituted-2-nitroanilines, *Aust. J. Chem.*, 11, 4981, 1958.

561. **Dyall, L.K., Evans, J.O.M., and Kemp, J.E.,** Kinetic studies of the reaction of 2-nitroaniline with phenyliodoso acetate in toluene solutions, *Aust. J. Chem.*, 21, 409, 1968.

562. **Dyall, L.K.,** Cyclization of 2-substituted anilines with alkaline hypohalite, *Aust. J. Chem.*, 37, 2013, 1984.

563. **Rastogi, R., Dixit, G., and Zutschi, K.,** Electrochemical synthesis of benzofuroxan: role of electrogenerated iodonium in the intramolecular cyclization of 2-nitroaniline, *Electrochim. Acta*, 29, 1345, 1984.

564. **Boulton, A.J. and Brown, R.C.,** The formation of formylbenzofurazan oxide from a nitroanthranil, *J. Org. Chem.*, 35, 1662, 1970.

565. **Boulton, A.J., Ghosh, P.B., and Katritzky, A.R.,** A general rearrangement of heterocycles, and novel syntheses of the benzotriazole, benzofurazan, and anthranil system, *Angew. Chem. Int. Ed. Eng.*, 3, 693, 1964.

566. **Senga, K., Fukami, K., Kawazawa, H., and Nishigaki, S.,** New syntheses of pyrido[3,2-*d*]pyrimidines, *J. Heterocycl. Chem.*, 19, 805, 1982.

567. **Chi, C., Chen, M., Liang, S., and Chen, D.,** Reactions of 3,5-dinitro-2-chlorobenzotrifluoride and 3,5-dinitro-4-chlorobenzotrifluoride with sulfur and nitrogen-containing nucleophiles, *J. Fluorine Chem.*, 38, 327, 1988.

568. **Tanasescu, I. and Macarovici, M.,** Condensation of 2-nitro-5-chlorobenzaldehyde with chloro- and bromobenzene with concentrated H_2SO_4, *Bull. Soc. Chim. Fr.*, [5], 4, 240, 1937.

569. **Drechsler, K.,** On the base $C_{13}H_9NO$ resulting from the action of aluminum chloride on *o*-nitrobenzylchloride and benzene, *Monatsh.Chem.*, 35, 533, 1914.

570. a. **Wohl, A.,** On the reaction between nitrobenzene and aniline in the presence of alkali, *Ber. Dtsch. Chem. Ges.*, 36, 4135, 1903; b. **Wohl, A.,** On azoxy derivatives, *Ber. Dtsch. Chem. Ges.*, 36, 4139, 1903.

571. a. **Maffei, S., Pietra, S., and Cattaneo, T.,** Phenazine halogenation, *Gazz. Chim. Ital.*, 83, 327, 1963; b. **Maffei, S., Pietra, S., and Rivolta, A.M.,** Synthesis of 2-phenazinecarboxylic acid, *Ann. Chim. (Rome)*, 42, 519, 1952.

572. a. **Serebryanyi, S.B.,** 1-Alkoxyphenazines, *Zhur. Obshchei. Khim.*, 20, 1629, 1950; b. **Serebryanyi, S.B.,** Alkyl derivatives of 1-alkoxyphenazines and their *N*-oxides, *Ukr. Khim. Zh.*, 34, 805, 1968, *Chem. Abst.*, 70, 11678y, 1969.

573. **Maki, Y., Sako, M., and Taylor, E.C.,** Novel synthesis of pyrimidopteridine 10-oxides, *Tetrahedron Lett.*, 4271, 1971.

574. **Pennings, M.L.M. and Reinhoudt, D.N.,** Synthesis and thermal isomerization of 2,3-dihydroazete 1-oxides, *J. Org. Chem.*, 47, 1816, 1982.

575. **Van Eijk, P.J.S.S., Overkempe, C., Trompenaars, W.P., Reinhoudt, D.N., Manniren, L.M., van Hummel, G.J., and Harkema, S.,** Stereoselective synthesis of *cis* and *trans* four-membered cyclic nitrones, *Rec. Trav. Chim. Pays Bas*, 107, 27, 1988.

576. **Van Elburg, P.A., Honig, G.W.N., and Reinhoudt, D. N.,** Chiral four-membered cyclic nitrones: asymmetric induction in the (4+2) cycloaddition of chiral ynamines and nitroalkenes, *Tetrahedron Lett.*, 6397, 1987.

577. a. **de Wit, A.D., Trompenaars, W.P., Pennings, M.L.M., and Reinhoudt, D.N.,** Reduction of the nitro group versus insertion in the C-O bond of 3-nitrobenzofuran by ynamines. Synthesis and x-ray crystal structure determination of a 1-benzoxepin and a quinoline 1-oxide, *J. Org. Chem.*, 46, 172, 1981; b. **Marcelis, A.T.M. and van der Plas, H.C.,** Cycloaddition of 5-nitropyrimidines with ynamines. Formation of 3-nitropyridines, *N*-5-nitropyrimidyl α-carbamoylnitrones, and 2,2a-dihydroazeto[2,3-*d*]- 3,5-diazocines, *J. Org. Chem.*, 51, 67, 1986.

578. a. **Benko, P., Berenyi, E., Messner, A., Hajos, G., and Pallos, L.,** Pyrido[4,3-*e*]-*as*-triazines, *Acta Chim. Acad. Sci. Hung.*, 90, 405, 1976; b. **Benko, P., Beremyi, D., Messmer, A., Hajos, G., and Pallos, L.,** Study of pyrido[4,3-*e*]-*as*-triazines, *Magy. Kem. Foly.*, 82, 183, 1976, *Chem. Abst.*, 85, 46593d, 1976; c. **Unangst, P.C. and Heindel, N.D.,** Preparation of 1,2,3-triazolo[4,5-*f*]quinoline *N*-oxides from 6-nitro-5,8-dimethoxyquinaldine, *Org. Prep. Proced. Int.*, 6, 295, 1974.

579. **Robertson, D.N.,** Improved synthesis of α-nitrostilbenes, *J. Org. Chem.*, 25, 47, 1960.

580. **Matelkina, E.L., Sopova, A.S., Perekalin, V.V., and Ionin, B.I.,** Reaction of *gem*-bromonitroethenes with nitroacetonitrile and ethyl nitroacetate, *Zh. Org. Khim.*, 10, 209, 1974.

581. **Johnson, C.R., Lockard, J.P., and Kennedy, E.R.,** *S*-Ethenylsulfoximine derivatives. Reagents for ethylenation of protic nucleophiles, *J. Org. Chem.*, 45, 264, 1980.

582. **Kaji, E. and Zen, S.,** Synthesis of 4-(β-D-ribofuranosyl)isoxazoline *N*-oxide derivatives, *Synth. Commun.*, 9, 165, 1979.

583. **Zen, S., Kaji, E., and Takahashi, K.,** Substituent effect on the ring transformation of 4-(di/trisubstituted phenyl)-3,5-bis(methoxycarbonyl)-2-isoxazoline 2-oxides, *Nippon Kagaku Kaishi*, 55, 1986.

584. **Takahashi, K., Kaji, E., and Zen, S.,** Synthesis of 4-substituted-3,5-bis(methoxycarbonyl)-2-isoxazoline 2-oxides, *Nippon Kagaku Kaishi*, 1678, 1983.

585. **Krasnaya, Z.A., Stytsenko, T.S., Prokof'ev, E.P., Yakovlev, I.P., and Kucherov, V.F.,** Reaction of enaminocarbonyl compounds with nitroacetic ester, *Izv. Akad. Nauk. SSSR, Ser. Khim.*, 845, 1974.

586. **Zen, S. and Kaji, E.,** Formation of 4-substituted-3,5-bis(methoxycarbonyl)isoxazoline *N*-oxides via *O*-alkylation of nitroacetate with *n*-alkyl halides, *Chem. Pharm. Bull.*, 22, 477, 1974.

587. **Magdesieva, N.N., Sergeeva, T.A., and Kyandzhestian, R.A.,** Selenonium ketylides in the synthesis of isoxazoline N-oxides, *Zh. Org. Khim.,* 21, 1980, 1985.

588. a. **Melot, J.M., Texier-Boullet, E., and Foucaud, A.,** Alumina supported potassium fluoride promoted reaction of nitroalkanes with electrophilic alkenes. Synthesis of 4,5-dihydrofurans and isoxazoline N-oxides, *Tetrahedron,* 44, 2215, 1988; b. **Melot, J.M., Texier-Boullet, F., and Foucaud, A.,** A convenient one-pot synthesis of 4-substituted 3,5-bis(alkoxycarbonyl)-4,5-dihydroisoxazole 2-oxides from aldehydes and nitroacetic esters via solid-liquid reaction system and subsequent deoxygenation, *Synthesis,* 558, 1988.

589. a. **Fukunaga, K.,** A new synthesis of 3,4,5-triaryl-4,5-dihydro-1,2-oxazole N-oxides, *Synthesis,* 55, 1978; b. **Fukunaga, K., Hamanaka, K., and Kimura, M.,** The formation of 3,4,5-tris(substituted phenyl)-4,5-dihydro-1,2-oxazole N-oxides in the oxidation of (substituted phenyl)nitromethanide anions with silver nitrate or peroxydisulfate, *Bull. Chem. Soc. Jpn.,* 52, 1543, 1979.

590. **Fridman, A.L., Gabinov, F.A., and Surkov, V.D.,** Mechanism of the reaction of diazo compounds with halotrinitromethanes and *gem*-dinitroalkenes, *Zh. Org. Khim.,* 8, 2457, 1972.

591. **Onishchenko, A.A., Chlenov, I.E., Makarenkova, L.M., and Tartakovskii, V.A.,** Reaction of trinitromethane and its halo derivatives with diazomethane, *Izv. Akad. Nauk SSSR, Ser. Khim.,* 1560, 1971.

592. **Ioffe, S.L., Makarenkova, L.M., Kashutina, M.V., Tartakovskii, V.A., Rozhdestvenskaya, N.N., Kovalenko, L.I., and Isagulyants, V.G.,** Formation of dinitrocarbene during trimethylsilylation of trinitromethane derivatives, *Zh. Org. Khim.,* 9, 905, 1973.

593. **van Velzen, J.C., Kruk, C., Spaargaren, K., and De Boer, T.J.,** Route from 1,3,5-dinitrobenzene to a dinitrotropane derivative, *Rec. Trav. Chim. Pays Bas,* 91, 557, 1972.

594. a. **Stetter, M. and Hoehne, K.,** On the course of the Michael addition of dihydroresorcin to nitroolefins, *Chem. Ber.,* 91, 1344, 1958; b. **Berestovitskaya, V.M., Sopova, A.S., and Perekalin, V.V.,** Synthesis of oxygen and nitrogen containing heterocycles from bromonitroalkenes and cyclic β-diketones, *Khim. Geterotsikl. Soedin,* 396, 1967.

595. a. **Kienzle, F., Fellman, J.Y., and Stadlwieser, J.,** On the formation of dihydro-1,2-oxazine N-oxides by intramolecular addition of a nitronate anion on a carbonyl group, *Helv. Chim. Acta,* 67, 789, 1984; b. **Gomez-Sanchez, A., Fernandez-Fernandez, R., Pascual, C., and Bellanato, J.,** Furan and pyrrole formation from 2-nitro-1-phenylpropene and acetoacetic esters, *J. Chem. Res. (S),* 318, 1986.

596. **Denmark, S.E., Dappen, M.S., and Cramer, C.J.,** Intramolecular [4+2] cycloadditions of nitroalkenes with olefins, *J. Am. Chem. Soc.,* 108, 1306, 1986.

597. a. **Denmark, S.E., Cramer, C.J., and Sternberg, J.A.,** Intermolecular [4+2] cycloadditions of nitroalkanes with cyclic olefins. Transformation of cyclic nitronates, *Helv. Chim. Acta,* 69, 1971, 1986; b. **Denmark, S.E., Cramer, C.J., and Sternberg, J.A.,** The sterostructures of [1,1′-bicyclohexyl]-2,2′-diones: a reassignment, *Tetrahedron Lett.,* 3963, 1986.

598. **Risaliti, A., Forchiassin, M., and Valentin, E.,** The reaction of cyclohexanone enamines with 1- and 2-nitropropene, *Tetrahedron,* 24, 1889, 1968.

599. **Bradamante, P., Pitacco, G., Risaliti, A., and Valentin, E.,** α-Nitro styrene, First reaction with enamines, *Tetrahedron Lett.,* 2683, 1982.

600. **Ferri, R.A., Pitacco, G., and Valentin, E.,** Nucleophilic behavior of 1-substituted morpholino ethenes, *Tetrahedron,* 34, 2537, 1978.

601. **Daneo, S., Pitacco, G., Risaliti, A., and Valentin, E.,** Bicyclic 1,2-oxazine N-oxides. Different behavior in ring fission between systems derived from 5- and 6-membered ring cyclic enamines, *Tetrahedron,* 38, 1499, 1982.

602. **Asaro, F., Pitacco, G., and Valentin, E.,** 1-Nitro-1-phenylpropene. 1,2-Oxazine N-oxides from aminocycloalkenes, *Tetrahedron,* 43, 3279, 1987.

603. **Nielsen, A.T. and Archibald, T.G.,** Synthesis of substituted 5,6-polymethylene-5,6-dihydro-4H-1,2-oxazine 2-oxides by reaction of enamines with nitroolefins, *Tetrahedron,* 26, 3775, 1970.

604. **Tohda, Y., Yamawaki, N., Matsui, H., Kawashima, T., Ariga, M., and Mori, Y.,** Synthesis and a novel fragmentation of 6-alkoxy-5,6-dihydro-4H-1,2-oxazine 2-oxide, *Bull. Chem. Soc. Jpn.,* 61, 461, 1988.

605. **Scheeren, H.W. and Frissen, A.E.,** Zinc chloride catalysis in cycloaddition between ketene acetals and electron-poor olefins; synthesis of highly substituted 1,1-dimethoxycyclobutanes, *Synthesis,* 794, 1983.

606. **Seebach, D. and Brook, M.A.,** Reversed stereochemical course of the Michael addition of cyclohexanone to β-nitrostyrene by using 1-(trimethylsilyloxy)cyclohexene-dichloro(diisopropoxy)titanium, *Helv. Chim. Acta,* 68, 319, 1985.

607. **Brook, M.A. and Seebach, D.,** Cyclic nitronates from the diastereoselective addition of 1-trimethylsilyloxycyclohexene to nitroolefins. Starting materials for stereoselective Henry reactions and 1,3-dipolar cycloadditions, *Can. J. Chem.,* 65, 836, 1987.

608. **Valentin, E., Pitacco, G., Colonna, F.P., and Risaliti, A.,** Unusual stereochemical course in nitroalkylation of biased enaminic system, *Tetrahedron,* 30, 2741, 1974.

609. **Risaliti, A., Forchiassin, M., and Valentin, E.,** *Erythro* configuration of the reaction products from cyclohexanone enamines and β-nitrostyrenes, *Tetrahedron Lett.,* 6331, 1966.

610. **Seebach, D. and Prelog, V.,** Specification of the steric course of asymmetric syntheses, *Angew. Chem. Int. Ed. Engl.,* 21, 654, 1982.

611. **Shin, C., Yonezawa, Y., and Yoshimura, J.,** Reaction of ethyl α,β-unsatured β-nitrocarboxylate with triethylphosphite, *Tetrahedron Lett.,* 3995, 1972.

612. **Gareev, R.D., Loginova, G.M., and Pudovik, A.N.,** Synthesis and properties of 2,2-dimethoxy-2-phenyl-3-isopropyl-5-oxo-1,5,2-oxaza-phosphol-4-ene, *Zh. Obshch. Khim.,* 46, 1906, 1976.

613. **Gareev, R.D., Savin, V.I., Il'yasov, A.V., Levin, Y.A., Goldfarb, E.I., Shermegorn, I.M., and Pudovik, A.N.,** Homolytic stages in the reactions of *o*-methyldiphenylphosphinites with 1-nitro-1-propene and phosphoranes formed from them, *Zh. Obshch. Khim.,* 51, 2137, 1981.

614. **Gareev, R.D., Pudovik, A.N., and Shermergorn, I.M.,** Mechanism of reactions of derivatives of phosphorus (III) acids with 1-nitro-1-alkenes, *Zh. Obshch. Khim.,* 53, 38, 1983.

615. **Grundmann, C. and Grünanger, P.,** *The nitrile oxides,* Springer-Verlag, Berlin, 1971.

616. **Snyder, H.R. and Boyer, N.E.,** Synthesis of furoxan from aryl methyl ketones and HNO_3, *J. Am. Chem. Soc.,* 77, 4233, 1955.

617. **Chang, M.S. and Lowe, J.U.,** Bis(cyclopropanocarbonyl)furoxan, *J. Org. Chem.,* 33, 866, 1968.

618. **Brophy, G.C., Sternhell, S., Brown, N.M.D., Brown, I., Armstrong, K.J., and Martin-Smith, M.,** Nitration of 3-acetyl- and 3-formyl-benzo[*b*]thiophenes, *J. Chem. Soc. (C),* 933, 1970.

619. a. **Peterson, L.I.,** The reaction of acetone with nitrogen tetroxide, *Tetrahedron Lett.,* 1727, 1966; b. **Tezuka, H., Kato, M., and Sonehara, Y.,** Mechanism of nitric acid oxidation of acetophenone to dibenzofurazan 2-oxide, benzoic acid and benzoylformic acid, *J. Chem. Soc., Perkin Trans. 2,* 1643, 1985.

620. **Kornblum, N. and Weaver, W.M.,** The reaction of sodium nitrite with ethyl bromoacetate and with benzyl bromide, *J. Am.Chem. Soc.,* 80, 4333, 1958.

621. **Otsuji, Y., Tsujii, Y., Yoshida, A., and Imoto, E.,** Preparation and reactions of benzoyl nitrile oxide from dimethylphenacylsulfonium bromide, *Bull. Chem. Soc. Jpn.,* 44, 223, 1971.

622. **Bagal, L.I., Stotskii, A.A., and Novatskaya, N.I.,** Nitration of methyl and ethyl dimethylacrylates by nitric acid and sulfuric acid mixtures, *Zh. Org. Khim.,* 3, 1201, 1967.

623. **Tzeng, D. and Baum, K.,** Reaction of hexanitroethane with alcohols, *J. Org. Chem.,* 48, 5384, 1983.

624. **Shimizu, T., Hayashi, Y., and Teramura, K.,** A new synthetic method for alkyl carbonocyanidate *N*-oxides, *Bull. Chem. Soc. Jpn,* 58, 2519, 1985.

625. **Kelley, J.L., McLean, E.W., and Williard, K.F.,** Synthesis of bis(arylsulfonyl)furoxans from aryl nitromethyl sulfones, *J. Heterocycl. Chem.,* 14, 1415, 1977.

626. **Sifniades, S.,** Nitration of acetoacetate ester by acyl nitrate. High yield synthesis of nitroacetoacetate and nitroacetate esters, *J. Org. Chem.,* 40, 3562, 1975.

627. **Krzhizhovskii, A.M., Mirzabekyants, N.S., Cherburov, Y.A., and Kumuyants, I.L.,** Reactions of polyfluoronitroalkanes containing an active hydrogen atom, *Izv. Akad. Nauk SSSR, Ser. Khim.,* 2513, 1974.

628. **Cherest, M. and Lusinchi, X.,** The action of acetyl chloride and of acetic anhydride on the lithium nitronate salt of 2-phenylnitroethane. Reactivity of the intermediate nitrile oxide as an electrophile or as a dipole, depending on the nature of the medium, *Tetrahedron,* 42, 3825, 1986.

629. **Harada, K., Kaji, E., and Zen, S.,** Synthesis of isoxazoles and isoxazolines from aliphatic nitro compounds, *Nippon Kagaku Kaishi,* 1195, 1981.

630. **Marx, M., Marti, F., Reisdorff, J., Sandmeier, R., and Clark, S.,** A stereospecific total synthesis of (±)biotin, *J. Am. Chem. Soc.,* 99, 6754, 1977.

631. **Shimizu, T., Hayashi, Y., Shibafuchi, M., and Teramura, K.,** Convenient preparative method of nitrile oxides by the dehydration of primary nitro compounds with ethyl chloroformiate or benzenesulfonyl chloride in the presence of triethylamine, *Bull. Chem. Soc. Jpn.,* 59, 2827, 1986.

632. **Conutouli-Argryropoulou, E.,** The reaction of triphenylphosphine with arylbromonitromethanes. Formation of aryl nitrile oxides, *Tetrahedron Lett.,* 2029, 1984.

633. **Hirai, K., Matsuda, H., and Kishida, Y.,** Syntheses and reactions of 2-substituted thiazolidines, *Chem. Pharm. Bull.,* 20, 97, 1982.

634. **Wade, P.A. and Pillay, M.K.,** Benzenesulfonylcarbonitrile oxide. 3. Useful new procedures for generation, *J. Org. Chem.,* 46, 5425, 1981.

635. **Hurd, C.D., Nilson, M.E., and Wikholm, D.M.,** Hydroximyl chlorides from nitrostyrenes, *J. Am. Chem. Soc.,* 72, 4697, 1950.

636. **Novikov, S.S., Khmel'nitskii, L.I., and Lebedeva, O.V.,** A study of conditions of reaction of nitrogen tetroxide with benzaldoxime. Composition of products and the equation of reaction, *Zh. Obshch. Khim.,* 28, 2286, 1958.

637. **Suzuki, N., Wakatsuki, S., and Izawa, Y.,** Revised structure for " a novel peroxide, 4,5-diphenyl-1,2-dioxa-3,6-diazine", *Heterocycles,* 16, 1195, 1981.

638. **Sandhu, J.S. and Mohan, S.,** Manganese dioxide oxidation of oximes, *J. Indian. Chem. Soc.,* 49, 427, 1972.

639. a. **Paul, R. and Tchelitcheff, S.,** Synthesis of some new mono- and bicyclic isoxazolines from vinyl acetate and various vinyl ethers, *Bull. Soc. Chim. Fr.,* 29, 2215, 1962; b. **Boulton, A.J., Hadjimihalakis, P., Katritzky, A.R., and Hamid, A.M.,** Isomerism in the oxadiazole series, *J. Chem. Soc. (C),* 1901, 1969.

640. **Trouchet, J.M.J., Baehler, B., Le-Hong, N., and Livio, P.F.,** Mechanistic data on the halogenation of sugar hydrazones and oximes, *Helv. Chim. Acta,* 54, 921, 1971.

641. **Oxenzider, B.C. and Rogic, M.M.,** Reactions of 2,2-dialkoxy ketone oximes with chlorine and bromine. Halogenation vs. Beckmann fragmentation, *J. Org. Chem.,* 47, 2629, 1982.

642. **Mukaiyama, T., Saigo, K., and Takei, H.,** Reactions of sulfonium and pyridinium salts with alkyl nitrite, *Bull. Chem. Soc. Jpn.,* 44, 190, 1971.

643. **Middleton, W.J.,** Trifluoroacetonitrile oxide, *J. Org. Chem.,* 49, 919, 1984.

644. **Fridman, A.L., Ismaiglilova, G.S., and Nikolaeva, A.D.,** Reaction of nitrodiazoketones with nitrogen tetroxide, *Khim. Geterotsikl. Soedin,* 7, 859, 1971.

645. **Engberts, J.B.F.N. and Engbersen, J.F.J.,** Reaction of aliphatic diazo compounds with dinitrogen trioxide. Facile route to 3,4-disubstituted 1,2,5-oxadiazole-2-oxides, *Synth. Commun.,* 1, 121, 1971.

646. **Dahn, H., Favre, B., and Leresche, J.P.,** The nitrosation of primary aliphatic diazocarbonyl compounds: formation of α-carbonyl nitrile oxides, *Helv. Chim. Acta,* 56, 457, 1973.

647. **Jagt, J.C., van Buuren, I., Strating, J., and van Leusen, A.M.,** Synthesis of C-sulfonylcarbohydroximoyl chlorides from α-diazosulfones and nitrosyl chloride, *Synth. Commun.,* 4, 311, 1974.

648. **Kissinger, L.W., McQuistion, W.E., Schwartz, M.N., and Goodman, L.,** Reactions of perfluoralkyldiazom-ethanes. I. Nitrogen dioxide, *Tetrahedron,* 19, Suppl. 1., 131, 1963.

649. **Birckenbach, L. and Sennewald, K.,** On the reaction of fulminic acid and its salts with halogens, *Liebigs Ann. Chem.,* 489, 7, 1931.

650. **De Sarlo, F., Guarna, A., Brandi, A., and Goti, A.,** The chemistry of fulminic acid revised, *Tetrahedron,* 41, 5181, 1985.

651. **Barbaro, G., Battaglia, A., and Dondoni, A.,** Kinetics and mechanism of dimerization of benzonitrile N-oxides to furazan N-oxides, *J. Chem. Soc. (B),* 588, 1970.

652. **Grundmann, C., Frommeld, H.D., Flory, K., and Datta, S.K.,** Dimerization of a sterically hindered nitrile oxide, dimesitylfurazan oxide, *J. Org. Chem.,* 33, 1464, 1968.

653. **Dondoni, A., Barbaro, G., Battaglia, A., and Giorgianni, P.,** Synthesis and reactivity of adamantane-1-carbonitrile N-oxide, *J. Org. Chem.,* 37, 3196, 1972.

654. **Panattoni, C., Clemente, D.A, Bandoli, G., Battaglia, A., and Dondoni, A.,** Structure of the asymmetric diarylfuroxans obtained by reaction of p-chloro- with p-methoxy-benzonitrile N-oxide, *J. Chem. Soc. Chem. Commun.,* 60, 1970.

655. **Huisgen, R.,** 1,3-Dipolar cycloadditions, *Angew. Chem.,* 75, 604, 1963; *Int. Ed. Eng.,* 2, 565, 1963.

656. **Lo Vecchio, G., Foti, F., Grassi, G., and Risitano, F.,** New perspective in the chemistry of nitrile oxides, *Tetrahedron Lett.,* 2119, 1977.

657. **De Sarlo, F. and Guarna A.,** Substituent effect on the rates of dimerization of aromatic nitrile oxides to 3,6-diaryl-1,4,2,5-dioxadiazines, *J. Chem. Soc., Perkin Trans.,* 2, 626, 1976.

658. **De Sarlo, F. and Guarna, A.,** Dimerisation of aromatic nitrile oxides catalysed by trimethylamine, *J. Chem. Soc., Perkin Trans. 1,* 1825, 1976.

659. a. **Morrocchi, S., Ricca, A., Selva, A., and Zanarotti, A.,** Catalysed dimerisation of nitrile oxides, *Gazz. Chim. Ital.,* 99, 165, 1969; b. **Caramella, P., Corsaro, A., Compagnini, A., and Marinone Albini, F.,** Fragmentation of nitrile oxides with triethylamine, *Tetrahedron Lett.,* 4377, 1983.

660. **Faragher, R. and Gilchrist, T.L.,** Imidoyl-substituted oxosulfonium ylides: preparation and reaction with nitrile oxides, *J. Chem. Soc., Perkin Trans. 1,* 1196, 1977.

661. **Gilchrist, T.L., Harris, C.J., Hawkins, D.G., Moody, C.J., and Rees, C.W.,** Synthesis of $1H$-1,2,4-triazole 2-oxides and annelated derivatives, *J. Chem. Soc., Perkin Trans. 1,* 2166, 1976.

662. **Lo Vecchio, G., Caruso, F., Risitano, F., Foti, F., and Grassi, G.,** Further studies on the reaction of benzonitrile oxide with isoxazol-5-ones in methanolic sodium methoxide: 1,3-oxazin-6-ones, imidazoles, and imidazole 3-oxides, *J. Chem. Res. (S),* 76, 1983.

663. **Foti, F., Grassi, G., Risitano, F., and Caruso, F.,** $1H$-1,2,4-triazole 2-oxides from O-benzoyl derivatives of N substituted benzamide oximes and benzonitrile oxide, *J. Chem. Res. (S),* 302, 1982.

664. **Shiraishi, S., Shigemoto, T., and Ogawa, S.,** Reaction of nitrile oxides with N-aryl-S,S-dimethylsulfimides, *Bull. Chem. Soc. Jpn.,* 51, 563, 1978.

665. **Bell, S.C., Sulkovski, T.S., Gochman, C., and Childress, S.J.,** 1,3-Dihydro-$2H$-1,4-benzodiazepin-2-ones and their 4-oxides, *J. Org. Chem.,* 2762, 1962.

666. **Walser, A. and Zenchoff, G.,** A cinnoline 2-oxide from nitrosobenzophenone, *J. Heterocycl. Chem.,* 13, 907, 1976.

667. **Pak, K. and Testa, A.C.,** Photoreduction of 2-nitroso-1-naphthol, *J. Photochem.,* 16, 223, 1981.

Chapter 4

REACTIONS

I. GENERALITIES

As stated in Chapter 1, the present discussion centers on the reactivity of typical heteroaromatic *N*-oxides, i.e., of the *N*-oxides of π-deficient (six-membered) heterocyclics. The chemistry of these compounds can be introduced by making reference to three main characteristics, as suggested by structural analogy with other compounds, though, as appears from the following, differences are possibly more apparent than similarities.

First, like the parent heterocyclics, these compounds are electron-poor aromatics, and this leads to the expectation of easy nucleophilic substitution of substituents in α and γ to the *N*-oxide group. However, both electrophilic and nucleophilic aromatic substitution occur on these compounds, and even in the latter class of reaction the scope is remarkably different from the parent heterocycles.

Second, the *N*-oxides contain the nitrone functionality, but as a part of an aromatic system. This strongly limits the importance of the typical nitrone reactions, i.e., nucleophilic addition and 1,3-dipolar cycloaddition, but such processes do take place whenever an ensuing rearrangement leads to stable end products.

Third, electrophilic attack at the oxygen atom is obviously facile, and the role of the quaternary salts (often prepared *in situ* as intermediates) is central in a great number of reactions leading in the end to functionalized heterocycles.

A quick overview, taking pyridine 1-oxide as an example, will indicate the typical chemistry of these compounds.

Thus, pyridine and congeners are activated toward nucleophilic substitution in positions α and γ by the stabilizing effect of the electronegative nitrogen atom on the intermediate anion, and correspondingly are deactivated toward electrophilic substitution, the more so because the electrophile first attacks the nitrogen atom, and thus substitution on the ring actually involves a dication and occurs in the β position (since in this case the dication is less destabilized) only under drastic conditions.

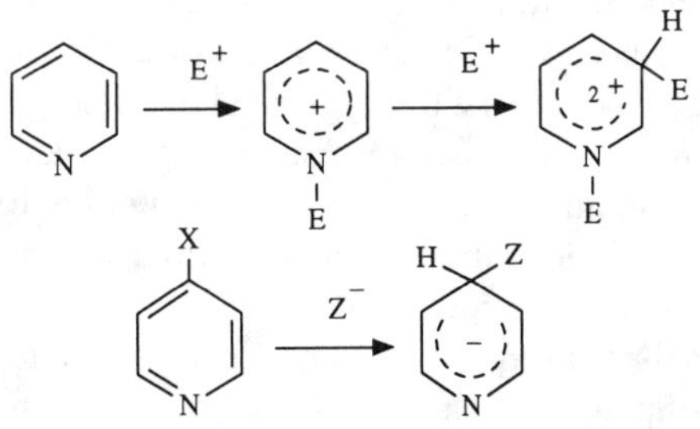

With the *N*-oxides the situation is more complex, since the mesomeric effect exerted by this group is both electron donating and electron withdrawing, and in addition there is an electron-withdrawing inductive effect. As a consequence, pyridine 1-oxide, though less reactive than benzene, is sufficiently reactive to react with at least some electrophiles with a γ (or α) orientation due to the stabilization of the intermediate cation by mesomeric forms such as **1**. Thus, while the effect of aza substitution on electrophilic substitution can be compared to the introduction of a nitro group (strongly deactivating, *meta* orienting), the effect of aza *N*-oxide substitution is rather to be compared to the introduction of a chloro atom (less strongly deactivating, *ortho-para* orienting; there is some analogy between pyridine 1-oxide and isoelectronic chlorobenzene). The double effect of the *N*-oxide function results also from spectroscopic properties (see Chapter 2).

On the other hand, the $\geq$N$\rightarrow$O group activates nucleophilic substitution of leaving groups in α and γ more strongly than the nitrogen atom in pyridine does, due to the elevated stabilization of the intermediate anion through mesomeric forms such as **2**. Actually, with strong nucleophiles attack at an unsubstituted (usually α) position takes place and leads either to substitution (through formal elimination of a metal hydride) or to ring cleavage (in some cases, followed by ring reclosure). Furthermore, abstraction of an α proton takes place under relatively mild conditions, and the stable anion **3** has been shown to add both to double bonds and to electron-poor centers (Scheme 1).

Reaction of electrophiles at the oxygen atom is, as predicted, a facile, though in several cases reversible, process. *N*-oxides form stable salts (of either 1:1 or 1:2 stoichiometry) by reaction with acids and are converted by the appropriate halides to the corresponding *N*-alkoxy as well as the (usually unstable) *N*-acyloxy and *N*-sulfonyloxy pyridinium salts. These compounds are extremely reactive with nucleophiles, giving rise to a varied and useful chemistry. Thus, addition of a nucleophile, generally at the α or γ position, cleavage of the N-O bond, and deprotonation leads to substituted pyridines. This class of reactions, properly termed *deoxidative nucleophilic substitution,* overcomes the problem of hydride elimination in the direct nucleophilic substitution of an unsubstituted position through simultaneous deoxygenation, and EX serves as an auxiliary reagent, which is easily eliminated as the anion EO$^-$ from the intermediate dihydropyridine. It must be noted that the 1,2- (or 1,4-) dihydro intermediate may rearrange before rearomatization and a β-substituted pyridine may be the end product. Furthermore, such dihydropyridines have been shown to react in position β with some electrophiles (electrophilic substitution under acylating conditions), and in other cases they may cleave to open-chain compounds. Finally, with other electrophiles, e.g., phosphorus (III) derivatives, the adducts undergo N-O bond cleavage to yield the deoxygenated pyridine (Scheme 2).

Interesting variations on this mechanism are obtained when an ambient bifunctional reagent is used and the initial electrophilic attack at the oxygen atom is followed by intramolecular nucleophilic cyclization. Cleavage of the N-O bond, rearrangement, and elimination then lead to α- or β-substituted pyridines (Scheme 3).

Addition of a dipolarophile to the nitrone function in heterocyclic *N*-oxides leads to a similar intermediate. Though 1,3-dipolar addition with alkenes is usually not an efficient process, since it involves the loss of the aromaticity, the reaction with alkynes and cumulenes is often efficient because the primary adduct rearranges to stable products. Examples are given in scheme 3, where it again appears that the initially formed 1,2-dihydropyridines may rearrange, and thus pyridines substituted in different positions, as well as various ring-fused pyridines, are obtained (Scheme 4).

Furthermore, many of the above-mentioned deoxidative nucleophilic substitutions, including some reactions with ambient additives, can be extended to α and γ alkyl groups. These reactions involve, at least formally, anhydro-bases, such as intermediate **4**, and offer a mild method of chain functionalization (Scheme 5).

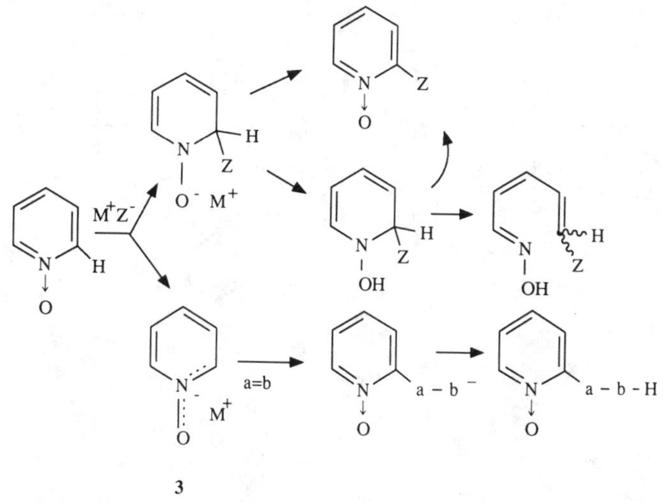

Scheme 1.

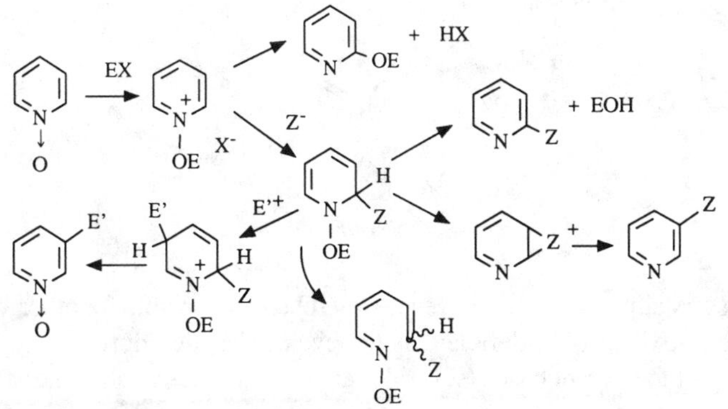

Scheme 2.

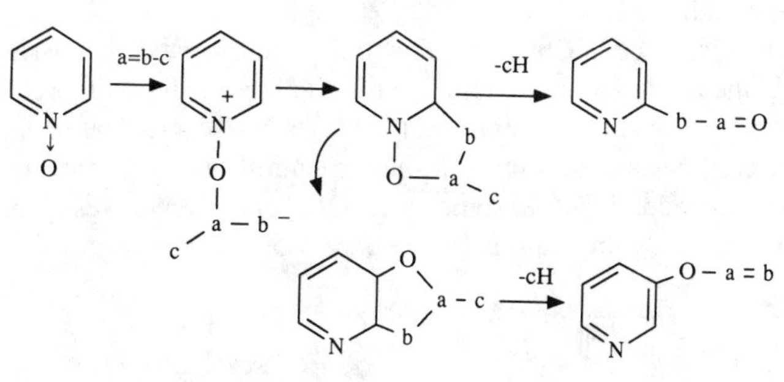

Scheme 3.

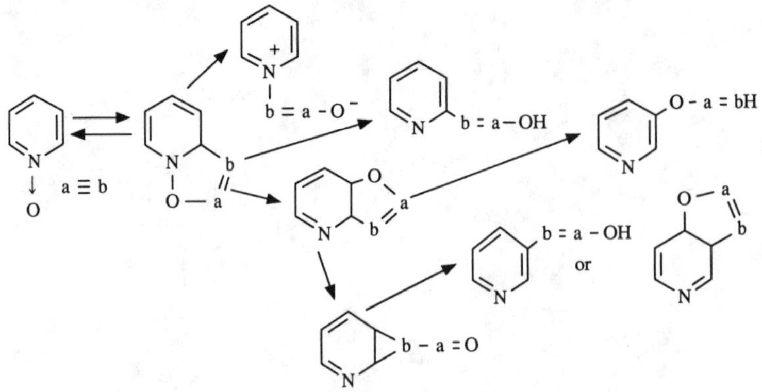

Scheme 4.

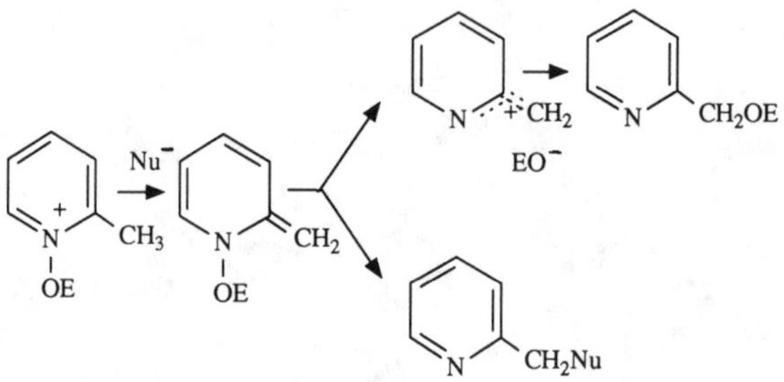

Scheme 5.

With the necessary adjustment the above general schemes also hold for other azine *N*-oxides. In summary, the *N*-oxides of π-deficient heterocycles, besides duplicating the (relatively limited) chemistry of the parent bases (several reactions that are simply the analogs of what is observed on the corresponding heterocycles are not mentioned in this introduction), undergo many peculiar reactions, which offer new methods for the direct introduction of a substituent in the ring or in the chain (conserving or not the *N*-oxide function), or, in some cases, for obtaining polyunsatured open-chain derivatives.

As will be further discussed in Chapter 5, due to this multifaceted chemistry, in addition to simple preparation, the *N*-oxides are extremely versatile intermediates in organic synthesis.

The situation is obviously different in the field of π-excessive (five-membered) heterocycle *N*-oxides. 1-Hydroxypyrroles and azapyrroles are in equilibrium with the corresponding *N*-oxide tautomer, which predominates in some cases. Of course, derivatives "blocked" in the *N*-oxide form are known, e.g., compound **5**, but then the aromatic character is lacking.

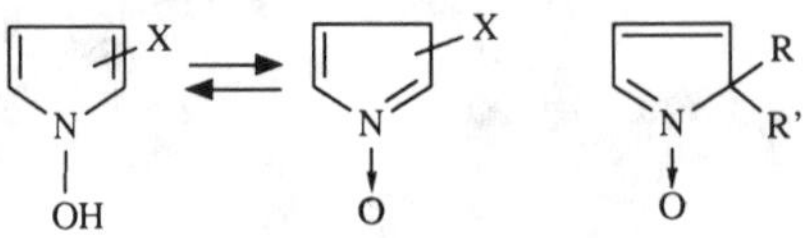

5

Some reactions that are at least formally analogous with those discussed above are also observed with the *N*-oxides of five-membered heterocycles (or with their benzo derivatives), e.g., deoxidative substitution initiated by the addition of an auxiliary electrophile, and this also holds for seven-membered heterocycles (Scheme 6).

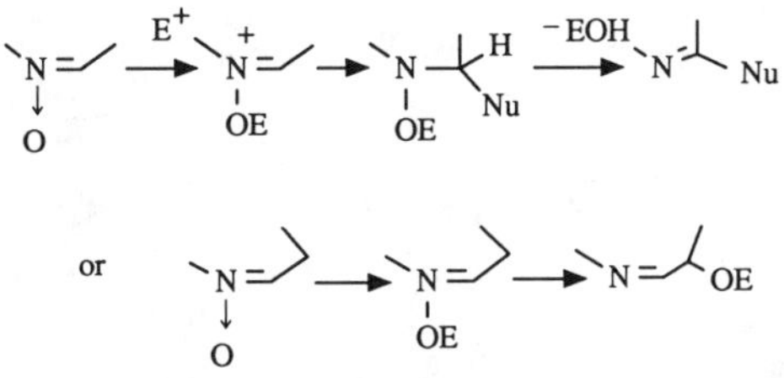

Scheme 6.

Particularly important among five-membered *N*-oxides are furoxans and benzofuroxans, which exhibit a rich chemistry of their own that is initiated both by electrophilic and by nucleophilic addition.

Reduced heterocyclic *N*-oxides containing the isolated nitrone function, e.g., compound **6**, behave very much like their open-chain counterpart, and nucleophilic addition and 1,3-dipolar cycloaddition now become the dominating reactions. The chemistry of these compounds will be only briefly considered for the sake of comparison.

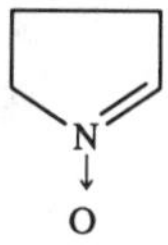

6

II. FORMATION OF QUATERNARY DERIVATIVES

A. FORMATION OF SALTS WITH PROTIC ACIDS

Heterocyclic *N*-oxides readily form salts with protic acids. These are usually crystalline materials, which can be used for purification purposes. The salts are in some cases the primary product obtained in the *N*-oxidation process (see Chapter 3), e.g., treatment of pyridine with perphthalic acid yields the phthalate of pyridine 1-oxide,[1] or else they are obtained by treatment of the preformed *N*-oxide with mineral[2-10] or carboxylic acids or acidic phenols,[11-12] as well as by anion exchange from other salts[1] or by decomposition of *N*-quaternary salts.[13]

Two types of adducts are obtained with many acids (HA), the "normal" *N*-oxide.HA and the "hemi" (*N*-oxide)$_2$.HA salts, and the result in some cases depends on the method of preparation. Thus, as an example, 2-picoline 1-oxide yields the normal salt on treatment with aqueous concentrated HCl or HBr, and the hemi salt on passing gaseous HCl or HBr in the benzene solution. However, only the hemiiodide is obtained by both methods.[4] In other cases, the hemi salt is obtained by recrystallization of the normal salt in the presence of an equimolecular amount of the *N*-oxide.[8-9] The structure of these salts, and in particular the nature of the hydrogen bond, have been the subject of extensive spectroscopic investigation, and further references are given in Chapter 2 (Table 1).

B. FORMATION OF *N*-ALKOXY QUATERNARY SALTS

As might be expected, heterocyclic *N*-oxides easily enter in S$_N$2 reactions with positively charged carbon atoms, and treatment with alkyl halides, sulfates, or sulfonates yields the corresponding *N*-alkoxy quaternary salts.[3,14-26] The halides are often hygroscopic and light sensitive, and for characterization it might be convenient to exchange the counterion, e.g., to perchlorate. Another effective alkylating agent is triethyloxonium fluoborate.[27] As for the *N*-aryloxy salts, these are obtained from aryldiazonium[28] or diaryliodonium[29-30] fluoborates; the

TABLE 1
Salts of Heterocyclic *N*-Oxides with Protic Acids

A. Salts of general formula N-oxide.HA

N-oxide	Substituent	Acid, HA	Mp,°C	Yield,%	Method of preparation	Ref.
Pyridine		HCl	180	75	From the phthalate on heating with aq HCl	1
		HNO$_3$	50–67		Add equim. acid to the warm EtOH soln., crystal. on cooling	2
		Phthalic	122–123			1
		Picrolonic	232–233	Quantit.	From pyridine with perphthalic acid	3
	(2-Me)	HCl	105–106	84	Treat with conc. aq HCl	4
	(2-Me)	HBr	144–145	93	Treat with conc. aq HBr	4
Quinoline		3,5-Dinitrobenzoic	144–145		Treat with stoichiometric acid in MeCN	11
	(2-Me)	HCl	203–205		Crystallize from conc. aq HCl	8
	(2-Me)	HBr	188		Crystallize from 10 parts 48% aq HBr	8
	(2-Me)	HI	147		Saturate an alcoholic solution with HI	8
Isoquinoline		HBr	149–150		Crystallize from conc.aq HBr	9
Acridine		HI	177			10

B. Salts of general formula (*N*-oxide)$_2$.HA

N-oxide	Substituent	Acid, HA	Mp,°C	Yield,%	Method of preparation	Ref.
Pyridine		HAsF$_6$	118–119		Treat with 60% aq HAsF$_6$	5
		HSbF$_6$	117–118		Treat with 60% aq HSbF$_6$	5
	(2-Me)	HCl	48–49	55	Treat with gas HCl in benzene	4
	(2-Me)	HBr	142–143	58	Treat with gas HBr in benzene	4
	(2-Me)	HI	145–146	89	Treat with conc. aq HI	4
	(2,6-Me$_2$)	HClO$_4$	185.5–187	54	Dissolve the 1-acetoxy perchlorate in water, crystallizes on cooling	7,13
Quinoline		HClO$_4$	188–189	40	Dissolve the 1-acetoxy perchlorate in water, crystallizes on cooling	7,13
	(2-Me)	HBr	180		Crystallize from equiv. weight of 48% HBr	8
	(2-Me)	HI	163		Crystallize an equiv. mixt. of *N*-oxide and normal salt from EtOH.	8

Isoquinoline	$HClO_4$	156.5–157.5	50	Dissolve the 2-acetoxy perchlorate in water; crystallize on cooling	9,13
	HBr	118–119[a]		Crystallize an equiv. mixt. of N-oxide and normal salt from EtOH	9
Acridine	HI	155			10

[a] Crystallizes with one molecule of water.

reaction is efficient only either when there is an electron-withdrawing substituent in the aryl group or, in the latter case, also when there is an electron-donating substituent in the *N*-oxide. Representative examples are collected in Table 2.

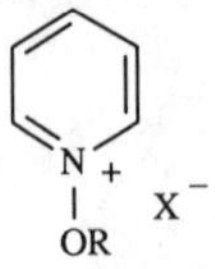

7 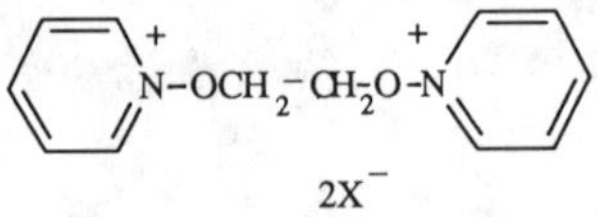   8

1-Ethoxy-4-methylpyridinium fluoborate. *23.0 (0.12 mol) triethyloxonium fluoborate are added to a solution of 10.9 g (0.1 mol) 4-methylpyridine 1-oxide in 30 ml dry chloroform. After some time an exothermic reaction brings the mixture to boil. At the end of this phase, the mixture is refluxed for 2 min more. The oil obtained on evaporation crystallizes on rubbing and cooling in ice-salt mixture. Filtering and washing with ether yields 20.5 g (0.091 mol, 91%) of the salt (mp 56—57°C after recrystallization from dichloromethane).[23]*

1,1'-(Ethylenedioxy)-bis-[pyridinium bromide] (formula 8). *Pyridine 1-oxide (4.75 g, 0.05 mol) and ethylene dibromide (4.7 g, 0.025 mol) are refluxed for 2 h on a steam bath, while the product begins to separate as large, colorless blades. After cooling, the crystalline mass is filtered off and dried (mp 173—174°C). Recrystallization from a small volume of ethanol gives 4.8 g (51%) of a product melting at 175°C.[26]*

Interesting intramolecular variations leads to dihydrooxazino- and dihydroisoxazolopyridinium salts starting from γ- or, respectively, β-bromoalkylpyridine 1-oxides.[31] A related carbocation attack is involved in the cyclization of 2-(1-benzoylmethyl)pyridine 1-oxide.[32]

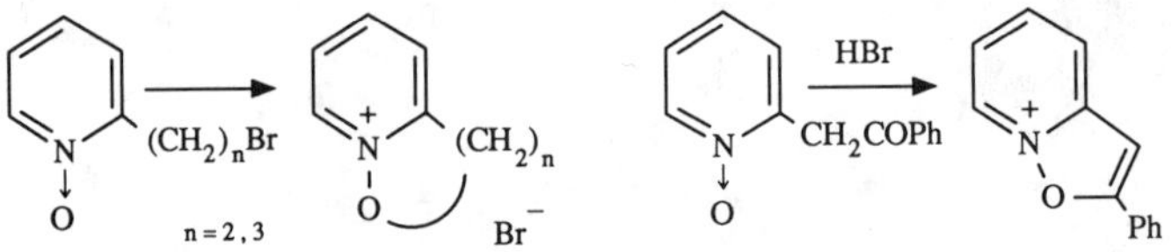

Yet another alkylation is observed in the reaction of 2,6-lutidine 1-oxide with styrene oxide in the presence of strong acids to yield a 1-(2-hydroxyyethoxy)pyridinium salt.[33]

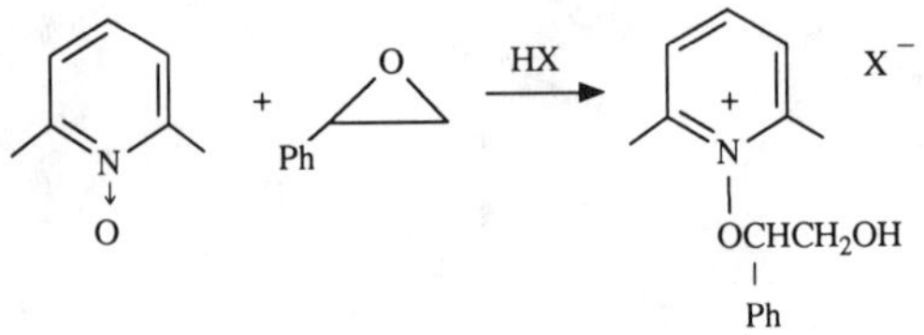

Competition between alkylation of the *N*-oxide group and of other functionalities present in the molecule is discussed in Section XIC2a according to the relative function and in connection with the tautomerism usually observed in such molecules.

C. FORMATION OF OTHER QUATERNARY SALTS

N-acyloxy salts are easily obtained by reaction with carboxylic anhydrides,[13,37] esters,[42] or acyl halides (not the fluorides).[34] The equilibrium constants for the formation of these salts in solution have been determined.[35-36] The counter-ion has an important role, the *N*-acyloxy derivatives behaving as soft acids and giving more stable salts with soft bases such as iodide.[34] The *N*-acyloxy, as well as the analogously obtained *N*-sulfonyloxy salts,[34] are highly reactive

TABLE 2
N-Alkoxy Heterocycles ⩾N$^+$-OR X$^-$ from the Corresponding *N*-Oxides

N-oxide	Substituent	R	Anion, X$^-$	Mp, °C	Yield %	Method of preparation	Ref.
Pyridine		Me	ClO$_4^-$	69–70	80	Heat on steam bath with Me$_2$SO$_4$, then treat with EtOH, HClO$_4$, AcOEt	18
		Me	TsO$^-$	86–90	74	Heat at 110°C for 5 h with MeOTs	3
		Me	I$^-$	90	97	Reflux with excess MeI, evaporate	20
		t-Bu	ClO$_4^-$	95–96	22	Treat with *t*-BuBr in MeNO$_2$, then with AgClO$_4$; filter and dilute with ether	19
		n-C$_9$H$_{19}$	I$^-$	87–80	55	Reflux with n-C$_9$H$_{19}$I in MeCN	21
	(4-Me)	Et	BF$_4^-$	56–57	91	Reflux with Et$_3$O$^+$BF$_4^-$	27
		CH$_2$I$_2$	Br$^-$	175	51	Reflux with CH$_2$BrCH$_2$Br in MeCN	26
		4-CNC$_6$H$_4$	BF$_4^-$	214–215	70	Treat with p-CNC$_6$H$_4^+$N$_2^+$·BF$_4^-$ at r.t. for 24 h	28
	(4-OMe)	Ph	BF$_4^-$	122–122.5	95	Reflux with Ph$_2$I$^+$BF$_4^-$ in MeCN	29–30
Quinoline		Me	ClO$_4^-$	110–111	57	Heat with MeOTs at 120°C for 8 h, then treat with EtOH, HClO$_4$, AcOEt	19
		Me	ClO$_4^-$	110–111	65	Heat with Me$_2$SO$_4$ at 95°C for 2 h, then treat with HClO$_4$ and AcOEt	19
		Et	BF$_4^-$	89–90	96	Reflux with Et$_3$O$^+$BF$_4^-$	27
Isoquinoline		Me	ClO$_4^-$	94–99	60	Heat with Me$_2$SO$_4$ at 95°C for 2 h, then treat with EtOH, HClO$_4$, AcOEt	19
	(3-Me)	Et	BF$_4^-$	88–89	66	Reflux with Et$_3$O$^+$BF$_4^-$	27
Phenanthridine		Me	MeSO$_4^-$	142	89	Reflux with Me$_2$SO$_4$ for 20 min	25

compounds and often are directly used for further reactions (see Section VI) rather than isolated as such. However, several salts of both classes of derivatives have been obtained and purified as crystalline materials (see selected examples in Table 3).[13,37-40] Likewise, *N*-sulfenyloxy-,[38] sulfiniloxy-,[39] and silyloxypyridinium salts[41] were obtained from pyridine 1-oxide and the appropriate reagents. Further derivatives, e.g., *N*-phosphoryl salts, have been characterized in solution[34] (Table 3).

III. DEOXYGENATION

One of the first reported observations about heterocyclic *N*-oxides is their enhanced stability towards deoxygenation with respect to aliphatic amine *N*-oxides. Thus, e.g., pyridine 1-oxide is stable to ferrous hydroxide, whereas aliphatic *N*-oxides are reduced under these conditions[43,44] (but other aromatic *N*-oxides are more reactive, see below). However, several reagents are available for the deoxygenation of heterocyclic *N*-oxides, and this process has been used for the identification of new *N*-oxides. Obtaining a heterocycle from the corresponding *N*-oxide is a significant preparative reaction in several cases:

1. Since the functionalization of the *N*-oxide is often easier than the corresponding reaction of the parent base, it may be convenient to go through the *N*-oxidation - reaction - *N*-deoxidation sequence (see examples in Sections IV and V).
2. In some cases the *N*-oxide is conveniently obtained through a cyclization reaction, and this offers a useful entry to the specific heterocycle (see examples in Chapter 3).
3. It may be useful to initially oxidize both the ring nitrogen(s) and other functionalities present in the molecule (e.g., a sulfur atom) and then to proceed to specific *N*-deoxygenation and preserve the other oxidized function (see an example in Section IIID).
4. In polyazaheterocycles, partial deoxidation of a polyoxide is in some cases a selective process.

In the present section, the available methods for deoxygenation are reviewed according to the reactive employed, and the cases of selectivity in *N*-polyoxides are discussed at the end. The mechanism involved ranges from electrophilic attack on the oxygen (e.g., with electrophilic derivatives of phosphorus and sulfur) to nucleophilic attack on the ring (e.g., with dithionite and mixed hydrides) (Scheme 7).

A. REACTION WITH PHOSPHORUS DERIVATIVES

Phosphorus (III) derivatives probably form the most consistently used class of reagents for the *N*-deoxygenation of heterocyclic derivatives. The reaction appears to involve an electrophilic attack at the oxygen atom, followed by cleavage of the N-O bond. This is indicated by the enhanced reactivity of electron-withdrawing substituted phosphorus derivatives [e.g., P(III) halides are far more reactive than phosphines],[45-46] as well as by the retarding effects of electron-withdrawing substituents on the rate of the (kinetically complex) deoxygenation of the heterocycle.[47]

The mechanism does not need to be univocal, however, as an example, a radical mechanism has been postulated for the peroxide accelerated *N*-deoxygenation of pyridine 1-oxide by triethylphosphite.[48]

Some typical procedures are reported in Table 4. As mentioned above, phosphines are less active; however, under drastic conditions (heating a triphenylphosphine-*N*-oxide mixture at 230°C or above), good results are also obtained in this case.[49] Furthermore, one can take advantage of the easy oxygen transfer to some metal derivatives in order to carry out the reaction under mild conditions in the presence of a catalytic amount of such products [e.g., Mo(IV) complexes].[50]

TABLE 3
Various Quaternary Derivatives $\geqslant$N$^+$-OY X$^-$ from the Corresponding N-Oxides

N-oxide	Y	X$^-$	Mp,°C	Yield %	Method of preparation	Ref.
Pyridine	Ac	ClO_4^-	125–127.5	79	Add ice-cold $HClO_4$ in Ac_2O to the N-oxide in Ac_2O/AcOH	13
	p-$NO_2C_6H_4$S	ClO_4^-	160–170		Treat with p-$NO_2C_6H_4$SCl, add $HClO_4$, wash with H_2O	38
	p-$NO_2C_6H_4$SO	Cl^-	98–100		Treat with p-$NO_2C_6H_4$SOCl, filter, wash with EtOH	39
	CF_3SO_2	$CF_3SO_3^-$	62–64	93	Treat with $(CF_3SO_2)_2O$ in CH_2Cl_2 at –20°C, wash with Et_2O	40
	Me_3Si	$CF_3SO_3^-$				41
Quinoline	Ac	ClO_4^-	45		Add ice-cold $HClO_4$ in Ac_2O to the N-oxide in Ac_2O/AcOH	13

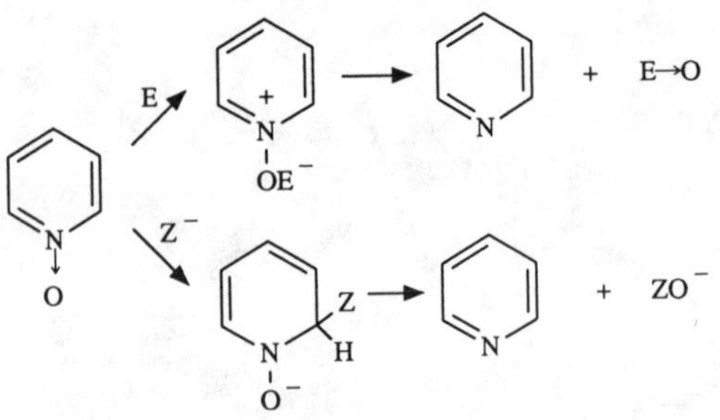

Scheme 7.

TABLE 4
Deoxygenation of Heterocyclic *N*-Oxides by Phosphorus Derivatives

N-oxide (substituent)	Reducing agent	Yield, %	Conditions	Ref.
Pyridine	PPh$_3$	90	Heat at 230°C for 35 min	49
	PPh$_3$	100	Heat at 40°C in benzene with PPh$_3$ and 0.02 mol(Et$_2$NCS$_2$)$_2$MoO	50
	P(OPh)$_3$	72	Heat in AcOEt on water bath for 4 h	55
	P(OEt)$_3$	70	48 h at r.t. in EtO(CH$_2$)$_2$O(CH$_2$)$_2$OEt[a]	48
	PCl$_3$	90	Heat in CHCl$_3$ on water bath for 30 min	53
	PCl$_3$	96	2 h at r.t. in AcOEt	53
	PBr$_3$	84	Add PBr$_3$ to the *N*-oxide in AcOEt at 0°C; heat 1 h on water bath	54
	P$_2$I$_4$	95	Add freshly prepared P$_2$I$_4$ to the *N*-oxide in CH$_2$Cl$_2$, reflux 10 min	71
	ArP(S)S$_2$P(S)Ar	65	1 h in benzene at 20°C	72
Quinoline	PPh$_3$	89	Heat at 270—283°C for 20 min	49
	P(OPh)$_3$	71	Heat in AcOEt on water bath for 4 h	55
	PCl$_3$	Quant.	Heat in CHCl$_3$ on water bath for 1 h	53
	PBr$_3$	76	Heat in AcOEt on water bath for 1 h	54
	NaH$_2$PO$_2$	83	Add aq NaH$_2$PO$_2$ to the *N*-oxide in THF in the presence of Pd/C and Na$_2$CO$_3$	73
Furazan (4,5-diphenyl)	P(*n*-Bu)$_3$	82	Heat in Et$_2$O at 35°C for 4 h	74
	P(OEt)$_3$	97	Heat at 150—160°C for 18 h	74
Benzofurazan(4,6-dimethyl)	PPh$_3$	65	Treat the furoxan in benzene with powdered PPh$_3$, filter[b]	68

[a]   In the presence of peroxides the reaction is faster.
[b]   The phosphine complex is obtained in this way; the free furazan is directly obtained in 24% yield when the benzofuroxan is refluxed with PPh$_3$ in xylene for 2 h.

The procedure used in most cases involves phosphorus trichloride or the corresponding bromide in chloroform or in other aprotic solvents. With these reagents, deoxygenation occurs on brief heating, or even at room temperature.[51-55] Many substituents, e.g., hydroxy, alkoxy, and amino, are unaffected by these reagents (but with 4-hydroxyquinoline 1-oxide, some 4-chloroquinoline is obtained, together with 4-hydroxyquinoline by treatment with PCl$_3$).[51,57] A sulfone also resists the action of PCl$_3$.[58a] However, a nitro group in α or in γ (but not in ß)[59,60] is frequently (but not necessarily)[56a] substituted by a chloro (or, respectively, a bromo) atom under these conditions.[61-64] *N*-Deoxygenation and substitution are subsequent reactions, as longer reactions times and higher temperatures favor the occurrence of the latter step, and it is thus generally possible to limit the reaction to deoxygenation.[52,65,66] Substitution is favored by (or requires) the presence of hydrogen halides. Thus, 3-methyl-4-nitropyridine 1-oxide yields only the corresponding pyridine when treated with PCl$_3$ in purified chloroform, even after heating, while reaction in ethanol-containing chloroform, and even better in HCl-saturated solvent, yields the corresponding 4-chloropyridine.[67]

4-Nitroquinoline. A solution of 4-nitroquinoline 1-oxide in 200 ml of chloroform is concentrated to one third of the original volume, and an ice-cold mixture of 25 g PBr$_3$ in 25 ml dehydrated chloroform is added, maintaining the temperature below 15°C. After 15 min, the mixture is poured on crushed ice (200 g), basified, and extracted with chloroform to yield, after work-up and distillation, 8.7 g (quantitative yield) of the desired product (boiling range 130—135°C at 5 mmHg).[52]

Noteworthy in the five-membered N-oxides series is the smooth deoxygenation of 4,6-dinitrobenzofuroxan to the corresponding furazan.[68]

Red phosphorus and iodine in boiling acetic acid[69] or phosphorus/hydrogen iodide[70] have been used for the deoxygenation of α-hydroxy (as well as α-thio) N-oxides. Phosphorus tetraiodide is a useful (and selective, Cl and NO$_2$ groups unaffected) deoxygenating agent.[71]

Other reagents that have been used include the sulfurated derivative **9**[72] and sodium hydrophosphite[73] (see Table 4).

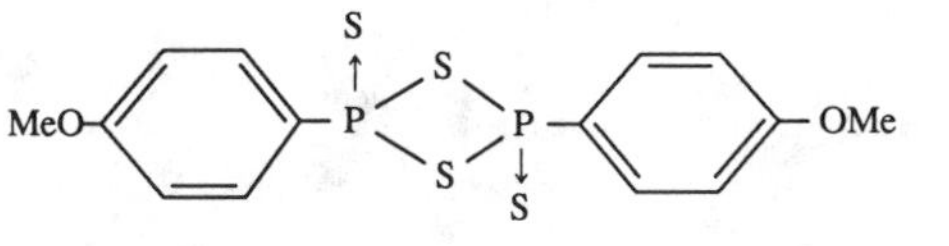

9

B. REACTION WITH SULFUR DERIVATIVES

It was early noted that heteroaromatic N-oxides are differentiated from their aliphatic counterparts by their resistance to reduction by sulfur dioxide or sodium sulfite at room temperature.[75] This difference is useful for selective deoxygenation (see Section IIIH). However, with a long reaction time[76,77] or at a higher temperature,[78] reduction by these reagents can be effected. As an example, bubbling SO$_2$ in boiling dioxane or water[78] or treating with a SO$_2$-tertiary amine complex in boiling tetrahydrofuran (a method that avoids the need of continuous saturation)[79] are useful preparative methods (furthermore, SO$_2$ is active onto the acyl adducts of N-oxides[80]). Under these conditions 4-nitropyridine 1-oxide reacts poorly, indicating the electrophilic nature of the reaction; however, 4-nitro-2,6-lutidine undergoes both deoxygenation and nitro group reduction on boiling with sulfite.[81] With some quinazoline 1-oxides, an important side process is rearrangement to 2-quinazolones.[77]

In keeping with the initial attack at the oxygen (see Scheme 7), the quaternary salts obtained from N-oxides and arenesulfenyl and sulfinyl chlorides[38,39,82] (but not those from arenesulfonyl chlorides, which react differently, see Section VIF2), as well as those from SOCl$_2$,[4] decompose on heating (apparently through a homolytic pathway) to yield the deoxygenated base. Sulfur monochloride reacts similarly.[83]

Aqueous sodium dithionite has been used with good results, at least with water-soluble N-oxides (e.g., quinoxalone, pteridinone, fervenulin derivatives).[84-90] The reaction with dithionite has been interpreted as a nucleophilic attack on the α-carbon, followed by two elimination steps.[90b] A nitro group is reduced with this reagent;[81] furthermore, with some 3-quinoxalinone 1-oxides over-reduction to tetrahydroquinoxalinones is observed.[84] 4-Pyrazolone 1,2-dioxides are selectively deoxygenated by Na$_2$S$_2$O$_4$ to 1,4-dihydroxypyrazoles (while complete reduction to 4-hydroxypyrazole is obtained with zinc in acetic acid)[91a] (Table 5).

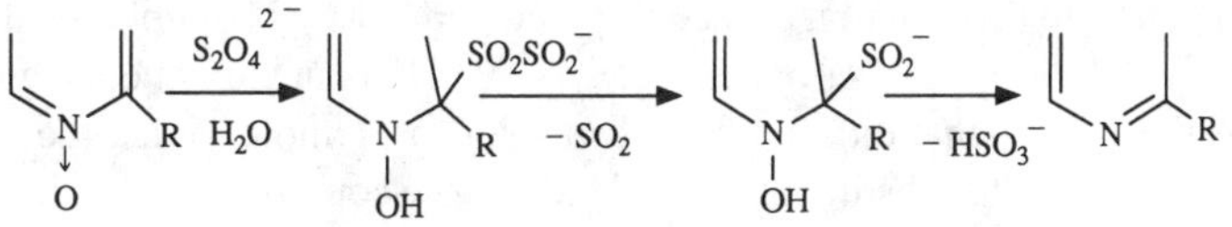

TABLE 5
Deoxygenation of Heterocyclic *N*-Oxides by Sulfur Derivatives

N-oxide (substituent)	Reducing agent	Yield	Conditions	Ref.
Pyridine	$SOCl_2$	92	Reflux in benzene	4
	SO_2	66	Pass a slow stream of SO_2 in refluxing dioxane	78
(5-Ethyl-2-methyl)	$NaHSO_3$	66	Reflux with 2 moles aq $NaHSO_3$ for 6 h	78
	$Et_3N.SO_2$	81	Add $Et_3N.SO_2$ to the THF soln. under N_2 reflux 2.5 h	79
	Me_2SO	94	Heat the *N*-oxide in DMSO with 6% H_2SO_4 at 190° C for 20 h	94
	$(nPrO)_2S$	96	Leave 3 h at r.t. in $CHCl_3$	93
	S_2Cl_2	80	Reflux in $CHCl_3$ for 1 h	83
	S	64	Add S to the *N*-oxide at 150°C over 100 min, heat at 135°C for 16 h	96
Quinoxaline (2-phenyl-3-benzoyl)	$Na_2S_2O_4$	92	Add dropwise aq $Na_2S_2O_4$ to the *N*-oxide in acidif. MeOH, precipitate with H_2O	86
Pyrazole[a] (3,5-diphenyl-4-oxo)	$Na_2S_2O_4$	91[b]	Treat with $Na_2S_2O_4$ in 25% EtOH for 2 h, precipitate with H_2O	91a
Benzimidazole (2,3-dimethyl)	SO_2		Pass SO_2 for 30 min in the $CHCl_3$ solution of the *N*-oxide, leave 3 h at r.t.	104

[a]   1,4-Dioxide.

[b]   3,5-Diphenyl 1,4-dihydroxypyrazole.

Sulfolenes (2,5-dihydrothiophene 1,1-dioxides)[91b] and trans-stilbene episulfoxide (via *in-situ* generated SO)[92] are useful deoxygenating agents. Dialkylsulfoxylates react with heterocyclic *N*-oxides in the same way as trialkyl phosphites (initial electrophilic attack at the oxygen atom) and yield the corresponding bases and sulfites.[93] With sulfoxides the reaction involves a similar intermediate and following cleavage of the C-S bond yields sulfonic acids (via the corresponding sulfinic derivatives), besides the deoxygenated base. The reaction is catalyzed by acids.[94]

Heating with sulfur deoxygenates pyridine 1-oxides,[95-97] and 4-nitropyridine is obtained from the corresponding 1-oxide when heated with sulfur in a nitrating mixture.[98] Other reagents that have been used are hydrogen sulfide,[99] sodium hydrogen telluride,[100] mercaptans[96] [the reduction occurs easily in the presence of catalytic amounts of the cluster $Fe_4S_4(SR)_4{}^{2-}$][101] thiophenol, thiourea,[96] carbon disulfide,[102] and diaryl disulfides.[103]

C. CATALYTIC HYDROGENATION

The *N*-oxide function is sensitive to catalytic reduction, and hydrogenation at room temperature and pressure (solvent: alcohols or acetic acid) in the presence of Raney[105-111,115] or Urushibara[112] nickel or (much more slowly) of palladium on charcoal[113] takes place in minutes or hours, and leads to the corresponding base in a quantitative or, at any rate, a high yield; much more drastic conditions are required for the hydrogenation of the heterocyclic nucleus, or of fused aromatic rings.[114] Substituents in the α position slow down the reaction, and some derivatives resist hydrogenation under these conditions due to steric hindrance, as with 2,6-lutidine 1-oxide,[105] or due to formation of an intramolecular hydrogen bond, as with 2-hydroxy and 2-carboxypyridine 1-oxide[105] (however, 2-hydroxyquinoline 1-oxide is deoxygenated over nickel[109] and hypoxanthine 1-oxide over platinum[116]).

Nickel catalysts not only influence the rate, but also the selectivity of the hydrogenation. Indeed, *N*-oxides of pyridine, quinoline, and diazine derivatives have been shown to undergo deoxygenation before hydrogenation of a conjugated C=C or N=N double bond,[111,117] dehalogenation,[107,108,112] and hydrogenolysis of benzyloxy[107,108,111,112] or acyloxymethyl groups.[111] Thus, interruption of the hydrogenation after the absorption of 1 mole yields the corresponding heterocycles, conserving the substituent intact. Under these conditions, a pyridinium ring is not reduced. Thus, with compound **10** absorption of 1 mol hydrogen leads essentially to deoxygenation[118] (Tables 6 and 7).

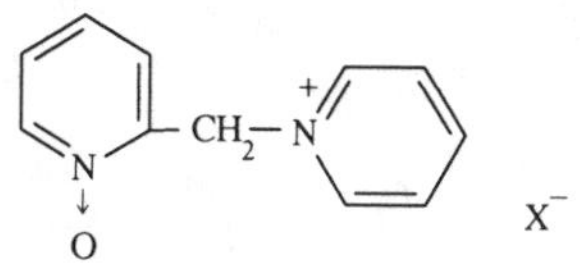

10

Furthermore, Raney nickel is also active with sulfur-containing N-oxides[120,121] (with some exceptions, e.g., with 4-thioethoxypyridine 1-oxide deoxygenation stops at ca. 50% conversion and the sodium salt of 4-mercaptoquinoline 1-oxide yields a mixture of quinoline and the disulfide 1,1'-dioxide under these conditions).[120] Although isolated failures have been reported,[119] the activity and selectivity of Raney nickel are general and some other catalysts, e.g., Raney cobalt and copper and copper chromite, appear to behave similarly; this has been rationalized on the basis of the affinity of the metals for oxygen.[111]

On the other hand, one can take advantage of the slow hydrogenation of the $\geqslant$N→O group in the presence of Pd/C for effecting the selective reduction of the substituent.[113] Indeed, saturation of a double bond, hydrogenolysis of C-O bonds, and dehalogenation can often be carried out under this condition, leaving the N-oxide unchanged (see Section XIC1). An exception is the azoxy group, which is reduced to azo before the N-oxide function with both types of catalysts.[117] As for the nitro group, this is usually reduced to an amino, while the N-oxide is deoxygenated,[56a,59,107,108,122,123], though again the use of Pd/C may lead to a partial reduction of the substituent without affecting the $\geqslant$N→O group.[124,125] Acids favor complete reduction, and hydrogenation has been often carried out in acetic acid/acetic anhydride[119] (this may lead to the amide instead of the amine).[123]

D. REACTION WITH DISSOLVING METALS AND LOW-VALENCY METAL DERIVATIVES

The reactions with dissolving metals, in particular with iron, zinc, or tin in acidic or basic medium,[128-135] or with sodium and alcohols,[136] as well as the reaction with Fe(II)[137] or Sn(II)[138] salts, have been largely used for the deoxygenation of N-oxides. Strong reducing agents, such as nickel-aluminum alloy in NaOH, completely reduce pyridine 1-oxide to piperidine.[139] Nitro groups are usually reduced to amino under these conditions, and partial reduction of the nitro group may precede that of the N-oxide. Thus azo or hydrazo derivatives that do or do not conserve the N-oxide function have been obtained in some cases.[122,140,141] Noteworthy is the resistance of sulfones;[58,142] thus, though in some cases it is possible to selectively oxidize a sulfide to sulfone with no significant N-oxidation, in the case of the cyclic pyridine derivative **11**, it is more expedient to oxidize to the disulfone N,N'-dioxide and then to carry out a selective N-deoxygenation[142] (Table 8).

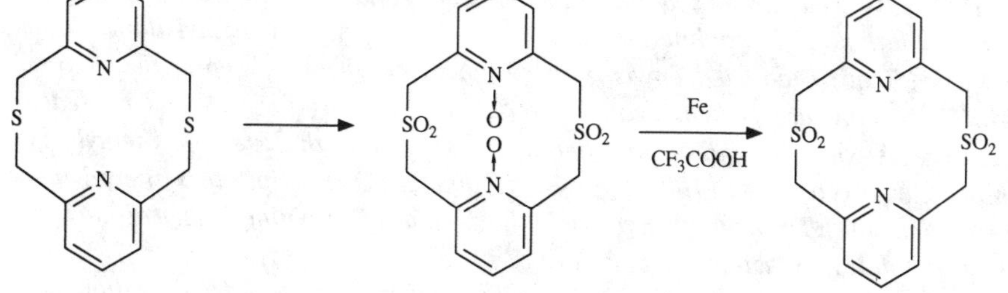

11

TABLE 6
Deoxygenation of *N*-Oxides by Hydrogenation in the Presence of Raney Nickel at Ordinary Temperature and Pressure

N-oxide	Substituent	Yield	Ref.
Pyridine	4-NO$_2$	90[a]	107
Quinoline		82	108
	4-NO$_2$	79[a]	108
	2-OH	83	109
Cinnoline[b]	4-NO$_2$	75[a]	126
Purine[c]	6-NH$_2$-7-Me	89	127

[a]   The 4-NH$_2$ derivative is formed.
[b]   1-Oxide.
[c]   3-Oxiole.

TABLE 7
Selective Deoxygenation of Substituted Heterocyclic *N*-Oxides by Hydrogenation in the Presence of Raney Nickel (After Absorption of One Molecule of Hydrogen)

N-oxide	Unaffected substituent	Yield, %	Ref.
Pyridine	2-styryl	Main product	117
	4,4'-azo	22	117
	4-PhCH$_2$O	92	107
	4-Cl	88	107
	4-PhS	76	120
	4-PhSO$_2$	91	120
Quinoline	4-PhCH$_2$O	81	108
	4-Cl	90	108

With cyclic nitrones, reduction may proceed beyond deoxygenation, e.g., 2,4,4-trimethylpyrroline 1-oxide yields the pyrroline with Zn/AcOH, but the main product with Sn/HCl is the pyrrolidine,[135] and 3,4-dihydroisoquinoline 2-oxide is reduced to 1,2,3,4-tetrahydroisoquinoline by Na/BuOH.[143]

Heating with metals,[144] cocondensation with metal atoms at low temperature,[145a] and reaction with metal carbonyls[145b] have also been shown to deoxygenate *N*-oxides. Several useful preparative methods that are effective under mild conditions have recently been reported and involve reaction with metal ions of low valency, typically Mo(III),[146] Cr(II),[147] and Ti(III),[148-150] as well as with the compositions obtained by hydride reduction of TiCl$_4$[151-154] and WCl$_6$,[155] or reaction with metals under milder conditions, e.g., with zinc in the presence of NaI and Me$_3$SiCl.[156] Deoxygenation with reducing salts is also useful from the analytical point of view.[157,158]

2-Picoline. To a mixture of sodium iodide (90 mmol) and zinc dust (60 mmol) in acetonitrile (30 ml), a solution of 2-picoline 1-oxide (30 mmol) and chlorotrimethylsilane (90 mmol) in acetonitrile (30 ml) is added dropwise with stirring over 1 h while keeping the temperature at 35—40°C. The mixture is heated at 55—60°C with stirring for 1 h and cooled to room temperature. A white precipitate is filtered and washed with 3× 30 ml diethyl ether. The combined filtrate is poured in 100 ml 5% NaOH, and the new precipitate is filtered and washed with 4 ×30 ml ether. The organic layer yields 88% of pure 2-picoline. The procedure is similar for other pyridines, pyrazines, and quinoline.[156]

Quinoline. 0.46 g (13 mmol) of lithium aluminum hydride are added to the yellow suspension obtained by dropping 2 ml (17 mmol) TiCl$_4$ in 50 ml anhydrous THF under nitrogen at 0°C, and *the mixture is stirred 15 min at room temperature, obtaining a black slurry. Quinoline 1-oxide (20 mmol) is added at 0°C, and stirring is continued for 15 min at room temperature. Decomposition in 50 ml water-50 ml 25% aqueous ammonium hydroxide and extraction with 5 × 50 ml diethyl ether yields the desired product (95%). The procedure is similar for various pyridines, acridine, and other heterocyclics.[152]*

5-Aminoisoquinoline. *To 193 mg 5-nitroisoquinoline 2-oxide in 6 ml AcOH-H$_2$O 1:1 5.1 ml (8 mEq) aqueous TiCl$_3$ are added in a single amount. The mixture is stirred for 7 min at 18°C and decomposes to yield 143.6 mg (98%) of the desired product.[149]*

E. REACTION WITH HYDRIDES

Heterocyclic *N*-oxides react with diborane and alkylboranes at room temperature, while the corresponding bases are usually inert under these conditions.[160] Determinations of the hydride consumed and the hydrogen evolved show that in pyridine 1-oxide addition across the nitrone function takes place and the adduct either decomposes to yield the deoxygenated heterocycle and hydrogen (this is the case with disiamylborane) or undergoes further addition to finally yield a hexahydro pyridine (with diborane or thexylborane; with 9-borabicyclo[3.3.1]nonane both paths are followed)[161,162] (Table 9).

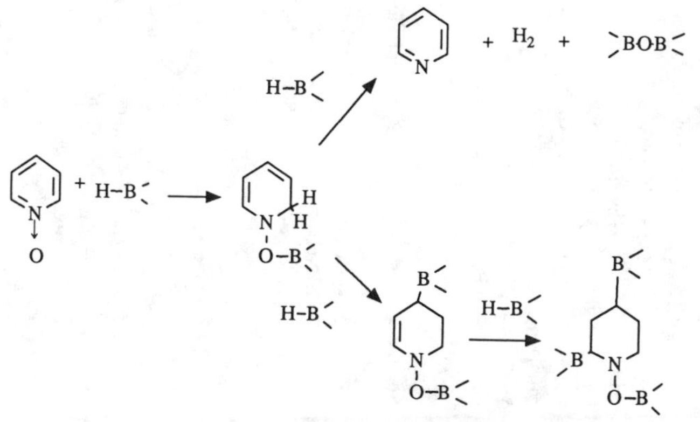

Diisobutylaluminum hydride leads to the 1,2-dihydro level.[163] Tris(*n*-butyl)tin hydride (in the presence of the radical initiator AIBN) deoxygenates efficiently the *N*-oxides via oxygen transfer to the R$_3$Sn· radical.[164]

Pyridine 1-oxide does not react with sodium boron hydride[165] (though some substituted derivatives do),[166] but quinoline 1-oxide is deoxygenated (with some ring reduction if the reaction is carried out in protic media).[167] The mechanism involves again nucleophilic addition on the ring, as shown by deuterium loss when starting form the 2-*d* derivative. (Partial) dehalogenation is observed under these conditions,[167] while reduction of a nitro group appears to precede attack on the ⩾N→O function.[126] Several mixed borohydrides have been tested with pyridine 1-oxide, the results varying from no reaction [e.g., with K(*i*-PrO)$_3$BH][168] to addition of three molecules in a similar manner as with B$_2$H$_6$ (e.g., with Li 9-boratabicyclo[3.3.1]nonane),[169] while Na$_2$BH$_2$S$_3$ deoxygenates selectively.[165] In contrast to the above-discussed hydride transfer mechanism, deoxygenation of crown ether **12** and of the weakly complexing analog **13** by potassium tris(*sec*-butyl)borohydride takes place via single electron transfer after complexation.[170]

TABLE 8
Deoxygenation with Metals, Dissolving Metals, and Metal Ions of Lower Valency

N-oxide	Substituent	Reagent	Conditions	Yield, %	Modified substituent	Ref.
Pyridine	(2-Me-4-NO$_2$)	Fe/AcOH	Heat 1 h at 100°C with Fe powder in AcOH	95	(4-NH$_2$)	128
	(3-Br-4-NO$_2$)	Fe/AcOH	Heat 1 h at 100°C with Fe powder in AcOH	80	(4-NH$_2$)	129
	(2-CN-4-NO$_2$)	Fe(II)	Distil at 300°C from Fe oxalate/Pb granulated	64		137b
		Fe(II)	Reflux 1 h with FeSO$_4$/NH$_4$OH	63	(4-NH$_2$-2-CONH$_2$)	137a
	(4-Me)	Fe$_3$(CO)$_{12}$	Stir overnight with Al$_2$O$_3$ adsorbed Fe$_3$(CO)$_{12}$ in hexane	86		145b
		Ni/Al	Add Ni/Al alloy to a 0.5 M NaOH soln. of the N-oxide, stir 18 h	59[a]		139
	(2,6-Cl$_2$)	Mo(III)	Add H$_2$O and Zn dust to MoCl$_5$ in THF, add the N-oxide and reflux 45 min	50		146
	(2,6-Me$_2$)	Cr	Condense with Cr at 27 K	90		145a
		Cr(II)	Treat with Cr(II) in HClO$_4$			147
		Ti	Add the N-oxide to the slurry obtained from TiCl$_4$-LiAlH$_4$ (or NaBH$_4$)	90		152
	(2,6-Me$_2$-4-NO$_2$)	Sn/HCl	Add Sn to the HCl soln. of the N-oxide at 100°C	80	(4-NH$_2$)	141b
	(4-MeO-3-Me)	Mg/NH$_4$OAc	Heat with Mg/NH$_4$OAc in MeOH			133
		Zn/H$_2$SO$_4$	Add Zn to the N-oxide in 2N H$_2$SO$_4$ at 60-70°C, leave 8 h	87		134
	(4-NO$_2$)	Zn/NaOH	Heat 3 h on water bath with zinc dust in 3N NaOH	[e]	(-N=N-)	140
	(2-Me)	Zn/NaI/Me$_3$SiCl	Add the N-oxide and Me$_3$SiCl in MeCN to NaI-Zn in MeCN at 35-40°C, heat at 55-60°C	88		156
	Hg(SiMe$_3$)$_2$		Add 1 mole of Hg(SiMe$_3$)$_2$ to the N-oxide in THF at 0°C			159
Isoquinoline	(5-NO$_2$)	Ti(III)	Add TiCl$_3$ to the N-oxide in AcOH/H$_2$O, stir 5 min	98	(5-NH$_2$)	149
Benzo[c]cinnoline[b]	(4-Br)	Sn(II)	Heat on water bath with SnCl$_2$/HCl in EtOH for 30 min	80		138
	(3-NH$_2$)	Na/Hg	Treat with Na amalgam in ethanol and boil for a few minutes	68		136
Phenazine	(1,6-Ph$_2$)	Fe	Heat at 300°C with Fe powder under CO$_2$	70		144c
1,2,4-Benzotriazine	(3-Cl)[c]	Zn/NH$_4$Cl	Stir a suspension with Zn dust/aq NH$_4$Cl at r.t. for 17 h	38		130
2 7,9,10-Tetra-azaphenanthrene[d]	(2-Ph)	Fe	Heat at 250°C with Fe dust	Quant.		144a
Benzimidazole		Zn	Heat with Zn dust at 240°C for 2 h	58		144b

a Complete reduction to piperidine.
b 6-Oxide.
c 1-Oxide.
d 9-Oxide.
e Mixture of 4,4'-azopyridine and its 1-oxide obtained.

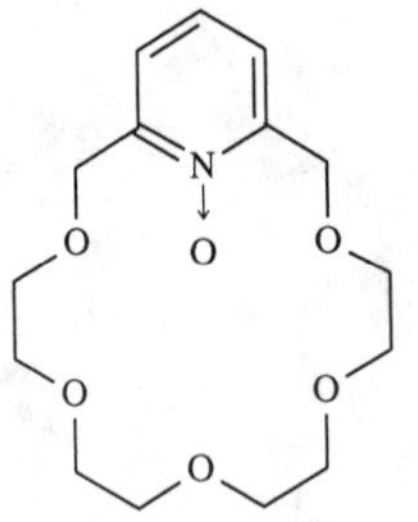

12 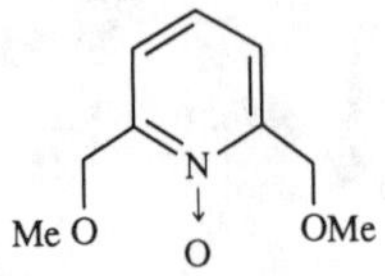   **13**

The stronger nucleophile LiAlH$_4$ has been found useful for the deoxygenation of various cinnoline 1-oxides,[171] though in other cases, e.g., with isoquinoline and phenanthridine *N*-oxides,[172] the reaction proceeds further to the dihydro level. With cyclic nitrones, addition to yield the corresponding hydroxylamine is usually observed.[135]

F. OTHER REAGENTS FOR THE DEOXYGENATION

1. Reaction with Disilanes, Distannanes, and Disulfides

Compounds containing weak X-X bonds (X=Si,Sn,S) include some of the best and the mildest reagents for the deoxygenation of *N*-oxides. This is the case for hexachlorodisilane.[173-175] Hexamethyldisilane is inactive towards pyridine 1-oxide, but the reaction occurs in the presence of fluorides[176] or butyl lithium,[177] the effective reagent being in both cases the anion Me$_3$Si$^-$. Hexabutyldistannane also does not react, but tetrabutyldichlorostannane reacts easily,[178] as do diaryldisulfides (in the case of unsymmetrical derivatives, oxygen transfer is to the sulfur atom, far from the electron-withdrawing substituted ring).[103]

Pyridine. To a solution of hexamethyldisilane (0.42 g, 2.9 mmol) in HMPT (2 ml) is added methyllithium (1.4 M in ether, 2 ml, 2.8 mmol) at 0°C. After 20 min stirring, pyridine 1-oxide (0.19 g, 2 mmol) in HMPT (0.95 ml) is added and the mixture stirred at room temperature for 8 h and then quenched with water and extracted with ether. The yield is 0.136 g, 86%.[177]

2. Reaction with Nitrogen Derivatives

Studies on the nucleophilic substitution of halogeno- and nitro-substituted *N*-oxides by ammonia, amines, and other nitrogen derivatives have shown that deoxygenation is often a bypath (see Section VA1). Some useful preparative deoxygenations by hydrazine and phenylhydrazine have been reported.[126,179] Sodium azide (at the best in carboxylic acids) deoxygenates benzofuroxans to the corresponding benzofurazans without affecting nitro substituents.[180]

3. Reaction with Carbon Derivatives

The main interest of this group of reactions lies in the oxidation and dehydrogenation of organic compounds by heterocyclic *N*-oxides. This may be a useful preparative method and is discussed in Chapter 5. From the point of view of the *N*-oxide reduction, one should mention the reportedly efficient reaction obtained on heating with alcohols and sodium hydroxide.[181] Interesting is the catalyzed deoxygenation by CO/H$_2$O.[182]

G. OTHER METHODS FOR DEOXYGENATION

1. Thermal Deoxygenation

Heterocyclic *N*-oxides are generally quite stable from the thermal point of view. The possibility that 2,2-bipyridine dioxide or its derivatives cleave thermally to yield bipyridine and singlet oxygen has been discounted.[183] However, for various *N*-oxides it has been reported that some deoxygenation takes place at elevated temperatures, and this must be taken into account,

TABLE 9
Deoxygenation of *N*-Oxides with Various Reagents

N-oxide	Substituent	Reducing agent	Yield	Method	Ref.
Pyridine	(4-Me)	Si_2Cl_6	60	Treat with Si_2Cl_6 in $CHCl_3$, leave 1 h at 25°C	173
		Si_2Me_6	90	Add 1.2 eqv Si_2Me_6 to the *N*-oxide in THF with 0.05 eqv Bu_4NF, the temperature rises at 35°C	176
Quinoxaline[a]	(2,3-Me$_2$)	Si_2Cl_6	80	Add Si_2Cl_6 to the *N*-oxide in $CHCl_3$ at 0–5°C, stir 3–4 h at room temperature	174
Pyrazole	(4,5-Me$_2$-2,2-Ph$_2$)	Si_2Cl_6	80	Add Si_2Cl_6 to the *N*-oxide in $CHCl_3$ at 25°C, stir 30 min	174
Pyridine		$(Bu_2ClSn)_2$	Quant.	Reflux 1 h in THF in a Schlenck tube	175
		p-$NO_2C_6H_4SSPh$	70[b]	Heat at 150–180°C in bromobenzene for 7 h	178
Quinoline	(3-Me-4-NO$_2$)	$PhNHNH_2$	Quant.[d]	Reflux in EtOH for 5 h	103
Benzo[c]cinnoline[c]		N_2H_4		Stir with conc N_2H_4,H_2O in EtOH until the *N*-oxide dissolves (1–3 d)	125
Benzofurazan	(4-NO$_2$)	NaN_3	63	Heat with 3 equiv. NaN_3 in AcOH at 140–150°C	179
					180

[a] 1,4-Dioxide.
[b] The 4-NH$_2$ derivative is formed.
[c] 5,6-Dioxide.
[d] 5-Oxide.

e.g., when analyzing the *N*-oxides by gas chromatography or mass spectrometry. In isolated cases, quantitative thermal deoxygenation is observed, e.g., with 2-nitrophenazine 10-oxide on subliming[184] or with benzo[*c*]cinnoline 5-oxide at 850°C.[185]

Easier thermal deoxygenations involve different pathways, as is the case with 2-(2'-hydroxyphenyl)pyridine 1-oxide (and, by inference, with the related alkaloid orellanine), where 1,5-oxygen shift from the *N*-hydroxy tautomer leads to the hydroperoxide.[186]

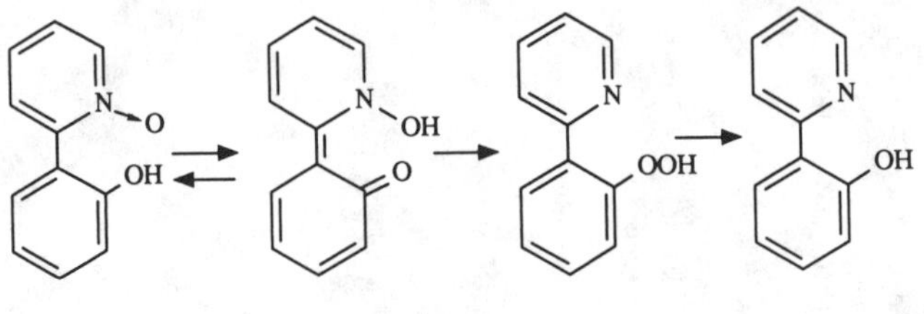

Furthermore, it must be taken into account that deoxygenation may be an important process under "oxidative" conditions (e.g., some carbostyril is formed from its *N*-hydroxy derivative on treating with peracetic acid,[187] and other deoxygenations when treating with peroxides have been reported[183,188]) or in attempted electrophilic reactions under severe conditions, e.g., in the case of pyridine oxide with a nitrating mixture at 240°C (and more easily with nitrosylsulfuric acid)[188] and with oleum above 270°C.[189]

2. Electrochemical Deoxygenation

The first (often reversible) reduction wave observed with heterocyclic *N*-oxides (see Chapter 2) corresponds to the formation of the radical anion, which has been characterized by epr and UV in a number of cases. At least for some derivatives, a second wave is discernible and the irreversible reaction occurring at this potential is *N*-deoxygenation. Thus, electrolysis at controlled potential may be useful preparatively, provided that further reduction of the heterocycle ring is avoided (this also depends on the proton availability in the medium)[192-197] (Table 10).

3. Other Methods

Photochemical deoxygenation is discussed in Section XB. Enzymatic deoxygenation has been studied for some substrates,[198] and particularly for carcinogenic derivatives such as 4-nitroquinoline 1-oxide (where reduction of the nitro group precedes that of the ≥N→O function).[199]

H. SELECTIVE DEOXYGENATION

Partial deoxygenation of *N*-polyoxides may be a selective and synthetically useful process. A first class of reactions exploits the already mentioned easier deoxygenation of aliphatic *N*-oxides in comparison with their aromatic counterpart with reagents such as SO_2[200-203] and CS_2[204] a process that has been used for the preparation of aromatic *N*-oxides of alkaloids from the easily obtained polyoxides.

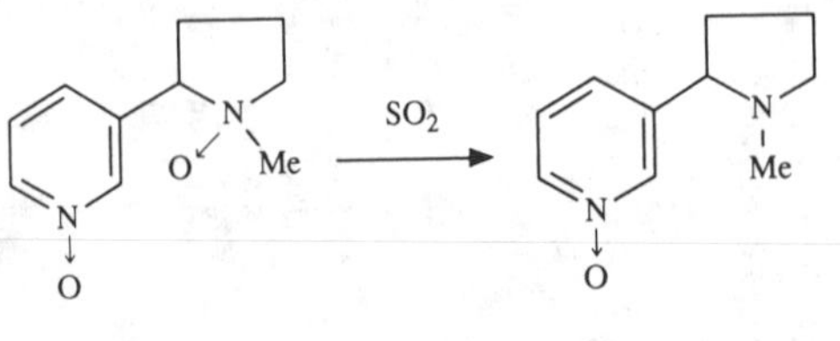

TABLE 10

Deoxygenation by Cathodic Reduction at Controlled Potential

N-oxide	Substituent	Yield	Conditions	Ref.
Pyridine		81	In MeOH at 65°C, Hg cathode	195
Quinoline	(2-Ph)		In DMF, Hg cathode	194
	(2-CN)		In DMF, Hg cathode	193
Quinoxaline[a]	(2,3-Me$_2$)		In DMF, Hg cathode	192
Purine[b]	(6-NH$_2$)	Good	In 0.1 N HClO$_4$, Pt cathode	196
Benzimidazole	(2-CN)	69	In 1 M HCl in MeOH, Hg cathode	197

[a] 1,4-Dioxide.
[b] 1-Oxide.

Another class makes recourse to the different reactivity of aromatic $\geqslant$N$\rightarrow$O functions. Thus, the two rings in 2,3′-diquinoline are sufficiently different to allow the preparation of the 1-oxide from the 1,1′-dioxide on treatment with a mild reagent like CS_2.[102]

The most thoroughly investigated molecules among polyazines are, understandably, quinoxaline 1,4-dioxides. Thus it has been observed that 3-methylquinoxaline 1,4-dioxides carrying an electron-withdrawing substituent in 2 (COOMe, COMe, CF_3, CONHR) yield the 4-oxide with both $P(OMe)_3$[205] and $Na_2S_2O_4$[87] (PCl_3 yields a mixture, with the 1-oxides as the main products).[205] The corresponding 3-unsubstituted amides again yield the 4-oxides with $Na_2S_2O_4$, but yield the 1-oxides with both $P(OMe)_3$[87] and PCl_3,[206] possibly due to the greater importance of the hydrogen bond between the $N(1)$-oxide and the amide in this case. 2-Phenylquinoxaline 1,4-dioxide yields the 1-oxide with SO_2 in methanol.[115] Selective deoxygenation of phenazine[207] and benzophenazine[208] dioxides on prolonged treatment with peracids has been reported.

IV. ELECTROPHILIC SUBSTITUTION

Substitution of the more electronegative nitrogen atom for a CH group in an aromatic molecule obviously decreases the reactivity toward electrophiles. Since N-oxidation adds a further electronegative atom, the reactivity would be expected to drop again. Actually, the situation is more complex, and three fundamental situations arise in the presence of the electrophile, leading to different end products.

1. The oxygen atom is protonated or complexed by the electrophile. The $\geqslant$N$^+$-OH (or $\geqslant$N$^+$-OE) group is deactivating just as the $\geqslant$N$^+$-H (or $\geqslant$N$^+$-E) group is deactivating (actually slightly more so) and *meta*-orienting. The N-oxide then reacts in the same manner as the corresponding nonoxidized heterocycle. This is what is observed in sulfonation.
2. The reaction with the electrophile involves the nonprotonated species present, though in a low amount, in equilibrium. The $\geqslant$N$\rightarrow$O group is again deactivating, but less so than the protonated group, and is *ortho-para*-orienting. This is what is observed for nitration in mixed acids under conditions compromising between the generation of sufficient NO_2^+ and permanence of sufficient free base, in bromination and in proton exchange at low acidity. As originally postulated by Ochiai, the orientation can be rationalized, considering the contribution of mesomeric formulae such as **1** (a reasoning similar to the usual rationalization of the halogenobenzenes reactivity; Scheme 8). On the basis of the calculated electron densities, the reactivity order of pyridine 1-oxide is position $2 > 4 \gg 3$. This is what is observed in deuteration, but nitration and bromination of pyridine and other azine N-oxides occurs preferentially at the γ position. This is probably an indication that electron transfer to the strong acceptor NO_2^+ (or Br_3^+) plays a determining role, as happens in the nitration of some aromatics.
3. The first reaction forms a σ bond with the oxygen, and the addition of a nucleophile leads to a dihydro heterocycle, usually a N-vinyl-hydroxylamine derivative, which in turn undergoes attack in β. Final elimination leads to a product of formal β-substitution and deoxygenation. This is the case for nitration or bromination in acylating media.

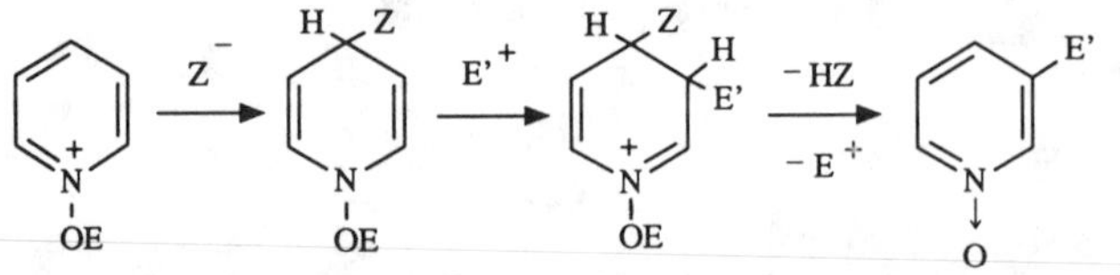

There is no reason to expect a corresponding activation of five-membered heterocyclic N-oxides, since in this case no stabilizing mesomeric formula such as **1** is possible for the

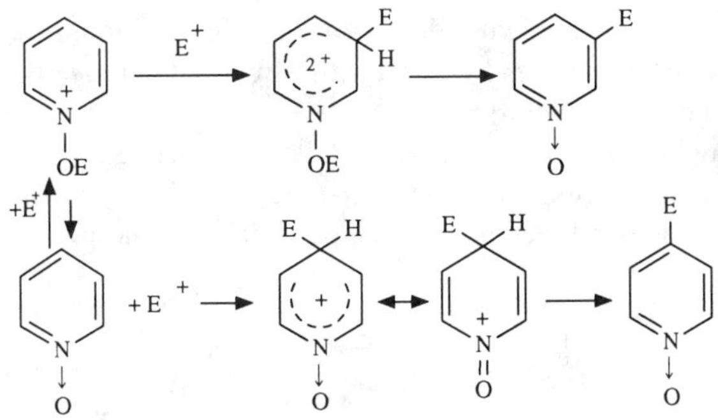

Scheme 8.

intermediate. However, in some cases a certain degree of activation has been noticed (e.g., in the bromination of triazole 1-oxides, see below).

A. ACID-CATALYZED H/D EXCHANGE

The deactivating and *meta*-orienting effect of the protonated *N*-oxide group towards electrophilic substitution, as well as the less deactivating (and *ortho-para* orienting) effect of the unprotonated group, are apparent in the course of the H/D exchange. Thus in pyridine, 1-oxide exchange in neutral D_2O solution is observed at positions 2 and 6 after 3 h at 180°C, but no exchange takes place after 2 h in 96% D_2O at 220°C.[209] Likewise, 3,5- and 2,6-dimethylpyridine 1-oxide are deuterated at the 2, 4, and 6 positions, and, respectively, at the 4 position via the free base (in the latter case at sufficiently low acidities), while 2,4,6-trimethyl- and 2,6-dimethylpyridine 1-oxide (the latter at high acidity) are deuterated at positions 3,5 through the conjugate acid.[210] Comparison of the exchange rate of 2,4,6-trimethylpyridine and its 1-oxide shows that the $\geqslant N^+$-OH group is slightly more deactivating than the $\geqslant N^+$-H group toward H exchange in position 3.[211,212]

The acid-catalyzed H/D exchange in quinoline and isoquinoline *N*-oxides takes place preferentially on the carbocyclic ring and involves the protonated species.[213,214] A study of quinoline 1-oxide in various D_2SO_4-D_2O mixtures and temperatures ranging from 150 to 250°C shows that the order of substitution is 8 > 5,6 > 7 > 3.[214] Conditions for preparative deuteration (and tritiation) of quinoline 1-oxide and some derivatives, including 4-nitroquinoline 1-oxide, have been determined,[215] as have been conditions for selective preparative deuteration in the α, 6, and 4,6 positions of 2-methylpyridine 1-oxide.[216]

When electron-donating substituents are present, the H/D exchange is easier and takes place either on the neutral or on the protonated species, depending on the substituents and on the conditions.[212,217-220] Also in the case of hydroxy derivatives, the $\geqslant N^+$-OH group has been shown to be somewhat more deactivating than the $\geqslant N^+$-H group.[218]

B. NITRATION IN MIXED ACID

Nitration is the most studied reaction among electrophilic substitution of *N*-oxides and is preparatively interesting, since it takes place under less stringent conditions and has a different orientation than with the nonoxidized heterocycles. Pyridine 1-oxide yields the 4-nitro derivative (65—95%) when treated with fuming H_2NO_3 in H_2SO_4 at 90°C. A small amount of 2-nitropyridine (1.5—4%) is the byproduct. The reaction was first reported by Ochiai[221] and later, independently, by den Hertog.[129] Detailed kinetic studies showed that nitration occurs on the free base, and the substitution of a $\geqslant N \rightarrow O$ for a $\geqslant CH$ group in a benzene ring reduces the rate of reaction by a factor of ca. 10^{-4} [222,223] (alternatively, formation of a complex may play a role).[266] Nitration under these conditions can be conveniently carried out on the crude *N*-oxide obtained by heating pyridine or its derivatives with H_2O_2/AcOH[224,225] and has been patented several times.[226-228]

Alkyl- and halogeno-substituted pyridine 1-oxides with position 4 free are nitrated in that position,[229-231] but steric hindrance is observed with the 3-*t*-butyl derivative, which is nitrated in α (with concomitant deoxygenation).[232]

Electron-withdrawing substituents may inhibit the reaction (e.g., 3-CN),[231] but picolinic acid is nitrated in 4.[233] 4-Alkyl pyridine 1-oxides resist nitration.[229,234] An interesting case is that of *N*-oxide **14**, which is easily nitrated in a vicinal position[235] (Table 11).

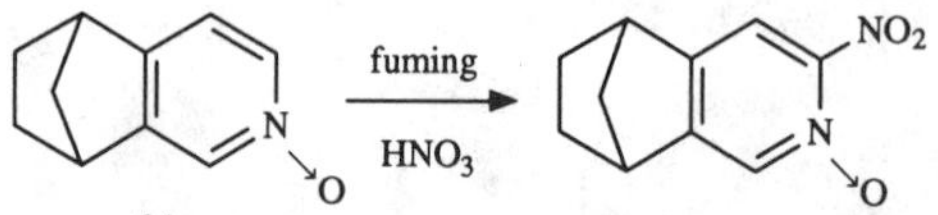

The directing effect of the N-oxide balances or overcomes that of the alkoxy group. Thus, both 2- and 3-alkoxypyridine 1-oxides are nitrated in 4,[128,236,237] unlike the corresponding pyridines; however, some substituted 3-alkoxypyridine 1-oxides are nitrated in 2 or 6.[238-241] Orientation by the hydroxyl group takes over and leads to the nitration in 2 of 3-hydroxypyridine 1-oxide [56b] (in 4 in the 2-methyl homologous and other α-substituted derivatives[242,243]), in 3,5 of the 4-hydroxy isomer,[244] and in 5 of *N*-hydroxy-2-pyridone.[245] Similarly, 2- and 4-dimethyl-amino- (as well as 2-methylamino-) pyridine 1-oxides are nitrated in *o*- and *p*- to the substituent.[246-248]

The deactivating effect of the *N*-oxide group is apparent in the nitration of the phenyl, rather than of the heterocyclic, ring in phenyl (as well as benzyl) pyridine 1-oxides.[20,249,250] The reaction of the 2-phenyl derivative has been shown to involve the protonated form (the protonated pyridine oxide, considered as a substituent, decreases the reactivity of the benzene ring and orients in *meta*[250]). On the other hand, the activation of the pyridine ring caused by *N*-oxidation is shown by the results from nitration of bipyridine mono-*N*-oxides, which leads to substitution at position 4 of the oxidized ring, provided, of course, that this position is free.[251,252]

The orientation in the benzopyridine *N*-oxides depends on their structure and the conditions. Treatment of quinoline 1-oxide with mixed acids at room temperature or below yields a mixture of 5- and 8-nitroquinoline 1-oxide, but at 70°C or higher a good yield of the 4-nitro derivative is obtained;[253-255] best conditions for nitration in 4 are a temperature of 70—80°C and the use of 80—85%, rather than 96%, sulfuric acid.[256] Analogous to the pyridine case, nitration in 5,8 involves the protonated form, and attack in 4 involves the free base. In contrast, isoquinoline 2-oxide is nitrated in 5[257] via the conjugate acid,[222] though under different conditions nitration occurs in 5, 6, and 8.[258] Acridine 10-oxide reacts already at 0°C to yield the 9-nitro derivative,[64] reasonably through the free base.

Pyrimidine[259] and pyrazine[260] 1-oxides are resistant to nitration, but pyridazine mono- and di-*N*-oxides are nitrated in γ to the ≥N→O function in medium to good yields.[261-264]

Cinnoline 1-oxide behaves similarly to quinoline 1-oxide, being nitrated in 4, with a small amount of the 5-nitro derivative also found at low temperature.[126] Cinnoline 2-oxide, like isoquinoline 2-oxide, is nitrated only on the carbocyclic ring; the 6-nitro derivative is formed via the free base (at low acidity and high temperature) and the 5- and 8-isomers are formed via the conjugate acid (at high sulfuric acid concentration and low temperature).[265,266a] A similar situation arises with benzo[*c*]cinnoline 5-oxide, nitration in 9 taking place with nitric acid,[267] while the main reaction in mixed acids involves position 4.[267,268]

Quinoxaline 1-oxide is resistant to nitration (though some hydroxy and alkoxy derivatives react),[269,270] and phenazine 5-oxide yields a mixture of the 3-nitro (major) and the 1-nitro derivatives.[184,271]

Among the results with heterocycle-condensed azine *N*-oxides, it is noteworthy that orientation to the 4 position is maintained, even in the presence of a thiophene ring in the case of thieno[2,3-*b*]pyridine 7-oxide.[272] Nitration of alloxazine 5-oxide derivatives occurs in the carbocyclic ring[273] (Scheme 9).

TABLE 11
Nitration

N-oxide	Substituent	Position (% yield)	Method	Ref.
Pyridine	(3-Et)	4(57)	Heat at 50°C with HNO_3 (d 1.5)/H_2SO_4; after that the exothermal reaction subsides, heat at 90–100°C for 3.5 h	232
	(2,3,6-Me_3)[a]	4(70)	Heat with HNO_3 (d 1.42)/H_2SO_4 at 60°C for 45 min and at 75°C for 1 h	229
	(2-Cl)	4(69)	Add HNO_3 (d 1.5)/H_2SO_4 to the *N*-oxide in H_2SO_4; 45 min at 20°C, 2 h at 90°C	230
	(2-COOH)	4(45)	Heat with HNO_3 (d 1.5)/H_2SO_4 at 120°C for 1 h	233
	(2-COOEt)	5(50)	Treat the *N*-oxide in $CHCl_3$ with $p\text{-}NO_2C_6H_4COCl$ at –3/–7°C, then with powdered $AgNO_3$. Stir 1 h at –3°C, 2 h at r.t., 2.5 h at 40°C, 4 h at 55°C	
	(2-OMe)	4(55)	Add HNO_3 (d 1.5)/H_2SO_4 to the *N*-oxide in H_2SO_4, heat 1.5 h at 75°C	284
	(3-OH)	2(61)	Add HNO_3 (d 1.5)/H_2SO_4 to the *N*-oxide in H_2SO_4 at 10°C, leave 24 h at r.t.	236
Quinoline	(2-Me)	5(27), 8(29), 4(1.5), 4, 8(6)	Treat with KNO_3/H_2SO_4 at –5/–10°C for 8 h	56b
	(2-Me)	5(11), 8(7), 4(51)	Treat with KNO_3/H_2SO_4 at 110–120°C for 1 h	254
		3(78)	Treat the *N*-oxide in cold $CHCl_3$ with $PhCOONO_2$, leave 15 d at r.t.	254
Pyridazine		3(60-70)	Treat the *N*-oxide repeatedly in cold anhyd. DMF with $NaNO_2$/TsCl	285
		4(22)	Treat HNO_3 (d 1.5) the *N*-oxide in H_2SO_4 at 130–140°C	291
		3(33)	Treat the *N*-oxide in cold $CHCl_3$ with PhCOCl, then add powdered $AgNO_3$ at –10°C, leave 4 h at –10°C and 4 d at r.t.	261
Cinnoline[b]		4(64)	Add the *N*-oxide to HNO_3/H_2SO_4 at 0°C, leave 8 h at r.t., 1 h at 50°C	295
Cinnoline[c]		5(8), 6(37), 8(35)	Heat with HNO_3 (60%)/H_2SO_4 (70%) at 80°C for 2 h	126
Benzo[c]cinnoline		5(18), 6(4), 8(58)	Heat with HNO_3 (60%)/H_2SO_4 (90%) at 80°C for 2 h	266a
Pyrazole	(1-Me)	9(69)	Add the *N*-oxide to stirred HNO_3 (d 1.5) at 50°C, leave 1 h	266a
		5(90)	Add HNO_3 (d 1.5) to the *N*-oxide in 66% H_2SO_4 at 0°C, leave 12 h at r.t.	267
Indole	(3-oxo-2-Ph)	5(34)	Add HNO_3 (d 1.42)/H_2SO_4 to the *N*-oxide in $CHCl_3$, shake 5 min	274
Benzotriazole[b]		7(50)	Add the *N*-oxide to HNO_3 (d 1.25) at 20°C	276 279

[a] In 45% mixture with the unreactive 2,4,6 isomer.
[b] 1-Oxide.
[c] 2-Oxide.

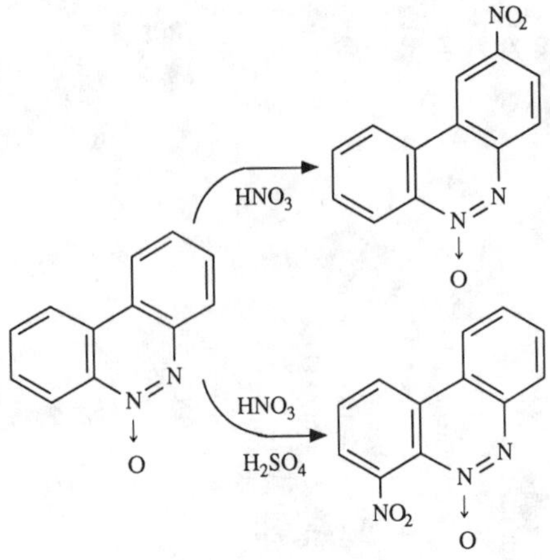

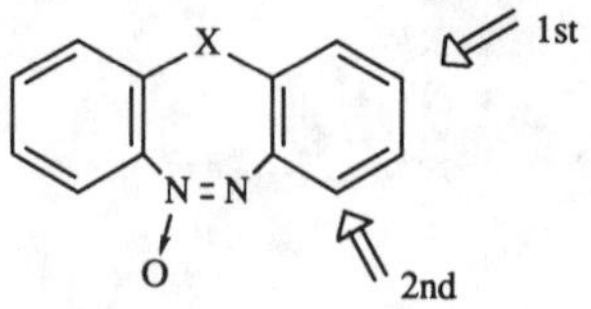

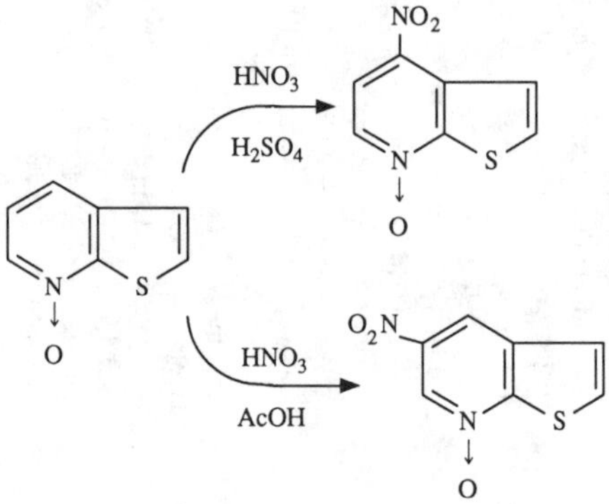

Scheme 9.

As for five-membered heterocycle N-oxides, in mixed acids 1-methylpyrazole 2-oxide is nitrated in 5, and 1,4,5-trimethylimidazole 3-oxide in 2, in both cases via the free base.[274,275a] 1-Phenyltriazole 3-oxide is nitrated in 4', not on the heterocycle ring.[275b]

2-Phenylisatogen is nitrated in the 5 position,[276] benzofuroxan and derivatives are readily nitrated in 4 (or in 7 when this position is occupied)[277,278] and benzotriazole 1-oxide in 7 (with 50% HNO_3) or in 5 (with HNO_3/AcOH, compare the following section).[279]

Seven-membered heterocycle N-oxides have also been studied. Thus the dibenzo derivatives **14'** and **14"** are nitrated as shown, and first and second position of attack are indicated: the orientation is the same observed with azoxybenzene.[280,281]

4-Nitropyridine 1-oxide (from pyridine). A mixture of pyridine (40 g), glacial acetic acid (100 ml), and 35% H_2O_2 (68 ml) are heated 3 h on a steam bath. 135 ml of the solution are distilled off under reduced pressure, 30 ml sulfuric acid are added, a further 20 ml are distilled off, and the residual solution is transferred to a three-necked flask. After the addition of 130 ml fuming sulfuric acid (30% SO_3), the solution is heated at 120°C, and 40 ml fuming nitric acid (d 1.52) are added in 1 h, the temperature remaining at 125—130°C; after a further 1 h stirring at this temperature and cooling, the solution is poured on 500 g crushed ice, neutralized with soda ash, and filtered to yield 49.5 g (70%) of the desired product.[282]

C. NITRATION UNDER ACYLATING CONDITIONS

The nitration of heterocyclic N-oxides takes a different course under conditions involving O-acylation, e.g., by treating with benzoylnitrate, and the nitro group now enters in β to the $\geqslant N \rightarrow O$

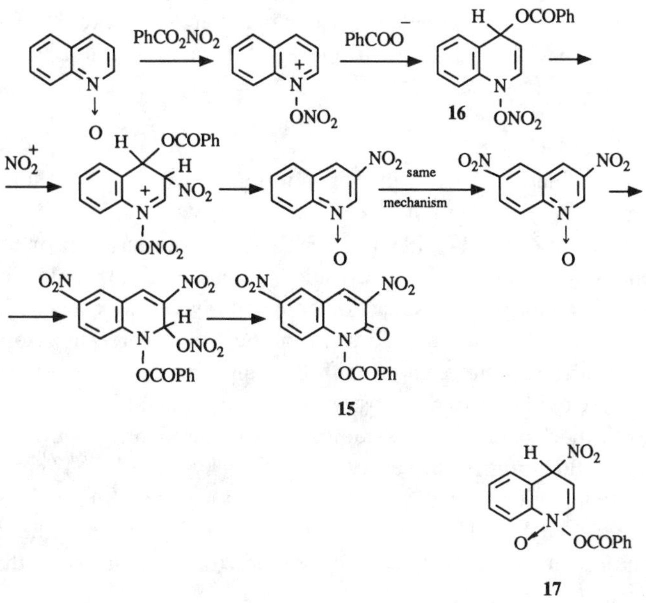

Scheme 10.

function. The reaction works poorly with pyridine 1-oxide, yielding 10% each of the 3-nitro and the 3,5-dinitro derivatives,[283] but works better with 3-cyano and 3-carboxyethyl pyridine 1-oxide, yielding 40—50% of the corresponding 5-nitro derivative.[284]

Nitration under these conditions was first reported and was more thoroughly investigated in the quinoline series. Reaction of quinoline 1-oxide with benzoyl nitrate yields the 3-nitro derivative (yield 50%).[285] With excess reagent a further nitration takes place in position 6 and the dinitro-2-oxo-derivative **15** is finally obtained (yield up to 51%).[286,287] The mechanism outlined in scheme 10 was proposed with 1,4-dihydro quinoline **16** as the key intermediate. The final step for the formation of **15** involves attack at position 2 followed by the elimination of nitrous acid. Nitration in 3 is also obtained by treating both 1-methoxyquinolinium methosulfate and the quinoline 1-oxide-boron trifluoride adduct with metal nitrates.[288] This supports the role of preliminary bonding at the oxygen atoms in this class of reactions. However, alternative mechanisms are possible.[266b,289] One proposal that has been made is that, in fact, nitration occurs at position 4 (see intermediate **17**) or 5, and then the nitro group rearranges at the 3 (or, respectively, 6) position[266b] (Scheme10).

At any rate, the use of acylating reagents (benzoyl chloride or acetyl chloride), followed by silver or potassium nitrate,[287] also in a related case of HNO_3 in AcOH (see page 140),[272] leads to nitration in 3 and then in 6, unless substituents are already present in these positions.[290] Substitution of tosyl for acyl chlorides in this reaction in unconvenient, but treatment of quinoline 1-oxide with $TsCl/NaNO_2$ in anhydrous DMF leads to nitration in 3, apparently via nitrosation and oxidation.[291]

Application of these methods to some derivatives causes reactions that are different from nitration, giving benzoylated products in some instances, with elimination or rearrangement of a substituent. Thus, 4-halogenoquinoline 1-oxides yield 1-benzoyloxy-4-oxo-3-halogenoquinolines, and 2-ethoxyquinoline 1-oxide yields 1-benzoyloxycarbostyril.[292] Furthermore, α-alkyl groups react competitively. Thus, 2-methylquinoline 1-oxide yields 20% of the 3-nitro derivative with benzoyl nitrate, but 2-(acetoxymethyl)quinoline and 2-cyanoquinoline 1-oxide (via the corresponding aldoxime acetate) are obtained with acethyl chloride and silver nitrate.[293] 4-Methyl and 2,4-dimethylquinoline 1-oxide are benzoylated (rather than nitrated) in 3.[294]

Nitration in 3 of pyridazine 1-oxides (in 5 when position 3 is blocked) is obtained under the same conditions,[295,296] and thieno[2,3-*b*]pyridine 7-oxide is nitrated in 5 when the reaction is carried out in acetic acid (see Scheme 9).[272]

D. HALOGENATION

The scope of bromination follows that of nitration, but forcing conditions are required. Bromination of pyridine 1-oxide in 90% sulfuric acid containing silver sulfate yields 8% 4-bromo and 3% 2-bromopyridine 1-oxide (reasonably via the free base) and in 65% oleum yields a mixture of mono- (largely the 3-bromo), di-, and tribromopyridine 1-oxide (via the conjugate acid or via the complex with SO_3).[297] A reaction of preparative value is obtained, however, when Br_2 and $Tl(AcO)_3$ in AcOH are used, good yields of the 4-nitro derivative from 3-picoline and some lutidines 1-oxides (but no reaction with the parent compounds) are observed.[298] This method might involve oxidation of the substrate by the metal cation, and thus be an example of electrophilic substitution via the *N*-oxide radical cation (compare the introductive section)

Strongly electron-donating substituents (amino[299] or hydroxy[243,245,300] groups) activate the nucleus toward bromination and direct the substitution *ortho*/*para* to them. Moderately activating substituents, such as alkoxy groups, promote bromination only when their *ortho*/*para* activation cooperates with the activation by the *N*-oxide function; thus, e.g., the 3 isomer is the only methoxypyridine 1-oxide that is brominated.[299]

Bromination in 4 of quinoline 1-oxide is obtained by treating the perbromide with bromine water at room temperature (on heating 3,4,6,8-tetrabromocarbostyyril is formed)[301] and is much better with the Br_2/Tl^{III} method.[299] Acridine 10-oxide is brominated in 9.[64] Diazine and 1,2,4-triazine *N*-oxides are brominated only in the presence of strongly activating groups, with a pattern similar to that described for pyridine 1-oxides.[264b,299,302,303] Useful preparative procedures for iodination and chlorination of some *N*-oxides have been reported, generally on activated substrates.[304,305] Interestingly, parent quinoline 1-oxide is iodinated in 5 by iodine in 20% oleum.[306]

As for five-membered *N*-oxides, triazole 1-oxides undergo bromination in 5 (while there is no nitration on the ring, see above),[275b] and benzofuroxans are easily halogenated (Table 12).[307]

3,5-Dibromo-2-(dimethylamino)pyrazine 1-oxide. 320 mg (2 mmol) of Br_2 are added to 139 mg (1 mmol) of 2-(dimethylamino)pyrazine 1-oxide in 30 ml CCl_4. After 30 min stirring, 151 mg (115 mmol) of NEt_3 are added and stirring is continued for 2 h. Evaporation and alumina chromatography yield the product, 196 mg (90%).[299]

E. BROMINATION UNDER ACYLATING CONDITIONS

Just as with nitration, bromination occurs in β to the *N*-oxide function when the reaction is carried out in the presence of acylating reagents, probably via an intermediate 1,4-dihydro adduct. Thus pyridine 1-oxide yields 35% 3,5-dibromopyridine 1-oxide when treated with bromine in the presence of $Ac_2O/NaOAc$.[308] Under the same conditions, quinoline 1-oxide yields 60% 3,6-dibromoquinoline 1-oxide and isoquinoline yields the 4-bromo derivative.[308,309] Bromination in β in the presence of acylating agents has also been reported for pyrimidine[310] and 1,5- and 1,8-naphthyridine[311] *N*-oxides.

F. OTHER ELECTROPHILIC SUBSTITUTIONS

Sulfonation of pyridine 1-oxide takes place only under severe conditions and yields the corresponding 3-sulfonic acid (with a small amount of the 2 and 4 isomers).[190] Thus, the course of the reaction is the same as with the parent pyridine and involves the protonated form. Likewise, 3-hydroxypyridine 1-oxide is sulfonated in 2 under the same conditions as the corresponding pyridine.[312] Chlorosulfonation of pyridine 1-oxide fails.[315]

Mercuriation of pyridine 1-oxide with mercury (II) acetate occurs mainly in 2, with a small amount of the 3 isomer (the yield of which increases when using $HgSO_4/H_2SO_4$).[313] With

TABLE 12
Halogenation

N-oxide	Substituent	Product (% yield)	Method	Ref.
Pyridine		2-Br(4), 4-Br(8)	Heat at 180° for 20 h with Br$_2$, 90% H$_2$SO$_4$, and Ag$_2$SO$_4$	297
		3-Br(18), 3,4-Br$_2$(4), 2,5-Br$_2$(5)	Heat at 220°C for 10 h in sealed tube with Br$_2$ and oleum (65% SO$_3$)	297
	(2,6-Me$_2$)	3,5-Br$_2$(35)	Reflux 3 h with Ac$_2$O, Br$_2$, and NaOAc in CHCl$_3$	308
	(3-OMe)	4-Br(49)	Heat at 70°C for 24 h with Br$_2$ and Tl(OAc)$_3$ in ArcOH	298
	(4-NMe$_2$)	2,6-Br$_2$(37), 4,6-Br$_2$(33)	Add Br$_2$ and K$_2$CO$_3$ to the N-oxide in dry CCl$_4$, stir 6 h	299
Quinoline		3-Br(77)	Add Br$_2$ and K$_2$CO$_3$ to the N-oxide in dry CCl$_4$, stir 6 h	295
		4-Br(32)	Shake the N-oxide perbromide with bromine water in sealed tube at r.t.	301
Acridine	(8-OH)	4-Br(66)	Heat at 50°C for 24 h with Br$_2$ and Tl(OAc)$_3$ in AcOH	298
		3,6-Br(60)	Reflux for 3 h with Ac$_2$O, Br$_2$, and NaOAc	308
		5,7-Cl$_2$(80), 5-Cl(7)	Add SO$_2$Cl$_2$ in CHCl$_3$ to the N-oxide in CHCl$_3$ at 5°C	304
		9-Br(45)	Heat with Br$_2$ in AcOH at 100°C for 2.5 h	64
Pyridazine	(3-OMe)	2,6-Br$_2$(37), 4,6-Br$_2$(33)	Add Br$_2$ and K$_2$CO$_3$ to the N-oxide in dry CCl$_4$, stir 6 h	299
Pyrazine	(3-NH$_2$)	2,6-I$_2$(97)	Heat with I$_2$ in DMSO at 80°C for 30 min	305a
Pyrimidine	(2-NH$_2$)	5-Br(88)	Add NBS to the N-oxide in CH$_2$Cl$_2$, stir for 48 h	295
	(4-Ph)	5-Br(40)	Heat with Br$_2$ in Ac$_2$O/AcOH at 80°C for 1-2 h	310
1,5-Naphthiridine[a]		6-Br(41), 3,6-Br$_2$(4)	Reflux with Ac$_2$O in CHCl$_3$ for 12 h, add Br$_2$ and NaOAc, reflux additional 12 h	311
1,2,4-Triazine[b]	(3-NMe$_2$)	6-Br(70)	Add Br$_2$ and K$_2$CO$_3$ to the N-oxide in dry CCl$_4$, stir 2 h	303a
	(3-NMe$_2$)	6-Cl(40)	Add the N-oxide to Cl$_2$ and NEt$_3$ in CHCl$_3$, stir 1 h	303a
Triazole	(3-Me)	5-Cl(84)	Add aq NaClO to the N-oxide in 4 N HCl in ice bath, stir 5 h at r.t.	275b
	(3 Me)	5-Br(98)	Add Br$_2$ at 0°C to the N-oxide in CHCl$_3$/aq Na$_2$CO$_3$; stir 3 h at 20°C	275b

[a] 1-Oxide.
[b] 2-Oxide.

quinoline 1-oxide, the reaction with various mercury salts has been found to occur predominantly in 8 (possibly because of coordination of the reagent with the oxygen atom), except that with mercury (II) acetate, which yields the 3 and 8 isomers in comparable amounts.[314]

Reaction of pyridine 1-oxide with benzyl chloride and $AlCl_3$ has been reported to yield the 3-benzylated (and deoxygenated) product[316] (although other authors reported unsuccessful attempts to apply the Friedel-Crafts reaction to pyridine N-oxide).[315]

Nicotinic acid 1-oxide is acylated in 2 when heated with acetic and propionic anhydrides.[317] This, however, is not an electrophilic substitution. Extensive investigation showed that the reaction probably involves deprotonation from the α position in the 1-acyloxy mixed anhydride to yield an ylide, and this cyclizes onto the carboxy group to give a furanone derivative as the primary product.[317c] A related process takes place with N-methylnicotinamide 1-oxide.[317b] These reactions are mechanistically related to the deoxidative acylation from the N-(β-acylalkoxy) salts discussed in Section VIC6.

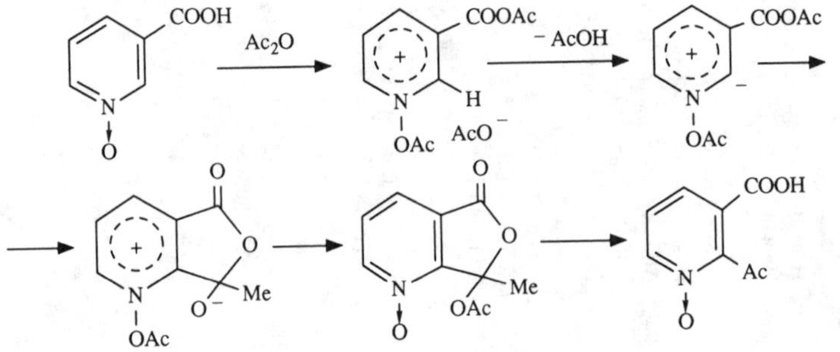

4-Phenyl-2,5-dimethyl pyridine 1-oxide is acylated on the phenyl group (in 4′),[318] and more generally pyridine and other heterocyclic N-oxides rearrange to lactams with acetic anhydride (see Section VIB).

Nitrosation[319,320] or coupling with diazonium salts[321-324] take place only on hydroxy- or aminosubstituted pyridine or quinoline N-oxides. Likewise, aminomethylation is a valuable preparative reaction with hydroxy-substituted pyridine,[323-326] pyridazine,[327,328] pyrimidine (a bis(dialkylamino) derivative is obtained from 5-hydroxy-4-phenylpyrimidine 1-oxide, while a mono derivative is formed from the nonoxidized substrate),[329] and quinoline[330] N-oxides, and in every case it involves a position that is *ortho* and/or *para* to the hydroxyl group (Table 13).

V. NUCLEOPHILIC SUBSTITUTION

Nucleophilic substitution in the N-oxides of azines presents no major difference from the general mechanistic scheme observed in heterocyclic chemistry. Thus leaving groups in α or γ to the ≥N→O function are usually substituted under mild conditions through the usual addition-elimination mechanism via a stabilized anion (other mechanisms, i.e., elimination-addition, stable "Meisenheimer" intermediate, $S_{RN}1$ have been found to occur in some cases, see below). Indeed N-oxidation remarkably increases the rate of such a process, so that it is often advantageous to carry out the substitution step on the N-oxide and then to deoxygenate. Some effect is also experienced in other positions, and in some cases a leaving group in β or in a condensed carbocyclic ring is sufficiently activated toward nucleophilic substitution. Furthermore, there are efficient methods for the direct introduction of a nucleophile at an unsubstituted position, with H⁻ as the formal leaving group (these are different from reactions involving direct substitution *and* N-deoxygenation, as discussed in Section VI), noteworthy among them is α-functionalization via the corresponding stable anion.

TABLE 13
Other Electrophilic Substitutions

N-oxide	Substituent	Product (% Yield)	Method	Ref.
Pyridine		3-SO$_3$H(42), 2-SO$_3$H(1), 4-SO$_3$H(2)	Heat at 230°C for 22 h with oleum (20% SO$_3$) and HgSO$_4$	190
		2 HgOAc(15), 3-HgOAc(3)[a]	Heat at 130°C for 2 h with Hg(OAc)$_2$ in AcOH	313
Quinoline		8-HgOAc (69), 7-HgOAc(5), 5-HgOAc(5)[a]	Heat at 130°C for 5 h with Hg(OAc)$_2$ in AcOH	314
Pyridine	(3-COOH)	3-CH$_2$Ph(42)[b]	Heat for 3 h with PhCH$_2$Cl and AlCl$_3$	316
	(3,5-(NH$_2$)$_2$)	2-Ac(27)	Reflux for 6 h with Ac$_2$O	317
		2-N=N-Ph(63)	Treat the N-oxide at 0°C in 10% HCl with PhN$_2$$^+Cl^-$, bring at pH5, leave 10 min at 0°C	321
Quinoline	(8-OH)	5-N=N-Ph(70)	Add PhN$_2$$^+Cl^-$ in HCl to the N-oxide in aq NaOH at 0°C, stir 2 h at 0°C	322
Pyridine	(3-OH)	2-CH$_2$NMe$_2$(81)	Add 30% aq CH$_2$O to the N-oxide and Me$_2$NH in water, leave at 20–25°C for 2 d	326
Pyridazine	(3-OH)	6-CH$_2$NMe$_2$(53)	Add CH$_2$O/Me$_2$NH to a suspension of the N-oxide in EtOH, stir for 5 h	327
Pyrimidine	(4-Ph 5-OH)	2,6-(morpholinomethyl)$_2$ (61)	Treat the N-oxide in benzene with paraformaldehyde and morpholine	329

[a] Calculated from the **corresponding** bromoderivative after substitution.

[b] Deoxygenated product.

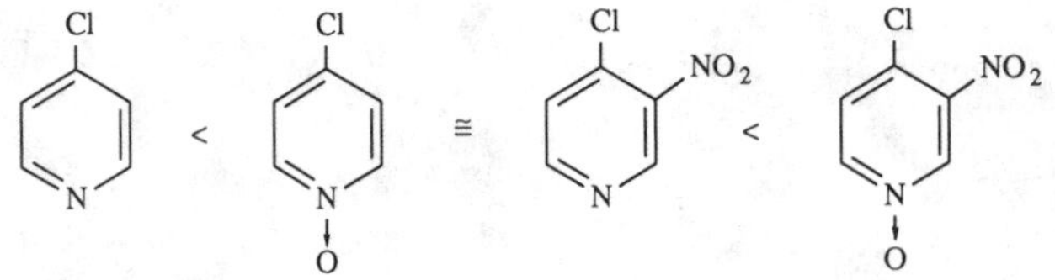

Scheme 11.

A. SUBSTITUTION OF A LEAVING GROUP ON THE HETEROCYCLIC RING

1. With Heteroatom-Centered Nucleophiles

N-oxidation enhances the electron-withdrawing effect of the nitrogen atom in six-membered heterocycles, and thus makes nucleophilic substitution in position α and γ easier. Since a nitro group is easily introduced by electrophilic substitution and halo *N*-oxides are available by *N*-oxidation of the easily obtained (see, for example, Section VIA) halo heterocyclics, substitution of these good leaving groups in *N*-oxides offers a useful entry to amino, hydroxy, alkoxy, mercapto, selenyl, sulfonyl, azido, and other derivatives, and these are easily deoxygenated, if desired, to the corresponding bases. A large number of straightforward nucleophilic substitutions involving, besides nitro and halogeno, leaving groups such as alkoxy, thioalkoxy, amino, sulfonyl, and other groups have been reported,[56a,130,331-356] and some examples are gathered in Table 14. Notice that in 2-benzoylpyridine 1-oxide, the substituent is displaced by alkoxides, while under this condition it is reduced in the pyridine.[355]

Kinetic studies of the reaction of halo- and nitro-pyridine[357-362] and quinoline 1-oxides[363] with amines and alkoxides clearly indicated the activating effect of the ≥N→O function, which is much higher than that caused by substitution of a ≥N or a ≥C-NO₂ for a ≥CH group in an aromatic ring (it should be observed, however, that the rate ratio for the reaction of different nucleophiles with nitroaromatics, heterocyclics, and the corresponding *N*-oxides is different, a fact that has been attributed to the retention of the negative charge of the intermediate on the heterocyclic ring in the latter two cases, whereas delocalization on the "external" nitro group is predominant in the first one).[362] The increase in the rate of nucleophilic substitution caused by oxidation of an *N*-heterocycle is similar to that caused by the introduction of a nitro group on the ring. Thus, substitution of the halogeno in 4-chloropyridine 1-oxide is as easy as in 4-chloro-3-nitropyridine, and is much easier than in 4-chloropyridine[361] (Scheme 11, Table 14).

The activation energy for substitution by the methoxy anion in **19** is 2.3 kcal/M lower than in **18**, and the reaction is two orders of magnitude faster at 50°C.[361] The order of the reactivity for the positions is 2 > 4 >> 3, and a nitro group is more easily substituted than a halogeno.[357,358] The order might differ for nonoxidized substrates, e.g., 2-chloropyridine 1-oxide reacts faster with methoxide than the 2-bromo derivative does, while the reverse is true for the corresponding pyridines.[360]

N-deoxygenation is often a minor pathway that competes with substitution, but in some cases a compound that is both deoxygenated and substituted is the main, or only, product, e.g., in the reaction of 2-methylthionaptho[2,1-*e*]1,2,4-triazine 4-oxide with hydrazine.[356] In hindered 2,3-dimethyl- (or 2,3,5-trimethyl-) 4-chloropyridine 1-oxide, substitution with amines under acidic conditions leads to deoxygenated 4-aminopyridines, apparently via elimination of HClO (only deoxygenation is observed under basic conditions).[369]

Acid catalysis has been found to be effective, in particular in the specific substitution of a chloro or bromo for a nitro group using hydrogen or acetyl halides,[56a,331,334,342,364,365] reasonably due to enhanced electron withdrawal by the ≥N⁺-OH or ≥N⁺-OAc groups. A sulfonyl group can likewise be substituted by chloro with hydrogen chloride and by OH on treatment with organic acids, followed by hydrolysis.[353] In some cases, POCl₃ causes substitution, probably through a

TABLE 14
Substitution of a Leaving Group by a Heteroatom Contered Nucleophile

N-oxide	Leaving group(s)	Unaffected or other group(s)	Entering group (yield)	Method	Ref.
Pyridine	2-NO₂		Cl	Treat with HCl. After that violent reaction subsides, reflux 30 min	56a
	2-Cl		F(32)	Reflux 48 h with finely ground KHF in DMSO	350
	2-Cl		OH(33)	Heat 5 h at 130–140°C with 20% aq HCl	336
	2-Cl		OEt(80)	Reflux for 30 min with NaOEt/EtOH	335
	2-Cl		OPh(64)	Heat for 4 h at 100°C with NaOPh/PhOH	336
	2-Cl		SEt(80)	Heat 3 h at 100°C with NaSEt/EtSH	336
	2-Cl	(3,4,5,6-Cl₄)	SH(80)	Saturate with H₂S ethanolic KOH at 0°C, add the N-oxide, leave 2 h at 10°C	379
	2,6-Cl₂	(3,4,5-Cl₃)	2,6(SH)₂(85)	Reflux 1 h the solution as above	379
	2-Cl		1-(2-Ph-4-NR₂ pyrrolidinyl) (46)	Reflux for 48 h with the amine in ethanol	349
	2-Br	(6-Me)	OH(77)	Heat on steam bath for 30 min with 5% aq NaOH	337
	2-Br		SeH(59)	Add the N-oxide to H₂Se satd. NaOEt/EtOH, reflux for 30 min	338
	2-Br		NCS(57)	Stir and reflux for 3 h with KCNS in EtOH	332
	2-Br	(4-NO₂)	NMe₂(75)	Reflux 5 h with Me₂NH/EtOH	389
	3-F	(4-NO₂-2,6-Me₂)	SEt(quant)	Leave 48 h with NaSEt in dioxane	393
	3-Cl		SMe(60)	Heat at 140°C for 5 h with NaSMe/MeOH and CuSO₄	373b
	3-Cl		SO₃H(63)	Heat for 10 h at 143°C with aq Na₂SO₃	339
	3-Br	(5-Br)	OMe(79)	Reflux for 30 min with KOH in MeOH	401
	3-Br	(5-OMe)	NH₂(95)	Heat for 5 h at 130°C with aq NH₃, and CuSO₄	401
	3-SO₂Ph		OEt(71)	Reflux with NaOH in EtOH	355
	4-NO₂		Cl(84)	Add the N-oxide in portions to stirred AcCl, reflux 3 h, leave overnight	331
	4-NO₂		Br(91)	Reflux for 17 h the suspension in 48% aq HBr	332
	4-NO₂		OEt(70)	Reflux 3 h with NaOH in EtOH	334
	4-NO₂	(2-Cl)	OMe(84)	Leave 24 h at r.T. with MeONa/MeOH	389
	4-NO₂	(3-Cl-2,6-Me₂)	SEt	Reflux for 3 h with NaSEt in EtOH	393
	4-Cl		NHMe(78)	Heat for 20 h at 140°C with aq MeNH₂	335
Quinoline	2-Cl		SH	Heat with NaHS in DMF	341

TABLE 14 (continued)
Substitution of a Leaving Group by a Heteroatom Contered Nucleophile

N-oxide	Leaving group(s)	Unaffected or other group(s)	Entering group (yield)	Method	Ref.
Quinoline	2-Cl		N_3(58)	Treat with NaN_3 in acetone at 25°C	340
	4-NO_2		Cl(95)	Heat on steam bath with conc. aq HCl	342
	4-NO_2		$OCH_2CH_2NH_2$(83)	Treat with $NaOCH_2CH_2NH_2$	344
	4-Cl		N_2H_3(67)	Treat with N_2H_4, H_2O in abs. EtOH	345
	4-NO_2		NH_2(90)	Heat for 2 h at 100°C with N_2H_4,H_2O in EtOH in the presence of copper	343
	4-Cl		SH(75)	Add the N-oxide to H_2S satd. NaOMe/MeOH reflux for 6 h	347
Isoquinoline	4-Cl		NHOH(63)	Reflux 5.5 h with NH_2OH in MeOH	346
	4-Br	(5-NO_2)	Piperidinyl(67)	Stir 12 h with piperidine in EtOH	377c
Acridine	9-Cl	(1-NO_2)	OPh(85)	Heat 30 min at 80°C with NaOPh/PhOH	348a
	9-OPh	(1-NO_2)	$NH(CH_2)_3NMe_2$(65)	Heat 30 min on steam bath with the amine in PhOH	348a
Pyridazine	3-OMe		OEt(65)	Heat 30 min at 50°C with NaOEt/EtOH	377b
	3-OMe		N_2H_3(26)	Reflux 3 h with N_2H_4. H_2O in EtOH	377a
	3-NO_2	(6-Cl)	OMe	Leave for 24 h at r.t. with NaOMe/MeOH	396
	3-Cl	(6-Cl)	OMe(80)[a]	Leave 1 h at r.t. with NaOMe/MeOH	397
	3-Cl	(6-Cl)	NHEt(54)[b]	Heat 4 h on steam bath with 70% aq $EtNH_2$	397
	4-NO_2		NH_2(94)	Treat with NH_3 in MeOH	394
	4-Cl	(3-OEt-6-NO_2)	N_3(51)	Heat for 5 h with NaN_3 in 80% EtOH	351
	6-Cl	(3-NO_2)	NHPh(67)	Reflux and stir with $PhNH_2$ in EtOH	396
	6-Cl	(3-Cl)	SH(72)	Add the N-oxide in dioxane to aq Na_2S, stir 10 h at r.T.	398
Pyrimidine	2-Cl	(4,6-Me_2)	Imidazolyl(71)	Reflux for 6 h with imidazole in MeCN	352
	5-Br	(4-Ph)	N_3(50)	Heat at 70°C with NaN_3 in DMSO	376
Pyrazine	3-Cl		NH_2(75)	Heat for 2.5 h at 115–120°C with aq NH_3	375
Quinoxaline[c]	2-SO_2Ph	(3-Me)	Cl(quant)	Dissolve in conc. aq HCl, heat a few min	353a
	2-SO_2Ph	(3-Me)	OH(95)[d]	Heat 30 min in AcOH	353a
1,5-Naphthyridine[c]	2-NH_2		OH(54)	Reflux for 45 min with 2 N NaOH	353b
Benzo-1,2,4-triazine[e]	3-Cl		$NHCH_2CH_2NMe_2$(70)	Leave 16 h at r.t. with the amine then reflux 1.5 h	130

				Conditions	
Naphto[2,1-e]-1,2,4-triazine[f]	3-SMe		N_2H_3(69)[g]	Reflux for 2 h with N_2H_4, H_2O in MeOH	356
Purine[h]	6-Cl		SH(73)	Heat and stir for 2 h at 60°C with $NH_4^+NH_2CS_2^-$	354a
	6-Cl	(2-NH_2)	NH_2(16)	Heat for 1 d at 90–100°C with 5M NH_3	354b
Furazan	4-NO_2	(3-Me)	OEt(85)	Mix the N-oxide in EtOH and NaOH/EtOH, stir 1 h at r.t.	407
	4-NO_2	(3-Me)	SEt(85)	Mix the N-oxide in Me_2CO and EtSH/NaOH in Me_2CO, stir 1 h at r.t.	407
Benzimidazole[e]	2-CN	(1-Me)	OH(63)	Reflux for 1 h with KOH in MeOH	409
Benzothiazole	2-CN	(7-NO_2-5-CF_3)	OH(66)	Add at 0°C aq NaOH to the N-oxide in 10% MeOH, stir 3 h at r.t.	410

a 7.5% of the regioisomer is also formed.
b 14% of the regioisomer is also formed.
c Dioxide.
d Sum of the products obtained after chromatography, i.e., 1-acetoxyquinoxalone (27%) and 1-hydroxyquinoxalone (68%).
e 1-Oxide.
f 4-Oxide.
g N-Deoxygenate product.
h 3-Oxide.

similar mechanism, e.g., it yields 4-chloropyridine 1-oxide from the nitro derivative if the reaction is carried out at 70°C,[366] (however, deoxygenative chlorination is usually observed with this reagent; see Section VIA). Carrying out a substitution under phase-transfer conditions may be advantageous (see below).[367] A catalytic effect of light on the substitution of a nitro group, both by hydrogen chloride[368a] and by amines,[368b] has been reported.

Reaction of 4-chloropyridine 1-oxide with 3-hydroxypyridine affords the ylide **20**.[370] In the field of purines, an important entry to 6-substituted derivatives is offered by substitution on the corresponding 6-chloro 3-oxides.[354]

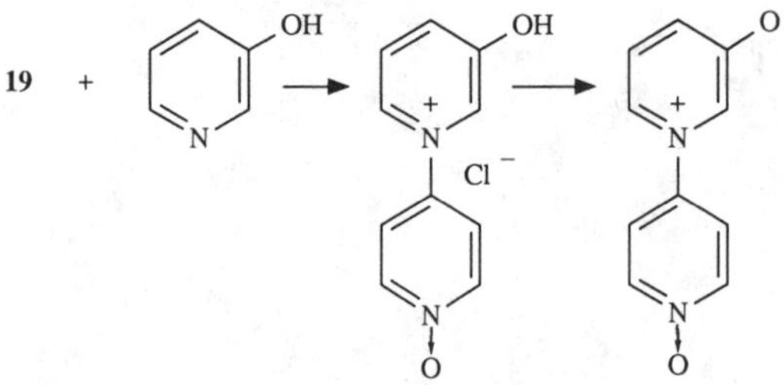

With bidentate nucleophiles, the product obtained may depend on experimental conditions. Thus, 2-thioethylamine reacts with 4-nitroquinoline 1-oxide, both at the sulfur and at the nitrogen atom, but in the latter case a secondary rearrangement is involved.[371] With methionine and cysteine, the corresponding S-derivatives are easily obtained, and these reactions have been studied also under physiological conditions, in view of the carcinogenic properties of the substrate.[372]

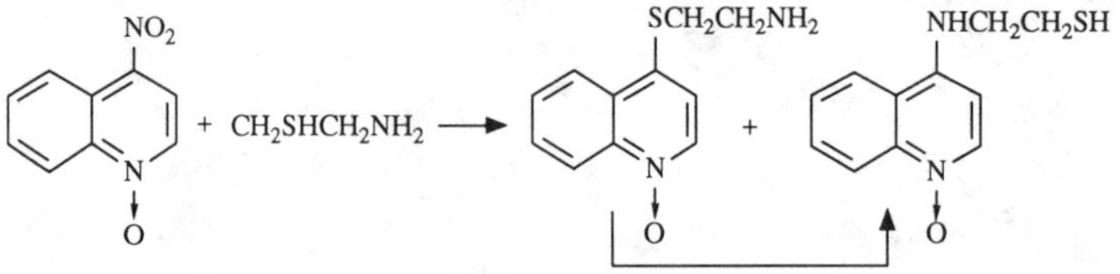

The inductive effect of the $\geqslant$N$\rightarrow$O function also activates the β position,[357,373-377] and, though more drastic conditions are required than for the substitution of an α or γ group, nucleophilic substitution on the N-oxide is also advantageous in this case. As an example, 3-chloropyrazine 1-oxide is aminated in 2.5 h at 115—120°C, while the nonoxidized substrate requires 16 h at 140°C.[375] In 4-bromo-5-nitroisoquinoline 2-oxide, amination in 4 is obtained under mild conditions.[377c] Substitution occurs easily in 3-fluoropyridine and quinoline 1-oxides.[374]

An elimination addition mechanism (via hetaryne) has been demonstrated to operate in the reaction of 3-chloro- and bromopyridine 1-oxides with potassium amide in liquid ammonia.[378] As opposed to the nonoxidized substrate, the 2,3- and not the 3,4-didehydro intermediate is formed, and finally yields 3-aminopyridine 1-oxide with traces of the 2-isomer.[378a,378b] 3-Bromoquinoline reacts analogously (with cycloaddition on the hetaryne as a side reaction; compare Section VIIIA2).[378b] A 3,4-pyridyne 1-oxide is formed from 3-bromo- and chloro-2-methylpyridine 1-oxides and yields the 4-amino derivative as the main product.[378a,378c] See Section VB for a discussion of proton abstraction from N-oxides.

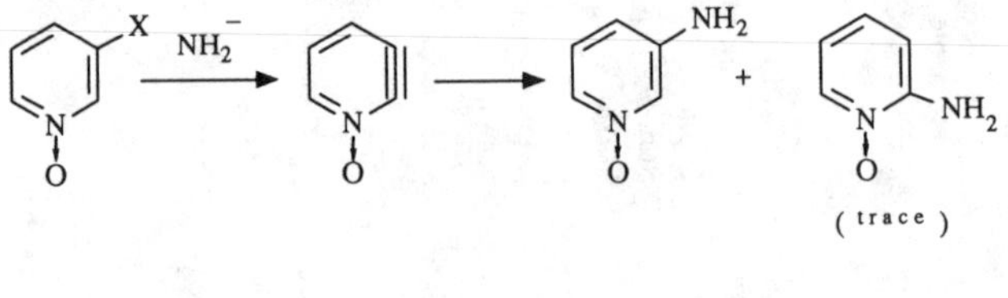

When more than one potential leaving group is present, position 2 is the most easily substituted.[379-382] Thus, pentachloropyridine 1-oxide is substituted at position 2, and further reaction leads to the 2,6-disubstituted derivative with S and N nucleophiles[380] (under the same conditions, reaction of pentachloropyridine ceases at monosubstitution). Substitution of a chloro in position 2 also takes place in the presence of a 4-arylsulfonyl[383] group (the opposite selectivity is observed with the nonoxidized substrate). Analogously, 2,4,6-trinitropyridine 1-oxide yields the 2-phenylamino and the 2,6-diazido derivatives.[384] As expected, a nitro group in 3 or 5 favors substitution of a 2-halogeno.[352,385]

With two different groups in α and γ, the course of the reaction depends on the nucleophile. Thus, with 2-halo-4-nitropyridine 1-oxide, amines substitute for the halogeno and alcohols substitute for the nitro group.[386-389] Bromo and chloro groups can be exchanged in 2-halo-4-nitroquinoline 1-oxide by treatment with phosphorus trihalides.[390]

Substitution of the halogeno by O, N, and S nucleophiles is easy in 3-fluoro-4-nitropyridine and quinoline 1-oxide, substitution of the nitro group being, as usual, observed with AcX, HX, or PX_3.[391,392] However, in 3-chloro-, bromo-, and iodo-4-nitro-2,6-dimethylpyridine 1-oxide, the nitro group is substituted.[393]

Competitive substitution in pyridazine 1-oxide changes considerably with the nucleophile used. Thus, in 3-alkoxy-4-nitropyridazine 1-oxide, both groups are substituted by NaOMe/MeOH,[264a] but only the 3-methoxy is substituted by ammonia or amines.[394] Similar behavior is observed with 3-alkoxy-2,6-dinitroderivatives (but a halogeno enters in 6 with HCl and HBr),[394,395] and in the 3-alkoxy-6-chloro-4-nitro derivatives NaSH, NaSMe, and NaN_3 substitute for the nitro group and amines substitute for the chloro.[395] In 3,6-dichloropyridazine 1-oxide, O and N nucleophiles attack in 3[397] and S nucleophiles in 6;[398] and again in 6-chloro-3-nitropyridazine 1-oxide, a methoxy group attacks in 3 and aniline in 6.[396] The reaction of some substituted 4-nitropyridazine 1-oxides with nucleophiles has been investigated from the mechanistic point of view, and it has been found that σ-complexes can be spectroscopically characterized and are involved in the observed substitution.[399,400] Furthermore, ring opening takes place on the treatment of some pyridazine 1-oxides with nucleophiles (see Section VIIA).

3,5-Dibromopyridine 1-oxide is converted to the diamino derivative by $NH_3/CuSO_4$ in an autoclave[321] or is transformed stepwise into 3-methoxy-5-aminopyridine 1-oxide.[401] Condensation between 3-chloro-4-nitropyridine 1-oxide and aromatic or heterocyclic dianions offers a useful entry to aza- and diaza-thianthrene, phenoxathiin, and phenoxaselenine systems[402-405]; the reaction often does not occur on the nonoxidized substrate. With O/S dianions the first step is substitution of the chloro by the S⁻ group. Condensation of 2,3-dichloroquinoxaline dioxide with 2-thioquinazolin-4-one yields product **21**.[406]

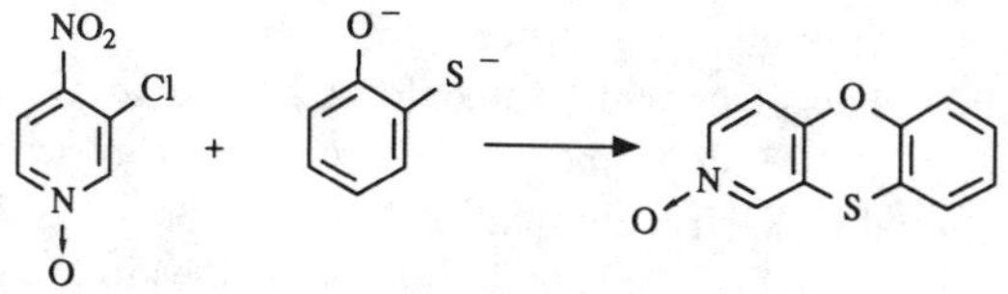

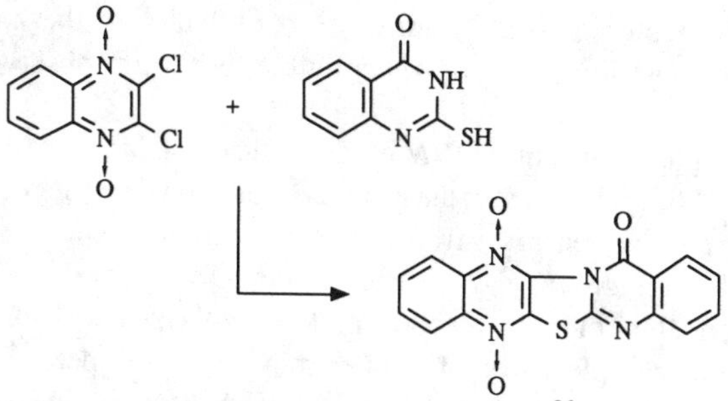

21

As for five-membered N-oxides, the easy substitution observed in nitro-[407] and chorofuroxans,[408] and the reactivity of a 2-cyano group in benzimidazole[409] and benzothiazole 3-oxides,[410] should be mentioned.

   1-Hydroxy-3-methoxy-2(H)-pyridone under phase-transfer conditions. 2-Chloro-3-methoxypyridine 1-oxide (5 mmol) is added to a dry suspension of powdered potassium hydroxide (20 mmol) and potassium carbonate (5 mmol) in toluene (50 ml) and benzyl alcohol (7.7 mmol). After adding tris(3,6-dioxaheptyl)amine (TDA-1, 0.5 mmol) the reaction is stirred at 120°C for 2 h. Toluene is decanted, water is added, and the solution is acidified to pH6. Concentration, extraction with absolute ethanol, and recrystallization give the product in a 69% yield.[367]

2. With Carbanions and Organometallics

   Nucleophilic substitution of a 4-chloro or nitro group by carbanions derived from malonic acid or nitrile, phenylacetonitrile, and pyridylacetonitrile has been found to take place with satisfactory yields in pyridine,[411] quinoline,[412,413] and pyrimidine[414] 1-oxides. In 3-bromo-4-nitropyridine 1-oxide, carbanions substitute for the halogen, and when acetoacetate is used, a secondary substitution leads to the furo[3,2-c]pyridine system.[415] With the quinazoline 3-oxide, **22** substitution is followed by cyclization to yield an isoxazoloquinazoline system.[416a]

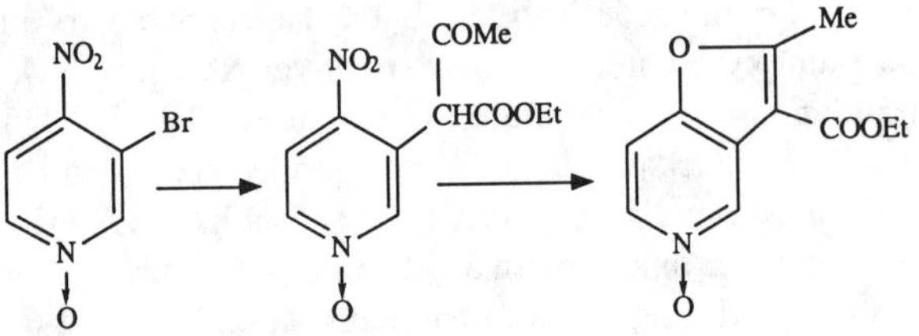

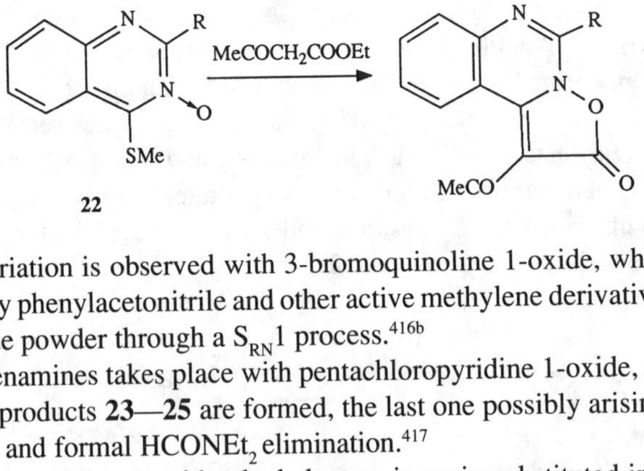

22

   An important variation is observed with 3-bromoquinoline 1-oxide, which is alkylated at room temperature by phenylacetonitrile and other active methylene derivatives in the presence of sodium hydroxide powder through a $S_{RN}1$ process.[416b]

   C-alkylation of enamines takes place with pentachloropyridine 1-oxide, e.g., from diethyl-aminocyclohexene products **23—25** are formed, the last one possibly arising by a competing path, cycloaddition and formal HCONEt$_2$ elimination.[417]

   In 3-fluoro-4-nitropyridine 1-oxide, the halogeno is again substituted in the reaction with derivatives of 3-aminocrotonic acid.[418] It must be noted that all of these reactions require the presence of the N-oxide group, as a poor or negligible reaction is observed with nonoxidized substrates.

   The reaction of Grignard reagents with N-oxides is discussed in more detail in Section VID3 and usually leads to addition. However, the product obtained when the reaction is carried out at room temperature or above often arises from formal substitution, which is true, of course, in the presence of a good leaving group, e.g., 2-chloro or nitro.[186,419]

   In this connection, the coupling reaction of halo N-oxides with organometallics in the presence of palladium complexes or of copper may be mentioned, though the reaction is mechanistically unrelated to the previously discussed substitutions and also occurs on nonoxi-

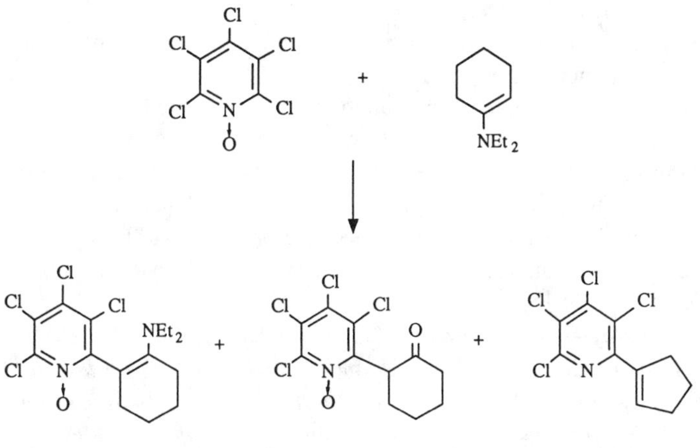

23   24   25

dized substrates. Good results are obtained in the alkynylation of 3- and 4-iodopyridine 1-oxides,[420] as well as in the alkylation,[421,422] alkenylation,[423] alkynylation,[424] and arylation[425] of 2-chloropyrazine 1- and 4-oxides and in the alkynylation of 3-chloropyridazine 1-oxides.[426] Dehalogenation is often a competing pathway and is the only reaction observed in the reaction of chloropyrazine oxides with formate.[427]

B. DIRECT SUBSTITUTION OF A HYDROGEN ATOM

1. Proton Abstraction

a. Hydrogen Exchange Under Basic Conditions

Hydrogen/deuterium exchange under basic conditions in heterocyclic N-oxides is strongly favored at the α position, the relative rate in pyridine 1-oxide being 2 >> 3 > 4.[428,429] This differs from the nonoxidized parent (4 > 3 > 2) and corresponds to the relative positional rate observed for proton exchange in N-alkylpyridinium salts and for decarboxylation of N-alkylpyridinecarboxylic acid betaines. This supports the intermediacy of a carbanion, and indeed the poor reactivity at the α position in pyridines has been explained on the basis of the destabilization of the developing carbanion by the lone pair of electrons on the nitrogen. This situation does not occur in N-oxides, and there the reactivity is determined by the larger electron density in the α position.[430] The reaction rate increases in diazine N-oxides, and the effect of the additional nitrogen is on the order $m > p > o$ with respect to the exchanging proton (thus, exchange in 2 in pyrazine 1-oxide is faster than exchange in 6 in pyrimidine 1-oxide, and this is faster than exchange in 2 in the same molecule).[431,432] Several kinetic and quantum-chemical investigations on various azine N-oxides fit with this scheme.[209,433-435] For a discussion of (competitive) deprotonation of the alkyl side chain, see Section XIB.

Prolonged reaction leads to complete exchange,[209] a useful application being the patented synthesis of pyridine-d_5 via base-catalyzed H/D exchange on the N-oxide.[436]

The available evidence excludes σ-(Meisenheimer) hydrated complexes having a significant role in H/D exchange, although is it well documented that adducts of this type (e.g., **26**) are formed under strongly basic conditions[431] or when electron-withdrawing substituents are present, e.g., with 3,5-dinitropyridine 1-oxide[283] and 4-nitropyridazine 1-oxide.[400]

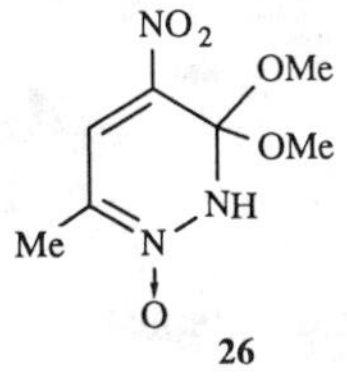

26

b. Substitution in α After Proton Abstraction

Support for the proton abstraction mechanism in the above-mentioned hydrogen-exchange reactions has been obtained by trapping of the carbanion. More importantly, the easy preparation of such anions has been exploited for the functionalization of the heterocycle by reaction with various electrophiles. Thus, alkali metal salts of the 2-pyridyl 1-oxide carbanion are generated by decarboxylating 2-picolinic acid 1-oxide[437] or directly by reaction of pyridine 1-oxide with strong bases such as *n*-butyl lithium[438] or sodium hydride,[439] and their stability is probably related to a chelating-directing effect of the oxygen lone pair. Treatment of these salts with elementary sulfur offers elegant and industrially significant access to 1-hydroxy-2-pyridinethione and derivatives[437,439-441] (Table 15).

1-Hydroxy-4-methyl-2-pyridinethione. 4-Picoline 1-oxide (2.18 g, 20 mmol) in 100 ml anhydrous THF at –65°C is flushed with N₂ and treated with 2.56 g n-butyllithium in hexane. After 1 h of stirring, sulfur (1.28 g) is added and the mixture is stirred for a further 30 min. Water addition and silica gel chromatography yield 1.01 g (39%) of the product .[441]

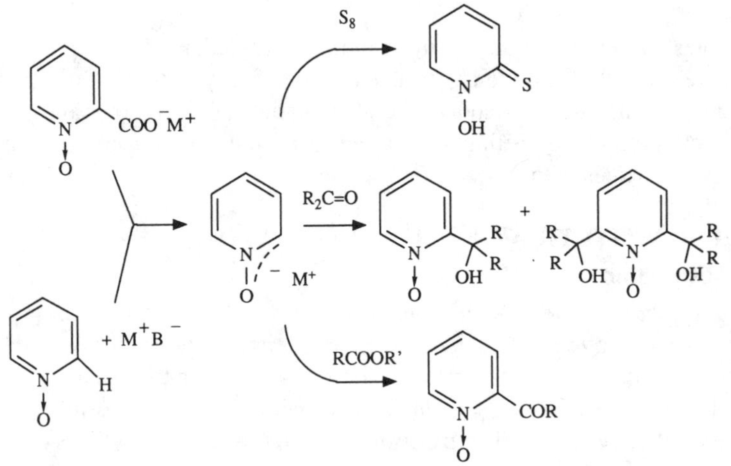

The reaction has been extended to many electrophiles. Thus, trapping with oxygen yields 1-hydroxy-2-pyridones,[441] with bromine and chlorine yields 2-halogeno 1-oxides,[440,442] and with mercuric chloride yields the 2-mercuriated derivatives.[442] Attack on a carbon-oxygen double bond yields hydroxyalkyl derivatives when ketones are used,[438,444] while carboxylic acids are obtained from CO_2[443] and ketones from esters;[443] related reactions take place with Schiff bases, amides, and nitriles.[443] Double addition (in 2,6) and dimerization accompany, to various extents, monosubstitution.

Yields by reaction of the previously prepared anion are mostly low, but they can be raised by *in situ* trapping. Thus, e.g., lithiation of pyridine 1-oxide in the presence of hexafluoroacetone yields 40% of the 2-alkylated and 50% of the 2,6-dialkylated derivative; likewise, lithiation of 4-*t*-butylpyridine 1-oxide in the presence of trimethylchlorosilane yields 99% of the disilylated derivative.[445]

The reaction is compatible with the presence of some substituents, e.g., 4-chloro and 4-methoxy groups are unaffected under these conditions, but in the presence of α- (and, to a lesser extent, γ-) alkyl groups, deprotonation and subsequent reactions take place competitively at the side chain (see Section XIB).

2. Reaction of Nucleophiles at an Unsubstituted Position in Activated N-Oxides

Direct substitution of a hydrogen without simultaneous deoxygenation (as opposed to deoxidative substitution, see Section VI) is possible in some cases 1) by reaction with strong nucleophiles, 2) when electron-withdrawing substituents activate an unsubstituted position of

TABLE 15

Functionalization by Attack at an Unsubstituted Position Under Retention of the *N*-Oxide Function

N-oxide	Substituent	Entering substituent (% yield)	Method	Ref.
Pyridine	(3,4-Me$_2$)	2-OH(10), 6-OH(14)	i) Stir for 1 h at –65°C with BuLi in C$_6$H$_{14}$ under N$_2$, ii) Bubble dry O$_2$ for 15 min at 20°C	441
		2-SH(22)	Heat at 80°C for 18 h a suspension of the *N*-oxide, LiH and sulfur in MeOCH$_2$CH$_2$OMe/MeOCH$_2$CH$_2$OH	441
	(4-Me)	2-SH(39)	i) Stir for 15 min at –65°C with BuLi in THF, ii) add excess sulfur and stir 30 min	441
	(4-*t*-Bu)	2-HgCl$_2$(13), 2,6(HgCl$_2$)$_2$(12)[a]	i) Stir for 1 h at –65°C with BuLi in THF under N$_2$ ii) Add HgCl$_2$ in THF dropwise	442
		2,6(SiMe$_3$)$_2$(99)	Treat with lithium tetramethylpiperidide and Me$_3$SiCl in Et$_2$O at 25°C	445
		2,6(CHOHMe)$_2$(30)	i) Stir for 15 min at –65°C with BuLi in THF, ii) And CH$_3$CHO and stir 1 h at –65°C	444
	(4-Cl)	2 COH(CH$_3$)$_5$(36), 2,6[COH(CH$_2$)$_5$]$_2$(21), 2[COH(CF$_3$)$_2$](40), 2,6[COH(CF$_3$)$_2$]$_2$(55)	i) Stir for 15 min at –65°C with BuLi in THF, ii) And (CH$_2$)$_5$CO and stir 1 h at –65°C Treat the *N*-oxide with (CF$_3$)$_2$CO and lithium tetramethyl piperidide in Et$_2$O at –78°C	444 444
	(3,4-Me$_2$)	6-Ac(65)	i) Stir for 15 min at –65°C with BuLi in C$_6$H$_{14}$ ii) add AcOEt in Et$_2$O and stir 1 h	445
	(4-Cl)	2-COOH(49a)	i) Stir for 15 min at –78°C with BuLi in abs. Et$_2$O, ii) Bubble CO$_2$ at –65°C for 3 h	443
Quinoline		2-OH(73)	Reflux for 1.5 h the *N*-oxide in benzene with Pb(OAc)$_4$ and CaCO$_3$	443
		2-OH(75)	Add simultaneously KOH and K$_3$Fe(CN)$_6$ solns. to the *N*-oxide in water at 5–10°C, leave overnight	460
		2-Me(40)	Stir for 5 h NaH in DMSO at 70°C under N$_2$, cool at r.t. add the *N*-oxide and stir for 2 h	459
		2-CH$_2$COMe(37)	Add the *N*-oxide to KO-*t*-Bu/Me$_2$CO in *t*-BuNH$_2$, at –10°C, stir for 2 h	447a 446a

TABLE 15 (continued)
Functionalization by Attack at an Unsubstituted Position Under Retention of the *N*-Oxide Function

N-oxide	Substituent	Entering substituent (% yield)	Method	Ref.
Quinoline	(4-Cl)	2-CH$_2$CO-t-Bu(80)	Add the *N*-oxide to KO-t-Bu/MeCO-t-Bu in t-BuNH$_2$ at −10°C, stir for 2 h	446a
	(4-NO$_2$)	3-CH(COOEt)$_2$(33)	Add NaOEt/EtOH to the *N*-oxide and CH$_2$(COOEt)$_2$ in abs EtOH, stir 12 h, filter the product as Na salt (36% yield)	451
Isoquinoline		2-CN(85)	Stir for 3 h at 75°C with aq KCN-K$_3$Fe(CN)$_6$	453
		1-Me(19)	Stir for 5 h at 70°C NaH and DMSO under N$_2$, cool at r.t., add the *N*-oxide and stir 2 h	447a
Acridine		1-CN(85)	Stir for 3 h at 75°C with aq KCN-K$_3$Fe(CN)$_6$	453
		9-CN(35)	Stir for 3 h at 70°C with KCN-K$_3$Fe(CN)$_6$ in 30% ETOH	453
Phthalazine[c]	(1-Ph)	4-CH$_2$COPh	Stir for 1.5 h at r.t. the *N*-oxide and PhCOMe with 50% aq NaOH	446b
Pyrrole	(4,5,5-Me$_3$)	2-OH(64)[b]	Stir for 2 h at 0°C the *N*-oxide in benzene with Pb(OAc)$_4$	463

[a] Determined as the corresponding bromides.
[b] 1-Acetoxy-4 5 5-trimethyl-2-pyrrolidone.
[c] 3-Oxide.

the substrate, 3) via elimination of an anion from the entering group ("vicarious" substitution), and 4) via the combined action of nucleophiles and oxidizers ("oxidative substitution").

a. Attack of Strong Nucleophiles

Quinoline 1-oxide is alkylated in position 2 by anions prepared from relatively weak carbon acids, e.g., methylketones and phenylacetonitriles with elimination of metal hydride. 3-Bromo- and 4-chloroquinoline 1-oxide are similarly alkylated in 2. No such reaction is observed when more stabilized carbanions, e.g., malonic acid derivatives, are used, though these react, of course, with acylated N-oxides, see Section VID1).[446a] A similar α-alkylation has been reported for a phthalazine oxide.[446b] Quinoline, isoquinoline, and benzo[f and g]quinoline (but not pyridine) N-oxides are methylated by reaction with the dimethylsulfinyl carbanion (see also Section VIIA2 for a competing reaction).[447] With methylthiomethyl-p-tolylsulfone, simple alkylation competes with vicarious substitution (see below).[448] A patent reports the successful ethynylation of pyridine and quinoline 1-oxides by treatment with sodium ethynylyde (compare Section VIIA).[449]

b. Attack on an Activated Substrate

Activation by a electron-withdrawing substituent causes, in some instances, substitution of a hydrogen rather than of the potential leaving group. Thus, the reaction of 3-nitroquinoline 1-oxide with cyanide in methanol involves attack of the activated position in 4 in preference to substitution of the nonactivated nitro group. The products are 3-methoxycynchoninonitrile 1-oxide and 3-methoxy-1-H-pyrazolo[4,3-b]quinoline-9-carboxamide, the latter resulting from attack at both activated positions (2 and 4) and cyclization.[450] Furthermore, while 4-nitroquinoline 1-oxide undergoes the expected substitution of the nitro group with heteroatom-centered nucleophiles (see Section VA1), its reaction with sodium diethyl malonate[451] and with enamines[452] proceeds through attack at nitro-activated position 3, rather than at position 4; in the latter case the product isolated arises from cyclization on the nitro group.

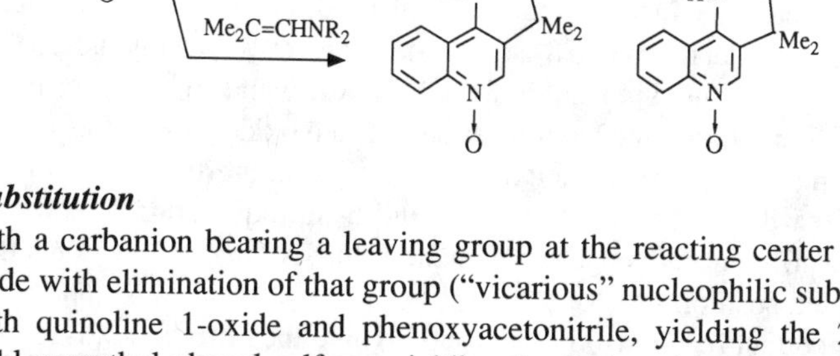

c. Vicarious Substitution

Reaction with a carbanion bearing a leaving group at the reacting center leads to an α-alkylated N-oxide with elimination of that group ("vicarious" nucleophilic substitution). This is the case with quinoline 1-oxide and phenoxyacetonitrile, yielding the 2-cyanomethyl derivative, or chloromethyl phenyl sulfone, yielding the 2-phenylsulfonylmethyl derivative. With methylthiomethyl p-tolyl sulfone both vicarious and normal substitution take place.[448]

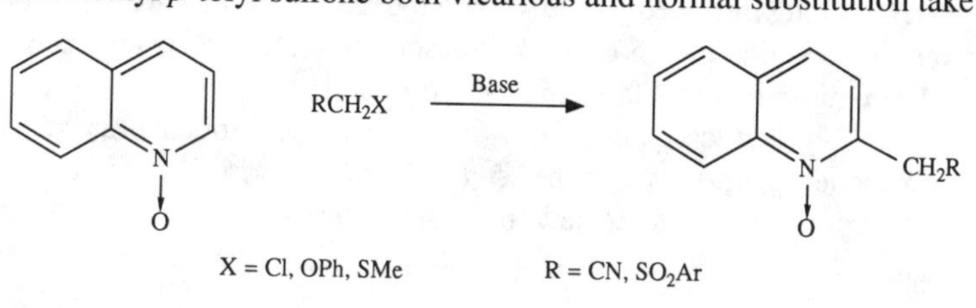

X = Cl, OPh, SMe   R = CN, SO₂Ar

d. Oxidative Substitution

The useful cyanation of acylated *N*-oxides (with related deoxygenation) is discussed in Section VIC1. Direct reaction with the *N*-oxides under oxidative conditions leads to the nitriles, without affecting the $\geqslant$N$\rightarrow$O function. Thus, treatment of bicyclic and tricyclic heterocyclic *N*-oxides with KCN and $K_3Fe(CN)_6$ ("oxidative cyanation") in protic solvents yields the corresponding nitriles (attack in 4 in quinoline, in 1 in isoquinoline, in 6 in phenanthridine, in 9 in acridine, and analogously in other *N*-oxides; monocyclic derivatives are unaffected, however).[453-455] Cyanogen bromide has been found to cyanate in α purine 3-oxide derivatives.[456]

In this group is also found oxidative amination, e.g., formation of the 5-amino derivative from 4-nitropyridazine 1-oxide upon treatment with $NH_3/KMnO_4$[457] and of the 5-amino from a similar reaction of 1,2,4-triazine 4-oxide,[458] as well as oxidative hydroxylation. The latter reaction occurs in the quinoline series by treatment with alkaline $K_3Fe(CN)_6$[459] or $Pb(OAc)_4$,[460-462] leading to good yields of 1-hydroxycarbostyril derivatives [with $Pb(OCOPh)_4$, the intermediate 1-acyloxy derivative is isolated]. Reaction with alkaline hydrogen peroxide similarly causes α-hydroxylation of phenanthridine 5-oxide[25] and of quinoxaline 1-oxide, in the latter case the reaction proceeds further to yield benzimidazole 1-oxides (compare with Section VIIA2).[790a] Hydroxylation of the ring is sometimes also caused by H_2O_2 under acidic conditions, i.e., under the conditions of preparation of *N*-oxides, (see Chapter 3).

As for cyclic nitrones, hydroxamic acids can be similarly prepared in the pyrroline series by reaction with lead tetraacetate[463] (as well as by $KMnO_4$, MnO_2, and $FeCl_3$, though the oxidation easily proceeds further with ring cleavage),[463-465] and in the azetine series, offering in this case a useful entry to *N*-acetoxy β-lactams.[466] 2-carboxypyrroline 1-oxides are decarboxylated by sodium hypobromide to again yield a hydroxamic acid, and 2-methylpyrroline 1-oxide (an analogue of a methylketone) undergoes a similar cleavage (haloformic reaction) under this condition, though the actually isolated product arises from further oxidation.[467]

In this connection, the nucleophilic attack by ozone on quinoline, isoquinoline, and phenanthridine *N*-oxides to yield the corresponding α-hydroxylated derivatives must be mentioned.[468] *N*-hydroxyphenanthridone and acridone are also obtained in the reaction of the *N*-oxides with potassium superoxide.[469] α-Hydroxylation is one of the reactions observed on heating quinoline 1-oxide with potassium *t*-butoxide and oxygen.[470]

C. NUCLEOPHILIC SUBSTITUTION ON A CONDENSED RING

In polycyclic heterocycle *N*-oxides, electron withdrawal toward the heterocyclic moiety also activates the carbocyclic ring for nucleophilic substitution, at least when mesomeric formulae with the negative charge on the oxygen are possible for the intermediate anion. Thus, fluorine is substituted by methoxide in 5-fluoroquinoline 1-oxide and by both methoxide and piperidine in the 7-fluoro isomer, but the 6- and 8-isomers do not react.[374] Analogously, a fluorine in 6 (and less easily in 8) is substituted in isoquinoline 2-oxides,[471] and so is a chlorine in phenazine 5-oxides.[472] 5- and 6-nitroquinoline 1-oxide react with methoxide, though the products are isolated in poor yield (compare with no reaction on the nonoxidized substrates),[342] and 2-nitrophenazine 10-oxide reacts easily.[184b] The reaction of 5- and 6- nitroquinoline 1-oxide with cyanide in methanol takes a different course (just as with the 3-nitro isomer, see Section VB2b), with attack at an unsubstituted position.[450]

The most important reactions in this group involve the nitro derivatives of benzofuroxan, benzimidazole 1-oxide, and related compounds. The acidic character of 4,6-dinitrobenzofuroxan (DNBF) in water has long intrigued scientists, and the metal salts that are easily obtained from it on treatment with alkaline bicarbonates were used for some time as explosives before the structure of salts of the anionic σ-adduct in position 7 was revealed.[473-475] In recent years, many such σ- (Meisenheimer) adducts from 4-nitro-[476,477] and 4,6-dinitrobenzofuroxan[478-480] (as well as from the corresponding furazans) and their 5- and 7-halogeno-[481,482] and alkoxy-substituted[483] derivatives with various nucleophiles have been characterized as crystalline salts or spectroscopically in solution. The attending kinetics have been extensively investigated, and the stability of these adducts is much higher than the analogous σ-adducts formed by polynitroaromatics. In some cases, double adducts of type **27** have been characterized.

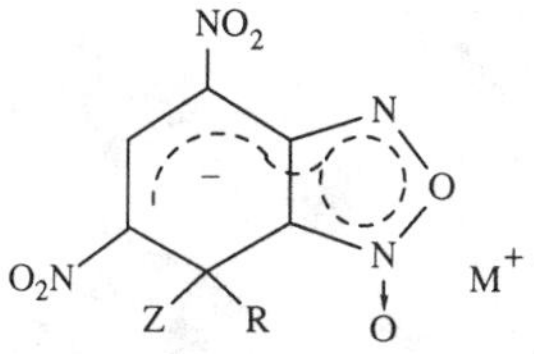

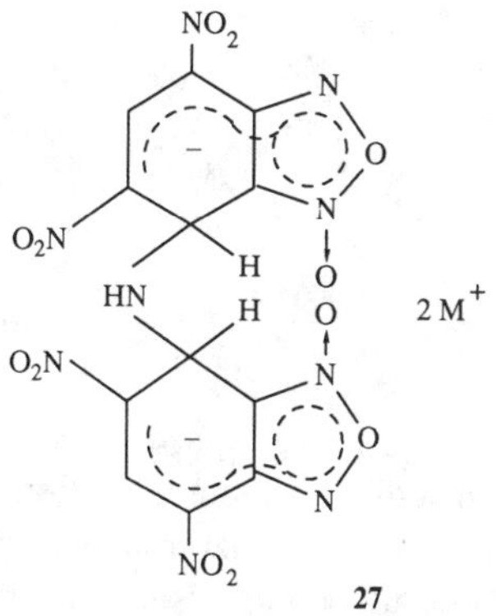

Even more interesting is the formation of carbon-carbon bonds. For example, with aromatic amines N σ-adducts are the kinetically preferred, but reversibly formed, adducts (particularly under base catalysis), and then give place to the thermodynamically more stable and isolable C-adducts.[484-486] Thus, from aniline, as well as N-methyl and N,N-dimethylaniline, zwitterionic adducts of type **28** are obtained in high yield (the corresponding metal salts are formed by treating with bicarbonates). 4-Substituted anilines react at the *ortho* position, and 2,4,6-trimethylaniline forms only a π-complex. Phenylaminobenzofuroxans are obtained from chloro derivatives of DNBF.[480]

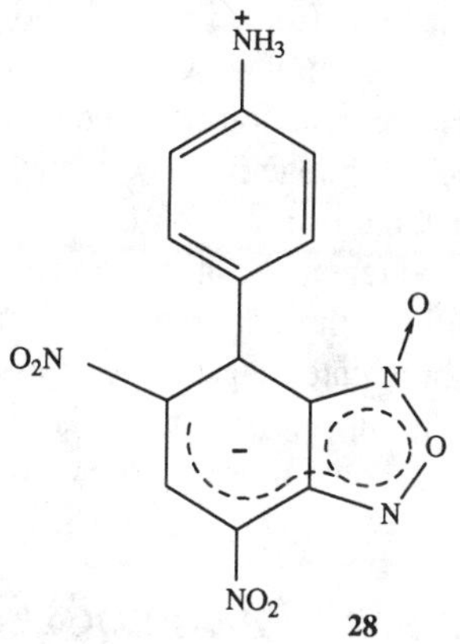

In view of the exceptional stability of the σ-adducts formed, DNBF and related compounds have been dubbed *super-electrophiles* (in contrast to "normal" electrophiles, such as polynitroaromatics),[487,488] and have been used to indicate the reactivity of weak carbon nucleophiles.

The (uncatalyzed) reactions of DNBF with ketones to yield (diastereoisomeric) 7-acylmethyl σ-adducts are typical and are analogous with β-diketones, in this case in the enolic form,[489] as well as with aromatic amines (including the "proton sponge", N,N,N′,N′-tetramethyl-1,8-diaminonaphthalene), phenols, and aromatic ethers.[484-488,490,491] With indene, an oxazine N-oxide is formed from the cyclization of an intermediate zwitterionic adduct.[492] With π-excessive five-membered heterocycles, attack takes place, as expected, in the α position,[493] and in β with 2,5-dimethylpyrrole and thiophene, as well as with 2-methylbenzofuran, but it is exclusively on the side chain with 2,5-dimethylfuran.[494]

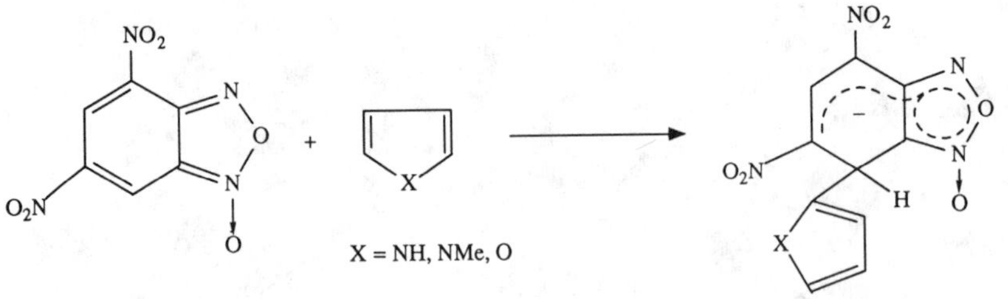

Not unexpectedly, covalent hydration also takes place in pyrimidofuroxans.[495] 4,6-Dinitro-2-(2,4,6-trinitrophenyl)-benzotriazole, an explosive, like most of the products discussed above, allows intramolecular comparison of the nucleophilic attack, either to yield a spectroscopically observable σ-adduct at position 7 of the benzotriazole moiety (both heteroatom-bonded, e.g., with thiols, and carbon-bonded, e.g., with *N,N*-dimethylaniline) or to yield displacement products through attack at the picryl moiety.[485,487,488,496,497b]

VI. DEOXIDATIVE NUCLEOPHILIC SUBSTITUTION

Since H⁻ is a poor leaving group, direct substitution of a hydrogen atom by a nucleophile is limited in scope in aromatic and heterocyclic chemistry, including the *N*-oxides, though some of these processes do occur satisfactorily (see Section VB). However, the *N*-oxide function offers a built-in hydride acceptor site, and easier processes, involving, at least from the stoichiometric point of view, elimination of a OH⁻ rather than of a H⁻, become possible and lead to the substituted deoxygenated heterocycle. Though in some cases such "deoxidative" nucleophilic substitutions take place by direct treatment of the *N*-oxide with the nucleophile, and then actually involve proton transfer from a ring carbon to the oxygen atom, in most instances these reactions are effected on the "activated" *N*-oxide, i.e., an *N*-acyloxy, -alkoxy, or -sulfonyloxy derivative prepared by reaction with the appropriate "auxiliary" electrophile, since such activation both enhances the reactivity with nucleophiles and offers a convenient oxygen acceptor site. Thus the mechanism is a variation of the normal addition-elimination mechanism, exploiting the enhanced reactivity of the oxygen atom in both steps (this has been referred to as an "abnormal" addition-elimination mechanism, S_NA, by some authors).[497a]

Variations include the use of bidentate reagents, with a site for electrophilic attack at the oxygen as well as the nucleophilic group, e.g., compounds containing a polarized double bond, such as imidoyl chlorides, or reagents containing an atom with high affinity for oxygen, e.g., silyl derivatives.

A. FORMATION OF HALOGENATED HETEROCYCLES

Inorganic halides cause, in some cases, a substitution reaction (e.g., 4-chloropyridine or quinoline 1-oxide from the corresponding nitro derivatives when treated with $POCl_3$), or on the contrary, they cause *N*-deoxygenation without affecting the substituents, or at least most substituents, as is usually the case with PCl_3 and PBr_3, (see Section IIIA). However, with P(V) and S(VI) halides, i.e., phosphorus oxychloride (or bromide) and pentachloride and sulfuryl chloride, both halogenation and deoxygenation take place. This process, sometimes referred to as the *Meisenheimer reaction,* since it was first reported by that author in 1926,[498] often gives satisfactory yields and is used largely for the synthesis of halogenated heterocycles.

Various pieces of evidence point to the intervention of both intramolecular (either concerted shift or recombination of the *O*-phosphorylated heterocycle-chloride ion pair within the solvent cage) and intermolecular (attack by a separated anion) mechanisms after an initial electrophilic attack on the oxygen atom, and their relative importance depends on the reagent chosen and the experimental conditions.[499,500] This leads to some change in the product distribution: with pyridine 1-oxide, 2- and 4-chloropyridine are formed in a 57:43 ratio with SO_2Cl_2,[1] 41:58 with PCl_5, and 68:32 with $POCl_3$.[500] It has been pointed out that small amounts of water-soluble ring-phosphorylated derivatives might be formed competitively with the chlorinated derivatives and be lost during the usual work-up (see Section VIG).[499] In the case of PCl_5, a different mechanism involving direct halogenation by PCl_4^+ might operate.[501]

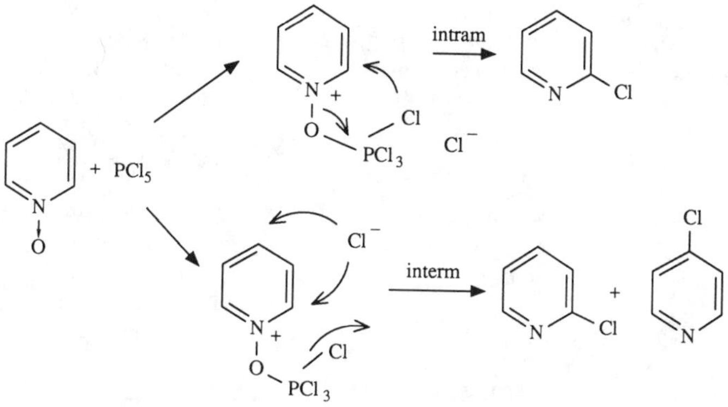

Substituent-induced regioselectivity may be important. Thus, 4-substituted pyridine 1-oxides are obviously chlorinated in 2,[500,502-504] except in the case of 4-cyanopyridine 1-oxide, which yields 3-chloro-4-cyanopyridine[505] (probably via attack at position 4, compare the reaction of the 4-nitro derivative below, followed by chlorine rearrangement or by addition of a second anion); with 4-cyano-3-ethylpyridine 1-oxide, the normal α orientation takes over again.[506] 2-Substituted pyridine 1-oxides react either in 4, as with the 2-methyl[500,507] (see, however, Section XIA2 for the competitive side-chain reaction) and the 2-phenyl[509] derivatives, or in 6, as with 2-chloro-[510] and 2-carboxypyridine 1-oxide.[511] A nitro and a carboxy group in 3 direct the attack to position 2[512] (in the latter case, this affords an industrial entry to 2-chloronicotinic acid).[513]

Some groups are substituted under the reaction conditions. Thus, while a 3-nitro group remains unchanged,[283] 4-nitropyridine 1-oxides invariably yield the corresponding 4-halopyridines ($POCl_3$, $POBr_3$, and HI/P have been used).[514,515] A 2- or 4-hydroxy group is likewise substituted;[516] notice that reaction under mild conditions ($POCl_3$ at <60°C) may result in OH/Cl exchange without deoxygenation,[516,517] while, on the other hand, these activated substrates undergo polychlorination under standard conditions.[516,517] As may be expected, amides are dehydrated to cyano derivatives during the reaction[503] (Table 16).

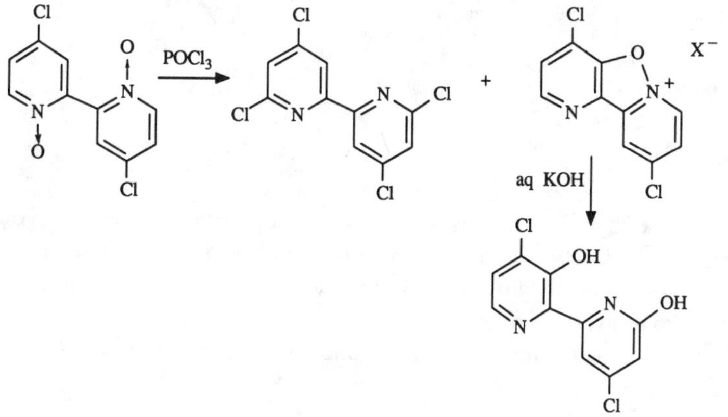

TABLE 16
Deoxygenative Halogenation

N-oxide	Substituent	Entering substituent[a] (% yield)	Conditions	Ref.
Pyridine		2-Cl(43), 4-Cl(20)	POCl$_3$, reflux 3.5 h	500
	2-Me	4-Cl(24)[b]	POCl$_3$, 100—140°C, 5h	507
	2-Ph	4-Cl(43)	SO$_2$Cl$_2$, reflux	509
	2-COOEt	6-Cl(69)	POCl$_3$/CHCl$_3$, 80°C, 7 h	511
	3-Me	2-Cl(30), 4-Cl(42), 6-Cl(25)	POCl$_3$, reflux 1.5 h	500
	3-COOH	2-Cl(41)	PCl$_5$/POCl$_3$, 115—120°C, 1.5 h	512a
	3-NO$_2$	2-Cl(64)	PCl$_5$/POCl$_3$, 150°C, 2.5 h	283
	4-Me	2-Cl(34)	POCl$_3$, reflux 1.5 h	500
	4-COOEt	2-Cl(70)	POCl$_3$, 130—150°, 2 h	502
	4-CONH$_2$	2-Cl(72)[c]	PCl$_5$/POCl$_3$, 120—130°C, 1.5 h	503
	4-CN	3-Cl(73)	PCl$_5$/POCl$_3$, 120—130°C, 1.5 h	505
	4-CN-3-Et	2-Cl(16), 6-Cl(16)	POCl$_3$/CHCl$_3$, reflux 5 h	506
Quinoline	2-Me	4-Cl(high)	SO$_2$Cl$_2$, reflux 2 h	522
	3-Me	2-Cl(78)	POCl$_3$, reflux 15 min	523
Isoquinoline		1-Cl(57), 3-Cl(5), 4-Cl(12)	POCl$_3$, reflux 2 h	526
Phenanthridine		9-Cl(83)	POCl$_3$, 100°C, 15 min	528
Pyridazine	3-OMe	6-Cl(55)	POCl$_3$, 30°C, 1.5 h	537
	3,6-Me$_2$	4-Cl(28)	POCl$_3$, 60—70°C, 30 min	538
	3,6-(OMe)$_2$	4-Cl(72)	POCl$_3$, r t., 4h	539
Pyrimidine[d]	4-Ph	2-Cl(50)[e]	POCl$_3$, 20°C, 5 h	540
Pyrazine		2-Cl(25)	POCl$_3$, reflux 15 min	541
	2-OMe	3-Cl(7), 6-Cl(54)	POCl$_3$, reflux 2 h	546
	2-COOMe	3-Cl(1), 6-Cl(88)	POCl$_3$, reflux 2 h	546
	3-OMe	2-Cl(35), 5-Cl(41), 6-Cl(5)	POCl$_3$, reflux 2 h	546
	3-COOMe	2-Cl(5), 5-Cl(82), 6-Cl(3)	POCl$_3$, reflux 2 h	546
	3-NH$_2$	2-Cl(51)	POCl$_3$, reflux 2 h	546
Pyrazine[f]		2,6-Cl$_2$(85)	POCl$_3$ reflux 45 min	542
Cinnoline[g]		4-Cl(50)	POCl$_3$, r.t., 45 min.	126
Phthalazine[c]	(1-Ph)	4-Cl(74)	POCl$_3$	560b
Quinazoline[g]	(4-iPr)	2-Cl(83)	POCl$_3$, r.t., 3 h	77
	(4-OMe)	2-Cl(55)	SO$_2$Cl$_2$ reflux 30 min	560c
Quinoxaline[g]		2-Cl(90)	POCl$_3$ reflux 15 min	556
Quinoxaline[f]		6-Cl(20)	AcCl, r.t., 1 h	557
		2,3-Cl$_2$	POCl$_3$, reflux 15 min	556
Phenazine		2-Cl(73)	POCl$_3$, reflux 3 h	560a
Pyrazole	(1-Me)	5-Cl(62)	POCl$_3$, 100°C, 30 min	275a
Benzimidazole	(1-Me)	2-Cl(75)	SOCl$_2$/CHCl$_3$, reflux 10 min	24
		2-Cl(58)	TsCl/CHCl$_3$, reflux 5 min	24

[a] Numbering of the position referred to the starting material.
[b] 26% of 2-chloromethylpyridine also obtained.
[c] 4-CN derivative.
[d] 1,3-Dioxide.
[e] 3-Oxide.
[f] 1,4-Dioxide.
[g] 1-Oxide.

Synthetically useful halogenations are also obtained from bi- and terpyridine polyoxides.[518,519] In 2,2′-bipyridine dioxides, the presence of a substituent in the 4 position favors a intramolecular cyclization to dipyridoisoxazolinium salts.[518] 2,7-Diazabiphenylene 2-oxide yields the 3-chloro derivative.[520]

The reaction occurs with quinoline 1-oxides under milder conditions than with the pyridines (usually an exothermal reaction is observed on mixing of the reagents, and refluxing for some minutes completes the reaction), and again leads to halogenation in 2 and 4;[521-524] a 4-nitro group is substituted.[66,344,366] Substituents in the benzo ring influence the product distribution. Thus, the ratio of 4-chloro to 2-chloro derivatives is 1.7 in the parent N-oxide, 0.6 in the 6-methoxy, and 3.5 in the 6-nitro derivatives.[521] 5-nitroquinoline 1-oxide yields some 3-chloro-5-nitroquinoline (20%), together with the 2- and 4-chloro isomers (35% and 10%).[525] Isoquinoline 2-oxide reacts mainly at position 1,[526] acridine 10-oxide at position 9,[527] and phenanthridine 5-oxides at position 6 (or 3 when this is occupied).[528] Pyrrolo- and imidazopyridine N-oxides are chlorinated in γ to the ≥N→O group in the pyridine ring.[529,530]

Among polycyclic derivatives, 1,5-, 1,6-, 1,7-, and 1,8-naphthyridine 1 oxides yield the 2-(from an intramolecular) and the 4-chloro derivatives (from an intermolecular pathway) in a ratio that can be explained by considering the inductive and mesomeric effect of the second nitrogen, with some reaction also taking place at the 3 position.[531] 1,5- and 1,6-Naphthyridine dioxides yield the 2,6-dichloro and, respectively, a mixture of 2,5-, 3,5- and 4,5- dichloro derivatives.[532,533] A single chlorinated product has been obtained from 1,8-phenanthroline 8-oxide[534] (the 7-chloro) and from 1,7-phenanthroline 7-oxide[535a] (the 8-chloro). 4,6-Phenanthroline 6-oxide reacts in 5 and the 4-oxide reacts mainly in 3.[535b]

3-Substituted pyridazine 1-oxides are chlorinated in 6[536,537] and are 3,6-disubstituted in 4.[538,539] 4-Phenylpyrimidine 1,3-dioxide yields 2-chloro-4-phenylpyrimidine 3-oxide.[540]

Much attention has been given to the Meisenheimer reaction of pyrazine N-oxides. The parent 1-oxide yields the 2-chloro[541] and the 1,4-dioxide the 2,6-dichloro derivative[542]; the reaction has been extended to a great variety of substrates, with the interesting result that reaction occurs both at the α and the β positions.[543-547] For example, 2-substituted pyrazine 1-oxide reacts mainly at position 6, with some reaction at position 3 when the substituent is electron donating, but with 3-substituted 1-oxides the reaction is mainly at position 5 when the group is electron withdrawing, and at both positions 2 and 5 with the 3-methoxy derivative. Various rationalizations have been advanced for this selectivity, e.g., dependence of the (intermolecular) attack on the π-electron density of the substrate,[544,545] or alternatively, formation of a dichlorodihydroadduct, and competitive dehalogenation to the observed products.[546] 3-Phenyl-1-hydroxy-2(1H)pyrazinone is chlorinated 5[548] (and the 3,5-diphenyl analogue in 6).[549] An interesting variation is the formation of 1-(2-pyrazinyl)pyrazinium salts, isolated as the chlorophosphate or hydrolyzed during work-up to 2-aminopyrazine in the reaction of pyrazine 1-oxide with POCl₃; from the NMR spectrum, the salt appears to be present in an equimolecular amount with 2-chloropyrazine before work-up.[543a]

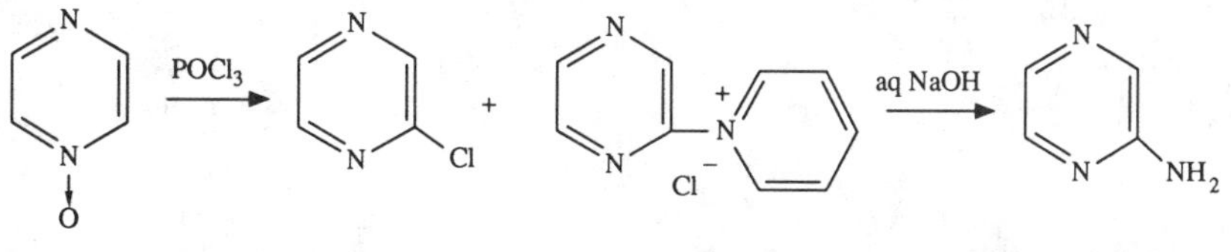

Chlorination of aminopyrazine N-oxides has been used largely as a key step for the synthesis of pteridines,[550,551] and chlorination of the appropriate pyrazine oxide offers a pathway for the synthesis of aspergillic acid.[552] 2,3-Diphenylpyrazine 1,4-dioxide yields the 5,6-dichloropyrazine, or, through partial reaction, the 6-chloropyrazine 1-oxide, which is easily transformed into the corresponding hydroxamic acid.[553] 2,5-Disubstituted pyrazine dioxides yield the 3,6-dichloropyrazines, and thus offer a convenient entry to diketopiperazines.[554,555]

With quinoxaline N-oxides, reactions at the heterocyclic and at the carbocyclic ring compete. Thus, on treatment with POCl₃, the 1-oxide yields 2-chloroquinoline and the dioxide yields the 2,3-dichloro derivative;[556] however, the dioxide[557] and some 3-hydroxy 1-oxides[558] unexpectedly yield 6-chloroquinoline derivatives when treated with acetyl chloride. 2,3-Diphenylqui-noxaline 1-oxide is chlorinated (as well as phosphorylated, see Section VIG) in 6,[499] 6-nitroquinoxaline 1-oxide reacts in 5, 7, and, to a minor extent, in 2, while the 5-nitro isomer reacts in 4 (in the first case the orientation of the two functions present is the same, in the latter the effect of the $\geqslant N \rightarrow O$ group predominates).[559]

Phenazine 5-oxide is chlorinated in 2.[560] Similar chlorinations of purine,[561] pteridine,[562-563] pyrimidopyridazine,[564] and flavine[565] N-oxides have been reported.

1-Methylpyrazole 2-oxide yields 5-chloropyrazole,[275a] and benzimidazole N-oxides yield the 2-chloro derivatives.[24]

2-Amino-3-cyano-5-methyl-6-chloropyrazine. A suspension of 3 g of 2-amino-3-cyano-5-methylpyrazine 1-oxide in 30 ml of DMF is treated dropwise with 6 ml of POCl₃ while maintaining the temperature at 80—90°C by cooling. The mixture is stirred at 80—90°C for an additional 10 min and poured in 300 ml water at 0°C. Filtration and recrystallization give 2.5 g (74%) of the product.[550a]

B. FORMATION OF HYDROXY AND ACYLOXY HETEROCYCLES

Heterocyclic N-oxides are easily O-acylated by anhydrides or halides of carboxylic acids, and salts of the N-acyloxy cations have been isolated in some cases (see Section IIC). However, usually the reaction proceeds further through a rearrangement to yield the α-acyloxy hetero-cycle, which in turn is easily hydrolized to the α-hydroxy heterocycle, or rather the tautomeric lactam; the entire reaction sequence formally corresponds to an isomerization process through migration of the oxygen atom. Thus, by boiling pyridine 1-oxide with acetic anhydride and anhydrous sodium acetate and pouring in water, the reaction mixture leads to a high yield of 2-pyridone.[566a] This procedure, often referred to as the *Katada reaction,* is of general application and has been extended to other carboxylic anhydrides (for less reactive N-oxides, trichloroacetic or trifluoroacetic anhydrides may be required, though this causes other reactions to occur in some cases, see below). Alternatively, the rearrangement is obtained by treatment with acyl (rarely sulfonyl) chlorides and then with bases (see below) or under related conditions, e.g., by heating with H_3PO_4/KH_2PO_4.[566b] Formation of lactams is often a side reaction in other deoxidative substitutions, e.g., cyanation, where anhydrides or other acylating reagents are used for activation. The reaction is somewhat analogous with the isomerization of aldonitrones to N-substituted amides (sometimes indicated as a Beckmann rearrangement; Table 17).

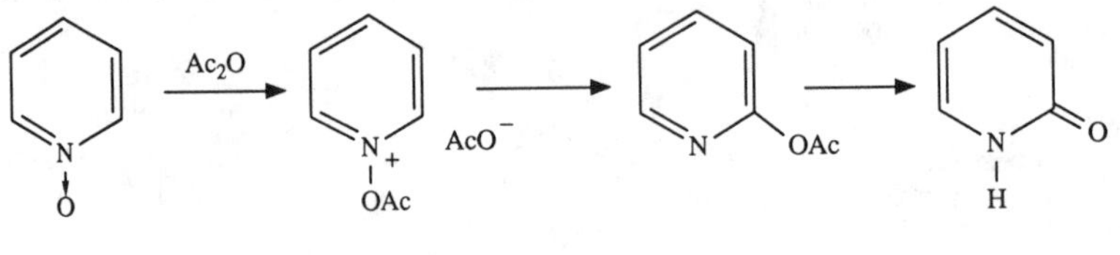

minor products

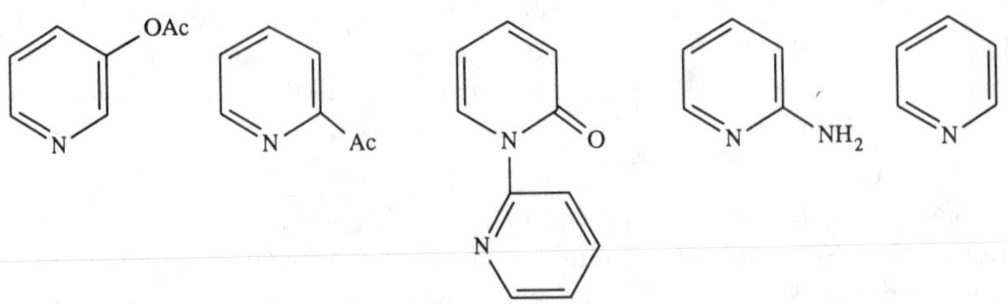

TABLE 17

Isomerization to Hydroxy Derivatives or Lactams[a] and Deoxidative Ring Acetoxylation

N-oxide	Substituent	Entering substituent (% yield)	Reagent	Ref.
Pyridine		2-OH(60)	Ac$_2$O, 140°C, 24 h	571
	(2-Cl)	2-OH(76)	Ac$_2$O, 130—140°C	510
	(2-OEt)	2-OH(62)	Ac$_2$O, 160—170°C	510
	(2-OPh)	2-OH(84)	Ac$_2$O, 160—170°C	510
	(3-Cl)	2-OAc(61)	Ac$_2$O, reflux, 4 h	578
	(3-NO$_2$)	2-OH(50)	Ac$_2$O, reflux, 24 h	59
	(3-NHAc)	2-OH(35)	Ac$_2$O, reflux, 4 h	579
Quinoline		2-OH(90)	Ac$_2$O, reflux, 3 h	582b
	(2-CN)	4-OH(83)	(CF$_3$CO)$_2$O, CH$_2$Cl$_2$, reflux, 5d	583
	(4-CN)	2-OH(64), 3-OH(36)	(CF$_3$CO)$_2$O, CH$_2$Cl$_2$, reflux, 12d	583
Isoquinoline		1-OH(64), 4-OH(8)	Ac$_2$O, reflux, 5h	587
Phenanthridine		6-OH(60)	Ac$_2$O/AcOH, reflux, 5h	25
		6-OH(60)	TsCl, reflux, aq K$_2$CO$_3$	25
1,6-Naphthyridine[b]		2-OH(10)	Ac$_2$O, 160°C	532b
1,6-Naphthyridine[c]		5-OH(23)	Ac$_2$O, 140—160°C	532b
4,6-Phenanthroline		5-OH(71)	Ac$_2$O, reflux	535b
Pyridazine	(3-OH)	6-OH	Ac$_2$O, reflux	589
Pyrimidine		4-OAc(38)	Ac$_2$O, 132°C, 3h	591
Pyrazine	(2-Ph)	3-OH(4), 6-OH(13)	Ac$_2$O, reflux	544a
	(3-Ph)	2-OH(13), 5-OH(16), 6-OH(4)	Ac$_2$O, reflux	544a
Quinoxaline	(3-Ph)	2-OH(13)	Ac$_2$O, reflux	115
Thiazole	(2,5-Ph$_2$)	4-OAc(95)	Ac$_2$O, 130°C	597a
Pyrazole[e]	(1-Me)	3-OAc[d]		596
Benzimidazole[f]	(1-Me)	2-OH(quant)	CHCl$_3$, reflux	24

[a] The hydroxy tautomer is indicated as the product.
[b] 1-Oxide.
[c] 6-Oxide.
[d] A mixture of 3-acetoxy-1-methylpyrazole and 2-acetyl-1-methyl-3-pyrazolone is obtained.
[e] 2-Oxide.
[f] 3-Oxide.

Mechanistic investigations[567] (see also Section XIA1 for the related reaction of alkylpyridine 1-oxides) involving kinetic studies,[568] observation of ^{18}O incorporation using labeled anydride and of a low isotopic effect in 2,6-d$_2$-pyridine 1-oxide,[569] as well as trapping experiments with anisole and benzonitrile[570] suggest a ionic mechanism, with a slow addition of acetate on the 1-acetoxypyridium cation, followed by fast elimination of acetic acid.

This kind of rearrangement is not the only process observed on treatment of the N-oxides with anhydrides. In several cases, ring opening or transformation (see Section VII), redox processes (Chapter 5), or reaction at the substituent (Section XIA1) occur.

From the synthetic point of view, the Katada reaction has the advantage of a nearly exclusive α-orientation, as opposed to the α,γ-orientation of the Meisenheimer reaction. Indeed, though quantitative acetoxylation in 2 was originally reported for pyridine 1-oxide,[566a] later studies showed that other products are formed in a low amount, including 3-acetoxypyridine, 1-(2-pyridyl)-2(1H)pyridone, 2-aminopyridine, and 2-acetylpyridine (15% in total).[571] At any rate, α selectivity overwhelmingly remains (see below for exceptions), and in some cases it is advantageous to prepare α-chloro heterocycles via the Katada reaction and the following substitution, rather than by direct reaction of the N-oxide with POCl$_3$, since in the latter case an isomeric mixture is obtained (this applies, e.g., with some pyridinylpyridine 1-oxides).[572a] Trichloroacetic anhydride in chloroform does not react with pyridine 1-oxide, but in acetonitrile an easy reaction occurs at room temperature and yields a complex mixture containing 2- and 4-trichloromethyl- and dichloromethyl-pyridine.[572b]

With 2- and 4-alkyl-substituted pyridine 1-oxides, reaction at the side chain, as discussed in Section XIA1, usually predominates over acetoxylation of the ring. Potential leaving groups in position 2 are eliminated. Thus, 2-chloro-, 2-alkoxy-, and 2-phenoxy-pyridine 1-oxide all yield 1-hydroxy-2-pyridone.[510] Likewise, 1-methoxypyridinium polychloropyridine methosulfates yield 1-methoxy-2-pyridones.[573] Under this condition, 2-carboxypyridine 1-oxide is decarboxylated to a mixture of pyridine 1-oxide and 2-pyridone, but with the corresponding esters, 6-carboxy-2-pyridones are formed.[510,574] With 2-acylpyridine 1-oxide, addition of the anhydride at the substituent (in the enolic form) takes place, instead of the usual rearrangement.[575]

3-Methylpyridine 1-oxide yields a mixture of 3- and 5-substituted 2-pyridones,[576,577] as does 3-carbomethoxypyridine 1-oxide,[574] but 3-halo-,[578] nitro-,[59] and amino-[579] pyridine 1-oxides yield only the corresponding 3-substituted 2-pyridones. With nicotinic acid 1-oxide the 2- and 6-pyridones are only minor products, the main pathway being acylation (see Section IVF). Anomalous eliminations take place in some cases. Thus, while 3-trimethylsilylpyridine 1-oxide yields, as expected, a mixture of the 3- and 5-substituted 2-pyridones,[580] in the heavily substituted derivative **29** the silyl group is eliminated to yield a 3-hydroxypyridine.[581]

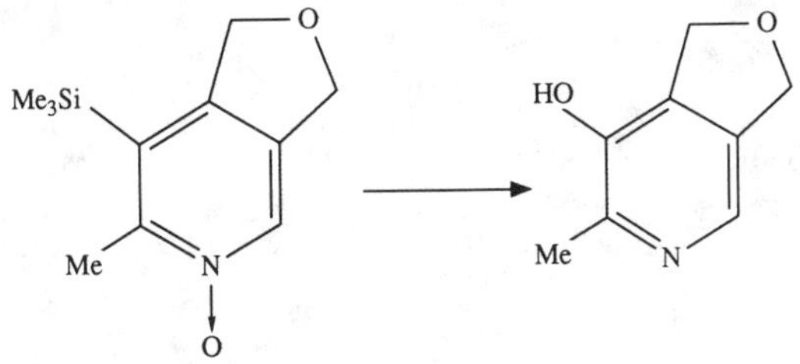

29

Carbostyrils are formed from quinoline 1-oxide by reaction with acetic anhydride[582,583] or by treating with acyl or sulfonyl chlorides and then with aqueous bases.[584,585] 2- and 4-Cyanoquinoline 1-oxides require trifluoroacetic anydride for reacting and yield the 4-quinolone in the former case and both the carbostyril and the 3-hydroxyquinoline in the latter.[583] Isoquinoline 2-oxide yields isocarbostyril along with a small amount of 4-hydroxyisoquinoline,[586,587] and acridone is obtained from acridine 10-oxide.[527] While pyrrolo[2,3-*b*]pyridine 7-oxide yields the expected lactam,[529] the corresponding pyrazolo derivative yields both the lactam and the 5-acetoxy derivative.[588]

As for diazine *N*-oxides,[589,590] noteworthy is that acetoxylation of pyrazine derivatives takes place both in α and β, just as is the case for chlorination (Section VIA).[543,544,555] Quinoxaline 1-oxides with a 2 position free behave normally.[115] The dioxide yields 1-hydroxyquinoxaline-2-one (and quinoxalin-2,3-dione on prolonged heating),[592,593] and in the corresponding 6-substituted derivatives, the hydroxamic function is formed exclusively *para* to the substituent when this is electron donating, and with the opposite regiochemistry when it is electron withdrawing.[593] Various purine 3-oxides have been found to rearrange to the 8-keto derivatives on treatment with anhydrides or dimethyl sulfate,[594] and useful rearrangements also take place in the pteridine series.[595]

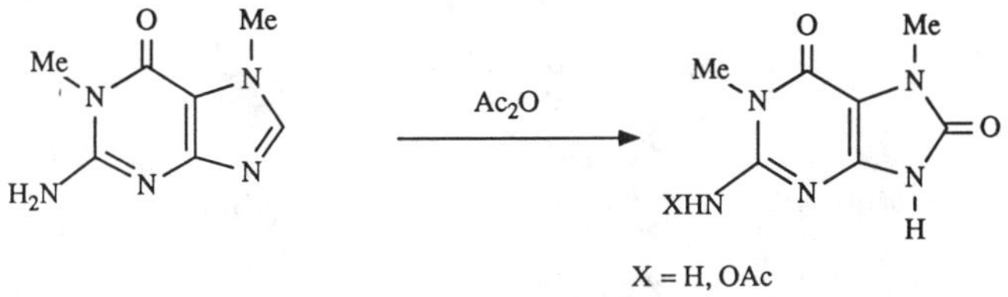

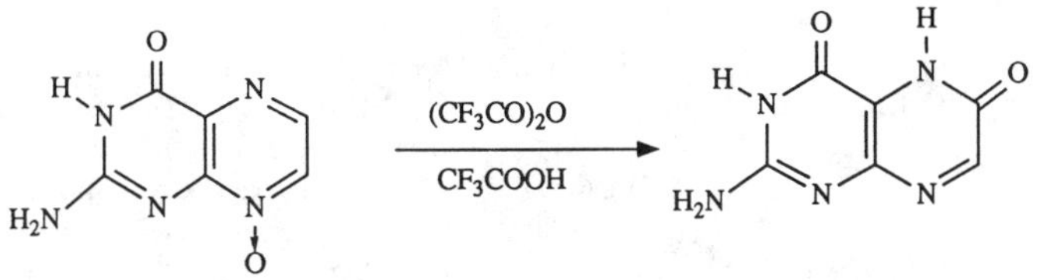

X = H, OAc

1-Methylpyrazole 2-oxide yields 3-acetoxypyrazole and 2-acetylpyrazol-3-one,[596] and from some thiazole 3-oxide 4,5-dihydro-4,5-diacetoxy adducts have been isolated, which are easily converted to the 4-acetoxythiazoles.[597a] Related reactions have been reported for imidazole and indolenine N-oxides.[597b,c] 1-Methylbenzimidazole 3-oxides yield the 3-acetylbenzimidazolone on heating with acetic anhydride, and rearrange to 1-methylbenzimidazolone on simply refluxing in acetone or chloroform.[24]

1,4-Benzodiazepine 4-oxides yield 3-hydroxydiazepines with acetic anhydride, apparently through an intramolecular rearrangement of the initially formed 4-acyloxy derivative,[598a] and similar reactions have been extensively used in this field.[598b,c]

2,4,4-Trimethylpyrroline 1-oxide yields the 1-acetyl-3-acetoxy-2-methylenepyrroline with acetic anhydride[599] and 3-benzoyloxypyrroline with benzoyl chloride.[600] These reactions of cyclic nitrones are more akin to the acetoxylation of alkyl substituted N-oxides discussed in Section XIA1 than to the ring acetoxylation of heteroaromatic derivatives.

Carbostyril. Quinoline 1-oxide (2 g) in 20 ml Ac$_2$O is refluxed for 2 h. The mixture is poured on 200 g ice, stirred for 3 h, and the product is filtered and recrystallized (1.8 g, 90%).[582b]

C. CYANATION, CARBAMIDATION, AND ACYLATION

Deoxidative cyanation is one of the most typical reactions of N-oxides and occurs under mild conditions. New methods of excellent synthetic value have been recently added to the classical reaction with benzoyl chloride/potassium cyanide.

1. Cyanation Under Acylating Conditions (Reissert-Henze Reaction)

Quinolines, isoquinolines, and some other polycyclic heterocycles yield α-cyano-N-acyldihydro derivatives (Reissert compounds) when treated with an acyl chloride and aqueous cyanide,[601] as first reported by Reissert in 1905.[602] In 1936 Henze found that quinoline 1-oxide yields 2-cyanoquinoline under the same conditions.[603] The mechanism is analogous, in that the cyanide adds to the heterocyclic cation to yield a dihydro adduct, but since in this case an N-acyloxy- rather than an N-acyl-dihydro derivative is formed, elimination is easier and the rearomatized nitrile is directly obtained. In view of the mild conditions and tolerance to substituents, the reaction has been largely used to gain access to α C-substituted N heterocycles, though there are limitations in its scope (see below). The reaction is conveniently carried out by treating the N-oxide with benzoyl chloride and aqueous potassium cyanide at room temperature, with some α-oxo or acyloxy derivatives being formed as by products (compare with Section VIB), and occasionally as the main products. The use of silver cyanide,[529,604] rather than alkali cyanide, has been shown to be advantageous in some cases. With tosyl chloride in place of benzoyl chloride, tosylated rather than cyanated products usually predominate (see Section VIF2; see Table 18).

TABLE 18
Deoxidative Cyanation

N-oxide	Substituent	Position of attack (% yield)	Reagents	Ref.
Pyridine		2(94)	Me$_3$SiCN, Me$_2$NCOCl, CH$_2$Cl$_2$, r.t., 1 d	634
		2(76)	NaCN, Me$_3$SiCl, Et$_3$N, DMF, 100°C, 18 h	635
	(2-Me)	4(18), 6(45)	i) Me$_2$SO$_4$, reflux, ii) KCN, r.t. 1 h	628
	(2-COOMe)	6(50)	i) Me$_2$SO$_4$, reflux, ii) KCN, r.t.	624
	(3-Cl)	2(85) 6(6)	Me$_3$SiCN, Et$_3$N, MeCN, reflux, 6 h	638
	(3-OMe)	2(73)	Me$_3$SiCN, Et$_3$N, MeCN, reflux, 6 h	638
	(3-OMe)	2(68)	i) Me$_2$SO$_4$, reflux; ii) KCN, r.t.	624
	(3-COOEt)	2(36), 6(51)	Me$_3$SiCN, Et$_3$N, MeCN, reflux 6 h	638
	(3-CN)	2(18), 6(28)	i) Me$_2$SO$_4$, reflux; ii) KCN, r.t.	624
	(4-COOMe)	2(69)	i) Me$_2$SO$_4$, reflux; ii) KCN, r.t.	624
	(4-NO$_2$)	2(55)	i) Me$_2$SO$_4$, 60°C; ii) KCN, r.t. overnight	627
	(4-Cl)	2(63)	PhCOCl, KCN r.t.	605
Quinoline		2(93)	i) Me$_2$SO$_4$, 100°C; ii) KCN, r.t., 4 h	21
		2(90)	NaCN, Me$_3$SiCl, DMF, 100°C, 3 h	635
	(4-OMe)	2(90)	PhCOCl, KCN, r.t.	606
	(3-NO$_2$)	2(81)	PhCOCl, AgCN, r.t., 2 d	604
	(8-OH)	2(72)	i) MeSO$_4$, heat; ii) NaCN, r.t.	631
Isoquinoline		1(95)	i) Me$_2$SO$_4$, 100°C; ii) KCN, r.t., 4 h	21
		1(81)	(EtO)$_2$POCl, Et$_3$N, DMF, 100°C, 3 h	636
	(5-OH)	1(53)	PhCOCl, KCN, r.t.	607
Phenanthridine		6(19)	PhCOCl, KCN, r.t., overnight	25
		6(94)	i)Me$_2$SO$_4$ reflux; ii) KCN, r.t., 20 min	25
Benzo[*f*]quinoline		2(74)	PhCOCl, KCN, r.t., 3 h	613
1,6-Naphthpyridine[a]		5(9)	PhCOCl, KCN, 15°C, 3 h	614
1,10-Phenanthroline		2(71)	PhCOCl, KCN, r.t., 30 min.	615
4,6-Phenanthroline[b]		2(71)	PhCOCl, KCN, r.t., 30 min.	623
4,6-Phenanthroline[a]		5(48)	PhCOCl, KCN, r.t.,	615
Thieno[2,3-*b*]pyridine		6(46)	PhCOCl, KCN, 25°C 16 h	617
Thieno[3,2-*c*]pyridine		4(84)	PhCOCl, KCN, 25°C 16 h	617
Pyridazine	(5-NH$_2$)	6(46)	Me$_2$SO$_4$, 4 h, 100°C; KCN, 10°C, 10 min	632
Pyrimidine		2(30), 6(45)	Me$_3$SiCN, Et$_3$N, MeCN, reflux, 10 min	640
	(2-Me)	6(71)	Me$_3$SiCN, Et$_3$N, MeCN, reflux, 2.5 h	640
	(4-OMe, 6-Me)	2(91)	PhCOCl, NaCN, r.t., overnight	620
Quinazoline[c]	(4-iPr)	2(33)[d]	PhCOCl, KCN, r.t., overnight	77
	(4-OMe)	2(87)	PhCOCl, KCN, r.t., overnight	560c
Quinoxaline	(3-Ph)	2(8)	PhCOCl, KCN, r.t.	115
Benzimidazole[e]	(1-Me)	2(high)	PhCOCl, KCN, r.t., 10 min	24
	(1,2-Me$_2$)	6(20)	PhCOCl, KCN, r.t., 5 min	104

[a] 6-Oxide.
[b] 4-Oxide.
[c] 1-Oxide.
[d] Along with 35% 2-(4-quinoxalyl)-2-methylpropionamide.
[e] 3-Oxide.

Various 2-cyanoquinolines,[603-606] as well as 1-cyanoisoquinolines,[257,607] are obtained in high yield from the respective *N*-oxides. In 2-substituted quinoline 1-oxides, the cyanide attacks at position 4, but in this case dihydro adducts are often obtained, along with rearomatized nitriles and/or benzoyloxyquinolines. Thus, 2-arylquinoline 1-oxides yield 1-benzoyloxy-4-cyano-dihydroquinolines, as well as 3-, 6-, and 8-benzoyloxyquinolines.[608,609] With 2-alkyl derivatives, an acyloxy group also enters into the side chain, and with 2,3- and 2,4-disubstituted quinoline 1-oxides, 2-hydroxy-3-acyloxy-4-cyano-tetrahydroquinolines or the corresponding 2-methoxy derivatives (from the use of water, or, respectively, methanol, as a solvent), or, when the group in 2 is a methyl, 2-methylenetrahydroquinolines are obtained.[610,611] Analogously, 1-substituted isoquinoline 2-oxides yield 4-acyl-1-cyanodihydroisoquinolines.[612]

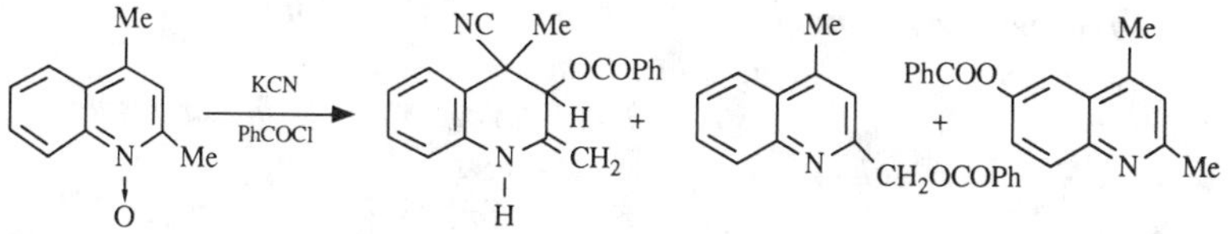

The Reissert-Henze reaction has been extended to various benzoquinoline,[613] phenanthridine,[25] quinoxaline,[115] phthalazine,[560b] quinazoline,[77,560c] naphthyridine,[614] phenanthroline,[615,616] pyrrolo- and thienopyridine,[529,617] as well as pyrazolopyrimidine *N*-oxides (in this case, with MeSO$_2$Cl/KCN in dioxane water).[618] 9-Methylguanine 7-oxide yields the 8-cyano derivative.[622]

However, the reaction works poorly with monocyclic *N*-oxides, pyridine 1-oxide, for example, being recovered unchanged under these conditions.[605] The exceptions are the 4-chloro[605] and the three trifluoromethyl[619] derivatives in the pyridine series and some pyrimidine 1-oxides.[620,621] As for five-membered rings, noteworthy is the easy reaction of 1-methylbenzimidazole 3-oxide to yield the 2-cyano derivative;[24] in 1,2-dimethylbenzimidazole 3-oxide the attack is on the benzo ring, at position 6.[104]

5-Benzoxyisoquinoline-1-carbonitrile. *10 g of KCN are added to a suspension of 10 g 5-hydroxyisoquinoline 2-oxide in 200 ml water, and 30 ml benzoyl chloride are added during 15 min (heat evolution). Extraction with chloroform and evaporation yields the product (9 g, 53%).*[607]

2. Cyanation via *N*-Alkoxy Derivatives

N-alkoxy heterocycles are more reactive with cyanide than are *N*-acyloxy derivatives. As mentioned above, pyridine 1-oxide and most of its derivatives are recovered unchanged when subjected to the Reissert conditions, but quaternarization by means of alkyl sulfates or iodides and treatment, usually without purification, with aqueous cyanide give high yields of cyano-pyridines.[16,17,21,624-629]

A drawback of this method, except, of course, with the 4-substituted derivatives, is that cyanation takes place in both the α and γ positions. Thus, for example, a mixture of 2-, 4-, and 6-cyano derivatives is obtained from 3-alkylpyridine 1-oxides; the isomeric distribution depends somewhat on the actual substituent and on the alkylating agent used.[21,625,626] Nevertheless, the reaction has a synthetic value in some cases, e.g., 64% of the 2-cyano derivative is obtained from 3-(2′-phenylethyl)pyridine 1-oxide and is easily separated.[629] Other substituents have a stronger orienting effect, e.g., 3-methoxypyridine 1-oxide yields only the 2-cyanated product.[624] The strong methylating agent FSO$_3$Me has been used with 3-hydroxymethylpyridine 1-oxide, and in this case addition of a second mole of cyanide (possibly on the anhydro base) leads to 2,6-dicyano-3-methylpyridine; the reaction has been applied to other 3-(α-hydroxyalkyl) derivatives for the synthesis of analogs of fusaric acid.[630]

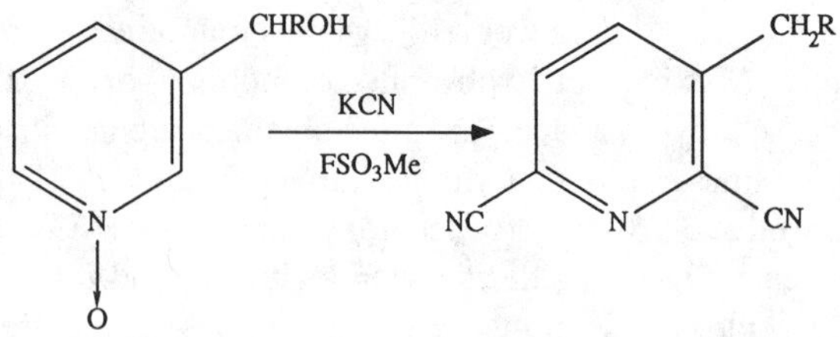

Cyanation via *N*-alkoxy derivatives has been extended to other azine *N*-oxides (with quinolines the reaction is almost exclusively at position 2)[16,21,536,628,631] and to benzimidazole *N*-oxides.[409,633]

2,6-Dicyano-3-methylpyridine. *10.72 g (94 mmol) methyl fluorosulfonate are added to 4 g (32 mmol) 3-hydroxymethylpyridine (slight exothermic reaction) and after 20 min of stirring, 10 g (154 mmol) KCN in 30 ml water are added during 5 min, while moderating the strong reaction with a cold water bath. Stirring for some hours, removing the solvent, and Kugelrohr distilling the residue yield 1.5 g (after recrystallization, 33%) of the product.*[630]

3. Cyanation with Silicon or Phosphorus Derivatives

A convenient modification of the Reissert-Henze reaction has been recently introduced that takes recourse to silicon or phosphorus cyanides, such as trimethylsilanocarbonitrile (TMSCN),[634,635] and its alkyl analogs, or to diethylphosphorocyanidate [DEPC,(EtO)$_2$POCN)].[636] This method also works beautifully with pyridine 1-oxides and, unlike the reaction via *N*-alkoxy derivatives, gives exclusively 2-cyanopyridines. Instead of using preformed TMSCN, an expensive and labile reagent, Me$_3$SiCl can be added to a solution of the *N*-oxide containing NaCN and NEt$_3$ in DMF.

3-Halo-, amino-, and alkoxypyridine 1-oxides yield exclusively the corresponding 2-cyanopyridines, and this selectivity has been attributed to the participation of the substituent n orbital, while in 3-aryl-, carboxy-, and cyano derivatives (there is disagreement about the 3-alkyl analogs), both α positions react.[637,638] The reaction has been used for the preparation of carboxypyridinyl alanine derivatives.[639] With DEPC, 4-nitropyridine 1-oxide reacts abnormally, yielding 3,4-dicyanopyridine.[636] These reagents also give useful results with quinoline, isoquinoline, phenanthridine, and pyrimidine (in this case both positions α to the ≥N→O function react),[635,636,640,641] and failures have been reported in only isolated cases.[642]

2-Pyridinecarbonitrile. Chloromethylsilane (11.42 ml, 50 mmol) is added during 1 h at 20—23°C to a stirred solution of pyridine 1-oxide (2.85 g, 30 mmol, NaCN (4.41 g, 90 mmol), and NEt$_3$ in 100 ml absolute DMF. The solution is heated at 100—110°C for 18 h and filtered. The filtrate is washed with ether, the combined organic phase is concentrated in vacuum, and the residue extracted with ether and concentrated again. The yield of the product is 76% (by vpc, since complete elimination of DMF is difficult).[635]

4. Direct Cyanation

Strongly activated (electron-withdrawing substituted) heterocycle *N*-oxides react directly with alkali cyanides. Thus, 2-, 3-, 4-, and 6-quinolinecarbonitrile 1-oxides undergo deoxidative cyanation (in the 2 or 4 position) in moderate yield when treated with potassium cyanide in DMSO[643a] (parent quinoline 1-oxide does not react appreciably under these conditions). 4-Isoquinolinecarbonitrile 2-oxide analogously yields 1,4-isoquinolinedicarbonitrile.[643b] Particularly noteworthy is the cyanation in the benzo ring observed with activated phenazine *N*-oxides. Thus, e.g., 2-nitrophenazine 10-oxide reacts smoothly with KCN to yield 1-cyano-2-nitrophenazine (in the presence of alcohols or amines, the nitro group is likewise substituted).[643c]

5. Carbamidation

As might be expected, carboxyamides are often obtained along with nitriles, using the methods discussed in the previous sections. Apart from this, direct carbamidation is obtained when the N-oxides are heated with formamide or N-methylformamide.[644a-644c] The reaction produces a high yield and involves the α, or, when this is unavailable, the γ position (2,4-dimethylquinoline 1-oxide yields the 5-amide and 2,3-dimethylquinoxaline 1,4-dioxide yields the 5- and the 6-amides). The mechanism has been formulated as a nucleophilic attack by the electron-rich ring carbon atoms onto the carboxylic function.[644a-644c] In view of the similarity with other deoxidative substitutions, it is also possible that the initial reaction involves the oxygen atom and that an intramolecular migration ensues.

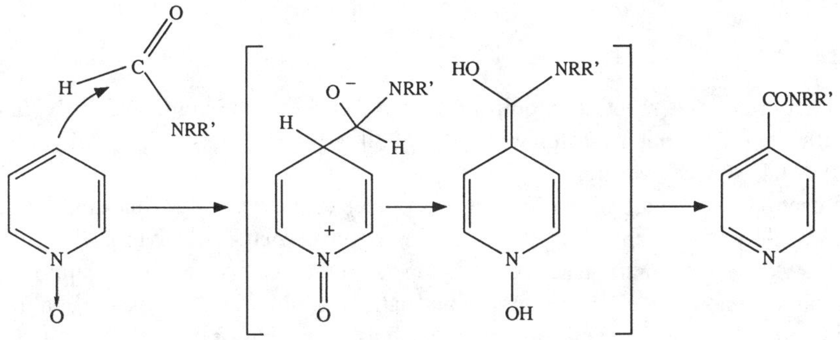

6. Acylation

Deoxidative acylation is the result of the treatment of N-(β-acylalkoxy)pyridinium, quinolinium, and isoquinolinium salts by bases.[644d,644e] The reaction involves an ylide intermediate formed by α-hydrogen abstraction and is mechanistically related to the acylation of nicotinic acid 1-oxide (Section IVF) and to the ring opening of the related β-carboxyalkoxy salts (see Section VIIA4, Scheme 15).

D. ALKYLATION

1. Reaction with Active Methylene Derivatives

Nitrones, and in particular cyclic nitrones such as pyrroline 1-oxides, are known to undergo aldol-like condensation in the presence of bases, and, more generally, to add carbanions. In contrast, aromatic heterocyclic N-oxides undergo related addition processes with strong nucleophiles only (compare with Sections VB2a and VIIA), but not with stabilized carbanions of the malonic ester type.

In exceptional cases, mostly connected with further activation by an electron-withdrawing group, active methylene derivatives react directly with N-oxides and cause deoxidative substitution. Thus, 2-nitro- and 2-methylsulfonylphenazine 10-oxides yield the corresponding 1-alkylphenazine with ketones, malonic acid derivatives, and nitromethane,[645] and 1-phenylphthalazine 3-oxide is alkylated in 4.[446b]

On the other hand, deoxidative alkylation with stabilized carbanions is a facile process when the reaction is carried out in the presence of acylating agents or on the preformed N-alkoxy salts. According to a procedure developed by Hamana,[646] in most cases acetic anhydride is used both as the acylating agent and as the solvent, and the reaction takes place at room temperature in good yield. These mild conditions make the process a useful entry to alkylated heterocycles, which provides an alternative to the reaction of carbanions with haloheterocycles. The reaction with ketones under acylating conditions is very slow[647a] and probably depends on the degree of enolization (e.g., quinoline 1-oxide is alkylated in 2 when treated with cyclohexanone and

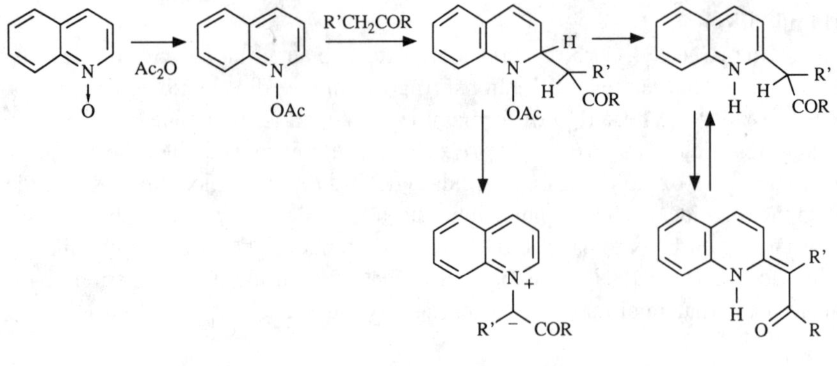

Scheme 12.

benzoyl chloride, the reaction requires 30 d at room temperature and only 6 d with acenaphthenone),[647a] but is faster when the *N*-alkoxy salts are used.[648] Alkylation with CH$_3$COCN takes place overnight. [647b]

The reaction with acylating auxiliary agents proceeds conveniently with stable carbanions. Indeed, a variety of conventional[649-656] (including trisubstituted derivatives such a nitromalonate),[649] as well as less common active methylene compounds, from acylpyridinium salts[657,658] to homophthalic anhydride[659] and esters of aromatic cyanohydrins,[660,661] have been used with good results, as have heterocyclic analogues such as 2-oxazolin-5-ones,[662] thiazolindiones,[663] pyrrolones, and oxindoles,[664] as well as butenolides.[665a] Although, as mentioned above, simple ketones react slowly[646] or not at all,[649] quinoline 1-oxide and 2-acetyl-1,3-cyclopentanedione yield products arising from attack at the methyl, not the methine, group[665b] (Scheme 12, Table 19).

The products obtained generally are the α-alkyl heterocycles, which in many cases are present as a mixture with the highly colored dihydroalkylidene tautomers.[651,653] Double substitution with formation of di(heterocycle)methylene derivatives has been observed in several instances.[649,665a,664b] Reaction may exceptionally take place at the γ position: thus, pyridine 1-oxide yields 3-(2,6-dioxocyclohexylidene)-2-(4-pyridyl)cyclohexanone with 1,3-cyclohexandione[650] and furthermore, thieno[2,3-*b*]pyridine 7-oxide, but not the isomeric [3,2-*b*] derivative, reacts in γ with ethyl cyanoacetate.[666-667] A particular reaction course is observed with 2-chloro- and 2-phenylquinoline 1-oxides and leads to cyclopropa[*c*]quinolines.[668]

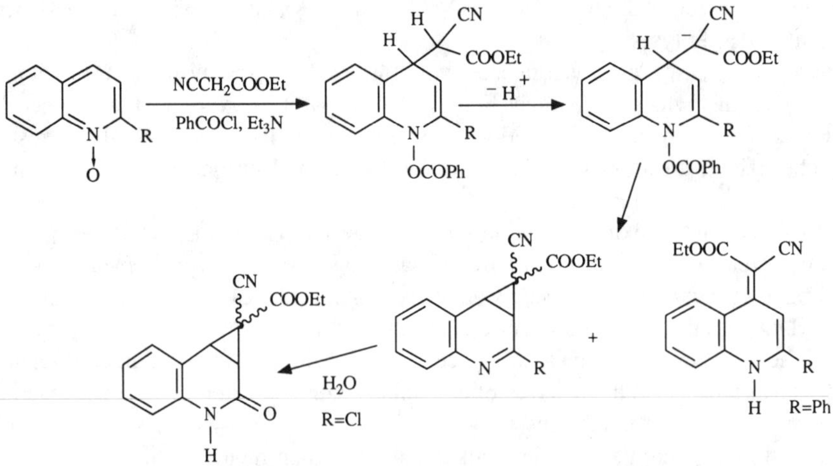

TABLE 19
Deoxidative Alkylation

N-oxide	Substituent	Alkylating agent	Auxiliary agent, conditions	Position alkylated (% yield)	Ref.
Pyridine		$CNCH_2COOEt$	Ac_2O, 30–40°C, 7 h	2(88)	649
		Dimedone	Ac_2O, 90°C	2(48)	650
	(2-Me)	1-Morpholinocyclohexene	PhCOCl, r.t., 3 d	2(71)	679a
		1-Morpholinocyclohexene	PhCOCl, r.t., 3 d	6(82)	679b
	(3-COOEt)	1,10-Dehydroquinolizidine	PhCOCl, CH_2Cl_2, r.t.	2(26)	683
		Indole	PhCOCl, $CHCl_3$, reflux	6	687
Quinoline		MeMg Br	i) EtBr; ii)MeMgBr, Et_2O, reflux	2(63), 4(12)	692
		Cyclohexanone	PhCOCl, $CHCl_3$, r.t., 30 d	2(24)	647a
		$MeCOCH_2COMe$	Ac_2O, r.t., 12 h	2(52)[a]	651
		Dimedone	Ac_2O, 90°C, 8 h	2(64)	650
		$CNCH_2COOEt$	Ac_2O, 30–40°C, 7 h	2(88)	649
		$CH_2(COOEt)_2$	Ac_2O, 30–40°C, 20 h	2(7+34[b])	649
		2-Phenyl-5-oxazolinone	Ac_2O, 90°C, 3 h	2(86)	662
		Oxindole	PhCOCl, $CHCl_3$, r.t.	2(60)	664b
	(2-Cl)	1-Morpholinocyclohexene	PhCOCl, $CHCl_3$, r.t., 3 d	2(70)	679a
		1-Morpholinocyclohexene	PhCOCl, $CHCl_3$, r.t., 3 d	4(54)	679c
		1-Morpholinoisobutene	PhCOCl, $CHCl_3$, r.t., 2 d	2(89)	681
		1-Ethoxycyclohexene	PhCOCl, $CHCl_3$, r.t.	2(58)	693
		Indole	PhCOCl, $CHCl_3$, r.t.	2(57)	685–687
Isoquinoline		MeMgBr	i) EtI; ii)MeMgBr, Et_2O, reflux	2(66)	23
		Dimedone	Ac_2O, 90°C, 8 h	1(100)	650
		2-Phenyl-5-oxazolinone	Ac_2O, 90°C, 3 h	1(71)	662
Benzimidazole	(1-Me)	1-Morpholinocyclohexene	PhCOCl, $CHCl_3$, r.t., overnight	1(66)	679a
		$CH_2(CN)_2$	i) MeI; ii) $CH_2(CN)_2$	2	633

[a] The acetonyl derivative is directly formed.
[b] 2,2-Bis(2-quinolyl) maleate.

Elaboration of the products thus obtained may involve decarbonylation[652] (spontaneous in some cases)[651] or conversion to α-aminomethyl-[662] and thiomehyl-heterocycles,[663] as well as to α-heterocyclic-substituted amino acids,[669] or condensation onto other functions present, as in the synthesis of aminofuropyridines.[654]

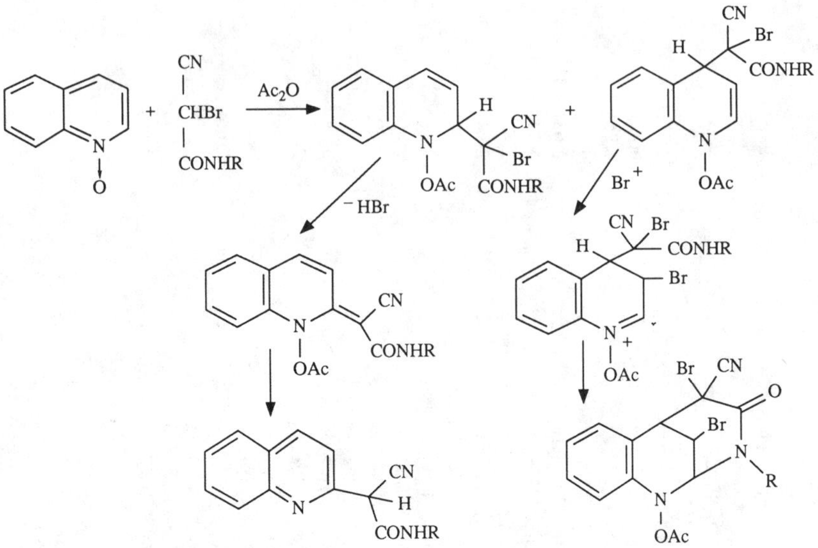

When a carbanion bearing a leaving group is used, the latter can be eliminated in preference to the *N*-acetoxy group, as appears in the complex reaction of quinoline 1-oxide with bromo derivatives of cyanoacetic acid and amide (involving attack at both the α and γ positions).[670]

With quinoline[671,672] and isoquinoline *N*-oxides,[671-673] *N*-ylides are formed in some cases (see Scheme 12). The competition between C and N functionalization depends on the structure, solvent, and temperature in a way that is not easy to rationalize. Thus, e.g., quinoline 1-oxide reacts with Meldrum's acid to give a 2- alkylated quinoline in neat Ac$_2$O and to give the ylide in Ac$_2$O/DMF or DMSO, in both cases at room temperature. However, with quinoline 1-oxide and barbituric acid, reaction at room temperature favors the first type of product and reaction at 90°C favors the latter. However, the reaction of 4-methoxyquinoline 1-oxide shows a temperature dependence in the opposite sense, and the directing effect of some other substituents is apparently uninfluenced by temperature.[671,672] The *N*-ylides, also formed in the reaction of benzo[*c*]cinnoline 5-oxide with malononitrile,[674] are, at any rate, secondary products, arising through rearrangement after primary attack at the α position.

Some reactions with vinilogous derivatives of carbanions have proven to be synthetically valuable. Thus, two molecules of pyridine (or quinoline) 1-oxide react with diethyl glutaconate to yield 2,3-annelated derivatives[675] and quinoline 1-oxide reacts with ethylidenmalonitrile or cyanophenylacrylonitrile and undergoes 1,2-annelation, thus offering a useful entry to quinolizine derivatives.[676]

Diketene has acylating and carbanion-like properties, and accordingly reaction with quinoline 1-oxide requires no auxiliary reagents and leads to 2-acetonylquinoline, in turn, adding further diketene to yield a pyrone derivative.[677] Related reactions are observed with isoquinoline and acridine *N*-oxides (also see Section VIIIA3).[678]

2-Quinolylacetone. *Acetylacetone (4 g, 40 mmol) is added dropwise over 30 min to a stirred, ice-cold solution of 5.8 g (40 mmol) of anhydrous quinoline 1-oxide in 8.2 g (80 mmol) acetic anhydride, and the resulting red solution is allowed to stand overnight at room temperature. Water (10 ml) is added and the mixture is heated at 50°C for 30 min and then evaporated in vacuum, making use of toluene azeotropes. The residue is triturated with 5 ml ether and recrystallized to yield the product (5.1 g, 68%).[651]*

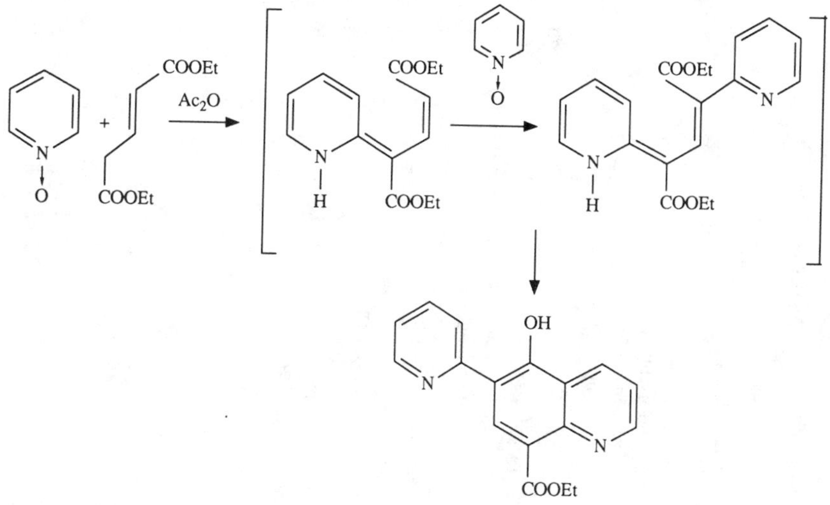

2. Reaction with Enamines and Enol Ethers

In the presence of acylating agents, heterocyclic *N*-oxides react easily with open-chain and cyclic enamines,[652,679-681] as well as with related compounds, such as 2-methylene-1,3,3-trimethyldihydroindole,[682] dehydroquinolizidines,[683,684] π-excessive heterocyclics (indoles,[685-688] pyrrole,[685,689a] 2-amino-, as well as 2-hydroxy-, isoxazoles[689b]), derivatives of β-aminoacrylic acid,[652] pyrazolones,[685,691] and *N,N'*-dimethylaniline.[690] Attack at the α position also predominates in this case, e.g., quinoline 1-oxide reacts with *N*-methylindole to yield the 2-(3-indolyl) derivative as the main product, along with some of the 4-isomer,[686] 2-methyl- and 2-chloquinoline 1-oxide, yielding, under analogous conditions, the corresponding 4-(3-indolenyl)quinolines.[679b,679c] However, pyridine 1-oxide reacts in position 4 with β-aminocrotonate (tosyl chloride is required for the activation)[652] (Table 20).

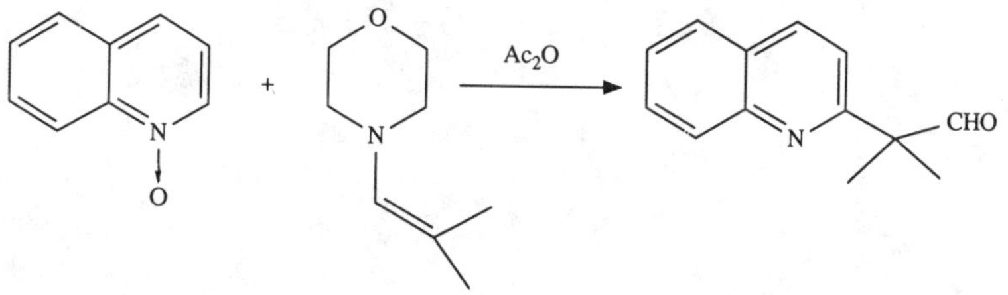

N-alkoxyquinolinium salts undergo a different reaction with enamines, and a new ring is formed.[692]

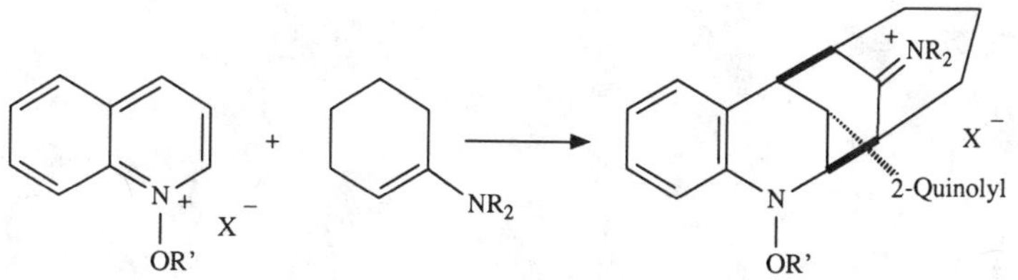

TABLE 20
Deoxidative Substitution with Amines, Imines, and Their Haloderivatives

N-oxide	Substituents	Conditions	Substituted heterocycle (% yield)	Ref.
Pyridine		PhNH$_2$, TsCl, CHCl$_3$, aq NaOH, r.t., 3 h	2-NPhTs(41)	699a
		PhNHTs, TsCl, CHCl$_3$, aq NaOH, r.t., 2 h	2-NPhTs(77)	699a
		2-BrC$_5$H$_4$N, dioxane, 110–120°C, 3 h	2-(2-oxo-1-pyridyl)(60)	716b
	(2-Me)	2-BrC$_5$H$_4$N, HBr,AcOH, PhMe, reflux, 3 h	6-(2-oxo-1-pyridyl)(29)	716a
		2-TsOC$_5$H$_4$N, 80–150°C, 5 h	2-(2-oxo-1-pyridyl)(20),	
			4-(2-oxo-1-pyridyl)(30),	718
			3(2-pyridyloxy)(20)	707a
Quinoline		PhC(Cl)=NPh, C$_2$H$_4$Cl$_2$, reflux, 10 h	2-NH$_2$(47)	698
		i) aq NH$_3$, TsCl, CHCl$_3$, 20°C, 1 h ii) aq HCl	2-NH$_2$(70)	700
		DMF, BF$_3$, TsCl, reflux, 1.5 h	2-NMe$_2$(30), 4-NMe$_2$(30)	714
		4-BrC$_5$H$_4$N, dioxane, 110–120°C, 3 h	2-(4-oxo-1-pyridinyl)(41)	
		PhC(Cl)=NPh, C$_2$H$_4$Cl$_2$, reflux, 4 h	2-NPhCOPh(30), 2-NHPh (33),	709a
			2-OCOPh(33)	
Isoquinoline		DMF, BF$_3$, TsCl, reflux 1.5 h	1-NMe$_2$(39)	700
		PhC(Cl)=NPh, C$_2$H$_4$Cl$_2$, reflux, 6 h	1-NPhCOPh(47), 4-Cl(18)	709a
		2-BrC$_5$H$_4$N, dioxane, reflux, 3 h	1-(2-oxo-1-pyridinyl)(65)	715
Pyridazine	(6-Me)	PhC=NPh+SbCl$_6$$^-$, CH$_2Cl_2$, reflux, 17 h	3-NHPhCOPh(18)	709a
Pyrimidine		PhC(Cl)=NPh, PhCl, reflux, 8 h	2-NPhCOPh(26)	709a
Benzimidazole	(1-CH$_2$Ph)	PhC(Cl)=NPh, CHCl$_3$, reflux, 22 h	2-NPhCOPh(92)	709a

A similar reactivity has been reported for enol ethers with quinoline and isoquinoline *N*-oxides (but not with pyridine 1-oxide).[693] Furan, thiophene, and anisole do not react under this condition.[693]

3-(2-Quinolyl)indole. Benzoyl chloride (0.8 g) is added dropwise to an ice-cooled, stirred solution of quinoline 1-oxide (0.8 g) and indole (0.7 g) in 10 ml chloroform, and the mixture is stirred for 1 h at room temperature and then refluxed for 2 h. The solution is shaken with K_2CO_3, evaporated, and the residue is chromatographed on alumina to give 0.9 g (67%) of the product.[686]

3. Reaction with Organometallic Derivatives

Grignard reagents are sufficiently strong nucleophiles for reacting directly with heteroaromatic *N*-oxides, with no need for activation, as discussed in Section VIIA, and the adducts thus formed often undergo ring opening. However, the reaction of Grignard reagents with *N*-alkoxy derivatives is itself valuable and gives good yields of the alkylheterocycles in the pyridine,[694a] quinoline,[23] and benzimidazole[633] series. *N*-alkoxypyridinium salts yield a mixture of 2- and 4-alkylated pyridines, but the *N*-(alkoxycarboxy) derivatives, prepared *in situ* by reaction with chlorocarbonates, react exclusively in α.[694b] A similar reaction has been reported using zinc organic derivatives.[694c]

The recourse to silyl derivatives is also fruitful in this case; thus, alkylation in α by benzyl and allyl silanes in the presence of catalytic amounts of tetrabutylammonium fluoride occurs in high yield, probably via addition to the oxygen atom and elimination of trialkylsilanol.[695a] Alkylation of the BF_3 *N*-oxide adducts by trialkylboranes occurs at the γ position, while reaction with the uncomplexed *N*-oxide occurs in α.[695b]

4. Reaction with Ylides

Several sulfur ylides have been shown to react with quinoline 1-oxide under acylating conditions to yield either the corresponding 2-quinolyl substituted ylides or products arising from further reaction of them.[696] Similar reactions take place with isoquinoline and quinoxaline, but not with pyridine *N*-oxides.[696a] A similar reaction of quinoline 1-oxide with a phosphor ylide has been reported.[646]

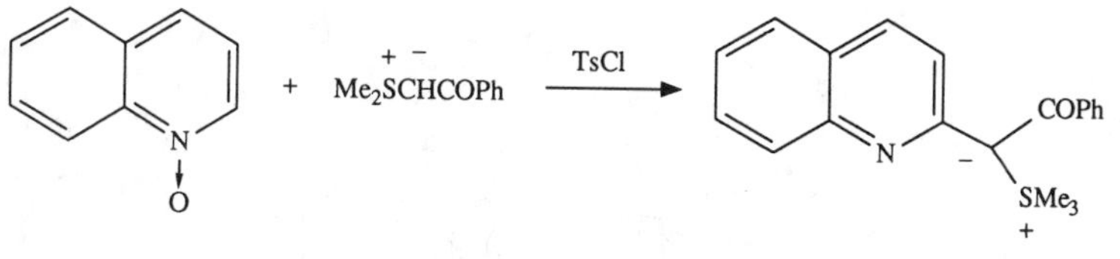

E. FORMATION OF A CARBON-NITROGEN BOND

1. Reaction with Amines and Amides

Amines add to nitrones, but in general are too weak nucleophiles to react with heteroaromatic *N*-oxides. Nucleophilic substitution of various leaving groups by amines is often, as mentioned in Section VA1, accompanied by deoxygenation to some extent. Direct amination at an unsubstituted position (with accompanying deoxygenation) is an efficient process with some electron-withdrawing substituted heterocyclic *N*-oxides, including derivatives of acridine,[697] phenazine,[643c] and pteridine.[550]

Activation by tosyl or benzoyl chloride is effective, and under this condition deoxidative amination is conveniently obtained with aqueous ammonia, amines, and their salts,[608,698,699] and also amides[700] and sulfonamides[699] react analogously. Similar reactions have been reported with *N*-alkoxy derivatives, e.g., with 1-alkoxybenzimidazolium salts.[633]

2. Reaction with Imines

The imine chromophores react with activated *N*-oxides, e.g., quinoline 1-oxide reacts with pyridine in the presence of tosyl chloride to yield the 1-(2-quinolyl) and (4-quinolyl)pyridinium salts.[702] Similar salts, which are useful intermediates for following nucleophilic substitution, are obtained from various pyridine, quinoline, and quinoxaline *N*-oxides (not from the hydroxy derivatives, which are sulfonylated under these conditions).[703-705] Xanthine and guanine 3-oxides yield 8-(1-pyridinium) ylides.[706]

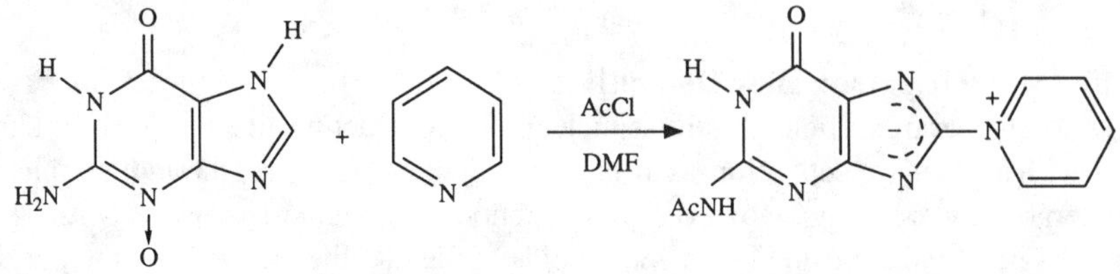

3. Reaction with α- and γ-Haloimines

Another general reaction with heterocycle *N*-oxides is α-acylamination by reaction with imidoyl chlorides, a typical case of a deoxidative substitution obtained with a single bifunctional reagent (see Scheme 5), rather than by using a nucleophile and an auxiliary electrophilic reagent.[707-709] The mechanism has been formulated as involving initial attack at the $\geqslant$N→O oxygen to yield the zwitterion **30** and the cycloadduct **31**, the latter further evolving intra- or intermolecularly to yield, in the case of pyridine 1-oxide, the acylamino derivatives **32**, 3-chloropyridine, pyridine, and the amide.

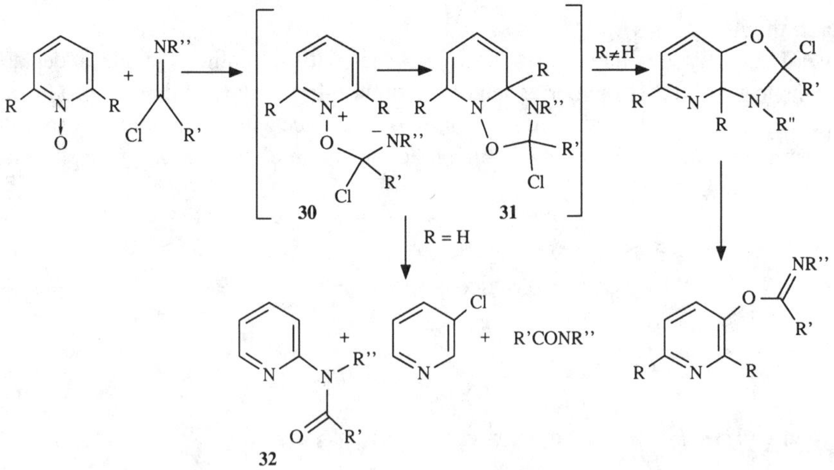

The reaction works with cyclic imidoyl chlorides (R′,R″ part of a ring), including saccharin pseudochloride, which yields *N*-(2-pyridyl)-saccharin. The reaction of pyridine 1-oxide has been extended to various *N*-phenylareneimidoyl chlorides and shows a complex dependence on the aryl group, e.g., the 2- and 3-nitrophenyl derivatives yield more 3-chloropyridine and less imidoyl derivative, but the 4-nitrophenyl derivative yields neither.[708]

As expected from the mechanism, the reaction does not take place with insufficiently basic *N*-oxides, e.g., 4-nitropyridine 1-oxide.[707a] A substituent in 3 may have an important directing effect, e.g., with 3-methylpyridine 1-oxide, acylamination in 6 greatly predominates; whereas with the 3-methoxy derivative, the reaction is mainly at position 2.[709b] With the *N*-phenylimidoyl chloride reactive covalently linked in a polymer, the reaction occurs with a lower yield than with nonpolymeric reagents, but one has the advantage of an easier work-up, and with 3-substituted pyridines the reaction is exclusively at the less-hindered position 6.[710a] With α-methyl derivatives, side-chain chlorination is observed.[710b] When both α positions are occupied, a 1,5 sigmatropic shift leads to a O(3-pyridyl)benzimidate, since dehydrohalogenation of the intermediate adduct to the amide is impossible.[709a] The reaction also takes place with pyridazine, pyrimidine, isoquinoline, and quinoline N-oxides, as well as with 1-benzylbenzimidazole 3-oxide.[707b,709a]

The amides thus obtained are easily hydrolyzed (indeed, a mixture of amide and amine is often obtained from chromatography). The reaction takes place similarly, though less conveniently from the preparative point of view, when nitrilium salts are used.[709a] However, extension to α-chloro aldoximes, which could provide 2-pyridylhydroxylamino derivatives, failed.[707a]

With a cyclic haloimine, 4-chloro-2H-1,3-benzoxazine, the reaction leads analogously to 3-(2-pyridyl)oxazines and hydrolysis of the latter compounds gives substituted 2-aminopyridines.[711] With 2- and 4-alkylpyridines, the reaction takes place both at the 3 position and on the chain. With 2- and 4-alkoxypyridine 1-oxides, the C-N bond is not formed, and ethers of 1-hydroxy-2- (or 4-) pyridone are obtained.[711,712a]

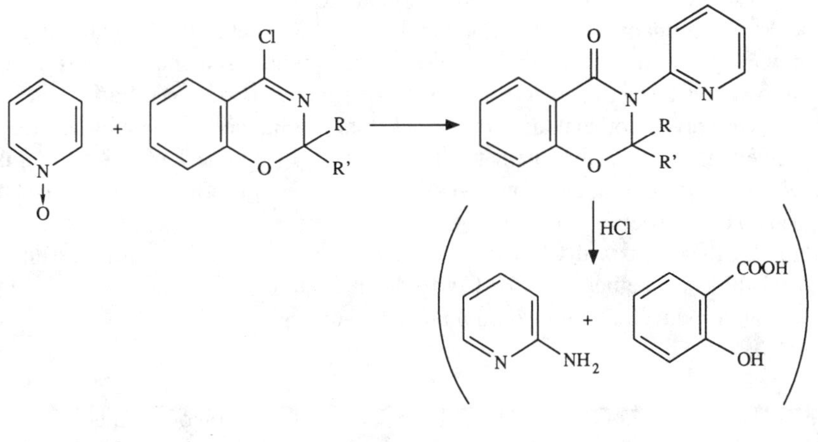

A polymer-linked benzoxazine has been used and has been shown to be a more selective reagent.[710a] With 4,5-dihydro-2-chloroimidazole, the 1-(2-quinolyl)-, and, remarkably, 1-(4-isoquinolyl)-2-imidazolones are obtained.[712b]

The reaction of N-oxides with chlorobenzoxazine (or with ammonia and acylating agents, Section VA1) is a useful method for the synthesis of primary heterocyclic amines and may be considered an alternative to the Tchichibabin reaction. Correspondingly, acylamination by imidoyl chlorides (or by amines/acyl chlorides, Section VA1) is a convenient source of secondary amines.

Heterocyclic N-oxides react with α- or γ-halogeno derivatives of azines, probably again through attack of the oxygen atom at the carbon-halogen bond and following rearrangement. Thus, 2-halogenopyridines and pyridine 1-oxides yield 1-(2-pyridyl)-2-pyridones,[713] and related reactions involving various N-oxides and 2- or 4-haloazines,[714-717] as well as 2- and 4-tosyloxypyridines,[718] have been reported.

1-(N-Benzoylanilino)isoquinoline. *A solution of isoquinoline 2-oxide (2.08 g, 14.4 mmol) and distilled N-phenylimidoyl chloride (2.76 g, 13 mmol) in 30 ml 1,2-dichloroethane is refluxed for 6 h, then washed with 10% aqueous Na$_2$CO$_3$ and chromatographed on silica gel to yield 0.33 g (18%) of 4-chloroisoquinoline, 1 g (39%) of benzanilide, and 1.6 g (47%) of the product.*[709a]

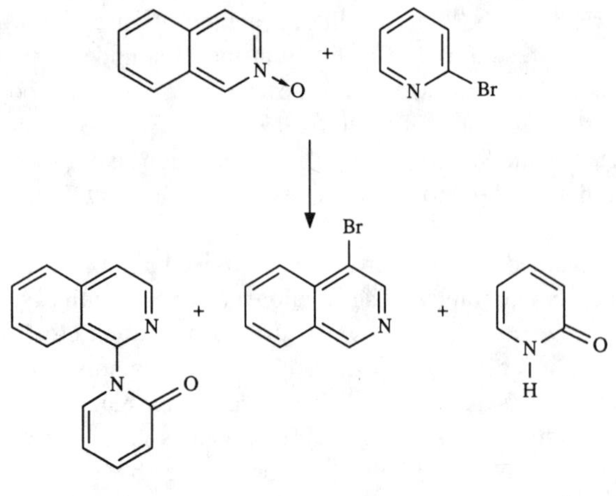

4. Formation of Carbamates

The *N*-oxides yield *N*-heterocyclic carbamates under two sets of conditions, i.e., by reaction with either cyanogen bromide or with alkali cyanate in alcohols. In the first case, quinoline 1-oxide yields predominantly the 8-quinolyl carbamate, reasonably through an intramolecular pathway, with minor amounts of the 2- and the 6-substituted isomers.[719] Yields are not high (much carbostyril is formed) and depend on the experimental conditions. Various alcohols can be used, but with solvents other than alcohols, only tars are obtained. Some substituted quinoline 1-oxides give better yields of carbamates,[720] but the only pyridine 1-oxide reacting is the 4-hydroxy derivative (yielding the 3-pyridyl carbamate).[721] From isoquinoline 2-oxide, both the 1- (major) and the 4-isoquinolyl carbamates are obtained.[721]

In contrast to the results with CNBr, the reaction of quinoline 1-oxide with potassium cyanate and benzoyl chloride in ethanol at −10°C affords the 2-quinolyl carbamate in a 70% yield. At reflux the main product is 2-ethoxyquinoline, however.[722]

5. Miscellaneous

Other reactions that should be mentioned, besides the formation of 2- and 4-nitroquinoline from the reaction of sodium nitrite and benzoyl chloride with the *N*-oxide,[291] include two methods for the synthesis of α-azidoazines, or rather of the tautomeric tetrazoles, which are valuable in view of the mild conditions required, i.e., treatment with tosylazide[723] and reaction with trimethylsilylazide in the presence of trimethylsilyl trifluoromethanesulfonate.[724]

F. FORMATION OF CARBON-SULFUR BONDS

1. Reaction with Thiols and Acylating Agents

The reaction of pyridine 1-oxide with thiols in the presence of acylating agents conveniently leads to 2- and 3-pyridyl sulfides.[725] The mechanism has been extensively investigated and is shown in scheme 13 (benzoyl and other acid chlorides or anhydrides are used as auxiliary reagents). As usual, nucleophilic attack involves the 1-acyloxypyridium salt and occurs in the α position. Then deprotonation leads to a 2-substituted pyridine, while detachment of the benzoate anion is followed by rearrangement via the episulfonium cation and preferentially yield a 3-substituted pyridine (the corresponding cation has no positive charge on the nitrogen). Competition between the two pathways is influenced by the acylating reagent used — both a higher total yield and a higher proportion of the β sulfide is obtained when it corresponds to a better leaving group — as well as by the presence of a base (triethylamine), which favors deprotonation and thus enhances the yield of the α derivative (except with *N*,*N*-dialkylcarbamoyl chlorides as acylating agents, since there the amine is formed *in situ* from the

TABLE 21

2- and 3-Alkylthiopyridine by Deoxidative Substitution

Substituents in pyridine 1-oxide	Mercaptan	Method	Products (% yield)	Ref.
	n-BuSH	Et$_2$NCOCl, C$_6$H$_6$, reflux 2 h	2-SR(68), 3-SR(1)	726
	n-BuSH	PhSO$_2$Cl, C$_6$H$_6$, NEt$_3$, reflux 2 h	2-SR(23), 3-SR(11)	726
	t-BuSH	Et$_2$NCOCl, C$_6$H$_6$, reflux 2 h	2-SR(15), 3-SR(16)	726
	t-BuSH	Ac$_2$O, reflux, 3 h	2-SR(43), 3-SR(19)	728
2-Me	t-BuSH	Ac$_2$O, 95°C, 3 h	5-SR(5), 6-SR(27)	728
3-Me	t-BuSH	Ac$_2$O, 95°C, 3 h	2-SR(30), 5-SR(24), 6-SR(12)	728
	t-BuSH	Ac$_2$O, NEt$_3$, 95°C, 3 h	2-SR(12), 5-SR(1), 6-SR(7)	731
4-Me	t-BuSH	Ac$_2$O, 95°C, 3 h	2-SR(29), 3-SR(11)	728
	PhSH	PhSO$_2$Cl, C$_6$H$_6$, reflux 2 h	2-SR(12), 3-SR(18)	726

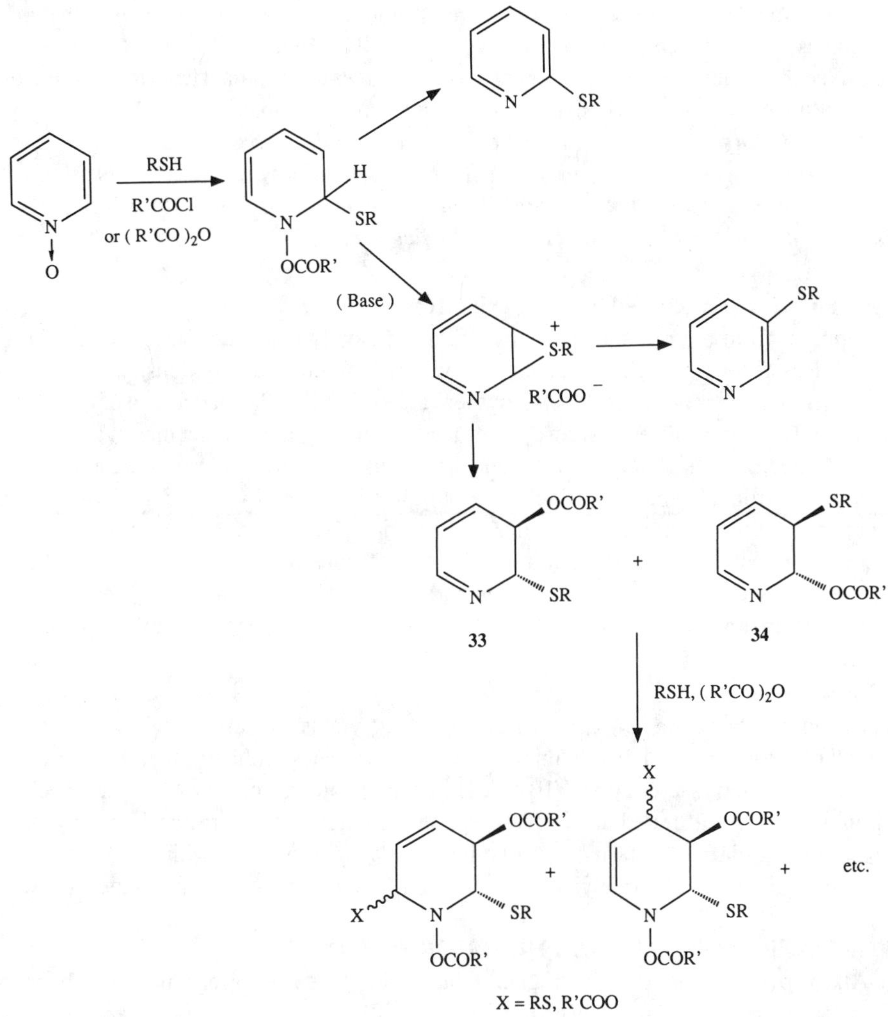

Scheme 13.

decomposition of the leaving anion).[726,727] Some representative examples are given in Table 21 (benzene as the solvent). Reaction with thiols is also observed starting from 1-alkoxypyridinium salts, though with a lower yield, and gives all three isomeric sulfides[22] (Scheme 13, Table 21).

The reaction has been extended to various aliphatic thiols, as well as to thiophenol. In substituted pyridine 1-oxides, the ratio between the yield of the 2- and the 3-sulfides increases with the increasing size of the pyridine substituent and of the alkyl group of the thiol.[725,728] With 2-methyl- and 2-phenylpyridine 1-oxide, products substituted at positions 6 and 5, but not 3, are obtained (again, an indication that both types of products arise from the same initial α-substituted dihydropyridine).[725,728] 3-Substituted pyridine 1-oxides undergo attack at both α positions, but nicotinic acid and nicotinamide 1-oxides react almost exclusively at position 2.[725,729-731]

If acetic anhydride is used both as an acylating reagent and as a solvent, low amounts of 1-acetyltetrahydropyridines are also obtained, and this kind of derivative (containing one to three sulfide groups and one or two acetoxy groups) increases when the thiol is added to the N-oxide in Ac_2O/NEt_3.[731-734] The products always have *trans* stereochemistry at C-2 and C-3, and accordingly the reaction is rationalized as involving *trans* addition of either the acetate ion or the thiol to the above-mentioned episulfonium intermediate to yield the dihydropyridines **33** and **34** (see Scheme 13). Following quaternarization and 1,2- or 1,4 addition by either of the nucleophiles, the observed products are yielded (not all isolated).

The expected 2-quinolyl- and 1-isoquinolyl- derivatives are obtained from the reaction of the corresponding N-oxides with S nucleophiles such as ethyl 2-thioacetate.[735-736]

2-Pyridyl-n-butylsulfide. To a stirred, ice-cooled solution of 9.5 g (0.1 mol) pyridine 1-oxide in benzene (previously azeotropically dried), (0.1 mol) N,N-diethylcarbamoyl chloride is added at once, followed by n-butylmercaptan (0.15 mol) during 5 min. The mixture is refluxed 2 h, evaporated, basified, extracted, and distilled to yield 12.6 g (68%) of the product.[726]

2. Reaction with Oxygenated Sulfur Derivatives

As appears from the previous sections, organic (usually aromatic) sulfonyl chlorides are often used as activating agents in deoxidative nucleophilic substitutions, and usually no sulfurated derivative is obtained. However, there are exceptions. Thus, reaction of the N-oxides with sulfonyl chlorides and potassium cyanide leads to sulfonylation rather than cyanation. With pyridine 1-oxide, attack is at positions 2 and 4 with alkylsulfonyl chlorides and at positions 3 and 4 with the aromatic analogs.[737] With quinoline 1-oxide, reaction is at 2, or 4 or 3 when the former position is not available[738,739] (with tosyl chloride/KOH attack at 3 also predominates on the parent compound),[566b] and isoquinoline 2-oxide gives a mixture of the 1-, 3-, and 4-substituted products.[740]

1-Alkoxybenzimidazolium salts react with bisulfite to yield 2-benzimidazolesulfonic acid.[633]

3. Miscellaneous

The reaction of acridine 10-oxide with acetyl sulfide yields thioacridone via an intermolecular attack of the strong nucleophile thioacetate ion (as opposed to the intramolecular reaction with the acetate ion, see Section VIB).[741] Thiolactams are formed from cyclic nitrones by reaction with some sulfurated derivatives, e.g., in the reaction of pyrroline 1-oxide with 1,1'-thiocarbonyldiimidazole[742]; the reaction of dibenz[c,e]azepine N-oxide with CS_2 is analogous.[743] N,N-Diethylcarbamoyl chloride yields both O- and S- derivatives with quinoline 1-oxide.[744]

G. FORMATION OF CARBON-PHOSPHORUS BONDS

N-Alkoxypyridinium, quinolinium, and related salts give the corresponding 2-phosphonic acids esters when treated with $NaP(O)(OR)_2$, and these are easily hydrolyzed to the free acids.[745-746] While P(V) chlorides usually cause deoxidative chlorination (Section VIA), treatment of 2,3-diphenylquinoxaline oxide with $POCl_3$ yields the 6-phosphonic acid ester.[499] This leads one to suspect that phosphonic acids are also formed in other cases as primary products on reaction with phosphorus halides (see Section VIA), though they may undergo secondary substitution to the observed chorinated derivatives.[499]

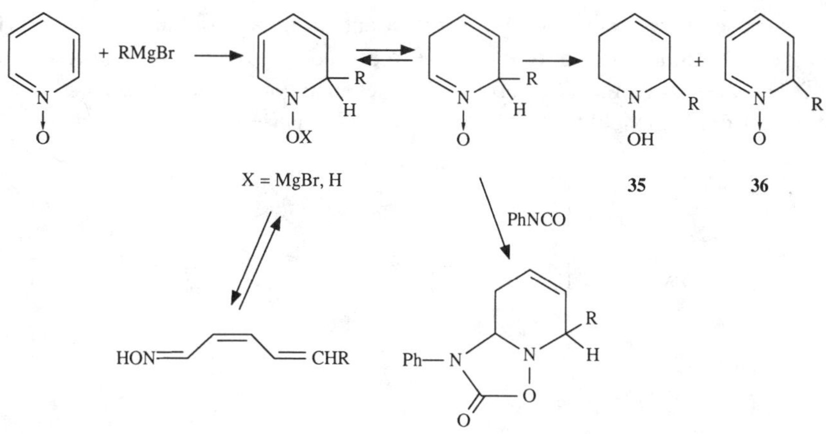

Scheme 14.

VII. REACTIONS INVOLVING ADDITION AND/OR RING CLEAVAGE OR CHANGE OF RING SIZE

Some important reactions in *N*-oxide chemistry involve a cleavage or a change in the heterocyclic ring, the most important group probably being the conversion of benzofuroxans to quinoxaline oxides or to benzimidazole oxides. It was thought to be useful to gather in a single chapter all reactions in which the ring is modified, though different mechanisms are involved. In general, the first step is addition, and therefore reactions leading to isolable adducts are also reported here. The difference with the previously discussed substitution processes, in which adducts are labile intermediates is clear-cut in some cases; in other cases, it is only a matter of convenience. In particular, several deoxidative susbtitutions (Section VI) also leads to isolable adducts. When the ring change or cleavage is to be imputed to a reaction of the substituent rather than of the $\geqslant$N$\rightarrow$O function itself (e.g., decomposition of azido or diazo derivatives), the reaction is discussed in Section XI under the heading corresponding to the relevant substituent.

A. ADDITION AND RING OPENING OF AZINE *N*-OXIDES

1. With Carbanions

Grignard reagents easily add to nitrones to yield α-substituted hyroxylamines, and these are dehydrated to imines or reoxidized to substituted nitrones, a reaction reported, e.g., for azetine,[747] pyrroline,[135,748-750] and benzodiadepine *N*-oxides.[751] The nitrone function is attacked in preference to conjugated carbonyl in β-oxopyrroline[748] and indoline oxides.[752,753]

Grignard reagents also react with heteroaromatic *N*-oxides. This is one of the few cases in which direct nucleophilic addition is also observed with these compounds, and complexation with the metal of the $\geqslant$N$\rightarrow$O function certainly has an important role. The reaction of pyridine 1-oxide with alkyl or phenylmagnesium bromide has been investigated repeatedly. When the reaction is carried out at $-40°$C in THF, an adduct is formed. This undergoes electrocyclic ring opening at $0°$C to yield under kinetic control the *E,Z,E* oxime (the *Z,Z,E* isomer is slowly formed on standing). Protonation of the adduct in the cold leads not to the expected 1,2-dihydro-1-hydroxypyridine, but to a 2,5-dihydropyridine 1-oxide. This compound can be trapped with phenylisocyanate or otherwise disproportionates to rearomatized **36** and the tetrahydro derivative **35**[754] (Scheme 14).

The pentadienyl oximes are directly obtained in yields of between 15—64% by reaction of various Grignard reagents with pyridine 1-oxide at $-20/0°$C.[754-757] Heating the oximes leads to a mixture of the 2-substituted pyridine and the corresponding 1-oxide,[754] and carrying out the reaction at room temperature results in the direct production of these products, and thus in formal (deoxidative) substitution; the ratio of 1-oxide to deoxidized pyridine depends on the conditions.[756,758,759] Other organometallic derivatives react analogously; thus ethynyl sodium reacts with pyridine 1-oxide to yield the unsaturated oxime **37**, as well as the double adduct **38**.[760]

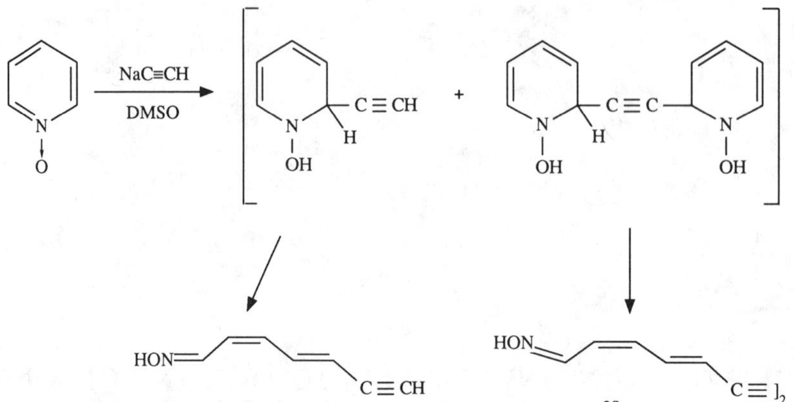

Scheme 14 is probably of general significance for the reaction of Grignard reagents with *N*-oxides. In several cases, open-chain compounds or secondary products arising from them are obtained. With pyridazine 1-oxides N_2O is lost, probably yielding a carbocation, and deprotonation or addition from the latter result in enynes or dienes, respectively. This procedures gives good yields of the *E*-enynes[761-763] and has been patented for the synthesis of terminal acetylenes.[764]

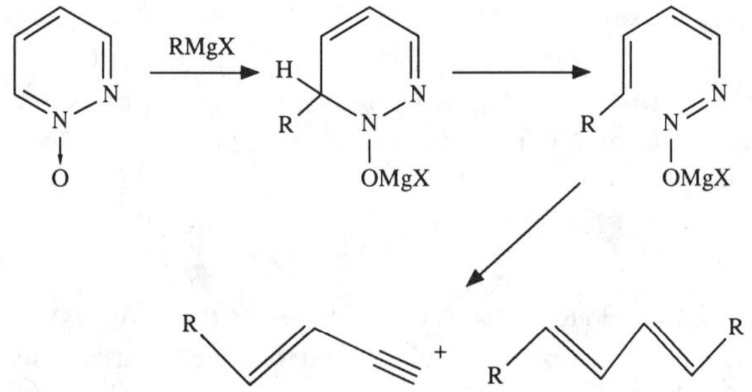

Ring opening and elimination has also been reported for condensed pyridazines oxides;[765,766] with cinnoline 2-oxide as an example, a dihydro adduct, an azobenzene derivative and products resulting from elimination are obtained.[766] The reaction with benzo-1,2,4-triazine 1-oxide similarly gives indole and azobenzene derivatives.[767]

In other cases, the starting heterocyclic ring remains in the final product, although the intermediacy of an open-chain oxime is likely, and a mixture of the α-substituted heterocycle and the corresponding *N*-oxide is obtained (reoxidation is effected by atmospheric oxygen or, better, by an added oxidizer, e.g., *p*-benzoquinone[768a] or lead dioxide[769]). This is the case with quinoline,[759,769-771] phthalazine,[560b,768] and quinoxaline[772] derivatives.

The behavior of α-substituted heterocyclic *N*-oxides is also interesting (compare with Section VA2). Thus 2-cyano- and 2-methoxyquinoline 1-oxide undergoes *ipso*-substitution with phenylmagnesium bromide to yield (after reoxidation) 2-phenylquinoline 1-oxide[769] (a related substitution of the cyano group by methyl magnesium bromide has been reported for 2-cyanopyrroline 1-oxide[773]). A similar ipsosubstitution is observed with some phenanthridine, and benzo[*f*] and [*h*]quinoline 1-oxides.[774] 2-Methyl and 2-phenylquinoline 1-oxides, on the contrary, add Grignard reagents to yield the α,α-disubstituted hydroxylamines, which are easily oxidized to the corresponding nitroxyl radicals.[769,775a] Analogously, 6-substituted phenanthridine 5-oxides yield the 6,6-disubstituted nitroxides[774] and acridine 10-oxides yield the 5-substituted nitroxides.[775b] For the reaction with Grignards in the presence of activating agents, see Section VID3.

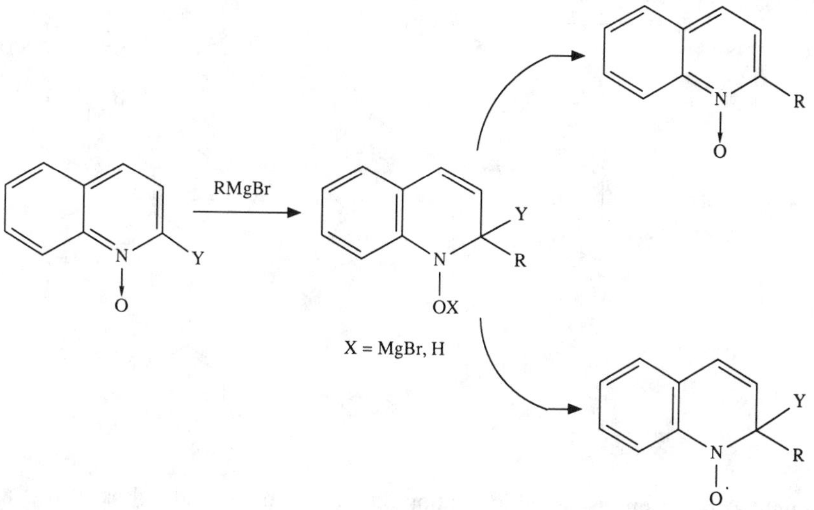

Apart from the exceptional case of Grignard derivatives and related organometallics, in general (cyclic) nitrones and heteroaromatic *N*-oxides show a different reactivity with carbanions and other nucleophiles, the former compounds usually undergoing easy addition, while aromatic *N*-oxides require activation, typically by acylation or alkylation, in order to react and then undergo deoxidative substitution, as has been discussed in Section VI. However, there are sparse reports of ring opening by reaction with stabilized carbanions, e.g., with fervenulin 4-oxide.[776a] 4-alkoxypyrimidine 1-oxides have been found to undergo addition and ring opening when treated with diketene and acetic anhydride and to finally yield oxazolopyridines.[776b]

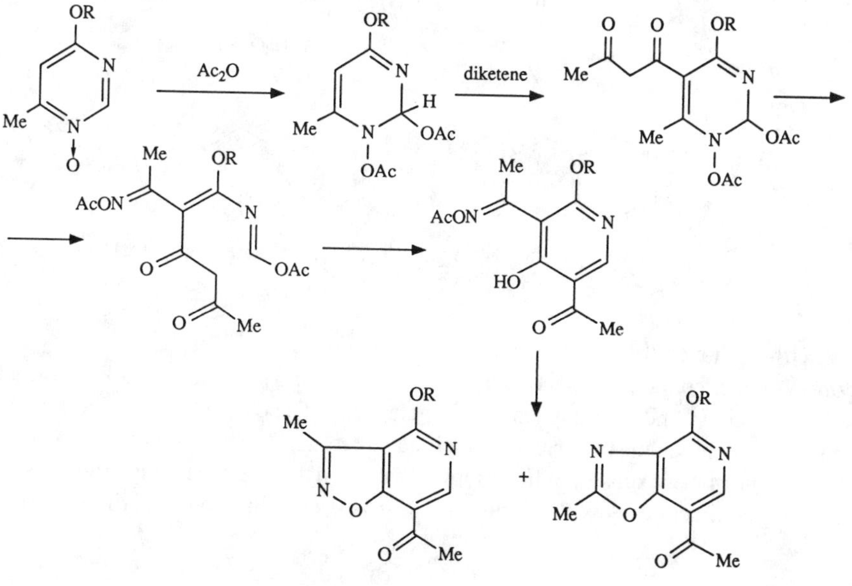

Interestingly, quinazoline 3-oxides react with ketones and active methylene compounds in the absence of a base to yield quinoline derivatives via addition and ring opening.[777] The same type of reaction has been observed in pyrido[2,3-d]pyrimidine 3-oxides.[778]

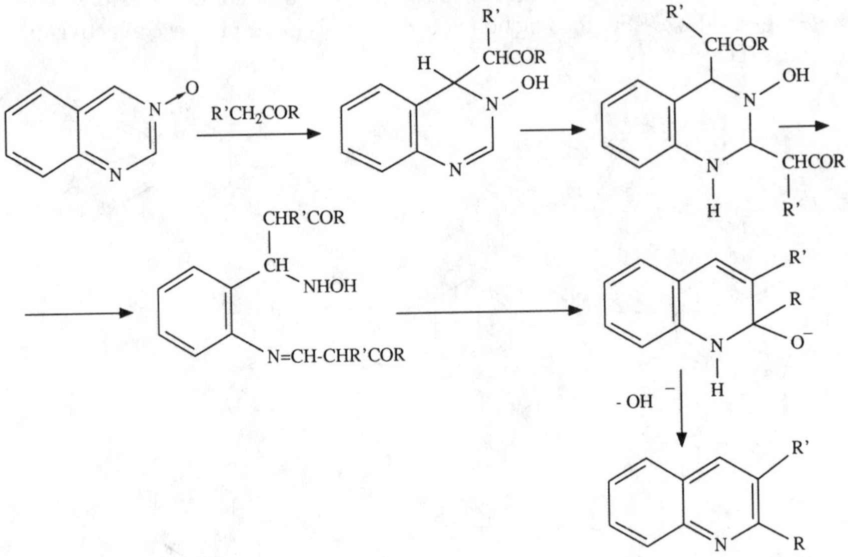

Although, as has been discussed in Section VB2a, the methylsulfinyl carbanion acts as a methylating agent with N-oxides,[447a] under different conditions (NaH at 70°C rather than t-BuOK at 20°C) a peculiar reaction leads to phenanthrene from both benzo[f] and [h]quinoline N-oxides, apparently via ring opening and reclosure; phenanthrene 1-d is formed when $(CD_3)_2SO$ is used.[447b] With the same anion, isoquinoline 2-oxide undergoes ring expansion to a benz[d]azepine.[779]

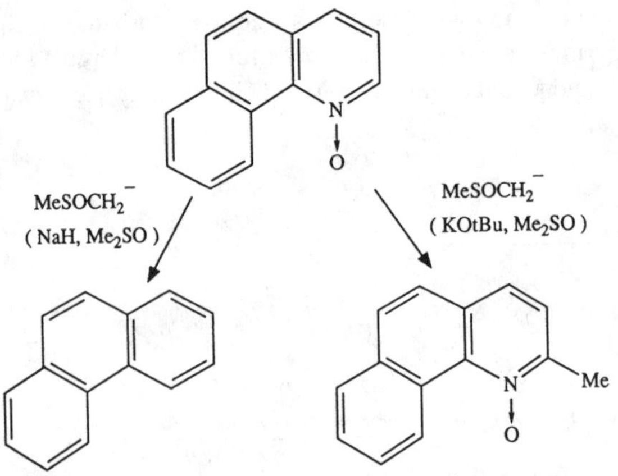

2. With Other Nucleophiles

Azine N-oxides containing a second nitrogen atom in the β position are liable to the addition of nucleophiles in α, and this may lead to ring cleavage. The case of pyrimidine 1-oxides, the "normal" nucleophilic substitutions of which have been discussed in Section VA, is typical. Studies on some of these substrates have shown that when there is no leaving group of sufficient mobility, ring contraction takes place and leads to an isoxazole; the course of the reaction further depends on the conditions, as is the case with 4-chloropyrimidine 1-oxides when treated with potassium amide in liquid ammonia.[780,781] Ring contraction to isoxazole is also observed in the reaction with hydroxylamine, the N-O group in the product in this case resulting from the nucleophile, not the pyrimidine $\geqslant$N→O, just as happens with the parent (deoxygenated) molecule.[782]

The pyrimidine ring has been found to be subject to cleavage under basic conditions in the N-oxides of some quinazolines,[783] pyridopyrimidines,[784] pteridines,[785] and purines.[786,787] Benzo-1,2,3-triazine 3-oxides are likewise cleaved under alkaline conditions,[788] and 1,2,4-triazine 2-oxides[789a] and benzo-1,2,4-triazine 1-oxides[789b] undergo ring contraction to triazoles.

A peculiar ring contraction is observed with 2-alkyl- and alkoxyquinoxaline 4-oxides, which yield 2-substituted-(1H)-benzimidazole 3-oxides when treated with hydrogen peroxide under basic conditions.[790] There is probably a mechanistic analogy with the "Beirut" synthesis (see Sections VIIB2, 3).

3. Under Acidic Conditions

The results are similar to those obtained under base catalysis. Some quinazoline 3-oxides form stable covalent hydrates under acidic conditions.[791] Pyrimidine 1-oxides are hydrolyzed in acidic medium and yield isoxazoles.[792] Ring cleavage is observed in purine derivatives[793] and in fervenulin 4-oxide.[794] In the latter case, the carbocation can be trapped with indole.

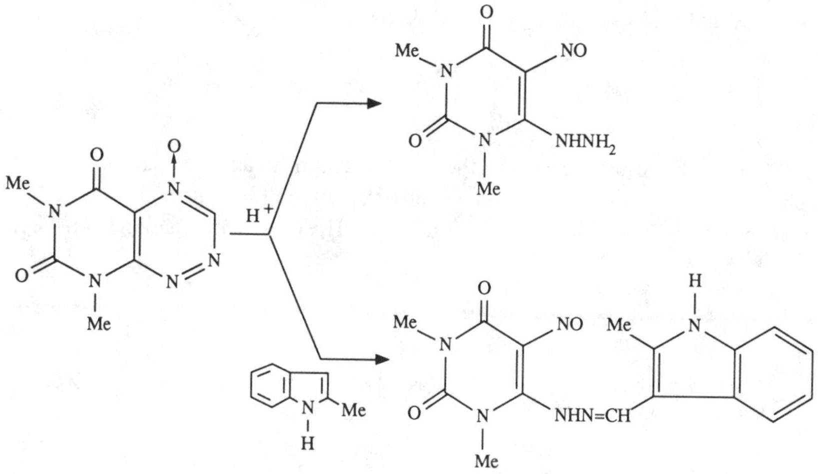

4. Via N-Alkoxy or N-Acyloxy Cations

O-alkylation is often carried out in order to "activate" N-oxides toward deoxidative substitution (Section VI). However, the interaction of N-alkoxy heterocycles with bases may lead to different reactions,[18,19] i.e., (i) proton abstraction from the alkyl group at position α to the oxygen, resulting in the formation of the deoxygenated heterocycle and a carbonyl derivative; (ii) displacement of the N-oxide from the alkoxy group; (iii) addition at the ring α-position, eventually resulting in either substitution or ring cleavage; (iv) proton abstraction from the ring α position to yield a ylide, a reaction that may again lead to cleavage to a carbonyl derivative, unless addition to the anionic site takes place competitively (see below); (v) proton abstraction from the alkyl chain at the β position to yield an alkene[795,796] (Scheme 15).

Thus, 1-methoxypyridinium salts undergo ring cleavage to glutaconaldehyde derivatives when treated with bases (in competition to attack at the alkyl group, leading to pyridine and formaldehyde).[18,19]

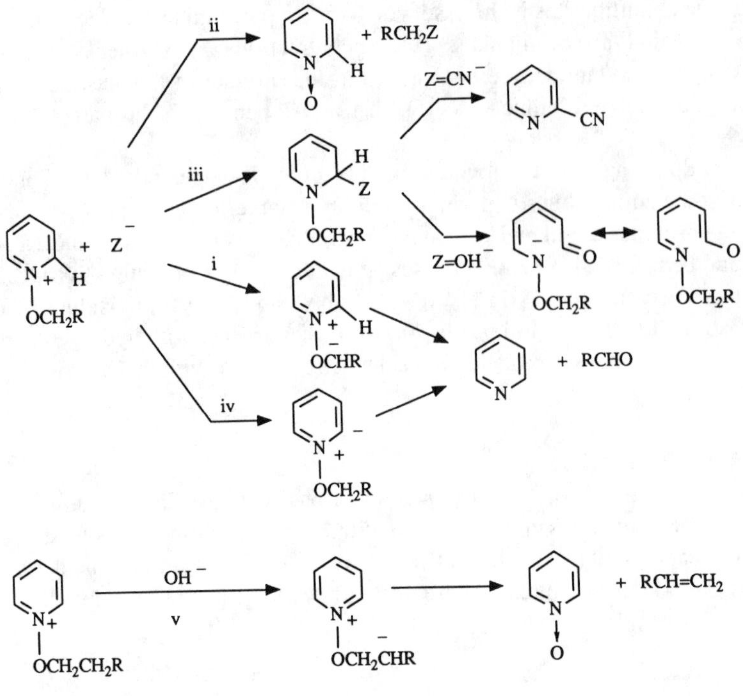

Scheme 15.

When there is a β-carboxy group in the *N*-alkyl chain, oxazolone derivatives are formed through proton abstraction-ring closure-addition of a nucleophile-ring opening (PARC-ANRO) mechanism, the key intermediate being an ylide formed by proton abstraction from the ring (path iv in Scheme 15).[795,796]

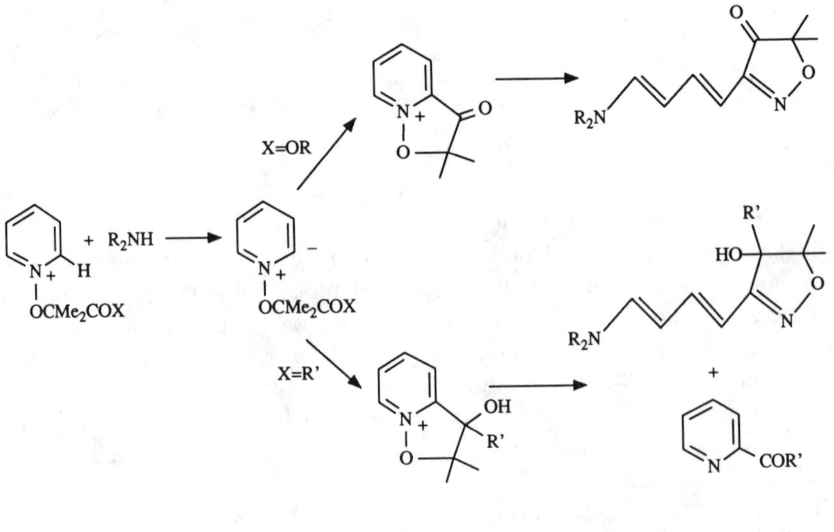

With a carbonyl rather than a carboxyl group, the oxazolopyridines are actually isolated. In this case, the ANRO sequence competes with a different mechanism, leading to 2-acylpyridine, and this is the only pathway with the quinolinium and isoquinolinium analogs (compare with Section VIC5).[644d-644e,796]

As might be expected, cleavage reactions are also observed on acylation, e.g., in the case of purine derivatives.[797]

5. By Thermal Decomposition

Thermal extrusion of N_2O from (poly)cyclic azoxy derivatives has been the subject of many preparative and mechanistic studies, which fall outside the scope of this book. With the heteroaromatic analogs, pyridazine 1-oxides, flash vacuum pyrolysis primarily affects the N-N bond, and the products are maleo-, fumaro-, and acrylonitrile, as well as indoles.[798]

B. ADDITION AND RING OPENING OF AZOLE N-OXIDES

1. By Thermal Decomposition

Several azole N-oxides are thermally unstable, e.g., 1-hydroxy- and 1-methoxyimidazole 3-oxides explode on attempted distillation,[799] and some nitro derivatives of benzofuroxan[474,475] and benzotriazole 1-oxide[496] are commercial explosives.

Of particular importance is the fragmentation of furoxans to yield nitrile oxides. This is a concerted process in the gas phase (probably preceded by rapid isomerization),[801,802] and its kinetics have been studied under various conditions.[801-804] Flash-vacuum pyrolysis of furoxans offers a useful entry to unstable nitrile oxides,[805] and more generally furoxans are used as precursors of nitrile oxides in cycloaddition reactions,[800,806-808] this method offering a particularly clean source of such reactive molecules, e.g., allowing one to carry out relatively slow cycloadditions, such as those involving trisubstituted olefins, with the use of a stoichiometric amount of reactive.[808] Furthermore, strained nitrile oxides formed from the furoxans rearrange, most easily in the presence of sulfur dioxide (probably through the intermediacy of thiadioxazoles), to isocyanates,[800,809,810,811b,811c] a reaction that also has industrial significance.[811a] De-oxidative thermolysis to form nitriles can also be carried out, e.g., in the presence of phosphites.[811d]

2. Conversion of Benzofuroxans into Quinoxaline Oxides and Related Reactions

In what has come to be known as the *Beirut reaction,* a benzofuroxan is condensed with a carbonyl derivative in the presence of a base, with an amine often giving better results, or with a preformed enamine, to yield a quinoxaline 1,4-dioxide.[812-817] The reaction usually takes place under mild conditions. Yields are variable, but are good in several cases, and the availability of the starting material, combined with the industrial interest of the products, has lead to much work in this field.

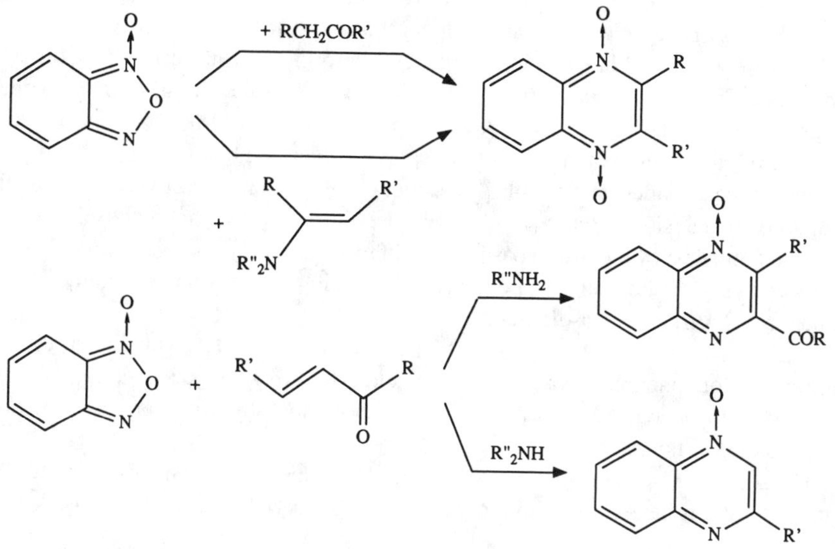

Different mechanistic pathways can be considered for the reactions, i.e., 1) cycloaddition of the enolate or the enamine onto the diazabutadiene *N*-oxide chromophore, 2) nucleophilic addition either at position 1 or at position 3, 3) nucleophilic addition onto the dinitrosobenzene present in equilibrium with the benzofuroxan. Hypothesis 1 is attractive and is probably inescapable in the reaction with alkynes or styrenes, but requires that the primary adduct has rearrangement pathways available, since in some cases products different from quinoxaline oxides are formed; path 2 is favored by analogy with other known nucleophilic additions on benzofuroxans and appears to be preferred by the majority of the authors; path 3 is supported by the fact that only benzofuroxans known to equilibrate in solution (through the dinitroso isomer, see Chapter 2) undergo this reaction, though some derivatives that equilibrate do not react. A substituent effect has been interpreted in favor of the various hypotheses, but unambiguous mechanistic evidence has probably not yet been reached. As has been noted, it is quite possible there is no unitary mechanism, and different alternatives are operative in various cases.

The process is not exclusive of benzofuroxans, since an analogous addition onto benzofurazans[818] (to yield quinoxaline mono-oxides) and benzo-2,1-isoxazoles[819] (to yield quinoline 1-oxides) has been reported, though more drastic conditions are required. Further complications are due to the likely importance of electron transfer pathways (the benzofuroxans are strongly oxidizers, and benzofurazans have been obtained during the reaction)[820] and thus to the possible intervention of radicals or radical ions in the process,[818] as well as to competitive ring opening by the base. In the presence of diethylamine, benzofuroxans react and nine products have been isolated, including quinoxaline 1,4-dioxide and 1-hydroxy-2-methylbenzimidazole 3-oxide from the reaction with Et-N=CHCH$_3$ formed *in situ*.[821]

In the following, the scope of the reaction, with attention to the products formed rather than to the mechanism, is briefly discussed.

With cyclic ketones, tetrahydrophenazine dioxides are obtained.[820] Enolizable imines and azines also react, but much better in the presence of a base.[818]

The actual product(s) obtained depend on the carbanion used, with many variations on the general scheme shown above. When a single proton is present at the reacting position of the ketone, 2-amino-2,3-dihydroquinoline 1,4-dioxides are obtained.[817] The base may cause substituent modification during the reaction. Thus, chloroacetaldehyde and a primary amine yield 2-alkylaminoquinoxaline dioxides.[823] Deacylation occurs easily (see below). Ring cleavage is observed with benzofuran-3-one, which yields 2-(2-hydroxyphenyl)-quinoxaline 4-oxide.[824a] Notice that in this case the product is one degree of oxidation lower than expected from a simple condensation. On the other hand, indole, a cyclic enamine, yields indole[2,3-*b*]quinoxaline dioxide, which is one degree of oxidation higher than expected.[820] Obviously, redox processes involving the reagents and/or some intermediates determine the final structure of the products.

Vinyl acetate in the presence of diethylamine yields, as expected, quinoxaline dioxide.[834b] β-Diketones, β-ketoesters, and amides react normally, and yield 2-acyl- (or carboxyalkyl, or carboxamido) 3-alkylquinoxaline dioxides.[825-829] (However, see below for a different course of the reaction.) The condensation with β-diketones can be conveniently carried out through molecular sieve catalysis.[829b] Benzoylacetonitrile yields 2-phenyl-3-cyanoquinoxaline dioxide,[829a] and with malonitrile a cyano group participates in the addition, yielding 2-cyano-3-aminoquinoxaline dioxide.[830,831] Formally analogous is the reaction with sodium cyanamide to yield 3-aminobenzo-1,2,4-triazine 1,4-dioxide.[832]

1,2-Cyclohexandione yields a mixture of 1-hydroxyphenazine 10-oxide and the corresponding dioxide.[833] Both phenol and hydroquinone yield 2-hydroxyphenazine dioxide, and from α-naphthol, benzo[*a*]phenazine dioxide and the corresponding 5-hydroxy derivative are obtained, possibly via quinones formed *in situ*.[820,831]

α,β-Unsaturated ketones appear to yield mono *N*-oxides rather than dioxides; thus, various aliphatic derivatives have been found to yield 2-acyl-3-alkyl 1-oxides with secondary amines and 2-alkyl-3-acyl derivatives with primary amines; in both cases deacylation is facile or spontaneous and affords 3-, or respectively 2-, substituted quinoxaline 1-oxides.[834,835] With cyclohexenone, deacylation leads to an open-chain amide.[835] Strict generalizations are difficult, however, as an example, cynnamaldehyde yields a mixture of the 2- and the 3-phenylquinoxaline 1-oxides in the presence of morpholine, while 4-phenylbut-3-en-2-one yields a dioxide, possibly because it is oxidized to the diketone under the reaction conditions.[834,835]

Dienamines,[836] and ynamines,[831] as well as acetylenes in the presence of amines,[837] also react to give quinoxaline dioxides. Anethole yields 2-methyl-3-(4-methoxyphenyl)-quinoxaline dioxide.[838] Phosphines have also been shown to participate into this reaction and act as a base. Thus, 1,4-benzo- and naphthoquinones react with benzofuroxan in the presence of phosphines to give the ylides **39**.[839]

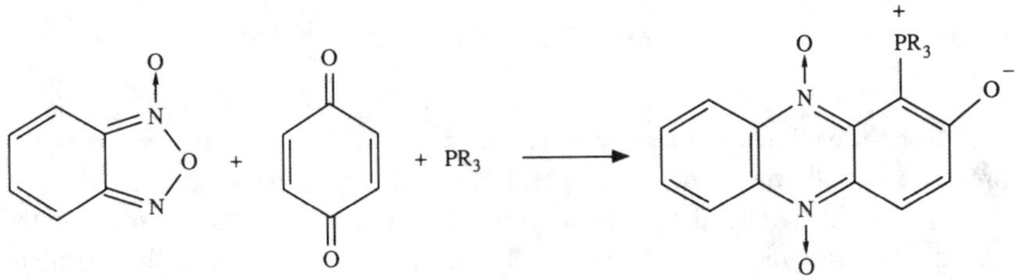

39

Besides quinoxaline and phenazine derivatives, some heterocycle-condensed pyrazine dioxides, e.g., pyrido[2,3-*b*]pyrazines[840] and thieno[2,3-*b*]pyrazines,[841a] as well as pyrido- and quinolinophenazines,[820] have been obtained by this method. Interestingly, pyrazino[2,3-*b*]quinoxaline 1,4-dioxides are obtained by simply refluxing a solution of furoxanoquinoxalines with aromatic alkenes or alkynes, no base being required.[841b]

2,3-Dimethylquinoxaline 1,4-dioxide. 54.4 g (40 mmol) of solid benzofuroxan are added in portions to a solution of 2-butanone (28.9 g, 40 mmol) in ammonia-saturated methanol. Ammonia is further passed through the dark-red solution, which is maintained at 40—50°C by occasional cooling. After 5 h the mixture is cooled at 0°C, and the precipitate is filtered and washed with MeOH. The yield is 68.0 g, 90%.[814]

3-Methyl-2-[2-methoxyphenyl]aminocarbonylquinoxaline 1,4-dioxide. Powder molecular sieve 3A (20 g) is added to a solution of 2 g (15 mmol) benzofuroxan and 3.05 g (15 mmol) o-acetoacetaniside in 50 ml methanol, and the mixture is rotary evaporated at 20°C. After 2 h at room temperature, the molecular sieve is added on top of a silica-gel column; elution with dichloromethane-chloroform 98-2 yields the product, with a 4.15 g, 87% yield.[829b]

3. Benzofuroxans to Benzimidazoles

As has been discussed above, many active methylene compounds add to benzofuroxans to yield quinoxaline dioxides; however, in other cases benzimidazoles are obtained. Thus, e.g., while malonitrile yields a quinoxaline, cyanoacetamides yield 2-carboxyamido-1-hydroxybenzimidazole 3-oxides.[829a] With barbituric acid, 2-carboxybenzimidazole is obtained.[842] In other cases, the reaction takes either course, depending on the conditions; thus, with β-ketoesters stronger bases and higher temperatures favor the benzimidazoles vs. the quinoxalines.[828a] Primary nitroalkanes,[843,844] and alkyl, aryl-[845] and trifluoromethyl-sulfonates[846] also yield 1-hydroxybenzimidazole 3-oxides. With α-benzenesulfonylacetophenone derivatives, both the sulfonyl and the benzoyl group are cleaved.[845] With diphenylnitrone, elimination of nitrosobenzene again leads to a 1-hydroxybenzimidazole 3-oxide.[847a] With secondary nitroalkanes 2*H*-2,2-dialkylbenzimidazole dioxides are obtained.[843,844]

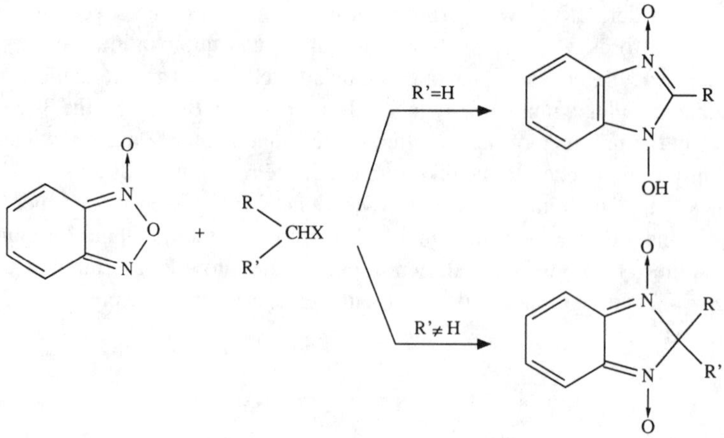

Another process leading to benzimidazole is the reaction of benzofuroxan with formaldehyde in the presence of sodium hydroxide to yield 1,3-dihydroxybenzimidazole-2-one;[845,847b] the reaction with formaldehyde and amines or with trialkylhexahydrotriazines leads to the corresponding imino derivatives.[842] These reactions are addition processes rather than condensations, as in the previous cases, and their mechanism is possibly unrelated; an electrophilic attack of the aldehyde onto the furoxan has been considered.[813]

4. Other Reactions

Nitrogen insertion and ring expansion to a 4-hydroxycinnoline 1-oxide take place when phenylisatogen is heated with ammonia (a different ring expansion to 3,4-dihydro-2-phenyl-quinazolin-4-one occurs with tetracyanoethylene).[848]

Furoxans are labile under basic conditions.[849-853] As an example, 4-phenylfuroxan rearranges to α-hydroxyimino-*anti*-phenylacetonitriloxide on simple dissolving in acetone, and this easily adds water to give the corresponding hydroxamic acid and 3-phenyl-1,2,4-oxadiazol-5-one,[849b] while the isomeric 3-phenylfuroxan is rearranged by bases to α-nitrophenylacetonitrile.[850b] 3-Methylfuroxan yields 2-isoxazolin-4-one oxime under basic conditions ("isoxazolinic", or Angeli's rearrangement), and isoxazoline,[851] pyrazoline,[853] and open-chain[852] derivatives are obtained from substituted furoxans.

Similarly, basic ring opening of benzofuroxans usually leads to *o*-nitrophenyl derivatives,[854-858] which in some cases is followed by ring-reclosure, as in the conversion of pyrimidinofuroxans into purines.[858]

Addition of Grignard reagents onto furoxans leads to dioximes or involves C-C bond cleavage to yield ketones and nitriles.[860,861] Chemical and electrochemical reductions of furoxans lead to open-chain derivatives.[859,860,862-864]

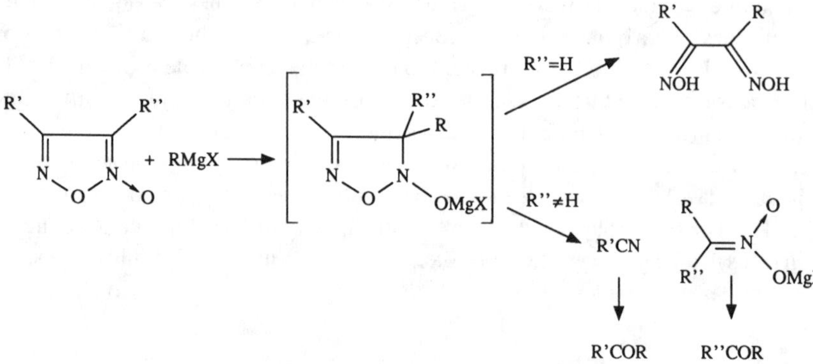

In the case of 4-substituted benzofuroxan, an important variation of these ring-opening ring-reclosure processes leads to new heterocycles (Boulton-Katritzky rearrangement, see the formula A=B is a nitro, nitroso, arylazo, acyl, or imidoyl group).[865-869]

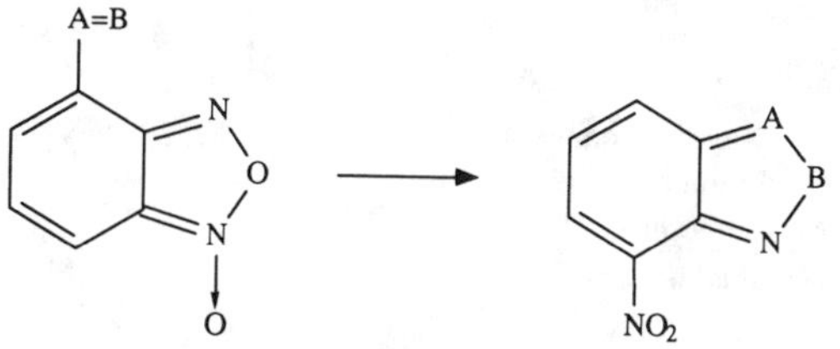

4-Oxopyrazolone 1,2-dioxides undergo ring cleavage both under basic (to yield an isoxazole in the case of the 3,5-dicarbomethoxy derivative)[870a] and under acidic conditions.[870b]

C. OTHER N-OXIDES

Various base-catalyzed ring contractions of benzodiazepine oxides have been reported.[871a-871d] For example, 3H-1,2-diazepine 2-oxides rearrange at room temperature to 3-alkenyl-3H-pyrazole 2-oxides,[871a] and a 2-amino-1,4-benzodiazepine 4-oxide has been found to rearrange to an indole derivative under basic conditions.[871b]

Although cyclic nitrones are generally neglected in this discussion, one should mention at least the theoretically interesting electrocyclic ring opening of azetine 1-oxide to unsaturated oximes.[871e,871f]

VIII. CYCLOADDITION

A. THE N-OXIDE FUNCTION AS A 1,3-DIPOLE

The rate of 1,3-dipolar cycloaddition processes in nitrones changes over a large interval depending on the steric and electronic characteristics of the reagents. In general, this process is much slower with heteroaromatic N-oxides, essentially due to the large loss of aromaticity that must be undergone in this case. As an example, isoquinoline 2-oxide reacts with ethyl crotonate at 100°C, 3.6×10^4 times slower than 3,4-dihydroisoquinoline 2-oxide, a cyclic nitrone, and this corresponds to enthalpy difference of 7—8 kcal/M. Phenanthridine 5-oxide reacts somewhat faster, since the loss of aromaticity is lower, but with pyridine 1-oxide no appreciable reaction takes place under these conditions[872] (Table 22).

Thus, thermodynamics favors the reagents rather than the adducts. However, when an alternative pathway is available for the primary adduct, convenient preparative reactions take place, and this is particularly the case when alkynes are used.

1. Reaction with Alkenes

Cyclic nitrones, such as pyrroline, tetrahydropyridine, dihydroisoquinoline N-oxides, easily add to olefins, yielding isoxazolines, which are in turn intermediates for a variety of transformations. This well-known and widely used reaction (in particular, for the synthesis of natural products) is not discussed here, and is adequately dealt with in recent treaties.[873]

On the other hand, only a limited number of cycloadditions between heteroaromatic N-oxides and alkenes have been reported. Thus, pyridine 1-oxide reacts with hexafluoropropene to yield 2-(1,2,2,2-tetrafluoroethyl)pyridine, reasonably through cycloreversion of the initially formed oxazolidine, as supported by the identification of some COF_2 in the effluent gas.[874] Pyridine 1-oxide apparently does not react with acrylic derivatives, but 3,5-lutidine 1-oxide and N-arylmaleimides stereoselectively yield the adduct **40**, rationalized as arising via a 1,5-sigmatropic shift from the initial exo adduct **41** (compare the reaction with alkynes, Section VIIIA2).[875,876]

TABLE 22
Rate Constants for the Cycloaddition of N-Oxides
to Ethylcrotonate in Toluene at 100°C

Substrate	k, $M^{-1}s^{-1}$
3,4-Dihydroisoquinoline 2-oxide	2.4×10^{-2}
Δ^1-Pyrroline 1-oxide	1.6×10^{-2}
Phenanthridine 5-oxide	1.3×10^{-6}
Isoquinoline 2-oxide	6.7×10^{-7}

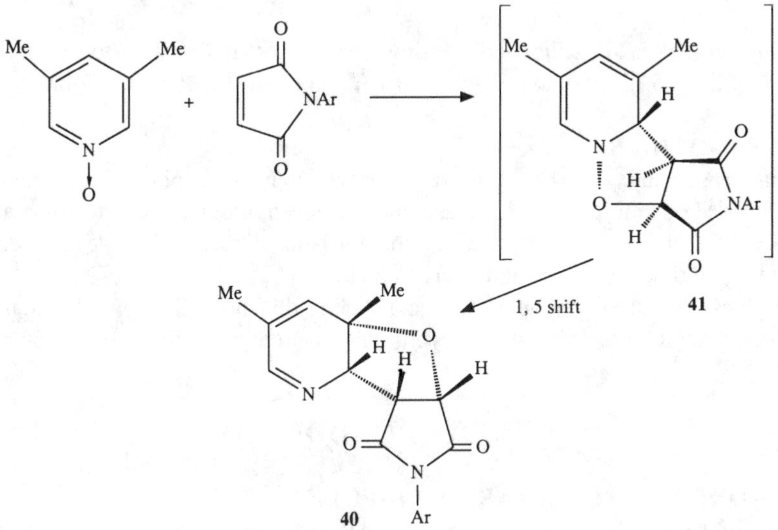

Reaction with acrylic derivatives has been reported for quinoline and isoquinoline N-oxides, but only in the presence of acetic anhydride or catalytic amounts of hydroquinone, and yields 2-, and respectively 1-, alkylated derivatives.[877,878] The reaction has been rationalized as involving the initial formation of an isoxazoline, the role of the additive being that of slowing down the polymerization of the acrylic derivative and possibly favoring irreversible decomposition of the isoxazoline to yield the end products. The reaction is complicated by bypaths initiated by acetic anhydride itself (compare with Section VIB).[879]

From phenanthridine 5-oxide and ethyl acrylate in DMF, the 6-(2-hydroxyalkyl) derivative **42** is obtained (notice that the regiochemistry is opposed to that observed with diphenylnitrone)[880] and a similar reaction takes place with naphthalene endoxide.[881] Isoquinoline 2-oxide analogously yields product **43** with thiophene S,S-dioxide.[882] The reaction of 1,2,4-triazine 4-oxides with 1-ethoxy-1-dimethylaminoethene to yield the 5-methyltriazines **44** can also be formulated as involving cycloaddition followed by reversion (when a 5-methyl group is present in the starting N-oxide, attack takes place on that group).[883] Benzo[c]cinnoline 5-oxide reacts with strained alkenes, such as E-cyclooctene; the initial adduct is again unstable and cleaves to an azomethine imine, and this is trapped by the alkene to yield the observed 1:2 adduct **45**.[884]

As for five-membered heteroaromatic derivatives, it might be noted that while 3-H pyrrole-3-one 1-oxides with three aryl substituents are stable purple compounds, attempted preparation of the less-arylated analogues yields the yellow dimers, and these revert to the monomers upon heating, an example of reversible 1,3-dipolar cycloaddition.[885] 3,3-Dimethylindolenine 1-oxide[886] and pyrazole-4-one 1,2-dioxides,[887] and among seven-membered N-oxides, 1,4-benzodiapine derivatives, e.g., diazepam oxide,[888] also yield condensed isoxazoline derivatives or products arising from their cleavage.

42

43

44

45

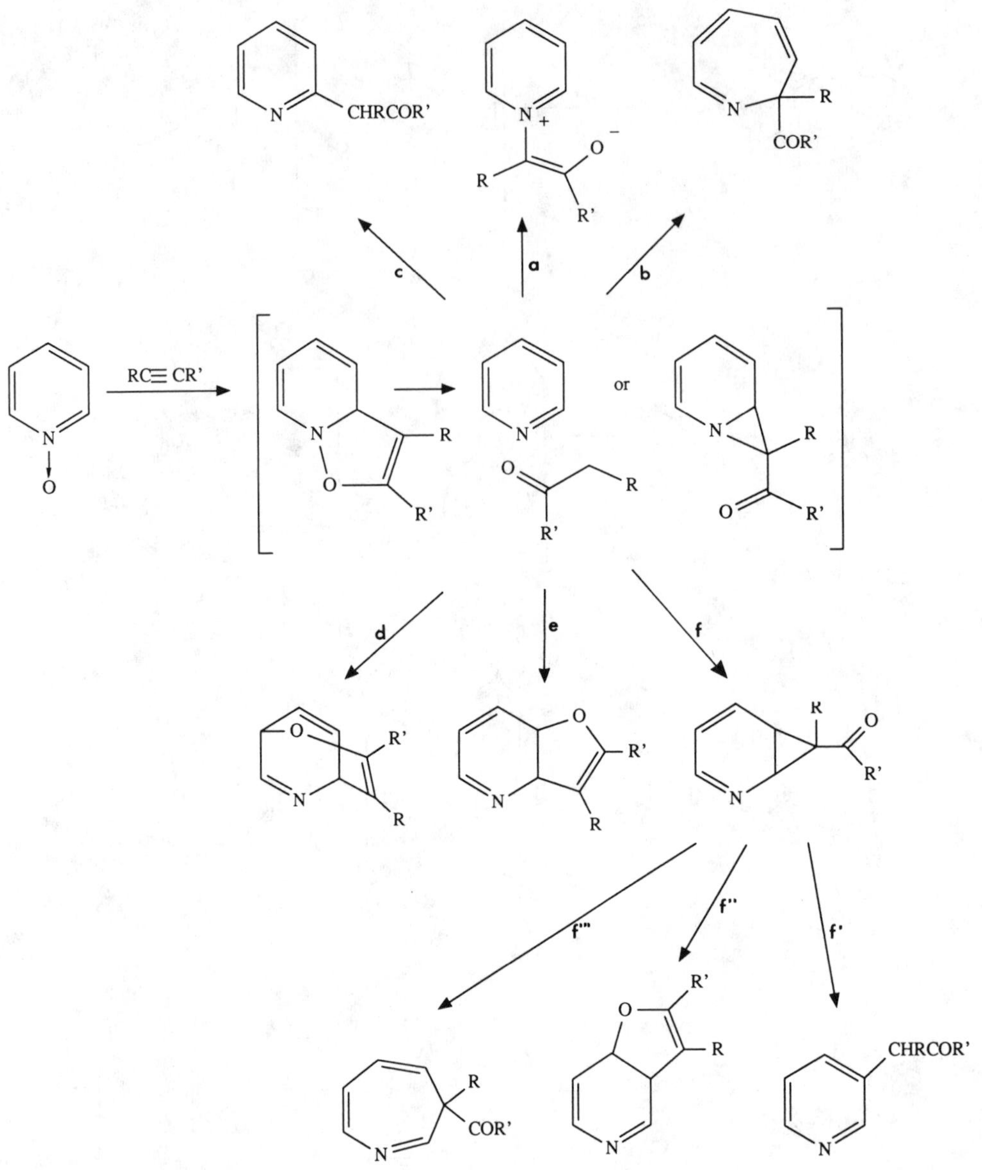

Scheme 16.

2. Reaction with Alkynes

The study of the reaction of heteroaromatic *N*-oxides with alkynes has revealed a multifaceted chemistry, and several processes of remarkable synthetic interest. Different types of products can be formed and have been rationalized as arising via various rearrangements from an initially formed isoxazoline.[889] Scheme 16 covers the main pathways that have been recognized in the case of six-membered *N*-oxides, taking pyridine 1-oxide as an example.

Thus, ring contraction to form a β-acyl aziridine (or, alternatively, cleavage to a tight-pair pyridine/α-oxocarbene) is followed by ring opening to yield a *N*-ylide (path a) or by ring enlargement to yield an azepine (path b). Cleavage of the N-O bond and rearomatization leads to a 2-alkyl derivative (path c), or alternatively, 1,3- or 1,5-sigmatropic shifts lead to 2,5 or, respectively, 2,3-dihydro heterocycles (paths d and e, in the latter case followed by rearomatization if there is a leaving group in 3). In several cases, however, the carbon-carbon bond migrates to position 3 through a rearrangement that, if concerted, is a $_\sigma 2_s + _\pi 2_a + _\pi 4_s$ symmetry allowed process (path f). In conclusion, *N*-ylides, 2- and 3-alkyl derivatives, as well as furo[3,2-*b*] and [3,2-*c*]pyridines (from the rearomatization of the corresponding dihydro derivatives in the presence of a suitable leaving group), are obtained in the various cases. Though the possible pathways are many, one of the products is often formed in high yield, and the direction can be usually predicted on the basis of the reagent structure, in particular, the presence of leaving groups that open a path for rearomatization (Scheme 16).

Thus, the reaction of phenylpropyonitrile or phenylpropiolates with pyridine 1-oxide yields the 3-alkylated derivatives as major products, along with minor amounts of the 2-alkyl derivative and the *N*-ylide.[889-892] 3,5-dichloropyridine 1-oxide yields a mixture of furo[3,2-*b*] and [3,2-*c*]pyridines, and only the latter type of product is obtained from 4-chloro-, methoxy-, or nitropyridine 1-oxides.[893-895] The reaction is quite complex, and in some cases important amounts of bis and tris adducts, including a pyridooxepin, have been isolated.[894]

Analogously with benzyne, 3-(2-hydroxyphenyl)pyridine is the major product; benzofuro[3,2-*b*]pyridines are obtained from 3,5-dihalopyridine 1-oxides and with 3,5-lutidine 1-oxide, and the reaction, when carried out under kinetic control (<45°C), yields products **46** and **47** according to paths c and d, and at a higher temperature yields the 2-alkyl derivative.[896]

With alkynylphosphonium salts, the phosphonium methylides **48** and **49** are obtained, and can be pyrolyzed to phenylethynylpyridines; here again, starting from 4-methoxy- or nitropyridine 1-oxide, a furo[3,2-*c*]pyridine is obtained.[897] A different course is followed with the sulfurated alkyne **50**, which reacts with pyridine 1-oxide to yield the spiroazepine **51**, in turn, rearranging to the substituted pyridine **52**.[898]

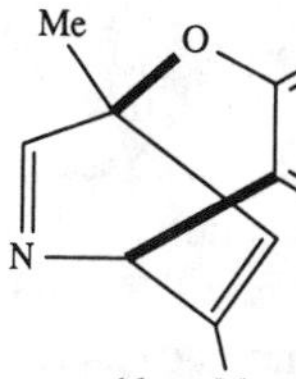

46   Me

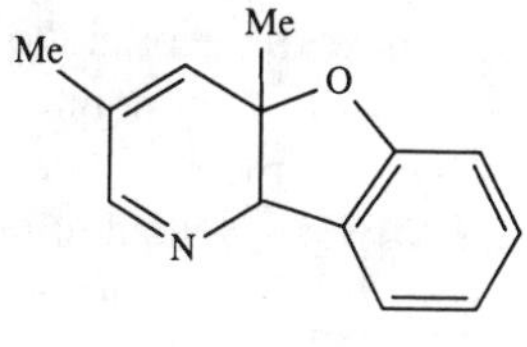

47

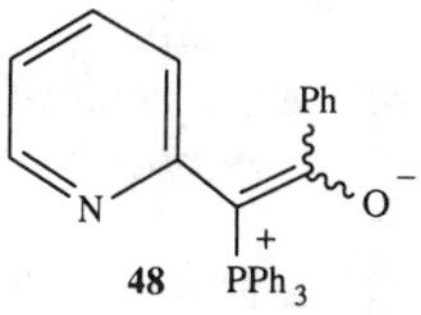

48   PPh$_3$

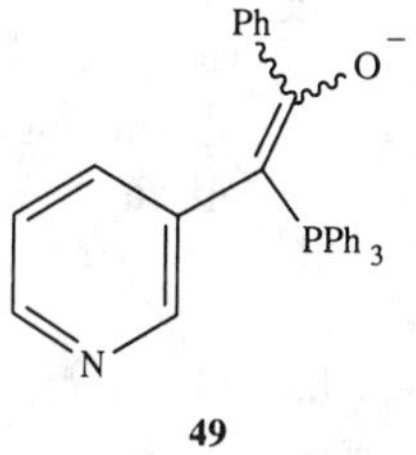

49

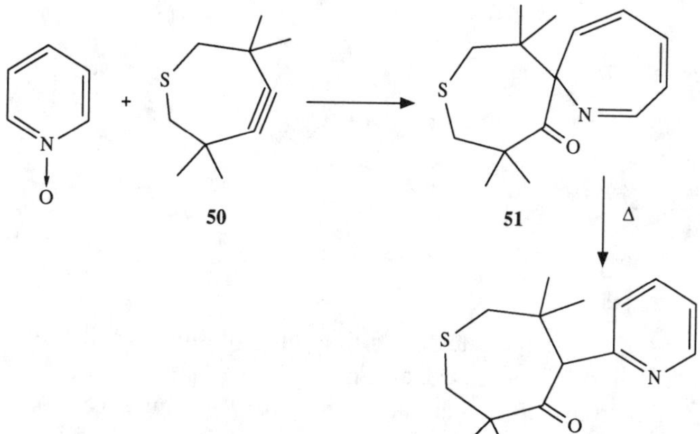

50

51   Δ

52

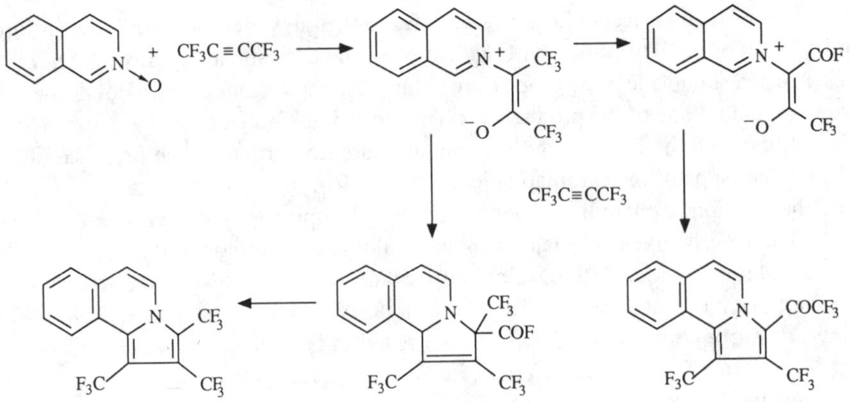

Scheme 17.

The reactions of quinoline 1-oxides are analogous and can be rationalized with the same considerations.[890,891,899-901] The reaction with propiolates has been reported to occur in the presence of acetic anhydride or hydroquinone.[878] Interestingly, 2-substituted quinoline 1-oxides yield 3-*H* 1-benzazepines when treated with dimethylacethylendicarboxylate (DMAD) (an example of path f''').[902] Quinoxaline 1,4-dioxides react similarly with dibenzoylacetylene.[903] With 2,3-dimethylquinoxaline 1-oxide (and 1,4-dioxide), "eliminative cycloaddition" involving the substitutuent gives pyrazolo[1,2-*a*]quinoxaline derivatives.[904]

With isoquinoline[901] (except in the reaction with propiolate/Ac₂O),[878] phenanthridine,[905,906] phthalazine,[907] and benzo[*c*]cinnoline[908,909] *N*-oxides formation of the *N*-ylides (path a in Scheme 16) or of products from their further evolutions appears to predominate (see Scheme 17).

With diazine, *N*-oxide cycloaddition is often followed by fragmentation. This is the case with pyridazine 1-oxide, in which nitrogen is eliminated after the addition of benzyne, and a benzoxepine is obtained;[910] however, with 3-hydroxypyridazine 1-oxide, 3-hydroxy-6-(2-hydroxyphenyl)pyrimidine is obtained.[911] Elimination of nitrogen from the pyridazine moiety also takes place in the reaction of 5-phenylpyrido[2,3-*d*]pyridazine 7-oxide with DMAD.[912] 1,3,4-Oxadiazin-6-one 4-oxides react with alkynes to yield butenolides after rearrangement and nitrogen elimination;[913] the analogous reaction with benzyne leads to 3-acylbenzofuranones and, by carbon dioxide elimination, to benzofurans.[914] In some pteridine 5-oxides, addition of DMAD is followed by ring contraction of the pyrazine moiety to yield purines and 7-deazapurines (the main reaction is simple deoxygenation of the substrate, however),[915] and related reactions have been reported for pyrimido[5,4-*e*]triazine 4-oxides, yielding 9-deazapurines.[916-918] 4-Phenyl-2-methylquinazoline 3-oxide adds DMAD and rearranges to yield a benzo-1,3-diazepine and a ring-cleaved product.[919]

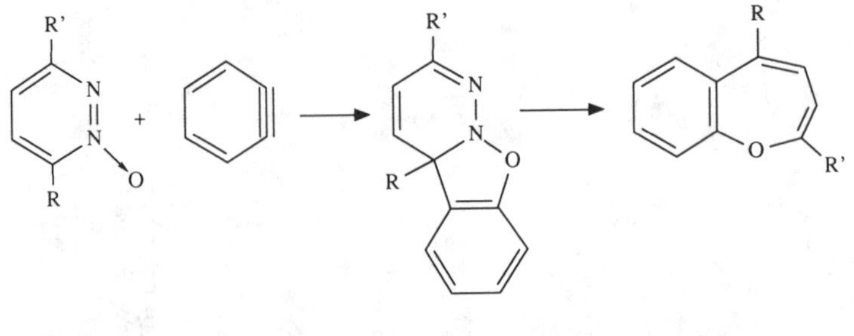

Several reactions of acetylenic esters with five- or seven-membered *N*-oxides have been reported, and lead either to condensed isoxazolines, as in the case of 2*H*-imidazole 1-oxides,[920] or to products resulting from their further transformations. Thus ring expansion to thiazines has been observed with some thiazole *N*-oxides,[921,922,927] and 1,2,3-triazine are obtained from 1,2,3-triazole 1-oxides.[923] With 2-aminothiazole 3-oxides, ring cleavage and condensation lead to fluorescing thiazolo[3,2-*c*]pyrimidones.[924] Further ring expansions occur in the reactions of pyrazol-4-one 1,2-oxides with DMAD (to yield pyrimidine 1-oxides),[925] and of isatogen with phenylacetylene and propyolic acid (to yield 4-hydroxyquinolines).[926] From 2-aminobenzo-1,4-diazepine 4-oxides, isoxazolobenzodiazepines and/or pyrazoloquinoxalines are formed.[928,929]

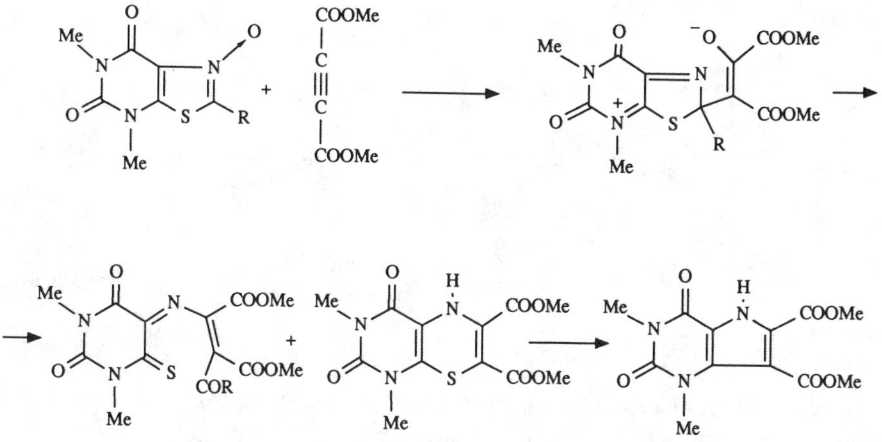

3-Methoxycarbonyl-2-phenylfuro[3,2-c]pyridine. 4-*Chloropyridine 1-oxide (0.3 g, 2.32 mmol) and methyl propyolate (0.30 g, 2.38 mmol) are refluxed in toluene (25 ml) for 2 h. Evaporation of the solution and silica-gel chromatography of the residue gives 0.415 g, a 71% yield.*[895]

3H-3-Methoxycarbonylcarbonyl-3-methoxycarbonyl-2-phenyl-1-benzazepine. A solution of 2-phenylquinoline 1-oxide and dimethyl acetylendicarboxylate (1.2 eqv) in dichloromethane kept at room temperature for 3 d gives an 88% yield of the product.*[902]

1,6-Dimethoxycarbonyl-2,5-diphenyldipyrrolo[1,2-a:2′1′a]quinoxaline. 2,3-*Dimethylquinoxaline 1,4-dioxide (1 g, 5.3 mmol) and dimethyl acetylendicarboxylate (8.5 g, 53 mmol) are heated together at 115°C for 3 d. The product is crystallized on cooling to 0.87 g, a 35% yield (after recrystallization).*[904]

3. Reaction with Isocyanates and Thioisocyanates

Heteroaromatic *N*-oxides react easily with phenylisocyanate and its derivatives. The reaction of pyridine 1-oxides has been repeatedly investigated,[930-937] and it has been firmly established by chemical,[933] as well as x-ray,[931,938] evidence that the product is an oxazolo[4,5-*b*]pyridine: apparently, the initially formed cycloadduct undergoes an efficient 1,5-sigmatropic shift. MO calculations show that the 2,3-dihydropyridine **53** is indeed more stable than the 1,2-dihydro isomer **54**. At a high temperature, the bicyclic derivatives **53** are converted into 2-anilino derivatives **55**,[934] and in several cases these are the only products isolated. With 3,5-dibromopyridine 1-oxide, dehydrobromination takes place and a 2-oxazolopyridine is obtained.[934] With 2-alkylpyridine 1-oxide, a further rearrangement of the adduct is followed by the addition of a second molecule of PhNCO to yield product **56**.[940] Kinetic studies and MO calculations support the concerted nature of the cycloaddition to PhNCO and explain the observed site selectivity.[939-941] A charge transfer complex between the reagents has been spectroscopically detected, and its role in cycloaddition might explain the solvent effect on the site selectivity.[943] The reaction has been extended to tosylisocyanate.[942]

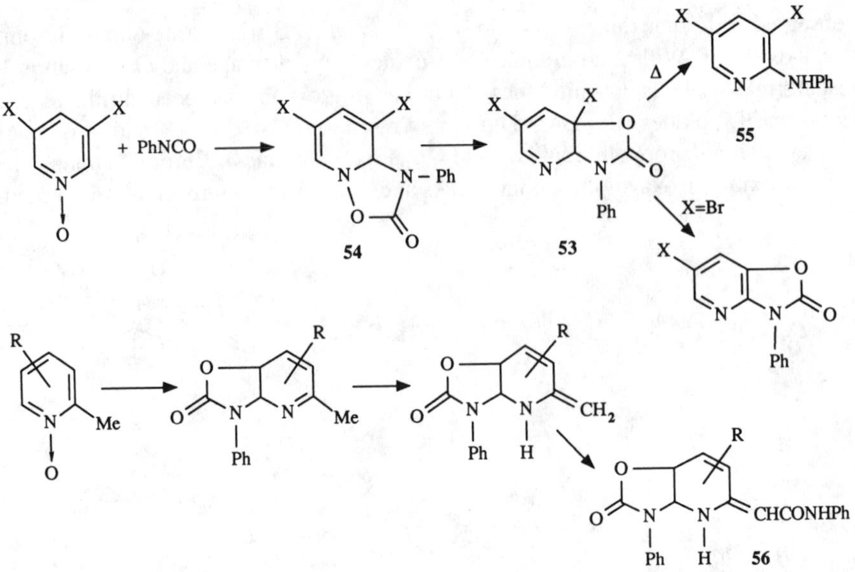

A similar pattern of reactivity has been found for the other N-oxides. Thus, α-anilino derivatives are obtained with quinoline, isoquinoline,[944,945] phenanthridine,[946] phthalazine,[560b] and quinoxaline[115] N-oxides with PhNCO and, in several cases, with PhNCS as well. 3-Bromo- and 3-nitroquinoxaline 1-oxide and PhNCO yield an oxazoloquinoline.[947] 2-Anilinopyrimidines have been obtained from the corresponding 1-oxides, including the 4-alkoxy derivatives, and PhNCO.[948] In contrast, 4-ethoxy-6-methylpyrimidine 1-oxide yields an oxazolo[4,5-d]pyrimidine with PhNCS, apparently through a variation of path d (Scheme 16). 2-Tosylaminoquinoline and 1-tosylaminoisoquinoline are, respectively, obtained from the corresponding N-oxides and N-sulfinyltosylamide (PhNSO).[949]

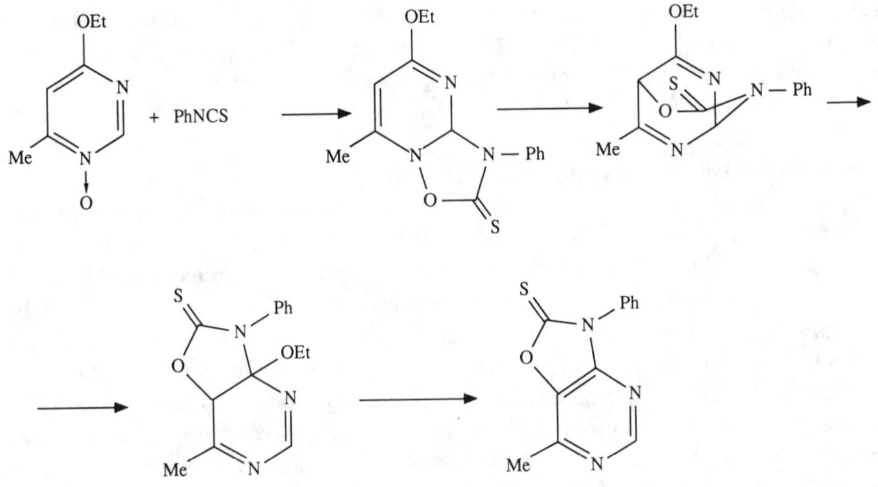

As for five-membered derivatives, 1-methylbenzimidazole 3-oxide yields 2-anilino-1-methylbenzimidazole with both PhNCO and PhNCS,[950] but with the 1,2-dimethyl analog the anilino group is found in position 6,[104] and 2-ethoxycarbonylbenzothiazole 3-oxide yields 1-arylimino(S^{IV})benzothiazole 2-carboxylate.[951] Interestingly, with thiazole 3-oxides elimination of CO_2 and cleavage of a C-S bond lead to bis(5-imidazolyl)disulfides.[952] Oxazole 3-oxides also cleave and add a second molecule of isocyanate to yield imidazole[4,5-*d*]oxazole derivatives.[953]

7-Methyl-3-phenyl-2,3-dihydrooxazolo[4,5-d]pyrimidine-2-thione. 4-Ethoxy-6-methylpy-rimidine 1-oxide (0.92 g, 6 mmol) and phenylisothiocyanate (0.81 g, 6 mmol) in 2 ml chloroform are kept at room temperature for 45 h. Dilution, acid and basic washing, silica-gel chromatography, and recrystallization from ether give 0.55 g (38% yield) of the product.

4. Reaction with Other Cumulenes

The reactions of heteroaromatic *N*-oxides with diketene leads to alkylation in 2 and has been discussed among deoxidative substitutions (Section VID1), but might also be considered as being initiated by 1,3-dipolar cycloaddition. The reaction of dichloroketene, which reacts with pyridine 1-oxide via both 1,3- and 1,5-dipolar cycloaddition (in the latter case, a 1,3-dipolar addition might, again, be the first step, and a following rearrangement might lead to the observed 3,4-dihydropyridine adduct, compare with, e.g., the reaction with phenylisothiocyanate),[948] is more complex. Final products are 2- and (mainly) 4-dichloromethylpyridine, as well as small amounts of two furo[2,3-*c*]pyridines.[954] Analogously, 2-dichloromethylquinoline and 1-dichloromethylisoquinoline are, respectively, obtained from the corresponding *N*-oxides.[954]

B. CYCLOADDITION ONTO HETERODIENES

Substitution of a -CH= group by a nitrogen lowers the energy of π orbitals, and the effect of the $\geqslant$N→O group is comparable. Thus azabutadiene *N*-oxides are electron-poor dienes and might be expected to be able to take part in inverse electron-demand Diels Alder cycloadditons. Indeed, 4 + 2 cyloaddition across a 2,3-diazabutadiene N-oxide chromophore has been observed in pyrazole derivatives,[955] and across a 1-azabutadiene *N*-oxide in acridine 10-oxide.[956]

The "Beirut" reaction between benzofuroxans and enols or enamines (Section VIIB2) may be, and has been, considered a related cycloaddition.

A different heterodiene, a cation with 1-oxa-2-aza substitution, appears to be involved in some interesting reactions. Thus, oxidation of 2-chloromethylpyridine 1-oxide derivatives leads to a cation that undergoes cycloaddition with olefins (primary adduct observed only by NMR), as well as dimerization to a dication.[957a] Furthermore, with 2-methylquinoline 1-oxides, oxidative [thallium(III) acetate or dichlorodicyano-*p*-benzoquinone] cycloaddition with DMAD takes place and yields 1,2-oxazinoquinoline derivatives.[957b]

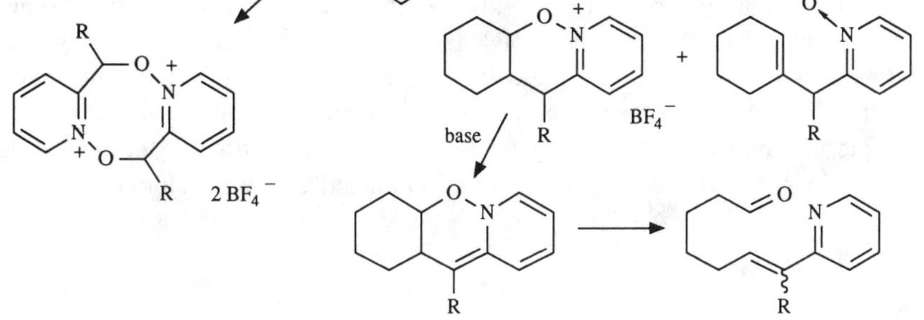

IX. HOMOLYTIC SUBSTITUTION

Radicals add to nitrones and yield nitroxide radicals.[958,959] Indeed, nitrones, 5,5-dimethylpyrroline 1-oxide being a common choice, are extensively used as spin traps[960] for the identification of the radicals formed during chemical reactions, as well as in biological systems,[961-965] though a warning about the limitation of the use of pyrroline oxides in this sense has been reported.[966]

Heteroaromatic *N*-oxides do not yield nitroxides with radicals (except, of course, if a substituent is involved, as is the case of 4-nitrosopyridine 1-oxide),[967] but rather undergo homolytic substitutions. Though this process has been less thoroughly studied than electrophilic substitution, it is not devoid of preparative interest. The total rate measured for the phenylation of pyridine 1-oxide at 0°C is 52 times higher than that of benzene. Under these conditions (electrochemical generation from phenyldiazonium fluoborate), a relatively high yield of pheylated pyridine 1-oxides is obtained (35%) and the 2-:3-:4- isomer ratio is 89:<1:10.[968] These results are in accord with calculated localization energies.[968] Various precursors and reaction conditions produce phenylpyridine 1-oxides in yields up to 45%, with the 2-isomer predominating in every case and competing minor deoxygenation.[968-971] Alkyl[968,972] and ketyl radicals[973-975] show a similar reactivity, yielding 2- and 4-substituted pyridine 1-oxide. However, 4-chloropyridine 1-oxide is dehalogenatated by metal ketyls, and a mixture of pyridine 1-oxide and 4,4'-bipyridine and its oxides is obtained.[976] 2-(2-Thiazolyl)-4-methylpyridine 1-oxide has been prepared by homolytic substitution.[971] Alkylation and arylation of quinoline and isoquinoline *N*-oxides occur exclusively at the α position.[977-979] Quinoline carboxamide 1-oxides are conveniently obtained by the reaction of quinoline 1-oxides with formamide or with derivatives initiated by *t*-butylhydroperoxide/Fe$_2$SO$_4$.[980]

X. PHOTOCHEMICAL REACTIONS

The $\geqslant$N$\rightarrow$O function is one of the few with which a monomolecular photochemical reactivity is associated, and products containing it are expected to photodecompose with reasonable efficiency. This, of course, introduces a major difference between azines (mostly only slightly photoreactive, and then usually through hydrogen abstraction, a reaction leading to useful preparative procedures only in a few cases) and azine *N*-oxides (mostly highly photoreactive and usually undergoing rearrangements of synthetic interest). This photoreactivity of heteroaromatic *N*-oxides is common with other compounds containing the $\geqslant$N$\rightarrow$O function, such as nitrones and azoxy derivatives, including, of course, cyclic compounds, e.g., pyrroline 1-oxides. Again, there is an important difference in the sense that irradiation of cyclic nitrones mostly causes electrocyclic ring closure to give an oxaziridine, or amides arising from it, while the photochemistry of aromatic *N*-oxides is much less simple and includes a large variety of ring rearrangements.[981-983]

Reported photochemical reactions fall into two groups, i.e., rearrangement and *N*-deoxygenation. There is now general agreement that the first process involves the first excited singlet state, and the latter one probably the triplet state. Analogy with the nitrones suggests that an oxaziridine is also the intermediate for the rearrangement of these molecules, and indeed many, but not all, of the observed reactions can be discussed in terms of a series of concerted (sigmatropic and electrocyclic) rearrangements.[981-984] MO calculations on the excited state show that, indeed, the bond order between the oxygen atom and the carbon atom in the α position is increased.[985]

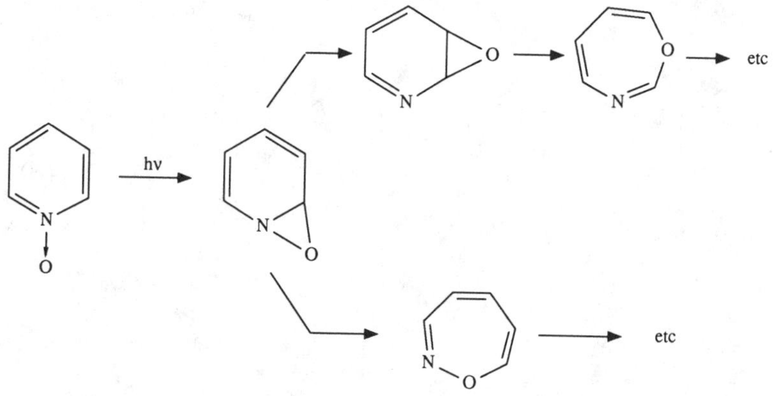

However, experimental evidence for the intermediacy of an oxaziridine is weak,[982,985-988] and, on the other hand, there are indications that in most cases no intermediate with a lifetime greater than a few nanoseconds is encountered before the formation of the end product.[989-991] Furthermore, while some trapping experiments are in accord with the chemistry expected from an oxaziridine,[987,992] other ones suggest a biradical[993,994] or a nitrene,[991] and in most cases no trapping is possible. Evidence from various techniques, such as ESR and ENDOR,[995,996] as well as the effect of a magnetic field[997,998] or of cations[999] or anions[1000a] on the reaction, have been variously interpreted in the favor of radicalic, ionic, or nitrenic intermediates. At this stage, it is probably better to state that excitation in the singlet state causes a major redistribution of electron density, in particular from the oxygen atom to the ring,[982,1001] and that this promotes complex bond rearrangements, often with a change of the ring size, the formation and trapping of intermediates being the exception, rather than the rule. Furthermore, the medium has a great importance, since free and H-bonded N-oxides react differently (protonated N-oxides tend to react less or not at all).[1000b]

As for deoxygenation, this is probably related to a diradicalic character of the triplet state, though other mechanisms are possible (see Section XB).

It appears convenient to subdivide the discussion according to the starting N-oxides, since the rearrangements observed are quite different in the various groups. Classification is made on the basis of what appears to be the primary product, even if this has not been isolated, since this makes discussion of the scope of the reaction easier.

A. REARRANGEMENT

1. Azabenzene N-Oxides

a. Oxygen Shift and Fragmentation

An important photoprocess for monocyclic azine N-oxides involves the shift of the oxygen atom from the nitrogen to the ring α position and cleavage of the N-C$_\alpha$ bond to yield open-chain unsatured derivatives. The typical example is pyridine 1-oxide, the irradiation of which leads mainly to "untractable tars", with only minute amounts of isolable products, such as 2-formylpyrrole.[999,1002-1004] However, when the reaction is carried out in basic aqueous solution, or in the presence of amines in both water and organic solvents, the anion 57 or, respectively, the aminonitriles 58 are obtained in a fair yield.[991,1005-1008] A mechanistic hypothesis is that the excited state rearranges directly to the zwitterion $\leftrightarrow$ nitrene intermediate 59, and this either transfers a proton to a base, yielding the above-mentioned anion, or otherwise polymerizes to tars or gives (in higher yield when complexed by metal ions) 2-formylpyrrole.[991] The same behavior is observed with alkylpyridine 1-oxides.

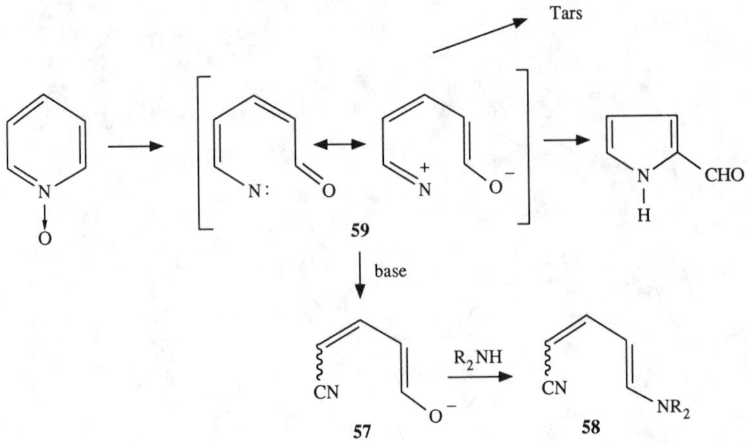

A similar cleavage occurs with pyridazine 1-oxides, the oxygen atom migrating exclusively to the carbon atom and not to the nitrogen, thus yielding the diazo derivatives **59′**, as has been indicated spectroscopically.[989] If the absorbed light flux is moderate and the temperature is low, **59′** rearrange to pyrazoles,[989] otherwise a second photochemical step leads to carbenes, and hence to furans or to cyclopropylketones[989,1009-1011] (these, in turn, give pyrroles when trapped with amines).[1009] 3-Aminopyridazine 1-oxide yields 3-cyanopropenal via the furan.[1010]

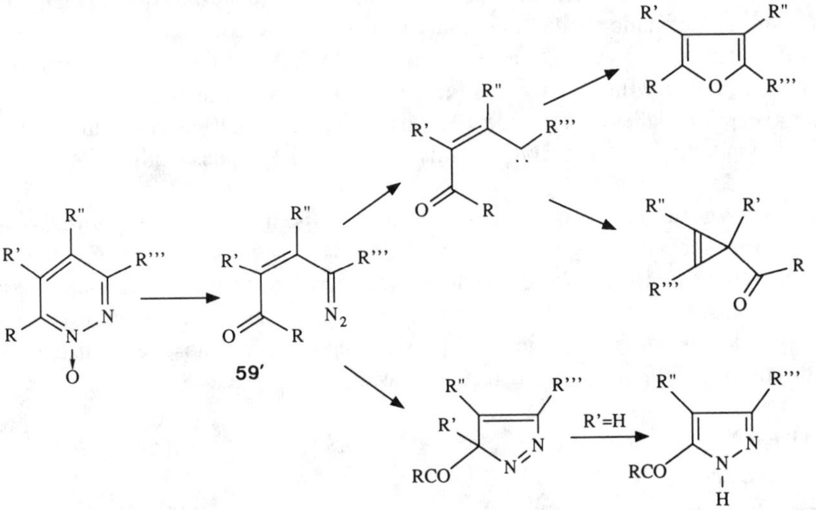

The cleavage is also selective (oxygen migration only to C_2) with pyrimidine 1-oxides and leads to 3-(acylamino)acrylonitriles.[1012-1013]

b. Ring Contraction

This is simply a variation of the previous mechanism, since here nitrene **59** cyclize onto the unsatured system. 2-Formylpyrroles are generally minor products with pyridine 1-oxide (but this is the main process with the 2,6-dicyano derivative),[1014] though their yield often increases in the presence of copper ions;[999] 3-monosubstituted pyridine 1-oxides yield only 4-substituted-2-formylpyrroles, while with the 2-substituted pyridine 1-oxides, the rearrangement occurs in both directions.[1003]

Pyrimidine 1-oxides selectively yield 4-acylimidazoles.[1015-1016]

c. Ring Enlargement and Related Secondary Processes

Ring enlargement is a process that is limited to some substituted derivatives. Tri- and tetraphenylpyridine 1-oxides rearrange in high yield to the corresponding 1,3-oxazepines,[1017] and the same type of products are obtained in lower yield from 2-cyanopyridine 1-oxides.[1014] The 1,3-oxazepine from 2,4-diphenylpyridine 1-oxide is too unstable to allow isolation,[1017] and in general these compounds are both thermally and photochemically labile, and rearrange easily to 3-hydroxypyridines or 1-acylpyrroles.[1018-1019] Since such derivatives are obtained in low yield from the photolysis of various pyridine 1-oxides,[1020] ring expansion is probably a relatively general, though minor, process. Similarly, 6-phenyl-1,2,4-triazine 4-oxides yield 3-phenyl-1,2,4-triazoles,[1021] and a small amount of imidazole is obtained from some pyrazine 1-oxides,[1022] in both cases probably via expansion to seven-membered, ring-contraction to N-acylazole, deacylation. On the contrary, the 2-acylpyrroles mentioned in the previous section are not formed via oxazepines (such a thermal process for oxazepines requires drastic conditions),[1019] but rather via nitrenes.

d. Other Processes

Some reactions that dominate the photochemistry of other *N*-oxides appear only as minor pathways for the monocyclic derivatives. This is the case for the rearrangement to lactams (observed with some pyridine[1002] and pyrazine 1-oxides,[1022] 2,5-dihydroxypyrazine being formed from the dioxide;[1023] the formation of pentachlorobutadienyl isocyanate from pentachloropyridine 1-oxide is presumed to involve the intermediacy of a 3*H*-2-pyridone)[1024] and for the ring contraction to 3-acylazoles (observed with 5-methoxypyrimidine 1-oxide).[1013]

Pyridazine 1,2-dioxide follows a different pathway and yields 3,4-dihydroisoxazole[4,5-*d*]isoxazole, apparently via initial N-N bond cleavage.[1025]

2. Azanaphthalene *N*-Oxides

a. Ring Enlargement

Ring enlargement to benzoxazepines (and aza analogs) is usually the main process from azanaphthalene *N*-oxides in aprotic solvents and offers a useful entry to this class of heterocycles (early papers attributed the structure of condensed oxaziridines to these compounds). Thus, a 50% yield of 3,1-benzoxazepine is obtained from the irradiation of quinoline 1-oxide and distillation.[1026-1027] In this case work-up under anhydrous conditions is required, otherwise this moisture-sensitive compound undergoes hydration to 2-hydroxy1-formylindoline and to indoles arising from it.[1026,1028] Reaction with amines leads to open-chain imidines.[1026]

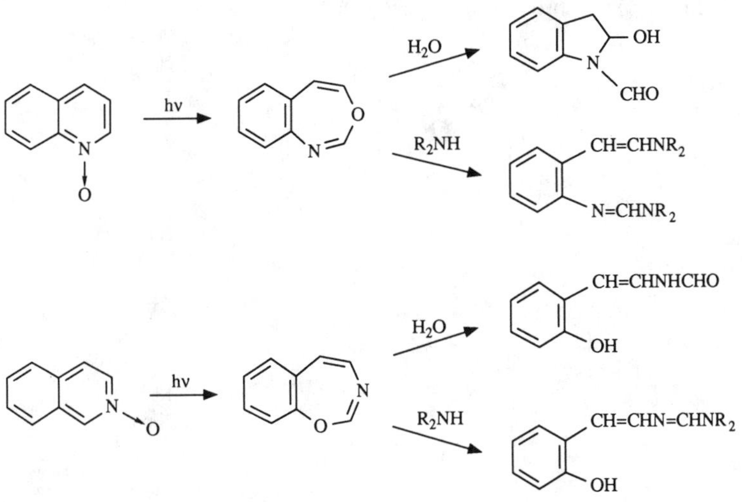

Starting from some quinoline 1-oxides with substituents in position 2 (Ph,[1029-1032] CN,[1032-1034] MeO,[1035] CF$_3$,[1036] PhCH=CH)[1037] or in other positions [3-Ph,[1030] 3,4-(COOMe)$_2$[1038]] much more stable benzoxazepines are obtained, in several cases both in protic and aprotic medium.[1030,1031,1035,1036] These compounds may be labile under acidic catalysis, e.g., 2-cyano-3,1-benzoxazepine is rearranged to 2-cyano-3-hydroxyquinoline on silica gel),[1034] or add nucleophiles, e.g., the same benzoxazepine yields 1-alkylaminocarbostyrils with amines.[992] Furthermore, some benzoxazepines have been shown to be themselves photochemically reactive and to yield 3-acylindoles.[1039-1041] In other cases, the latter compounds are the products isolated from the photolysis of the *N*-oxides,[1042] and, whether oxazepines are intermediates or not, this is a useful synthesis of some substituted indoles.

1,3-Benzoxazepine, which is expected from the photorearrangement of isoquinoline 2-oxide, has not been isolated, and apparently polymerizes rapidly. However, the isolation of derivatives of 2-(2′-hydroxyphenyl)ethylenamine arising from hydration or, in better yield, amine addition, supports its formation as the primary photoproduct.[990,1026] Analogous to the quinoline case, isoquinoline 2-oxides with some substituents in 1 (Ph,[1033,1043,1044] CN,[1043-1044] MeO,[1035] CF$_3$[1036]), and likewise papaverine *N*-oxide,[1045] yield stable 1,3-benzoxazepines. In other cases, treatment with acids of the irradiation mixture containing the oxazepine affords indoles (in particular, 4-hydroxyindoles)[1046] in fair yield.

2- and 3-methylquinoxaline 1-oxide photorearrange to 3,1,5-benzoxadiazepine,[1047] as do the 2- and 3-phenyl derivatives,[1048,1049] though the parent compound does not.[1047] 4-Phenylquinazoline 3-oxide rearranges to 4-phenyl-1,3,5-benzoxadiazepine,[1050] but ring enlargement is not observed with other diazanaphthalene *N*-oxides.

3,1-Benzoxazepine. *A mixture of 12 g (66 mmol) quinoline 1-oxide dihydrate and 1.3 l benzene is azeotropically dried in a Pyrex immersion-well irradiation cylindrical vessel while flushing with nitrogen. After cooling, while stirring and passing nitrogen, a 400-W medium-pressure mercury arc is inserted and irradiation is carried out for 3 h. Evaporation of the solvent at room temperature and reduced pressure, extraction with cyclohexane of the residue, and bulb-to-bulb distillation in a Kugelrohr apparatus yields 4.8 g (50% yield) of the product.[1027]*

b. Ring Cleavage

Two types of cleavage are observed in this class of derivatives. The first one involves oxygen migration to the α position and cleavage of the N-C$_a$ bond, which is analogous to the process observed with azabenzene *N*-oxides. Thus, 1,4-diphenylphthalazine 2-oxide yields diazoketone **60**, and this decomposes to diphenylisobenzofuran.[989] Analogously, benzo-1,2,3-triazine 3-oxides lose nitrogen to yield anthranils,[1051a] which are also obtained in low yield from 4-methylcinnoline 1-oxide through formal HCN loss.[1051b]

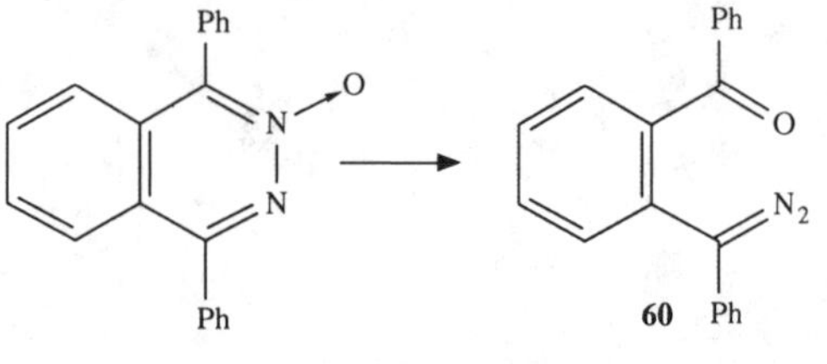

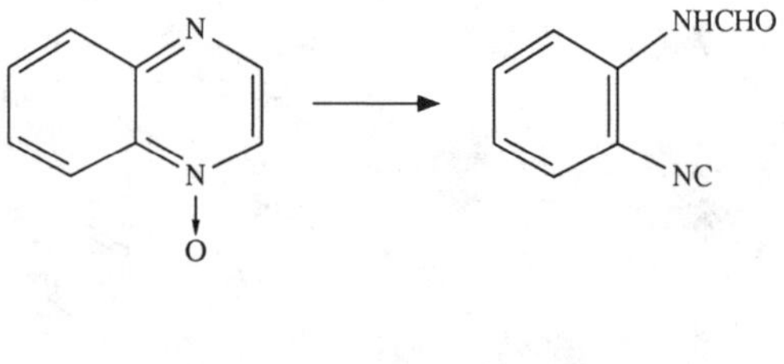

The second process, observed with quinoxaline 1-oxide,[1047] some of its derivatives,[1047,1052] and 3-methoxyquinoline 1-oxide,[1035] involves oxygen migration to the β position and cleavage of the C_α-C_β bond to yield an isonitrile (here seven-membered rings are not intermediates in the reaction).

c. Rearrangement to Lactams

With some exceptions (see below), rearrangement to lactams is the main process in protic solvents, although it usually accounts for only 10% of the process in aprotic media. Thus, quinoline 1-oxides yield carbostyrils in a 70—100% yield.[1053-1055] Substituents in 2 migrate either to the 1 or the 3 positions. With the 2-deuterio derivative, migration to C_3 predominates in water, but both processes occur in aprotic media;[1056] a methyl group migrates to both positions,[1053,1057] and with chloro and tolylthio derivatives only 3-substituted carbostyrils are obtained.[981] As mentioned in Section XA2a with some substituents in 2 (Ph, CN, MeO, CF_3), ring enlargment also predominates in protic solvents, and rearrangement to lactams has little or no importance.

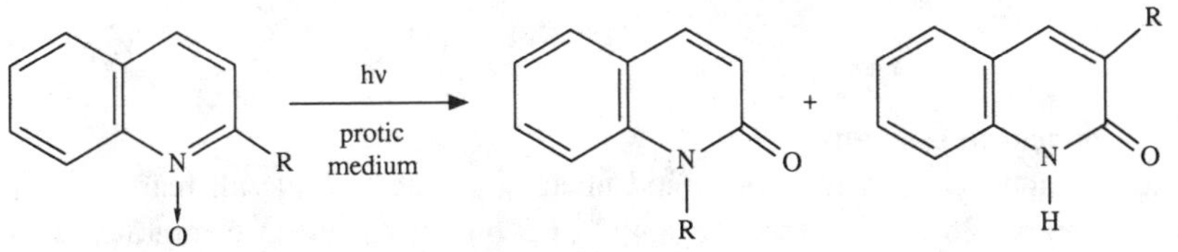

Isocarbostyrils are obtained from isoquinoline 2-oxides[990,1031,1055] (except those from the 1-OMe, Ph, CN derivatives), and alkyl groups in position 1 migrate to the nitrogen atom.[990,1045]

Quinoxaline 1-oxide yields quinoxalone (both 1- and 3-alkyl derivatives, starting from the 2-substituted N-oxide).[1047] The dioxides yield quinoxalin-3-one 1-oxides and then quinoxaline-2,3-diones.[1058] In 1,6-naphthyridine 1,6-dioxide, the chromophore in 1 rearranges preferentially, yielding the 2-oxo 6-oxide.[1059]

d. Other Reactions

There are a few reactions that are not encompassed by the previously mentioned general processes. Thus, benzo-1,2,3-triazine 3-oxide gives 3-phenilindazole via the 2-nitroso derivative,[1051a] and 4-methylcinnoline 2-oxide yields 3-methylindole, possibly also via a nitroso derivative, in addition to 3-methylindazole and 3-methylbenzofuran, probably via oxygen transfer and N-C_α bond cleavage (compare with Section XA1a).[1051b]

A characteristic reaction of some quinoxaline 1,4-dioxides is rearrangement to 1-acylben-zimidazol-2-ones, a reaction that apparently does not occur stepwise via quinoxalin-3-one 1-oxides, since these products have a different photochemistry.[1058,1060,1061]

3. Azaphenanthrene N-Oxides

a. Ring Enlargement

Both 1- and 4-azaphenanthrene N-oxides rearrange to the expected naphthoxazepine upon irradiation in aprotic solvents.[1062] The analogous ring expansion to a dibenzoxazepine is not observed with parent phenanthridine 5-oxide, but is important with the 6-aryl and cyano derivatives.[994,1064] These products undergo easy hydrolysis to 2-hydroxy-2′-aminobiphenyl derivatives, but interestingly in the case of 6-cyanophenanthridine 5-oxide, irradiation in EtOH leads to N-ethoxyphenanthridone (possibly via the oxaziridine),[1064] and in benzene/2,3-dimethylbutene to the addition of the alkene and formation of a dibenzoxonine,[993,994] the latter reaction pointing to an intermediate of diradicalic character.

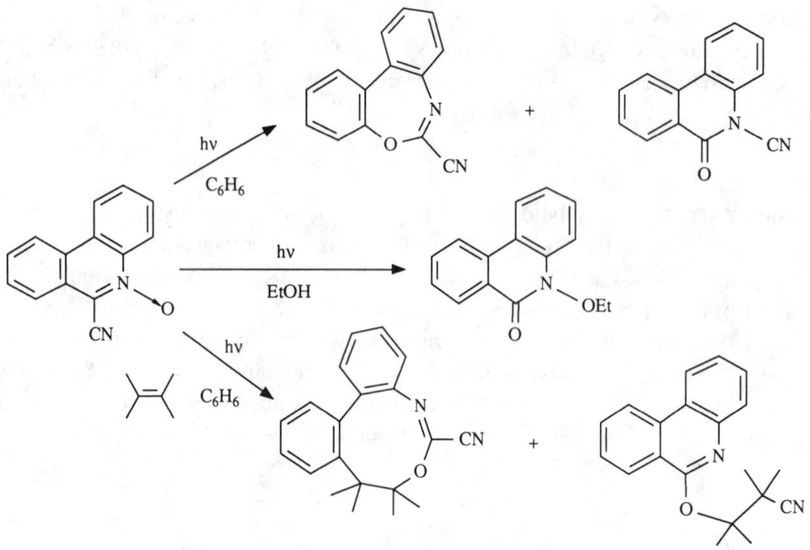

b. Rearrangement to Lactams

1- and 4-azaphenanthrene *N*-oxides and phenanthridine 5-oxide all rearrange to lactams in protic solvents. In the last case, an alkyl or a phenyl group in 6 migrates to the nitrogen.[994,1055,1063-1065] Chiral groups lose their configuration during migration, or are cleaved, yielding parent phenanthridone, when corresponding to a stable radical, e.g., benzhydryl.[994,1065] Similar rearrangements have been exploited for the synthesis of perlonine and analogs.[1066-1068]

4. Azaanthracene *N*-Oxides

a. Ring Enlargement

In acridine and phenazine *N*-oxides, expansion of the ring to the 1,3-oxazepine system would involve loss of the aromaticity of two rings. This is probably why such a process is much less important than in the azanaphthalene or phenanthrene series, though formation in low yield of a dibenzoxazepine (or oxadiazepine) has been observed from some acridine,[984a,1069] or, respectively, phenazine *N*-oxides[1070] (this is a reversible photochromic reaction with the dioxides[1071]).

More important is ring expansion to the 1,2-oxazepine system, i.e., to the valence tautomers of the oxaziridines that are often invoked as intermediates in the photochemistry of *N*-oxides. Dibenzo-1,2-oxazepines are stable products from the photolysis of 9-cyano- and 9-choroacridine 10-oxide[1072] and might also be formed as intermediates from other acridine oxides,[1072,1073] as shown by the isolation of dibenzoazeto[1,2-*b*]oxazolo derivatives (arising from secondary electrocyclic rearrangement) and of alkoxydibenzo-1,4-oxazepine (arising from nucleophilic addition and rearrangement).[984a,1074] 1,2-oxazepines might also be intermediates in the processes discussed in the two following sections.

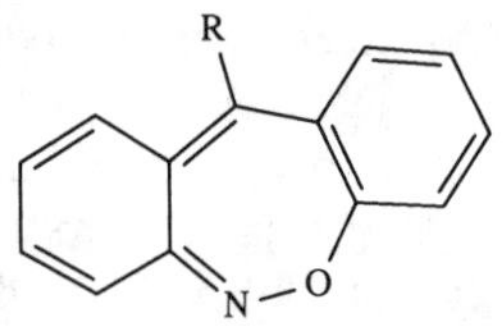

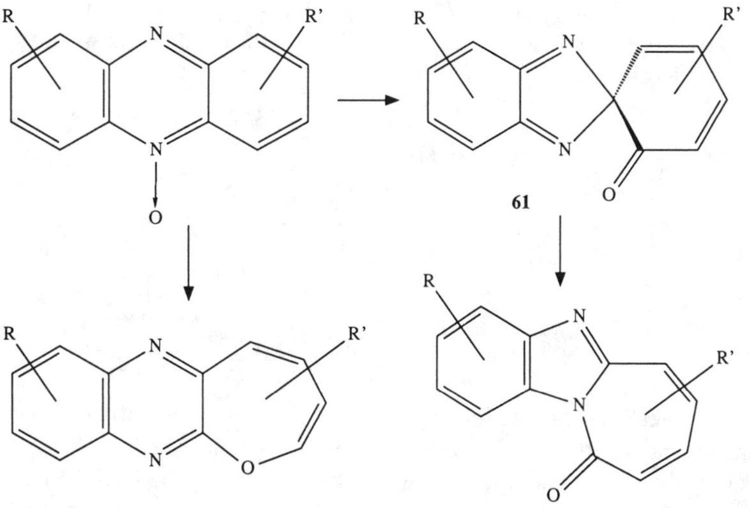

Scheme 18.

b. Ring Contraction

Ring contraction is the main process observed upon photolysis of acridine and phenazine N-oxides in aprotic solvents and leads to 1-, 2-, and 3-acylindoles,[984a,1069,1073] and, respectively, 1- and 3-acylbenzimidazoles.[1070,1075,1076] The study of substituted derivatives shows that oxygen migration is accompanied by rotation of one of the carbocyclic rings, and thus through an intermediate such as 61[984a,1069,1075] (in turn, this may arise form a nitrene, formally produced by N-O bond cleavage of a 1,2-oxazepine, and this is an analogy with the process discussed in Section XA1a) (Scheme 18).

A different ring contraction observed with phenazine 5-oxides leads to 5-(2-benzoxazolyl)-2,4-pentadienenitriles and involves cleavage of the C_α-C_β bond.[1070]

c. Migration of the Oxygen Atom to a Carbocyclic Ring

In polar solvent, the main products from acridine and phenazine oxides are usually oxepino[1,2-b]quinolines, or, respectively, quinoxalines (see Scheme 18). Contrary to the previously discussed rearrangement, in this case the molecule remains planar during the reaction, and the oxygen atom migrates to the two neighboring peri atoms.[984a,1069,1075,1077]

5. Benzanthracene N-Oxides

In benz[a]acridine[1078] and benzo[a]phenazine 7-oxides,[1079] the main process leads to 1,3-oxazepine (or 3,1,5-oxadiazepines, respectively), as in the azanaphthalene series, but other rearrangements that are typical of the azaanthracene N-oxides (ring contraction to acylindoles or benzimidazoles, oxygen migration to a carbocyclic rings) are also observed. With benzo[a]phenazine 12-oxide, the latter type of process predominates.[1078]

6. Other Azine N-Oxides

Deoxygenation is usually the main photoprocess with purine N-oxides, but some rearrangements are observed. These reactions are usually carried out in aqueous medium, and their courses are pH dependent; with tautomeric compounds, such as the amino and hydroxy derivatives, the reaction should, in some cases, be attributed to N-hydroxy heterocycles, rather than to N-oxides or to ions. Observed rearrangements include the formation of lactams,[1080-1083] oxygen migration from N_1 to N_3,[1084] and cleavage of the pyrimidine ring.[1085]

Rearrangement to lactam has been observed as a minor pathway in some pteridine deriva-tives,[1086] as have both ring cleavage and lactam formation in formycin **62**, a pyrazolo[4,3-*d*]pyrimidine nucleoside.[1087]

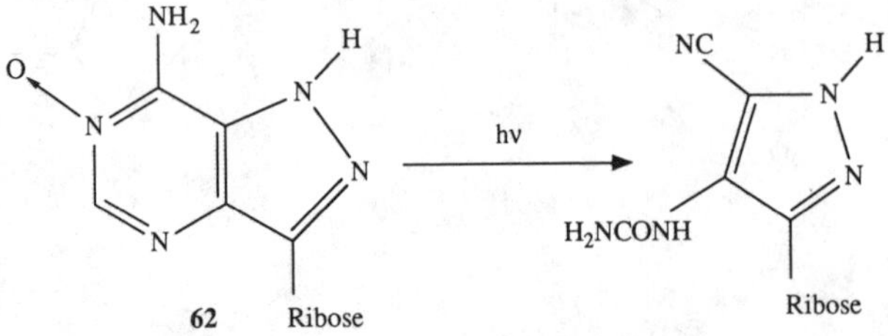

7. Azole *N*-Oxides

The photochemical behavior of azole *N*-oxides is partially reminiscent of that of nitrones, and though oxaziridinoazoles have not been isolated, they may be intermediates in the formation of what are usually the main photoproducts, i.e., oxoazoles. Thus 1-alkyl-1,2,4-triazole 4-oxides yield the corresponding 5-triazolones,[1088] and 1-alkyl (or 1,2-dialkyl-) benzimidazole 3-oxides yield 1-alkyl (or 1,3-dialkyl-) benzimidazol-2-ones.[1089,1090] In other cases, ring cleavage takes place, as in the formation of phenylendiamine derivatives from the photolysis of some 1,2-dialkylbenzimidazole 3-oxides in aprotic solvents[1090] (differently from the rearrangement men-tioned above in protic medium), of acyldiimines from 1*H*-tetraarylimidazole 3-oxides,[1091] and of *o*-nitrosoanilides from 1-hydroxybenzimidazole 3-oxides.[1092] Benzothia-[1093] (or selena-) diazole *N*-oxides[1094] are reversibly photolyzed to open-chain transients by cleavage of the N-S (or N-Se) bond, the end result being isomerization to the *S*-oxide in the first case and reclosure to benzofurazan in the latter. Positional isomerization is also observed with some ben-zofuroxans[1095] (in other cases, these molecules are cleaved to nitrile oxides, and these can be trapped),[1096] as well as in 1,2,3-thiadiazoles (from 2- to 3-oxide)[1097] and benzotriazoles (from 1- to 2-oxides).[1098]

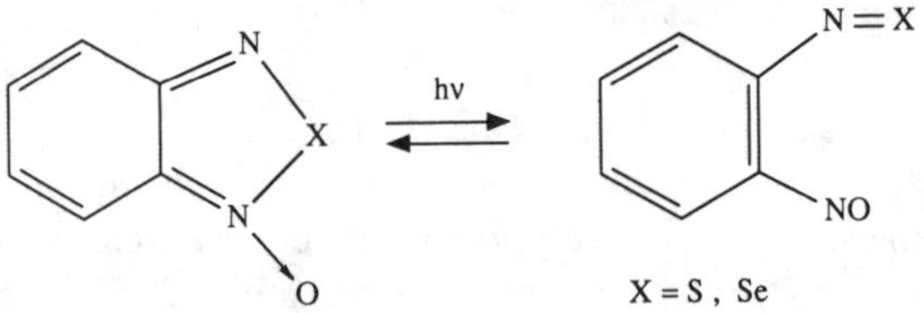

B. DEOXYGENATION

Two mechanistic hypotheses have most often been considered for *N*-deoxygenation, i.e., (1) that it is due to the triplet state of the *N*-oxide, while the singlet rearranges, and (2) that a labile intermediate (e.g., an oxaziridine) originates both processes, in that it either transfers the oxygen atom to a suitable acceptor or reacts further to give the rearranged products mentioned in Section XA.[981,982] The evidence presented in favor of the two possibilities is not definitive, and, at any rate, not of general validity, and it may well be that both pathways, as well as other ones, are involved in the various cases. Some reactions fit well with the triplet hypothesis. Thus 2- and 4-benzoylpyridine 1-oxide are deoxygenated (with low quantum yield) via a triplet of radicalic character.[1099] Furthermore, deoxygenation is more important for monocyclic than for polycyclic *N*-oxides, and still more with α or γ diaza *N*-oxides, e.g., in pyridazine[1009-1011] and pyrazine 1-oxides[1022] deoxygenation accounts for 15—50%, as compared with ≤5%, as typically observed with pyridine 1-oxides; a high yield of deoxygenation is also observed, e.g., with cinnoline,[1051b] benzo[*c*]cinnoline,[981] phenazine[1070] (in general, only in deoxygenated solvents, particularly in alcohols), and purine *N*-oxides.[1080-1085] In all these cases, it is likely that the lowest triplet is an

$n\pi^*$ state of radicalic character rather than a $\pi\pi^*$ state. Some spectroscopic characterization of the reaction of the triplet is available.[1100-1102] In other cases, e.g., with 6-cyanophenanthridine[987] 5-oxides, it would appear that an intermediate is formed, and then it transfers oxygen.

From the preparative point of view, it may be noted that a general method for deoxygenation (in 70—95% yield) is irradiation in the presence of boron trifluoride[1103] or triphenylphosphine,[1104] and in some cases amines are also effective.[1105] This might be useful, e.g., for the selective reduction of the N-oxide function in the presence of nitro groups.[1105,1106]

The main interest of this photochemical process is in the oxidation of various substrates by irradiation in the presence of heterocycle N-oxides, as will be discussed in Chapter 5.

XI. REACTION INVOLVING A SUBSTITUENT

A. DEOXIDATIVE NUCLEOPHILIC SUBSTITUTION INVOLVING A CARBON-LINKED SUBSTITUENT

Deoxidative nucleophilic substitution of heterocyclic N-oxides has been discussed in Section VI, and it has been shown that initial attack is usually at the α position, though later rearrangements may lead to differently substituted end products. When an alkyl group is present in the α or γ positions, the course of these reactions is diverted in most cases, and attack to the side chain, rather than to the nucleus, results (formally an intramolecular redox process in that the alkyl group is oxidized while the oxygen of the $\geqslant$N→O group is lost). In the following subsections, the main classes of these processes are discussed.

1. Hydroxylation

As first reported by Boekelheide,[1107] heating 2-picoline 1-oxide with acetic anhydride results in reaction at the side chain, yielding 2-acetoxymethylpyridine, and to a lesser degree on the nucleus, yielding 3- and 5-acetoxy-2-methylpyridine as minor products. For optimal yield the reaction is carried out at 70°C; over this temperature, neither the total yield nor the isomer ratio change appreciably.[1108] Other carboxylic anhydrides can be conveniently used, e.g., butyric anhydride gives a higher yield of the 2-acyloxymethyl derivative.[1108] With trifluoroacetyl anhydride, the reaction occurs at room temperature,[1109] and this is also the case when acetyl chloride (1 equivalent) is added to acetic anhydride.[1108] The raw acetate obtained after evaporation of the anhydride can be hydrolyzed by acids to the hydroxymethyl heterocycles. Acyl chlorides[70,1110] have also been used (though other reactions may occur, see below). Ketene in sulfuric acid has been used as an equivalent of acetic anhydride[1111,1112] (Table 23).

The mechanism of this reaction has been the subject of much work and some controversy (Scheme 19).[567] That a 1-acetoxypyridinum salt is first formed is indicated by the beneficial effect of better acylating agents in preparative experiments (see before) and has been supported by the isolation of the corresponding picrate (from the reaction of picoline 1-oxide with picryl acetate) and its conversion to the acetoxymethyl derivative with triethylamine.[37] Rate-determining, and irreversible, proton abstraction then leads to intermediate 63 (ion pair $\rightleftarrows$ anhydrobase), which can be envisaged as the precursor of both side-chain and nuclear (β) acetoxylated derivatives. This intermediate does not accumulate, and its conversion to pyridine 64 is not a concerted rearrangement, but rather involve a tight ion (or radical) pair, as shown, inter alia, by the oxygen scrambling in the acetoxy group when the reaction is carried out with ^{18}O-labeled acetic anhydride.[568,113-115]

Some pieces of evidence in favor of the heterolytic cleavage of intermediate 63 are the reactions observed when a group susceptible to rearrangement via cation is present in position 2 (e.g., a cyclopropyl group, see page 210),[1115,1116] as well as by the trapping of picolyl cations with anisole (2% yield, o:m:p ratio 57:4:39),[1117] and by the increased rate observed when an electron-donating substituent is present in position 4.[1118] A radical pathway is excluded by the

TABLE 23
Hydroxylation of Alkyl Side Chain Accompanied by *N*-Deoxygenation

N-oxide	Substituent	Substituent affected	Method	Ref.
Pyridine	2-Me	2-CH$_2$OAc(71)	Ac$_2$O, 70°C, 5 h	1108
		3-OAc(13), 5-OAc(16)		
	2,3-Me$_2$	2-CH$_2$OH(70)	Ac$_2$O, reflux 2 h	1128
	2,4-Me$_2$	2-CH$_2$OAC(75)	Ac$_2$O, reflux 15 min	1126
	2,5-Me$_2$	2-CH$_2$OAc(73)	Ac$_2$O, reflux 2 h	1127
	2,6 Me$_2$	2-CH$_2$OAc(71)	Ac$_2$O, 100°C, 10 h	1107,1129
Quinoline	2-Me	2-CH$_2$OH(66)	Ac$_2$O, reflux 1 h	1136a
	2,4-Me$_2$	2-CH$_2$OH(30),		
		4-CH$_2$OH(6), 3-OH(2)	Ac$_2$O, 100°C, 2 h	1137
Isoquinoline	1-Me	1-CH$_2$OAc(51), 4-OAc(34)	Ac$_2$O, C$_6$H$_6$ reflux 2 h	1138
Pyridazine	3-Me-6-OMe	3-CH$_2$OAc(50)	Ac$_2$O, reflux 2 h	1140b
Pyrimidine	4,6-Me$_2$	4-CH$_2$OAc(30)	Ac$_2$O, 15 min at 100°C	229
Pyrazine	2-Me	2-CH$_2$OH(26)	Ac$_2$O, reflux 30 min	260,1141
	2,5-Me$_2$	2-CH$_2$OAc(25)	Ac$_2$O, reflux 3 h	260,1141
Pyrazine[a]	2,3-Me$_2$	2,3-CH$_2$OH(58)	Ac$_2$O, 110°C 20 min	1142
Quinoxaline	2,3-Me$_2$	2-CH$_2$OAc(57)	Ac$_2$O, reflux 2 h	1143
Quinoxaline[a]	2,3-Me$_2$	2,3-(CH$_2$OAc)$_2$(54)	Ac$_2$O, reflux 4 h	1143
Purine[b]	6-Me	6-CH$_2$OAc(33)	Ac$_2$O-dioxane, 65°C, 1.5 h	1145b

[a] 1,4-Dioxide.

[b] 1-Oxide.

lack of effect of radical traps and by the formation of good yields of 2-acyloxypyridines, even when the acyloxy radical, if formed, would be expected to decompose quickly, e.g., using phenylacetic and trichloroacetic anhydride (though minor radicalic paths are indeed more important in this case).[1119,1120] On the other hand, free acyloxy radicals are involved in the formation of side products (e.g., in the 2-picoline 1-oxide/acetic anhydride reaction, CINDP signals are observed only for methyl acetate and ethane, see Scheme 19).[1121] Of radical origin is also the 1,2-dipyridyl ethane obtained from 2,6-lutidine 1-oxide.[1122]

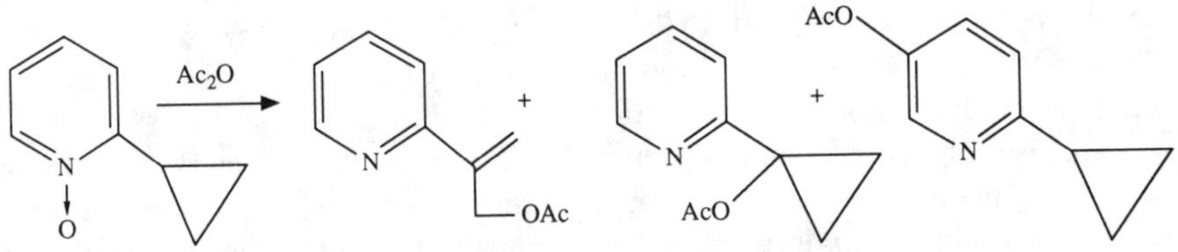

4-Picoline 1-oxide undergoes a similar side-chain acyloxylation, and mechanistic investiga-tions lead to similar conclusions, though there is indication that the ion pair is "less tight" in this case, and free radicals play a more important role (e.g., anisole traps the 4-picolyl cation to a larger extent than the 2 isomer,[1117] there is some CINDP signal for the formation of 4-acetoxymethylpyridine,[931,1123] a small amount of 4-ethylpyridine is formed[1124]). Preference for attack at the α position is clearly shown in the reaction of 2,4-lutidine 1-oxide, which mainly yields the 2-acetoxymethyl derivative, with minor amounts of the 4-acetoxymethyl and the 3-acetoxy isomers.[1125]

Side-chain acetoxylation of *N*-oxides, in particular in the case of 2-picoline 1-oxides, is a convenient source of hydroxymethylpyridines[514,1107-1109,1125-1131] and of alkenylpyridines (e.g., 1,2-dipyridylethylene *N*-oxide from the corresponding ethane *N, N'*-dioxide).[1132] The reaction is compatible with the presence of electron-withdrawing groups, e.g., 3-NO$_2$[1133] (however, with 4-nitro-2-picoline 1-oxide, partial substitution to yield the 4-pyridone occurs under the reaction conditions),[1125] 3-CONH$_2$,[1134] 6-COOH,[1135] and has largely been used with heavily functional-

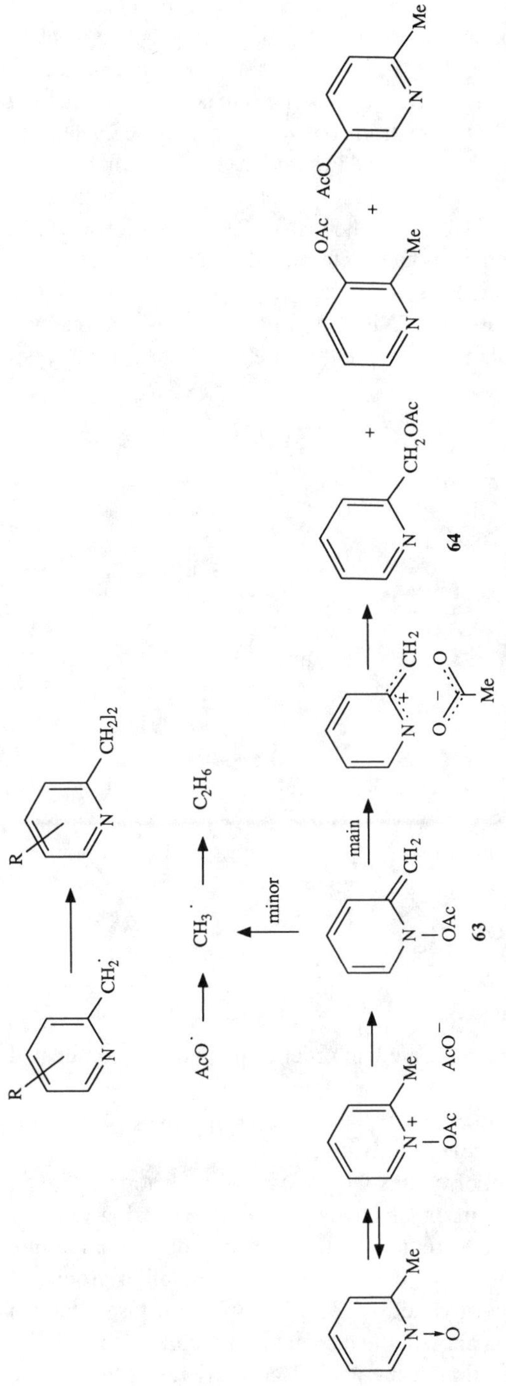
AcO˙
CH3˙
C2H6
CH2]2
minor
main
CH2OAc
OAc
AcO
AcO⁻
Me
63
64

ized pyridine derivatives as a key step in the synthesis of new heterocyclics (see also Chapter 5).

In quinoline 1-oxides, a 2-methyl and, less easily, a 4-methyl group are likewise acetoxylated.[1136-1138] The mechanism is analogous to the pyridine case.[1113,1114,1136-1138] In 1-methylisoquinoline 2-oxides, ring (in position 4) and side-chain acetoxylation are both observed.[1138,1139] Related reactions have been reported for other N-oxides, e.g., in the pyridazine,[1140a,1140b] pyrimidine,[1140c,1140d] pyrazine,[544,1141,1142] quinoxaline,[1143,1144] purine,[1145a,1145b] and pteridine series.[1145c]

As expected, hydroxylation similarly takes place when the alkyl group is part of a saturated ring fused to the heterocyclic N-oxide.[1144,1146-1149] Competition between attack to the nucleus, to the linear side chain, and to a saturated cycle has been studied in some cases.[1146,1147] In the series of compounds **65**, acetoxylation of endocyclic CH_2 decreases with increasing n; acetoxylation of the methyl group follows the reverse order.[1147] The stereochemistry of acetoxylation has been studied on the deuterated pyridinophane **66**, and it has been determined that proton abstraction and introduction of the substituent occur on the same side[1150-1152] (this need not bear a consequence for the mechanism in the general case). With β,γ-fused N-oxides, the reaction occurs on the methylene in γ (see, e.g., formula **67**),[1153] but in the case of compound **68**, the β position is acetoxylated.[1154]

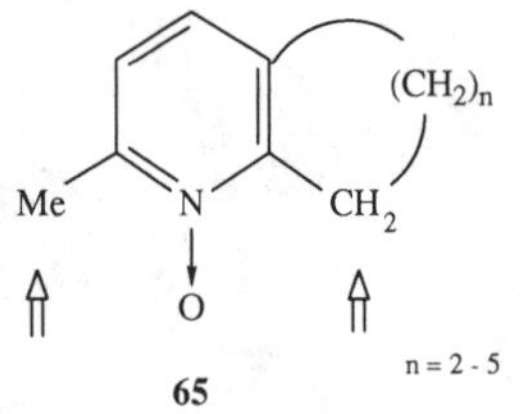

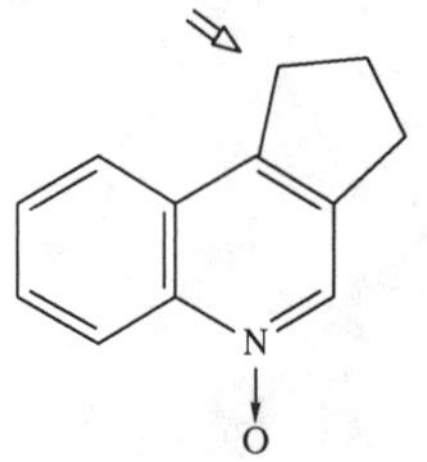

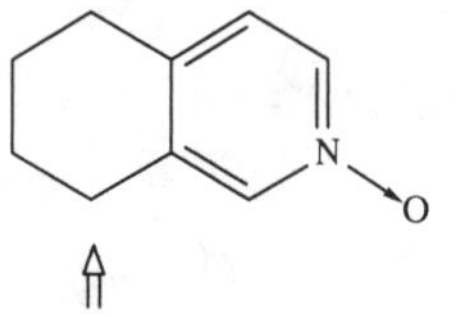

The alcohols thus formed have often been dehydrated in order to prepare the conjugated fused heterocycles.[1155,1161]

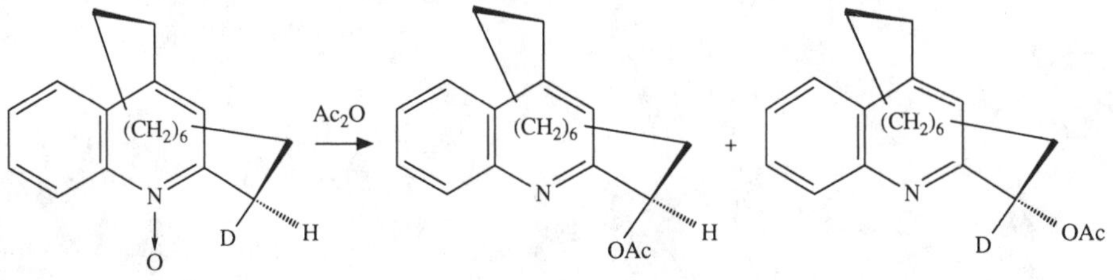

Among five-membered heterocyclic N-oxides, the reaction has been extended to pyrazole,[1163] triazole,[1164] thiazole,[1165] and benzimidazole[104] derivatives.

Oxidation under mild conditons of the alcohols yields the corresponding ketones,[1162] e.g., in the useful synthesis of 3-hydroxypyridin-4-carboxyaldehyde, a vitamin B_6 analog, from 4-methyl-3-hydroxypyridine.[1162b] Another method to get to a higher oxidation degree in the side chain is to carry out the rearrangement on N-oxides having functionalized alkyl groups as substituents.[532,1107,1166] Thus, starting from 2-pyridylmethyl acetate 1-oxide (in turn obtained from the N-oxidation of the acetate obtained as above), the diacetate is prepared and is easily hydrolyzed to the aldehyde.[1107] However, reaction at the methyl group is faster than at the acyloxymethyl, e.g., 6-methyl-2-acetoxymethylpyridine 1-oxide gives 2,6-bis (acetoxymethyl)pyridine, rather than the aldehyde diacetal.[1107] Analogous reactions take place with aminoalkyl derivativatives.[1166b] The rate of acetoxylation of various 2-CH_2X pyridine 1-

oxide derivatives (X=Cl, OAc, SAc, CN, OMe, NMe_2, NHAc, OPO_3H_2, NH_2) has been measured and has been found to correlate with the electron-withdrawing character of the group X, with the cyanomethyl group giving the fastest, and the methoxymethyl group the slowest, reaction.[1167] The use of side-chain acetoxylation for deblocking a protecting group is discussed in Chapter 5.

Heating 2,3-bis(hydroxymethyl)quinoxaline 1,4-dioxide in DMSO yields a cyclic semiacetal mono-N-oxide,[1168a] and cyclic derivatives are likewise obtained from bis(aminomethyl) derivatives (prepared *in situ* from the halomethyl analogs).[1168b,1168c]

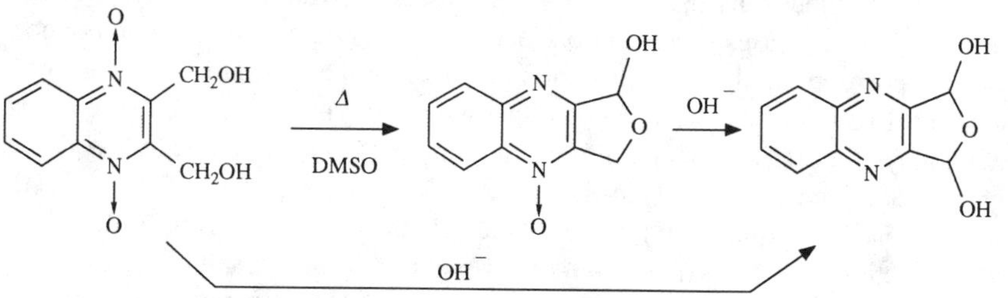

Related internal redox processes occur under different conditions; thus 4-α-phenyl-α-hydroxymethylpyridine 1-oxide (and less readily the alkyl analogs) yields the acylpyridine in the presence of bases, and a similar reaction takes place with 6-hydroxymethylphenanthridine 5-oxide under acidic conditions.[1169] At a higher degree of oxidation, the acetal of 2-quinoxaline-carboxyaldehyde 1-oxide yields the N-deoxygenated carboxylic acid with bases (both the acid and the decarboxylated quinoxaline are obtained from the aldehyde dihydrate).[1170]

An intermolecular redox process occurs when 4-benzylpyridine 1-oxide is heated in DMSO, with disproportionation to a mixture of 4-benzylpyridine (29%), 4-benzoylpyridine (41%), and 1,2-diphenyl-1,2-bis(4-pyridyl)ethane (9%).[1171a]

The pyrolysis of 2-acylmethylpyridine 1-oxide in the presence of $FeCl_3$ yields 2-pyridinecarboxyaldehyde.[1171b] Irradiation of the 2-benzylpyridine 1-oxide complex leads to hydroxylation of both the methylene and the phenyl (*ortho* position) groups with N-deoxygenation.[1171c]

2. Halogenation

As has been seen above, trichloroacetic anhydride and chloride easily give the trichloroacetylmethyl derivatives with methyl-substituted N-oxides. These derivatives readily decompose to chloromethylheterocycles.

A general method for deoxidative side-chain chlorination is the reaction of α and γ-alkyl N-oxides with aryl-[4,104,1173-1177] and alkylsulfonyl chlorides[1178,1179] (in the absence of these substituents, sulfonylation in β takes place, see Section VIF2). However, the reaction of 2-methylquinoline 1-oxide with tosyl chloride in the presence of sodium *p*-toluensulfinate leads to side-chain sulfonylation,[1177b] and 4-methylquinoline 1-oxide yields the bis(2-quinolyl)ether with tosylchloride-triethylamine.[1177c]

Occasionally, side-chain chlorination also takes place with $POCl_3$ (compare with Section VIA for the general reaction),[1180,1181] e.g., 4-chloromethylpyridine is the main product from 4-picoline 1-oxide.[1180]

3. Formation of 1,2-Bis(Heterocycle)Ethanes and Related Reactions

As seen in Section XIA1, radicalic decomposition of alkyl-substituted N-acyloxy salts is only a side pathway in most cases. However, a good yield of the 1,2-disubstituted ethanes arising from the recombination of the heteroarylmethyl radicals is occasionally obtained,[1122,1171] e.g., ethylendiamine derivatives are quantitatively obtained from the reaction of 4-acylaminopyridine 1-oxide with acetic anhydride.[1182,1183]

Pyrolysis of α-alkylheterocycle *N*-oxides also yields alkyl radicals. Thus, flash vacuum pyrolysis of 2-picoline 1-oxide affords the picolyl radical, and the end products obtained include 2-hydroxymethylpyridine, as well as bis(2-pyridyl)-methane and -ethane.[1185] 2-Benzylpyridine 1-oxide yields pyrido[1,2-*a*]indole, and both 2-(*o*-methylbenzyl)pyridine and 2-benzyl-3-methylpyridine 1-oxides yield benzo[*g*]quinoline.[1184,1185]

4. Other Substitutions

Virtually all deoxidative substitution processes occurring on the ring can likewise involve α- and γ-alkyl groups, though, as discussed in Section VI, this appear to be the exception rather than the rule, since in many cases such reactions occur at other ring positions or do not occur at all when α and γ positions are blocked (except, of course, for the above discussed acetoxylation).

Examples of reactions on the side chain involve sulfonylation (e.g., 2-methylquinoline 1-oxide with tosylchloride),[1177b] reaction with carbanions (4-acylaminomethylpyridine 1-oxide with active methylene compounds in the presence of acetic anhydride),[1186-1188] reaction with cyanide (in 4-isopropylquinazoline 3-oxide, attack is both at the ring and at the tertiary position),[77] reaction with thiols (from 1-alkoxy-4-methylpyridinium salts[1180] or from 4-acylaminomethylpyridine 1-oxide in the presence of Ac$_2$O),[1166b,1190] reaction with pyridine (from picoline 1-oxide-tosyl chloride),[1189] and reaction with imidoyl chlorides (from picoline 1-oxides).[1191] There is some indication for the radical character of some of these rearrangements.[1191]

A peculiar reaction is the conversion of of α-hydroxymethyl *N*-oxides to the *N*-deoxygenated phenylhydrazones by reaction with phenylhydrazine (not acting as an oxidant) in the presence of bases, a procedure that allows, *interalia*, the preparation of 6-methylpyridine-2-carboxyaldehyde from the 6-methyl-2-hydroxymethyl 1-oxide,[1192] a result not obtained with the Boekelheide reaction (see Section XIA1).

Another reaction that may be mentioned in this connection, since it again involves electrophilic attack at the oxygen (as well as at the substituent) is the conversion of pyridine and quinoline-2-carboxyaldeheide 1-oxides into carbonylcarboxyamides (see, e.g., compound **69**) by reaction with cyclohexylisonitrile.[1193]

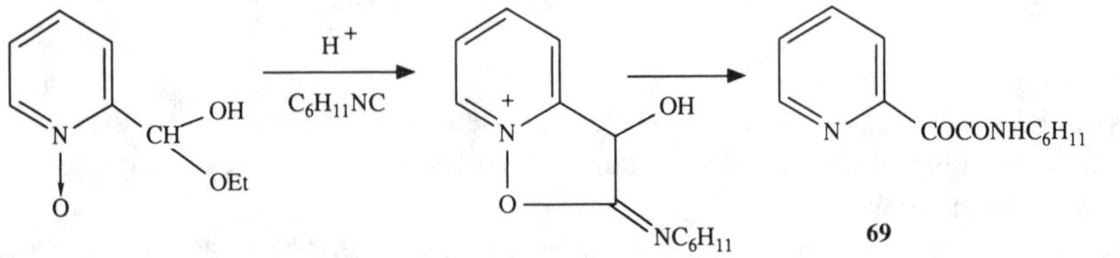

B. REACTIONS AT CARBON-LINKED SUBSTITUENTS LEAVING THE *N*-OXIDE FUNCTION UNCHANGED

In contrast to the previous section, the reactions discussed below fit into the general frame of azaheterocyclic chemistry, and are therefore considered more briefly. However, in several cases *N*-oxidation has a significant effect, both on the reaction rate and on selectivity with respect to the parent compound, and thus the synthetic interest of such processes should not be underestimated.

1. Alkyl Groups

As is well known, alkyl groups adjacent to the heterocyclic ring in α- and γ-substituted azines show a significant acidity ("carbonyl analogy"), and several reactions based on the attack of the corresponding carbanion onto various electrophiles have been developed. *N*-oxidation reinforces the electron-withdrawing effect and therefore is expected to enhance such acidity. Indeed, the reactivity of *N*-oxides in this respect is often midway between that of azine and that of azinium salts. As a result, some reactions that work poorly on the parent heterocycles are synthetically useful with the *N*-oxides. Furthermore, while γ-alkyl groups are often the most reactive ones in the first case, reaction on the α groups generally predominates in the latter.

a. Proton Exchange and Metalation

Base-catalyzed deuterium exchange usually takes place readily with α-(and less easily γ-) alkyl azine *N*-oxides. As an example, exchange at the side chain on treatment with Na_2CO_3, or Et_3N in D_2O or with NaOD, is faster than at the ring with 2-picoline 1-oxide and with both 2- and 4-benzylpyridine 1-oxides, but not with 4-picoline 1-oxide,[1194] and similar results have been obtained with other *N*-oxides.[1195-1198] Metalation by strong bases follows the same trend, e.g., treatment of 2,4-dimethylquinoline with lithium butyl yields the 2-lithiomethyl derivative, which rearranges to the thermodynamically more stable 4-lithiomethyl reagent, whereas with the 1-oxide only the 2-lithiomethyl derivative is formed.[1199,1200] The oxygen atom is involved in the coordination, and these metalated reagents are conveniently used for side-chain alkyla-tion,[1201-1203] acylation,[1203] or carboxylation.[1204]

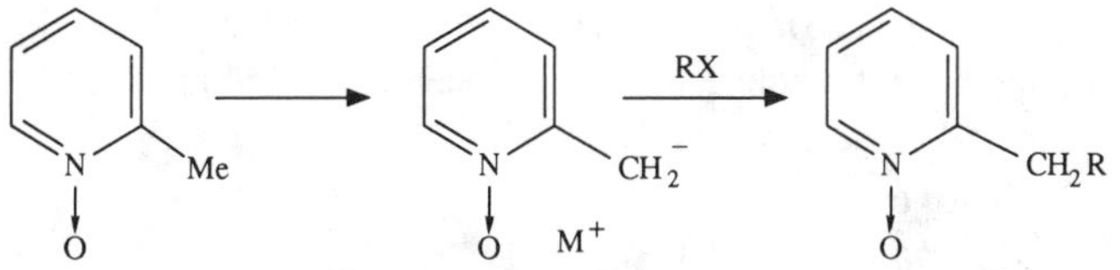

b. Condensation to Form Carbon-Carbon Bonds

Condensation of α- and γ-methyl *N*-oxides with aldehydes under basic conditions leads to the corresponding styryl derivatives,[833,1021,1205-1216] or to the carbinols.[1213,1214] The conditions required for the reaction of pyridine 1-oxides are more severe than for the corresponding methiodides, but less so than for the parent heterocycles.[1208] A methyl in position 3 is reactive when further activated by a 4-nitro group.[1208] Benzyl alcohols can also be used, since they are oxidized to the aldehydes by the *N*-oxide.[1215]

Claisen condensation with carboxylic esters (oxalates, benzoates),[1217] orthoesters,[1218] and othoamides has also been exploited for the side-chain acylation of *N*-oxides, these substrates again showing an enhanced reactivity with respect to the parent heterocycles. The Mannich reaction with formaldehyde precursors and amines is successful with 2-picoline,[1220a] and 6-methylpyrimidine 1-oxides[1216] and 2,3-dimethylquinoxaline 1,4-dioxide.[1220b]

c. Formation of Carbon-Nitrogen Bonds

Treatment of 2- and 4-picoline 1-oxides,[1221,1222] as well as related alkylazine[1223,1224] and benzimidazole *N*-oxides,[104] with alkyl nitrites in the presence of strong bases yields the oximes via the carbanion. Treatment of 6-methyl-4-nitropyridazine 1-oxides with acetyl chloride convert them to the 4-chloro derivatives (see Section VA1), and the nitrous acid liberated attacks the methyl group, again yielding the oxime.[1225a] Reaction with alkyl nitrates in metal amide/liquid ammonia yields nitromethyl derivatives;[1225b] 2-picoline 1-oxide yields 2-cyanopyridine, reasonably via dehydration of the oxime.[1226a]

Heating 2-picoline 1-oxide with iodine and pyridine (King reaction) yields a 1-[(1-oxido-2-pyridyl)methyl]pyridinium salt.[118,1226b]

d. Other Reactions

Other reactions of alkyl-substituted *N*-oxides follow expected pathways. Examples are bromination,[1178] oxidation of methyl groups to aldehydes by selenium dioxide[1127,1208,1231a,1231b] (but in dioxane oxidative demethylation is obtained with some nitro derivatives),[1231c] or to acid (or ketones from ethyl derivatives) by reaction with permanganate,[1231d,1231e] chromic acid,[1231f,1231g] or electrogenerated superoxide anion.[1231h]

2. Aromatic Rings

As discussed in Section IV, heterocycle N-oxides are much less susceptible to electrophilic substitution than are carbocyclic aromatics. Thus, phenyl- and benzyl-substituted azine N-oxides are sulfonated or nitrated in the carbocyclic,[20,249,250] not in the heterocyclic, ring, even when the latter is activated by electron-donating substituents.[1232,1233] In 4- and 5-phenylfuroxans, nitration again occurs in the benzene ring.

A fused aromatic ring is not considered as a substituent in the present discussion, but rather the chemistry observed is referred to the entire heteroaromatic system. Nevertheless, it appears appropriate to mention here some reaction of benzofuroxans, in which the heterocyclic part of the molecule modifies the reactivity of the carbocyclic ring, rather than introducing a new chemistry. Thus, in nitrobenzofuroxans the electron-withdrawing effects of the substituent and the heterocyclic ring combined make these molecules behave as electron-poor olefins, and indeed both Diels-Alder[1235] and 1,3-dipolar cycloaddition with diazomethane (leading finally to cyclopropanation)[1236] take place in positions 4,5. Attack of nucleophiles to the carbocyclic ring in benzofuroxans and benzotriazole 1-oxides has been discussed in Section VIC.

3. Halomethyl Derivatives

a. Nucleophilic Substitution

The electron-withdrawing effect of the heterocyclic moiety enhances the rate of nucleophilic substitution (a S_N2 process) of halomethyl N-oxides, in comparison with the nonoxidized substrates[1237,1238] (see, e.g., the σ^- values obtained for pyridine derivatives and the corresponding 1-oxides in Table 24). Furthermore, the effect is extended to an unconjugated position in the alkyl chain, at least in suitable rigid substrates (e.g., in compound **70** N-oxidation brings about both an increase in the solvolysis rate and in sterochemical control with respect to the parent molecule).[1239] The reaction of chloromethylpyridine and quinoline 1-oxides with carbon and heteroatom-centered nucleophiles is often usefuly synthetically[1240-1242] (Table 24).

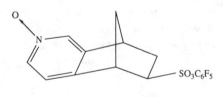

70

b. Nucleophilic Substitution with Ring Enlargement

2-Halomethylquinazoline 3-oxides undergo the expected substitution of the halogen atom with many nucleophiles.[1243-1245] However, with nitrogen-centered nucleophiles (amines, hydrazine),[1245-1252] and in some cases also with hydroxide,[1248] a different pathway leads to 1,4-benzodiazepines 4-oxides with incorporation of the side-chain carbon. Apparently the base initially attacks the ring (the presence of the N-oxide function is essential, probably because of the stabilizing effect on the intermediate adduct), and then ring opening and reclosure onto the halomethyl group lead to the product.[1245,1248] Oximes (compare formula **71**) have actually been isolated under the reaction conditions and have been shown to be further transformed into the diazepines.[1248] In a related reaction, 1,5-benzodiazocine 5-oxides are obtained by treatment of 2-[(1-hydroxyalkyl)methylamino]-3H-1,4-benzodiazepine 4-oxides with a strong acid or base.[1253]

TABLE 24

σ⁻ Values for the Nitrogen Moiety Within the Pyridine Ring, as Derived
from the Reaction of the Chloromethyl Derivatives with Methoxide

Position	≥N	≥N→O	≥N⁺-Me
2-CH₂Cl	1.00	1.50	2.49
3-CH₂Cl	0.60	1.23	1.15
4-CH₂Cl	1.17	1.67	2.16

Adapted from Liveris, M. and Miller, J., *J. Chem. Soc.*, 3486, 1963; and Jarvis, B. B. and
Marien, B. A., *J. Org. Chem.*, 42, 2676, 1977.

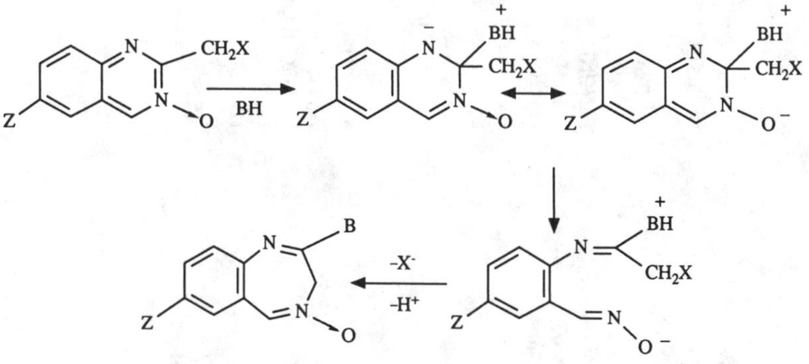

71

4. Other Carbon-Linked Substituents

In general, these derivatives display the expected chemistry. The activating effect of the
heterocyclic N-oxide moiety on reactions involving nuleophilic attack is important, e.g.,
acylation of alcohols, which is also seen on 2-(2-pyridine)ethanol,[1254] nucleophilic opening of
an oxirane ring,[1255] substitution in alkylpyridinium derivatives,[118] hydration and substitution
reactions of aldehydes[1256-1258] and related nitrones,[1259] reactivity of phosphoranes,[1260] and
solvolysis of esters[1261] and trifluoromethyl derivatives.[1262]

Saturation of a conjugated double bond precedes the reduction of the N-oxide function if the
hydrogenation is carried out with palladium on charcoal as the catalyst.[113]

C. REACTION OF HETEROATOM-LINKED SUBSTITUENTS

1. Redox Processes

As mentioned in Section IIIC, hydrogenation of several substituents is competitive with, and
when Pd/C is used as the catalyst precedes, N-deoxygenation. This offers an access to amino
heterocycle N-oxides from the easily available nitro derivatives[81,265,378d,538,1140b,1263,1264,1266,1267,1271]
(as well as to the aminophenyl-substituted N-oxides from the nitrophenyl analogs,[1268] and, under
suitable conditions, to the azo di-N-oxides).[1266] Other catalytic hydrogenations include conversion of azo to hydrazo,[117] reduction of halogenated derivatives,[1140b,1225a,1269] and hydrogenolysis
of alkoxy groups. Reduction of the azoxy group precedes attack at the ≥N→O function[117] (Table
25).

Methods for the chemical reduction of the nitro group to azoxy,[122,1263,1266,1272,1273] azo,[140,1263]
hydrazo,[141a] hydroxylamino,[167,1263,1265] and amino groups[267,1273,1274b] with no N-deoxygenation are
available.

TABLE 25
Reduction of Substituents

N-oxide	Substituent affected	Other substituents	Product (% yield)	Conditions	Ref.
Pyridine	4-NO$_2$		4-NH$_2$(85)	H$_2$, Pd/SrCO$_3$, MeOH	1263
	4-NO$_2$	2,6-Me$_2$	4,4'-azoxy(47)	Zn/AcOH, 16°C	1266
	4-NO$_2$	2,6-Me$_2$	4,4'-azo(80),		1266
	4-NO$_2$	5-Et-2-Me	4-NH$_2$(7)	H$_2$, Pd/C, 10% HCl	81,1266
	4-NO$_2$		4-NH$_2$(quant)	H$_2$, Pd/C, EtOH	1267
	4-NO$_2$		4-NHOH(quant)	PhNHNH$_2$, EtOH	1274a
	4-NO$_2$		4,4'-azoxy(60)	N$_2$H$_4$,H$_2$O, Cu, EtOH, r.t.	1273
	4-NO$_2$		4-NH$_2$(quant)	N$_2$H$_4$,H$_2$O, Cu, EtOH, reflux	1273
	4-NO$_2$		4,4'-azo(60)	As$_2$O$_3$, aq NaOH, reflux 2 h	140
	4-NO$_2$		4,4'-azo(60)	Sn(OH)$_2$, aq NaOH, reflux 2 h	140
	3-NO$_2$	2,6 Me$_2$	3-NH$_2$(quant)	H$_2$, Pd/Norit, H$_2$O	378d
	4,4'-azo		4,4'-hydrazo(78)	H$_2$, Pd/C, MeOH	117
	2-OCH$_2$Ph		2-OH(69)	H$_2$, Pd/C, EtOH	1270
	4-OCH$_2$Ph		4-OH(68)	H$_2$, Pd/C, EtOH	1270, 1271
Quinoline	4-NO$_2$		4 NHOH(25)	NaBH$_4$, EtOH	167
Pyridazine	4-NO$_2$	6-Me	4-NH$_2$(58)	H$_2$, Pd/C, MeOH	1140b
	4-NO$_2$	3,6-Me$_2$	4-NH$_2$(82)	H$_2$, Pd/C, EtOH	538
	6-Cl	3-OMe	6-H(38)	H$_2$, Pd/C, EtOH, aq NH$_3$	1269
	3-Cl	6-Me	6-H(90)	H$_2$, Pd/C, MeOH, aq NH$_3$	1140b
Cinnoline[a]	6-NO$_2$		6-NH$_2$(65)	H$_2$, Pd/C, EtOH	265
Benzo[c]cinnoline[b]	2-NO$_2$		2-NH$_2$(68)	N$_2$H$_4$, Ni Raney, reflux, 30 min	267

a 2-Oxide.
b 6-Oxide.

The microbiological reduction of 4-nitroquinoline 1-oxide is of obvious interest in view of the carcinogenic properties of these derivatives,[1275] and so also is the oxidation of products, such as the 4-hydroxylamino 1-oxide, to which the toxic effect is more directly connected.[1276]

The oxidation of 4- or 5-amino derivatives to the nitro compounds has synthetic interest in the furoxan series.[1277,1278] Other useful oxidations involve conversion of thiols to disulfide,[1279] sulfinyl, and sulfonyl derivatives.[1280]

2. Competition Between Substituents and the *N*-Oxide Group

As discussed in Chapter 2, the tautomeric equilibrium in α- and γ-amino, hydroxy, and mercapto heterocyclic *N*-oxides generally favors the *N*-oxide form in the first case and the *N*-hydroxy ketone or thione tautomer in the others. The related chemical reactions are discussed in the present section.

a. Functionalization

Treatment of 1-hydroxy-2- and -4-pyridones (tautomers of hydroxy *N*-oxides) and related compounds from other heterocyclic series with alkylating agents yields, in most cases, the corresponding 1-alkoxy derivatives,[1281-1288] though a certain amount of alkoxy 1-oxides may be formed (e.g., from 1-hydroxy-4-pyridone and quinolone with diazomethane).[1283,1284,1286] 2- and 4-pyridinamine 1-oxides undergo rate-determining *O*-acylation, followed by rapid isomerization to the amides.[1289-1291] 2- and 4-acylaminopyridine 1-oxides yield the 1-methoxy-2-amino tosylates by reaction with methyl tosylate[1291] (Table 26).

With 2-aminopyrimidine 1-oxide, acylation occurs both at the oxygen and the nitrogen atom, alkylation occurring only at the first and reaction with aldehydes only at the latter atom.[1292] 4-Aminoquinazoline 3-oxide gives a mixture of 3-methoxy-4-imino and 3-methoxy-4-(*N*-methyl)imino-quinazoline with diazomethane,[1293] and adenine 1-oxides yield the 1-alkoxy derivatives.[1294-1296]

1-Hydroxypyridine-2-thiones are both alkylated[1297] and acylated[1298] at the sulfur atom, and among the three potential reaction sites, the sulfur atom is preferred in the alkylation of 6-amino-1-hydroxypyridine-2-thiones.[1299]

b. Rearrangement

2-Alkoxypyridine[1300,1301] and quinoline 1-oxides[1302] rearrange thermally to 1-alkoxypyridones, or, respectively, quinolones. The allyloxy, as well as benzyloxy, derivatives react at a lower temperature than the methyl analogs. More precisely, it has been shown that with 2-allyloxypyridine 1-oxides at low temperature, the reaction leads to the 1-alkoxypyridones arising through group migration without double-bond shift, whereas at a higher temperature, the main process is a Claisen rearrangement, leading to 3-allyloxy-1-hydroxy-2-pyridones with double-bond shift in the chain.[1301] Mechanistic studies, including CIDNP measurements,[1303-1305] the use of chiral substituents,[1305] and the effect of pressure,[1306] show that with stable radicals as the migrating groups, the process involves a radical pair, whereas a concerted sigmatropic shift (suprafacial) is important in other cases. 2-Aryloxypyridine 1-oxides also rearrange to 1-aryloxy-2-pyridones, in this case through intramolecular nucleophilic substitution with the Meisenheimer complex **72** as the intermediate.[1307] 6-Alkylpyridazine 1-oxides likewise rearrange thermally[1308] and 4-alkoxyquinoline 1-oxides yield 1-alkoxy-4-quinolones when heated in the presence of boron trifluoride[1309] (Table 27).

TABLE 26

Alkylation and Acylation of Tautomeric N-Oxides

N-oxide	Substituent	Other substituents	Reacted group (% yield)	Conditions	Ref.
Pyridine	2-OH		1-OMe(77)	NaOEt/TsOMe/EtOH	1281
	4-OH		1-OMe(68)	NaOEt/TsOMe/EtOH	1281
	4-OH		1-OMe, 4-OMe	CH_2N_2	1283,1284
	4-OH	$3,5\text{-}I_2$	1-OCH₂COOH(85)	CH₂ClCOOH, aq NaOH 90–95°C	1282
	2-NH₂		2 NHCOPh(51)	PhCOCl, C_5H_5N	1290
	2-NH₂		2 NHAc(58)	Ac_2O, MeCN	1290
	2-NH₂		2 NHCONHPh(45)	PhNCO, MeCN	1290
	2-SH		2-SCH₂OEt(69)	CH₂ClOEt, Me_2CO, reflux	1297
	2-SH		2-SCOCH₂Ar(80)	ArCH₂COCl, C_6H_6, C_5H_5N	1298a
	2-SH		2-SMe(87)	Me_2SO_4, NaOH	1299
	2-SH	6-NH₂-3Ph-5-COOEt	2-SMe(87)	EtI, KOH, EtOH, reflux	1285
Quinoline	2-OH		1-OEt(80)	CH_2N_2	1286
	4-OH		1-OMe, 4-OMe	Me_2SO_4, NaOH	1288
Pyridazine[b]	3-OH	6-Me	2-Me(41), 3-OMe(5)	MeI, Ag_2O, 100°C	1287
	6-OH	3-OMe	1-OMe(quant)	PhCH₂Cl, NaOMe, MeOH	1287
	6-OH	3-OMe	1-OCH₂Ph(65)	CH_2N_2, MeOH, Et_2O	1293
Quinazoline[a]	4-NH₂		3-OMe-4(=NH), (17), 3-OMe-4(=NMe)(15)	MeI, $AcNMe_2$	1296
Purine[b]	6-NH₂		1-OMe(93)		

[a] 3-Oxide.

[b] 1-Oxide.

TABLE 27

Rearrangement of Alkoxy Heterocyclic N-Oxides

N-oxide	Substituent	Product (% yield)	Conditions	Ref.
Pyridine	2-CH$_2$CH=CH$_2$	1-(OCH$_2$-CH=CH$_2$)-2-(=O) (83)	100°C, 3.5 h	1300
	2-OCH$_2$Ph	1-OCH$_2$Ph-2(=O) (92)	100°C, 2.5 h	1300
	2-OMe	1-OMe-2-(=O) (89)	140°C, 1.5 h	1300
Quinoline	4-OCH$_2$Ph	1-OCH$_2$Ph-2-(=O) (28)	BF$_3$, DMF, reflux	1309
	4-OCH$_2$CH=CH$_2$	1-(OCH$_2$CH=CH$_2$)-2-(=O) (25)	BF$_3$, 150°C	1309

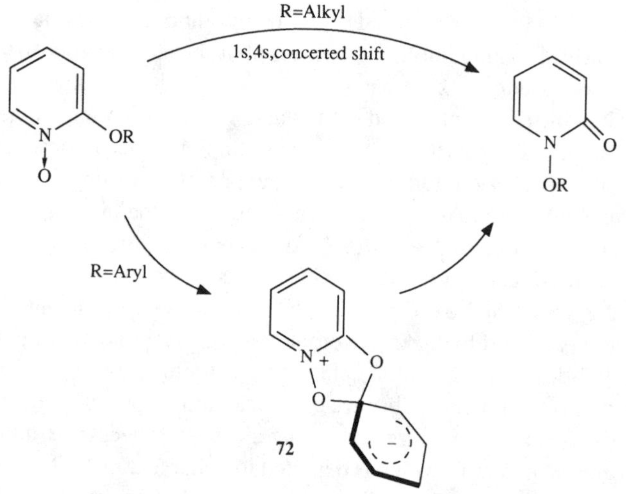

72

As for five-membered derivatives, 1-methoxy-3,4,5-trimethylpyrazole rearranges to the 3-methoxy 1-oxide isomer,[1310] and 2- and 3-acylbenzotriazole 1-oxides and 1-acyloxybenzotriazole are interconverted.[1311-1312]

c. Reaction of Alkoxy N-Oxides with Halides

2-Alkoxypyridine[1313-1315] and quinoline 1-oxides,[1313] as well as 6-alkoxypyridazine 1-oxides,[1308,1316,1317] react with alkyl, acyl, and sulfonyl halides to yield 1-alkoxy- (or acyloxy-, or sulfonyloxy-) 2- (or, respectively, 6-) oxo heterocyclics. This is a practical synthesis of cyclic hydroxamic acids esters, which are, in turn, useful intermediates in synthesis (see Chapter 5).

3. Reactions of Diazonium Salts

It was early noticed that the α and γ amino derivatives of pyridine and quinoline 1-oxides are easily diazotized, and the salts obtained are more stable than those formed from the parent compounds, due to the contribution of mesomeric structures such as **73**.[1318,1319] The role of free and protonated N-oxides in the diazotization has been evaluated.[1320,1321]

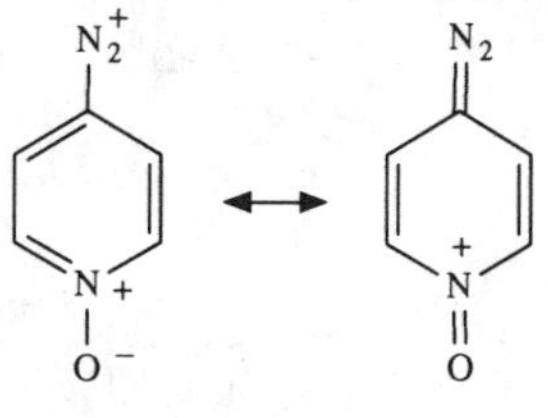

73

The chemistry of these derivatives is analogous to that of the diazonium salts obtained from carbocyclic aromatics, e.g., the Sandmeyer reaction[1318,1319,1322-1325], and coupling.[233,1290,1326] Indeed, diazotization of aminoazine *N*-oxides in hydrochloric or bromic acids offers a useful access to the corresponding halo derivatives. With some β-amino derivatives, electrophilic substitution competes with diazotization, and polyhalo derivatives are obtained.[1323-1325]

4. Reactions of Azido Derivatives

The decomposition of azides of six-membered heterocyclic *N*-oxides follows the same pattern as the corresponding non-oxidized derivatives, which is again that of arylazides,[1327] except for the α derivatives (see below), and involves intermediate nitrenes. Thus photolysis of 4-azidopyridine 1-oxide in acetone yields 4,4′-azopyridine 1-oxide via the stable nitrene triplet.[1328,1329] Similarly, 4-azidoquinoline 1-oxide is photolyzed in acetone to the azo derivative and yields 4-aminoquinoline 1-oxide on thermolysis in hydrogen-donating solvents.[1328-1330] Related reactions have also been observed with 3- and 5-azidoquinoline 1-oxide.[1331] Decomposition of various azides under strongly acidic conditions has been shown to yield halo- or alkoxyaminoquinoline and isoquinoline *N*-oxides, apparently via singlet nitrene - fused azirine,[1332,1333] and aminosulfides are analogously obtained from irradiation in ethylmercaptan.[1334]

The thermal decomposition of α-azido *N*-oxides occurs at a lower temperature and takes a different course. Thus, 2-azidopyridine 1-oxide reacts smoothly at 85—95°C in degassed benzene to give 2-cyano-1-hydroxypyrrole.[1335,1336] Thermolysis in methanol yields 2-cyano-pyrrole and 3-methoxy-2,3-dihydro-2-pyrrolone, and correspondingly, reaction in aniline affords 3-anilino-2,3-dihydro-2-pyrrolone phenylimine (compare formula **75**). In view of the mild conditions employed and of the absence of typical nitrene products, the reaction has been envisaged as a concerted nitrogen elimination-ring cleavage process, yielding the unsaturated nitroso nitrile **74**, the chemistry of which is depicted in Scheme 20.[1336] The reaction is formally analogous to the pyrolysis of 2-azidopyridine, which yields 2-cyanopyrrole and glutaconitrile (similar results are obtained with other α-azido heterocycles).[1337] However, in the latter case, much more drastic conditions are required (480°C), in part due to the fact that in the solid state and in solution the tautomeric tetrazole, and not the azide, is present. Some variations are observed with 3-substituted derivatives. Thus 3-methyl- and 3-bromo-2-azidopyridine 1-oxide yield 3*H*-pyrrole 1-oxides (which in turn, easily react with phenyl isocyanate)[1342] and tetra-chloro-2-azidopyridine 1-oxide is converted into a 6*H*-1,2-oxazine (and this is hydrolyzed to an oxazinone)[1341] (Scheme 20).

2-Azidopyrazine 1-oxide analogously yields 2-cyano-1-hydroxyimidazole,[1335] and 1-hy-droxy-5-cyanopyrazole and 3-cyanopropenal are obtained from pyridazine 1-oxide.[1338] Several bicyclic derivatives behave as expected, e.g., 4-methyl-2-azidoquinoline 1-oxide, and 2-azidoquinoxaline 1-oxide and 1,4-dioxide, from which indole and benzimidazole derivatives are obtained.[1339,1340] However, parent 2-azidoquinoline 1-oxide reacts differently and yields 2-cyanoisatogen and 2,2′-dicyanobis[indole], together with 2-aminoquinoline 1-oxide.[1339] This has been rationalized as involving an interaction of the intermediate nitroso nitrile with a

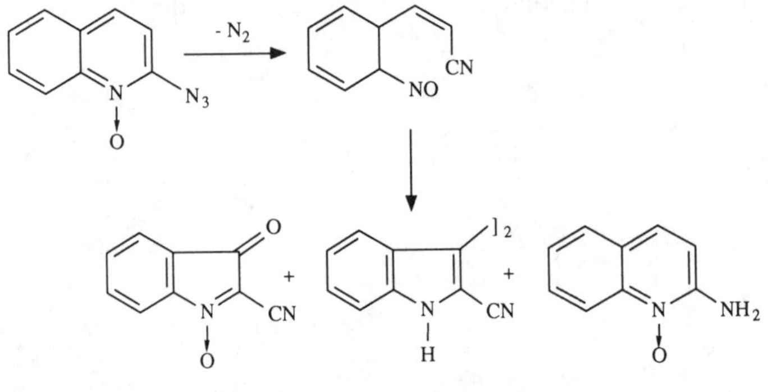

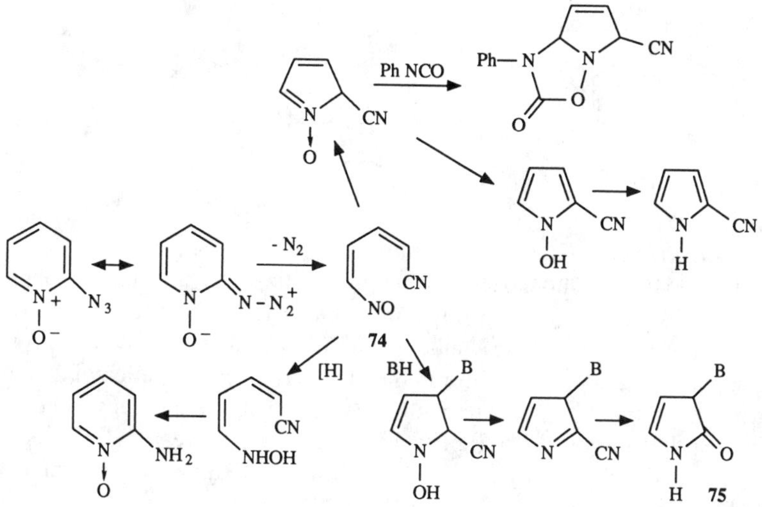

Scheme 20.

molecule of an *N*-oxide (either the starting material or rearranged, see Scheme 20). 2-Azido-3-methylquinoxaline dioxide yields the expected 2-*H* benzimidazole dioxide under suitable conditions, but at a higher temperature rearrangement to a benzo-2,1,4-oxadiazepine 4-oxide takes place.[1340]

Other reported reactions of azido derivatives include the preparation of cyanotriazenes by the addition of cyanides[1343,1344] and of iminophosphoranes by reaction with triphenylphosphine.[1345]

5. Formation of a New Heterocyclic Ring

An important, though heterogeneous, group of reactions includes the formation of a new heterocyclic system, exploiting either the reactivity of the ring positions or the $\geqslant$N→O function itself.

a. Reactions Not Involving the N-Oxide Function

Elaboration of substituents, most often amino groups, through conventional chemistry may lead to the formation of a new heterocyclic ring fused to the existing one(s), and thus of a new bi- (or poly-) cyclic *N*-oxide. As a typical example, purine[1346] and pteridine *N*-oxides[1346-1349] are conveniently prepared from aminopyrimidine *N*-oxides.

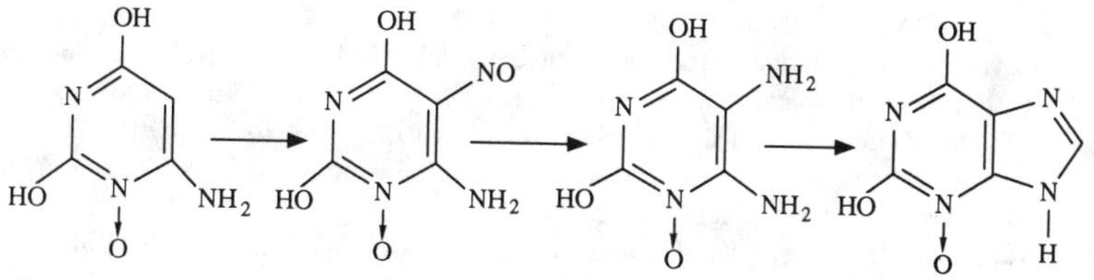

Other syntheses lead to pyrazolopyridines from 4-hydrazinonicotinic acid 1-oxides[1350] and to pyrrolo-,[1351] pyrazolo,[1352] as well as pyrido-pyridazines[1353] from aminopyridazine 1-oxides.

A ring nitrogen can be one of the starting points for the synthesis of a new ring in polyazine *N*-oxides, as in the preparation of triazolopyridazines from hydrazopyridazine 1-oxides[1354] and of pyrroloquinolinium salts from 2-cinnamoylquinoxaline 4-oxide.[1355] 1-Hydroxypyridinethione undergoes a cyclizative dimerization under irradiation and yields a diazathianthrene derivative.[1356]

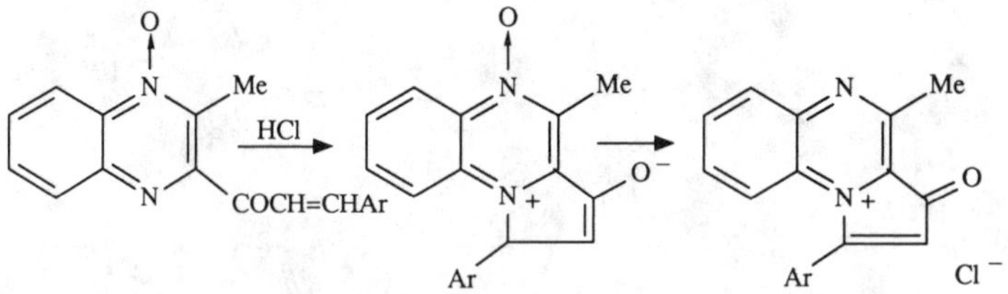

b. Reactions Involving the N-Oxide Function

Obviously, the *N*-oxide function itself can enter in a cyclization reaction, affording a new *N,O*-heterocyclic fused with the starting ring. An example is the intramolecular cyclization of haloalkyl derivatives to yield cyclic *N*-alkoxy salts, as discussed in Section IIB. Another is the formation of pyrido[1,2-*b*]oxazolone (or thione) by thermal condensation of 2-ethoxycarbon-ylaminopyridine 1-oxide,[1357,1358] or alternatively by reaction of 2-aminopyridine 1-oxide with phosgene[1359] (or thiophosgene).[1360] Similar reactions have been reported in the quinoline,[1361] isoquinoline,[1361] pyridazine,[1362] pyrimidine,[1363,1364] triazine,[1365] quinoxaline,[830] and benzo-diazepine and -diazocine[1366] series.

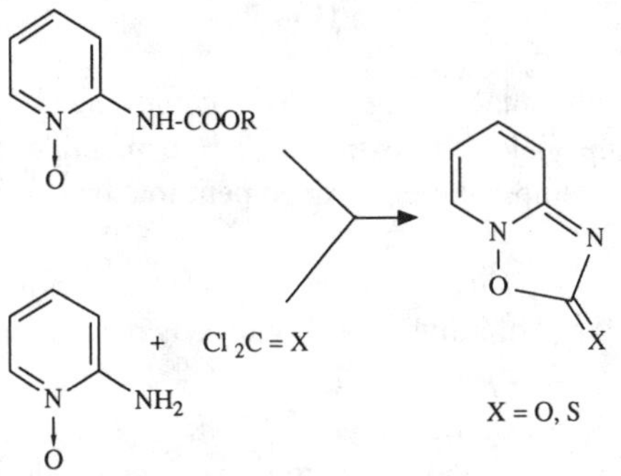

REFERENCES

1. **Bobranski, B., Kochanska, L., and Kowalewska, A.,** Reaction of sulfuryl chloride on pyridine oxide, *Ber. Dtsch. Chem. Ges.,* 71, 2385, 1938.
2. **Pinto, A.D. and Massabni, A.C.,** Solid adducts of organic bases with inorganic acids, *Eclectica Quim.,* 5, 17, 1980.
3. **Katritzky, A.R.,** The preparation of some substituted pyridine 1-oxides, *J. Chem. Soc.,* 2404, 1956.
4. **Vozza, J.F.,** Reactions of 2-picoline 1-oxide with reactive halides, *J. Org. Chem.,* 27, 3856, 1962.
5. **Cook, D.,** Short hydrogen bonds in some pyridine 1-oxide salts, *Chem. Ind. (London),* 607, 1963.
6. **Hadzi, D.,** Infrared spectra and structure of crystalline adducts of some phosphine, arsine and amine oxides, and sulphoxides with strong acids, *J. Chem. Soc.,* 5128, 1962.
7. **Szafran, M.,** IR spectra of normal and abnormal salts of *N*-oxides in crystalline state, *Bull. Acad. Pol. Sci., Ser. Sci. Chim.,* 11, 111, 1963.
8. **Szafran, M.,** IR spectra of crystalline quinaldine *N*-oxide salts, *Bull. Acad. Pol. Sci., Ser. Sci. Chim.,* 11, 497, 1963.
9. **Szafran, M.,** IR specta of crystalline isoquinoline *N*-oxide salts, *Bull. Acad. Pol. Sci., Ser. Sci. Chim.,* 11, 503, 1963.
10. **Szafran, M.,** IR spectra of salts of acridine *N*-oxide with mineral acids, *Bull. Acad. Pol. Sci., Ser. Sci. Chim.,* 12, 289, 1964.
11. **Szafran, M.,** Infrared spectra of salts of quinoline *N*-oxide and its methyl derivatives with dichloroacetic and benzoic acids, *Rocz. Chem.,* 42, 1469, 1968.

12. **Szafran, M. and Dega-Szafran, Z.,** Hydrogen bonding in salts of heterocyclic N-oxides with haloacetic acids. I. IR spectra, *Rocz. Chem.,* 44, 793, 1970.

13. **Muth, C.W. and Darlak, R.S.,** N-acetoxy pyridinium perchlorates, *J. Org. Chem.,* 30, 1909, 1965.

14. **Feely, W., Lehn, W.L., and Boekelheide, V.,** Alkaline decornposition of quaternary salts of amine oxides, *J. Org. Chem.,* 22, 1135, 1957.

15. **Coats, N.A. and Katritzky, A.R.,** Some reactions of 1-methoxypyridinium salts and a color test of N-oxides, *J. Org. Chem.,* 24, 1836, 1959.

16. **Okamoto, T. and Tani, H.,** Synthesis of 2- and 4-cyanopyridines, *Chem. Pharm. Bull.,* 7, 925, 1959.

17. **Tani, H.,** Reactions of N-alkoxypyridinium derivatives, *Chem. Pharm. Bull.,* 7, 930, 1959.

18. **Eisenthal, R. and Katritzky, A.R.,** The ring opening of N-methoxypyridinium perchlorate by hydroxide ion, *Tetrahedron,* 21, 2205, 1965.

19. **Katritzky, A.R. and Lunt, E.,** Reactions of N-alkoxy-pyridinium and -quinolinium cations with nucleophiles, *Tetrahedron,* 25, 4291, 1969.

20. **Hands, A.R. and Katritzky, A.R.,** Mononitration of 2-,3-, and 4-phenyl- and 2- and 4-benzyl-pyridine 1-oxide, *J. Chem. Soc.,* 1754, 1958.

21. **Feely, W.E. and Beavers, E.M.,** Cyanation of amino oxide salts. A new synthesis of cyanopyridines, *J. Am. Chem. Soc.,* 81, 4004, 1959.

22. **Bauer, L. and Dickerhofe, T.E.,** Substitution of 1-alkoxy-, 1-acyloxy-, and 1-sulfonyloxypyridinium salts by mercaptans, *J. Org. Chem.,* 29, 2183, 1964.

23. **Cervinka, O., Fabryova, A., and Matouchova, K.,** Reactions of Grignard reagents with quaternary salts of quinoline and lepidine N-oxides, *Collect. Czech. Chem. Commun.,* 28, 535, 1963.

24. **Takahashi, S. and Kano, H.,** Reactivity of 1-methylbenzimidazole 3-oxide, *Chem. Pharm. Bull.,* 12, 783, 1964.

25. **Hayashi, E. and Hotta, Y.,** Phenanthridine 5-oxide, *Yakugaku Zasshi,* 80, 834, 1960.

26. **Augustinsson, K.B. and Hasselquist, H.,** O-Substituted pyridine N-oxide derivatives 2. A new type of bis quaternary ammonium derivatives, *Acta Chem. Scand.,* 17, 956, 1963.

27. **Reichart, C.,** Preparation of N-ethoxy-pyridinium and -quinolinium salts, *Chem. Ber.,* 99, 1769, 1966.

28. **Abramovitch, R.A., Inbasekaran, M.B., Kato, S., and Singer, G.M.,** Reaction of pyridine 1-oxides and N-iminopyridinium ylides with diazonium salts. N-aryloxypyridinium salts and their base-catalyzed rearrangement, *J. Org. Chem.,* 41, 1717, 1976.

29. **Abramovitch, R.A. and Inbasekaran, M.N.,** Preparation and novel rearrangernents of N-aryloxy pyridinium salts. A [3,5]shift leading to pyrido[2,3-b]benzofurans, *Tetrahedron Lett.,* 1109, 1977.

30. **Iijima, H., Endo, Y., Shudo, K., and Okamoto, T.,** Phenoxenium ions. Identical intermediates in the acid catalyzed solvolysis of N-tosyl-O-hydroxylamines and in the thermolysis of N-aryloxypyridinium salts, *Tetrahedron,* 40, 4981, 1984.

31. **Boekelheide, V. and Feely, W.,** Amine oxides. Cyclic quaternary salts and their decomposition, *J. Am. Chem. Soc.,* 80, 2217, 1958.

32. **Connor, D.T., Young, P.A., and von Strandtmann, M.,** Synthesis of 2-arylisoxazolo[2,3-a]pyridinium bromides via acid catalyzed rearrangements of 1-aryl-2-(2-pyridinyl)ethanone N-oxides, *J. Org. Chem.,* 42, 1364, 1977.

33. **Marmer, W.N. and Swern, D.,** The reaction of aromatic amine oxides with styrene oxide, *Tetrahedron Lett.,* 531, 1969.

34. **Mizuno, Y. and Endo, T.,** Interaction of 2-picoline 1-oxides with acylating and phosphorylating agents. A case of product distribution control, *J. Org. Chem.,* 43, 747, 1978.

35. **Rybachenko, V.I., Chotii, K.Y., Kozhevina, L.I., and Titov, E.V.,** Study of the equilibrium of N-acetyloxypyridinium salt formation, *Zh. Fiz. Khim.,* 58, 1341, 1984.

36. **Rybachenko, V.I., Chotii, K.Y., and Titov, E.V.,** Molecular spectroscopy of symmetry fragmentary exchange in N-oxypyridinum salts, *J. Mol. Struct.,* 80, 251, 1982.

37. **Traynelis, V.J. and Pacini, P.L.,** The mechanism of the reaction of 2-alkylpyridine N-oxides with acetic anhydride, *J. Am. Chem. Soc.,* 86, 4917, 1964.

38. **Oae, S. and Ikura, K.,** The reactions of p-nitrobenzenesulfenyl chloride with pyridine and α-picoline N-oxides, *Bull. Chem. Soc. Jpn.,* 38, 58, 1965.

39. **Oae, S. and Ikura, K.,** The reaction of heterocyclic N-oxides with p-nitrobenzenesulfinyl chloride, *Bull. Chem. Soc. Jpn.,* 39, 1306, 1966.

40. **Chen, Z.C. and Stang, P.J.,** Reaction of pyridine N-oxides with triflic anhydride: formation of N-sulfonyloxy and dipyridinium ether salts. *Tetrahedron Lett.,* 3923, 1984.

41. **Bassindale, A.R. and Stout, T.,** The interaction of electrophilic silanes (Me$_3$SiX, X=ClO$_4$, I, CF$_3$SO$_3$, Br, Cl) with nucleophiles, The nature of silylation mixture in solution, *Tetrahedron Lett.,* 3403, 1985.

42. **Traynelis, V.J., Gallagher, A.I., and Martello, R.F.,** Reactions of 2-picoline N-oxide with phenyl acetates, *J. Org. Chem.,* 26, 4365, 1961.

43. **Ochiai, E.,** A new classification of tertiary amino oxides, *Proc. Imp. Acad. (Tokyo),* 19, 307, 1943.

44. **Baumgarten, P. and Erbe, H.,** Hydroxylamine and its O-sulfonic acid and on the pyridine N-oxide-O-sulfonic betaine, *Ber. Dtsch. Chem. Ges.,* 71, 2603, 1938.

45. **Cadogan, J.I.G.,** Oxidation of tervalent organic compounds of phosphorus, *Q. Rev., Chem. Soc.,* 16, 208, 1962.
46. **Ramirez, F. and Aguiar, R.,** in *Abst. 134th Meet. Am. Chem. Soc.,* Chicago, 1958, 42-N.
47. **Emerson, T.R. and Rees, C.W.,** The deoxygenation of heterocyclic *N*-oxides. Part III. Kinetics of their reactions with phosphorus trichloride in chloroform, *J. Chem. Soc.,* 2319, 1964.
48. **Emerson, T.R. and Rees, C.W.,** Deoxygenation of pyridine *N*-oxides with triethylphosphite: a novel oxygen- and peroxide-catalyzed reduction, *Proc. Chem. Soc., London,* 418, 1960.
49. **Howard, E. and Olszewski, W.F.,** The reaction of triphenylphosphine with some aromatic amine oxides, *J. Am. Chem. Soc.,* 81, 1483, 1959.
50. **Lu, X., Sun, J., and Tao, X.,** A new deoxygenation reaction catalyzed by molybdenum complex, *Synthesis,* 185, 1982.
51. **Hamana, M.,** Reduction of aromatic tertiary amine oxides with phosphorus trichloride, *Yakugaku Zasshi,* 71, 263, 1951.
52. **Nakayama, I.,** 4-Nitroquinoline, *Yakugaku Zasshi,* 71, 1088, 1951.
53. **Hamana, M.,** Reaction of pyridine 1-oxide, quinoline 1-oxide and dimethylaniline *N*-oxide with phosphorus trichloride, *Yakugaku Zasshi,* 75, 121, 1955.
54. **Hamana, M.,** Reduction of aromatic tertiary amine oxides by phosphorus tribromide, *Yakugaku Zasshi,* 75, 135, 1955.
55. **Hamana, M.,** Reduction of aromatic tertiary amine oxides by triphenylphosphite, *Yakugaku Zasshi,* 75, 139, 1955.
56. a. **Brown, E.V.,** Preparation and reaction of 2-nitropyridine-1-oxides, *J. Am. Chem. Soc.,* 79, 3565, 1957; b. **Lewicka, K. and Plazek, E.,** Nitration of 3-hydroxypyridine *N*-oxide, *Recl. Trav. Chim. Pays-Bas,* 78, 644, 1959.
57. **Hamana, M.,** Reaction of 4-substituted pyridine and quinoline 1-oxides having an active hydrogen with phosphorus trichloride, *Yakugaku Zasshi,* 75, 130, 1955.
58. **Foulger, N.J., Wakefield, B.J., and MacBride, J.A.H.,** Synthesis and fragmentation of oxidation products of bis(tetrachloro-4-pyridyl)sulfide and hexachlorodipyridothiophenes; hexachloro-2,7-diazabiphenylene, *J. Chem. Res. (S),* 124, 1977.
59. **Taylor, E.C. and Driscoll, J.S.,** 3-Nitropyridine 1-oxide, *J. Org. Chem.,* 25, 1716, 1960.
60. **Ochiai, E. and Kaneko, C.,** A new nitration of quinoline *N*-oxide, *Pharm. Bull.,* 5, 56, 1957.
61. **Talik, T. and Talik, Z.,** 3-Fluoro-4-nitropyridine *N*-oxide. I. Substitution reactions of the fluorine and the nitro group, *Rocz Chem.,* 38, 777, 1964.
62. **Talik, Z.,** Substitution reactions of halogens and the nitro group in 2-halo-4-nitropyridine *N*-oxides, *Bull. Acad. Pol. Sci., Ser. Sci. Chim.,* 9, 561, 1961.
63. **Talik, Z. and Talik, T.,** Reaction of 2-halo- and 3-halo-4-nitropyridine *N*-oxides with phosphorus trihalides, *Rocz. Chem.,* 36, 417, 1962.
64. **Acheson, R.M., Adcock, B., Glover, G.M., and Sutton, L.E.,** The bromination and nitration of acridine 10-oxide, *J. Chem. Soc.,* 3367, 1960.
65. **Hamana, M.,** Reaction of 4-nitropyridine 1-oxide with phosphorus trichloride, *Yakugaku Zasshi,* 75, 123, 1955.
66. **Hamana, M.,** Reaction of 4-nitroquinoline 1-oxide with phosphorus trichloride, *Yakugaku Zasshi,* 75, 127, 1955.
67. **Ross, W.C.J.,** The properties of some 4-substituted nicotinic acids and nicotinamides, *J. Chem. Soc. (C),* 1816, 1966.
68. **Read, R.W., Spear, R.J., and Norris, W.P.,** Synthesis of 4,6-dinitrobenzofurazan, a new electron-deficient aromatic, *Aust. J. Chem.,* 36, 1227, 1983.
69. **Micetich, R.G. and MacDonald, J.C.,** Metabolites of *Aspergillus sclerotiorum* Huber, *J. Chem. Soc.,* 1507, 1964.
70. **Dornow, A., Marquardt, H.H., and Paucksch, H.,** Reaction of α-chlorooximes, II., *Chem. Ber.,* 97, 2165, 1964.
71. **Suzuki, H., Sato, N., and Osuka, A.,** A mild deoxygenation of heteroaromatic *N*-oxides with diphosphorus tetraiodide, *Chem. Lett.,* 459, 1980.
72. **Joergensen, K.A., Shabana, R., Scheibye, S., and Lawesson, S.O.,** Reactions of 2,4-bis(4-methoxyphenyl)-1,3,2,4-dithiadiphosphetane-2,4 -disulfide with compounds containing the N→O function, *Bull. Soc. Chim. Belg.,* 89, 247, 1980.
73. **Boyer, S.K., Bach, J., McKenna, J., and Jagdmann, E.,** Mild hydrogen-transfer reductions using sodium hypophosphite, *J. Org. Chem.,* 50, 3408, 1985.
74. **Grundmann, C.,** Specific reduction of furoxans to furazans, *Chem. Ber.,* 97, 575, 1964.
75. **Ochiai, E., Ishikawa, M., and Zai-Ren, S.,** Synthesis of aromatic tertiary amine *N*-oxides, *Yakugaku Zasshi,* 65, 72, 1945.
76. **Risaliti, A.,** Preparation and properties of the *N*-oxides of 2-phenylquinoline, *Ric. Sci.,* 24, 2351, 1954.
77. **Hayashi, E. and Higashino, T.,** 4-*i*-Propylquinazoline 1-oxide and 4-*i*-propyl-2-quinazolinecarbonitrile, *Chem. Pharm. Bull.,* 12, 43, 1964.

78. **Daniher, F.A. and Hackley, B.E.,** Deoxygenation of pyridine *N*-oxides with sulfur dioxide, *J. Org. Chem.*, 31, 4267, 1966.

79. **Olah, G.A., Arvanaghi, M., and Vankar, Y.D.,** Deoxydation of pyridine *N*-oxides with trimethyl(ethyl)amine-sulfur dioxide complexes, *Synthesis*, 660, 1980.

80. **Hamana, M. and Funagoshi, K.,** Reduction of acyl adducts of aromatic *N*-oxides, *Yakugaku Zasshi*, 80, 1027, 1960.

81. **Evans, R.F. and Brown, H.C.,** 4-Nitro-2,6-1utidine 1-oxide, *J. Org Chem.*, 27, 1665, 1962.

82. **Oae, S. and Ikura, K.,** The reaction of 4-substituted pyridine *N*-oxides with *p*-nitro benzenesulfenyl chloride and *p*-nitrobenzenesulfinyl chloride, *Bull. Chem. Soc. Jpn.*, 40, 1420, 1967.

83. **Begum, S. and Shakir, N.,** Deoxygenation of aromatic *N*-oxides with sulfur monochloride, *Pak. J. Sci. Ind. Res.*, 17, 198, 1975.

84. **Habib, M.S. and Rees, C.W.,** The reduction of 3-hydroxyquinoxalino-2-carboxylic acid and derivatives with sodium dithionite, *J. Chem. Soc.*, 3384, 1960.

85. **Tennant, G.,** A new synthesis of 2-hydroxyquinoline 4-oxides, *J. Chem. Soc.*, 2428, 1963.

86. **Issidorides, C.H. and Haddadin, M.J.,** Benzofurazan oxide. II. Reactions with enolate anions, *J. Org. Chem.*, 31, 4067, 1966.

87. **El Abadelah, M.M., Sabri, S.S., and Tashtoush, H.I.,** Selective monodeoxygenation of quinoxaline-amino acid and ester dioxides, *Tetrahedron*, 35, 2571, 1979.

88. **Tennant, G. and Yacomeni, C.W.,** New synthetic route to 2-dialkylaminopteridin-7(8*H*)-ones and their 5-oxides, *J. Chem. Soc., Chem. Commun.*, 819, 1975.

89. **Taylor, E.C. and Jacobi, P.A.,** Unequivocal total synthesis of L-*erythro*-biopterin, *J. Am. Chem. Soc.*, 96, 6781, 1974.

90. a. **Ichiba, M., Nishigaki, S., and Senga, K.,** Synthesis of fervenulin 4-oxide and its conversion to the antibiotics fervenulin and 2-methylfervenulone, *J. Org. Chem.*, 43, 469, 1978; b. **Haddadin, M.J., Zahr, G.E., Rawdah, T.N., Chelhot, N.C., and Issidorides, C.H.,** Deoxygenation of quinoxalines *N*-oxides and related compounds, *Tetrahedron*, 30, 659, 1974.

91. a. **Freeman, J.P., Gannon, J.J., and Surbey, D.L.,** The synthesis and structure of 3,4-diazacyclopentadienone derivatives, *J. Org. Chem.*, 34, 187, 1969; b. **Kaneko, C., Hayashi, R., Fujii, H., and Yamamoto, A.,** 3-Sulfolene as an alternative reagent for sulfur dioxide, *Chem. Pharm. Bull.*, 26, 3582, 1978.

92. **Bonini, B.F., Maccagnani, G., Mazzanti, G., and Pedrini, P.,** Deoxygenation of pyridine 1-oxides with sulfur monoxide generated from trans-2,3-diphenylthiirane 1-oxide, *Tetrahedron Lett.*, 1799, 1979.

93. **Kagami, H. and Motaki, S.,** Deoxygenation of amine *N*-oxides or *C*-nitroso compounds by dialkyl sulfoxylates, *J. Org. Chem.*, 43, 1267, 1978.

94. **Biffin, M.E.C., Bocksteiner, G., Miller, J., and Paul, D.B.,** Mechanism of deoxygenation of heterocyclic *N*-oxides and tertiary amine *N*-oxides, *Aust. J. Chem.*, 27, 789, 1974.

95. **Takeda, K. and Tokuyama, K.,** Reactions of derivatives of pyridine and quinoline 1-oxides with sulfur-liquid ammonia solution, *Yakugaku Zasshi*, 75, 620, 1955.

96. **Relyea, D.I., Tawney, P.O., and Williams, A.R.,** Deoxygenation of pyridine 1-oxides, *J. Org. Chem.*, 27, 477, 1962.

97. **Hisano, T. and Koga, H.,** Reaction of 2-picoline 1-oxide and aniline in the presence of sulfur, *Yakugaku Zasshi*, 90, 552, 1970.

98. **Kurbatov, Y.V.,** Selective reduction of the nitrogen→oxygen group in 4-nitropyridine *N*-oxide, *Khim. Geterotsikl. Soedin*, 701, 1981.

99. **Kikugawa, K., Suehiro, H., Yanase, R., and Aoki, A.,** Chemical transformation of adenosine into 2-thioadenosine derivatives, *Chem. Pharm. Bull.*, 25, 1959, 1977.

100. **Barton, D.H.R., Fekih, A., and Lusinchi, X.,** Selective reduction of the nitrogen-oxygen bond in *N*-oxides and in nitrones by sodium hydrogen telluride, *Tetrahedron Lett.*, 4603, 1985.

101. **Itoh, T., Nagano, T., and Hirobe, M.,** Clusters [Fe₄S₄(SR)₄]²⁻ as novel catalysts in organic reductions, *Chem. Pharm. Bull.*, 34, 2013, 1986.

102. **Noda, H., Minemoto, M., Narimatsu, K., and Hamana, M.,** Preparation and properties of 3,2′-diquinolyl *N*-oxides, *Yakugaku Zasshi*, 95, 1078, 1975.

103. **Ikura, K. and Oae, S.,** The reaction of diaryl disulfides with pyridine *N*-oxides, *Tetrahedron Lett.*, 3791, 1968.

104. **Takahashi, S. and Kano, H.,** The reactivity of 1,2-dimethylbenzimidazole 3-oxides, *Chem. Pharm. Bull.*, 14, 1219, 1966.

105. **Jerchel, D. and Melloh, W.,** Purification of methylpyridines through their *N*-oxides, *Liebigs Ann. Chem.*, 613, 144, 1958.

106. **Hayashi, E., Yamanaka, H., and Shimizu, K.,** A new deoxygenation reaction of aromatic heterocyclic *N*-oxides, *Chem. Pharm. Bull.*, 6, 323, 1958.

107. **Hayashi, E., Yamanaka, H., and Shimizu, K.,** Reduction of 4-substituted pyridine 1-oxide derivatives, *Chem. Pharm. Bull.*, 7, 141, 1959.

108. **Hayashi, E., Yamanaka, H., and Shimizu, K.,** Reduction of 4-substituted quinoline 1-oxide derivatives, *Chem. Pharm. Bull.*, 7, 146, 1959.

109. **Hayashi, E., Yamanaka, H., and Higashino, T.,** Reduction of hydroxamic acid type *N*-oxides, *Chem. Pharm. Bull.,* 7, 149, 1959.

110. **Yamanaka, H.,** Catalytic reduction of 4-benzyloxy-6-methylpyrimidines and related compounds, *Chem. Pharm. Bull.,* 7, 158, 1959.

111. **Mitsui, S., Sakai, T., and Saito, H.,** Selectivity of catalysts in hydrogenation of pyridine derivatives, *Nippon Kagaku Zasshi,* 86, 409, 1965.

112. **Hayashi, E., Yamanaka, H., and Iijima, C.,** Catalytic reduction of quinoline 1-oxide and 4-substituted pyridine 1-oxides with Urushibara nickel, *Yakugaku Zasshi,* 80, 839, 1960.

113. **Katritzky, A.R. and Monro, A.M.,** The hydrogenation of some pyridine 1-oxides, *J. Chem. Soc.,* 1263, 1958.

114. **Ishii, T.,** Pressure-hydrogenation of 4-nitro-, 4-hydroxy- and 4-methoxyquinoline 1-oxide, *Yakugaku Zasshi,* 72, 1317, 1952.

115. **Hayashi, E. and Iijima, C.,** 2-Phenylquinoxaline 4-oxide, *Yakugaku Zasshi,* 82, 1093, 1962.

116. **Taylor, E.C., Cheng, C.C., and Vogl, O.,** Hypoxanthine-1-*N*-oxide, *J. Org. Chem.,* 24, 2019, 1959.

117. **Hayashi, E., Yamanaka, H., Iijima, C., and Matushita, S.,** Reduction of 4,4'-azopyridine 1,1'-dioxide, 4,4'-azoxypyridine 1,1',-dioxide, and 2-styrylpyridine 1-oxide, *Chem. Pharm. Bull.,* 8, 649, 1960.

118. **Hamana, M., Umezawa, B., and Noda, K.,** The reduction of 1-[(1-oxide-2-pyridyl)-methyl]pyridinium salt, *Chem. Pharm. Bull.,* 11, 694, 1963.

119. **Middleton, R.W. and Wibberley, D.G.,** Synthesis of imidazo [4,5-*b*] and -[4,5-*c*]pyridines, *J. Heterocycl. Chem.,* 17, 1757, 1980.

120. **Hayashi, E., Yamanaka, H., and Iijima, C.,** Catalytic reduction of heterocyclic amine oxides with Raney nickel.V., *Yakugaku Zasshi,* 80, 1145, 1960.

121. **Wagner, G., Pischel, H., and Smidt, R.,** Preparation of *S*-β-D-glucosides of mercaptopyridine *N*-oxides and mercaptoquinoline *N*-oxides, *Z. Chem.,* 2, 86, 1962.

122. **Ochiai, E. and Suzuki, I.,** Reduction of 4-nitro-2-picoline 1-oxide, *Yakugaku Zasshi,* 67, 158, 1947.

123. **Itai, T. and Ogura, H.,** Syntheses of 3-picoline 1-oxide derivatives, *Yakugaku Zasshi,* 75, 292, 1955.

124. **Ochiai, E., Ohta, H., and Nomura, H.,** 4-Hydroxyaminoquinoline *N*-oxide, *Pharm. Bull.,* 5, 310, 1957.

125. **Kawazoe, Y. and Araki, M.,** Reduction of 3- and 8-substituted 4-nitroquinoline 1-oxides, *Chem. Pharm. Bull.,* 16, 839, 1968.

126. **Suzuki, I., Nakashima, T., Nagasawa, N., and Itai, T.,** Nitration of cinnoline 1-oxide, *Chem. Pharm. Bull.,* 12, 1090, 1964.

127. **Taylor, E.C. and Loeffler, P.K.,** 7-Methyladenine 3-*N*-oxide, *J. Org. Chem.,* 24, 2035, 1959.

128. **den Hertog, H.J., Kolder, C.R., and Combé, W.P.,** The directing influence of the *N*-oxide group during the nitration of derivatives of pyridine *N*-oxide, *Rec. Trav. Chim. Pays-Bas,* 70, 591, 1951.

129. **den Hertog, H.J., and Overhoff, J.,** Pyridine *N*-oxide as an intermediate for the preparation of 2- and 4-substituted pyridines, *Rec. Trav. Chim. Pays-Bas,* 69, 468, 1950.

130. **Jiu, J. and Mueller, G.P.,** Syntheses in the 1,2,4-benzotriazine series, *J. Org. Chem.,* 24, 813, 1959.

131. **Kröhnke, F. and Shäfer, M.,** Preparation of 4-nitropyridine and of its molecular compounds with phenol, *Chem. Ber.,* 95, 1098, 1962.

132. **Loudon, J.D. and Wellings, I.,** Substituent interactions in *ortho*-substituted nitrobenzenes, *J. Chem. Soc.,* 3470, 1960.

133. **Hahn, W.E. and Lesiak, J.,** Magnesium deoxidation of quinoline and quinoxaline *N*-oxides and their derivatives in methanol solutions of ammonium acetate, *Pol. J. Chem.,* 59, 627, 1985.

134. **Profft, E. and Schulz, G.,** Highly active anesthetics derived from falicain: β-[4-alkoxy(and 4-alkyl)-3-pipecoline]propiophenones and thienones, *Arch. Pharm.,* 294, 292, 1961.

135. **Bonnet, R., Brown, R.F.C., Clark, V.M., Sutherland, I.O., and Todd, A.,** The preparation and reactions of Δ¹-pyrroline 1-oxides, *J. Chem. Soc.,* 2094, 1959.

136. **Arcos, J.C., Arcos, M., and Miller, J.A.,** Synthesis of nitro and amino derivatives of benzo[*c*]cinnoline, *J. Org. Chem.,* 21, 651, 1956.

137. a. **Talik, T. and Plazek, E.,** 4-Aminopicolinic acid and its amide and 2-cyano-4-aminopyridine, *Roczniki Chem.,* 35, 463, 1961; b. **Abramovitch, R.A. and Adams, K.A.H.,** A new method for the deoxygenation of aromatic *N*-oxides, *Can. J. Chem.,* 39, 2134, 1961.

138. **Barton, J.W. and Thomas, J.F.,** The synthesis and reactions of some benzo[*c*]cinnoline oxides, *J. Chem. Soc.,* 1265, 1964.

139. **Lunn, G. and Sansone, E.B.,** Facile reduction of pyridines with nickel-aluminium alloy, *J. Org. Chem.,* 51, 513, 1986.

140. **den Hertog, H.J., Henkens, C.H., and Van Roon, J.H.,** Reduction of 4-nitropyridine-*N*-oxide in alkaline medium, *Rec. Trav. Chim. Pays-Bas,* 71, 1145, 1952.

141. a. **Ochiai, E. and Katada, M.,** Reduction of 4-nitroquinoline 1-oxide, *Yakugaku Zasshi,* 64, 206, 1944; b. **Suszko, J. and Szafran, M.,** Reduction of 4-nitro-2,6-1utidine and its *N*-oxide, *Roczniki Chem.,* 39, 709, 1965.

142. **Martel, H.J.J.B. and Rasmussen, M.,** Sulfur dioxide extrusion from di-2-pyridylmethylsulfones: synthesis of *trans*-1,2-di-2-pyridyl-ethene and [2.2] (2,6) pyridinophane, *Tetrahedron Lett.,* 3843, 1971.

143. **Schmitz, E.,** 3,4-Dihydroisoquinoline *N*-oxide, *Chem. Ber.,* 91, 1488, 1958.

144. a. **MacBride, J.A.H.,** Synthesis of 2,7-diazabiphenylene by thermal extrusion of nitrogen from 2,7,9,10-tetraazaphenanthrene, *J. Chem. Soc., Chem. Commun.,* 359, 1974; b. **Stacy, G.W., Ettling, B.V., and Papa, A.J.,** Reactions of benzaldehyde with *o*-nitroaniline, *J. Org. Chem.,* 29, 1537, 1964.

145. a.**Togashi, S., Fulcher, J.G., Cho, B.R., Hasegawa, M., and Gladysz, J.A.,** Deoxygenation and desulfurization of organic compounds via transition metal atom condensation, *J. Org. Chem.,* 45, 3044, 1980; b. **Alper, H. and Gopal, M.,** Preparation of azo compounds and amines by triirondodecarbonyl or molybdenum hexacarbonyl on alumina, *J. Org. Chem.,* 46, 2593, 1981.

146. **Polanc, S., Stanovnik, B., and Tisler, M.,** Selective deoxygenation of *N*-oxides and reduction of nitrocompounds by molybdenum (III) species, *Synthesis,* 129, 1980.

147. **Zipp, A.P. and Ragsdale, R.O.,** Kinetics and mechanism of the chromium(II) reduction of pyridine *N*-oxides, *J. Chem. Soc., Dalton Trans.,* 2452, 1976.

148. **Everton, T., Zipp, A.P., and Ragsdale, R.O.,** Reduction of pyridine *N*-oxides by titanium(III), *J. Chem. Soc., Dalton Trans.,* 2449, 1976.

149. **Somei, M., Kato, K., and Inoue, S.,** Titanium(III)chloride for the reduction of heteroaromatic and aromatic nitro compounds, *Chem. Pharm. Bull.,* 28, 2515, 1980.

150. **Akers, H.A., Vang, M.C., and Updike, T.D.,** Reduction of specific disulfides with titanium(III) chloride, *Can. J. Chem.,* 65, 1364, 1987.

151. **Kano, S., Tanaka, Y., and Hibino, S.,** The reaction of heteroaromatic amine oxides with titanium chloride (TiCl₄)/sodium borohydride (NaBH₄), *Heterocycles,* 14, 39, 1980.

152. **Malinowski, M.,** A facile deoxygenation of unfunctionalized aromatic *N*-oxides, *Synthesis,* 732, 1987.

153. **Malinowski, M. and Kaczmarek, L.,** A selective deoxygenation of halogen containing heteroaromatic *N*-oxides, *Synthesis,* 1013, 1987.

154. **Malinowski, M. and Kaczmarek, L.,** A convenient preparation of 4-pyridinamine derivatives, *J. Prakt. Chem.,* 330, 154, 1988.

155. **Tang, W., Li, J., and Chen, T.H.,** Reaction of low-valent tungsten-composition and reactions with sulfoxide and pyridine *N*-oxides, *Huaxue Xuebao,* 45, 472, 1987.

156. **Morita, T., Kuroda, K., Okamoto, Y., and Sakurai, H.,** A new method for deoxygenation of heteroaromatic *N*-oxides with chlorotrimethylsilane/sodium iodide/zinc, *Chem. Lett.,* 921, 1981.

157. **Siggia, S.,** *Quantitative Organic Analysis via Functional Groups,* 3rd ed., Wiley, New York, 1963, 526.

158. **Brooks, R.T. and Sternglanz, P.D.,** Titanometric determination of the *N*-oxide group in pyridine *N*-oxide and related compounds, *Anal. Chem.,* 31, 561, 1959.

159. **Becker, H.P. and Neumann, W.P.,** Reductive silylation of azo compounds and *N*-heterocycles with bis(trimethylsilyl)mercury, *J. Organometal. Chem.,* 37, 57, 1972.

160. **Brown, H.C. and Subla Rao, B.C.,** A new powerful reducing agent—Sodium borohydride in the presence of aluminium chloride and other polyvalent metal halides, *J. Am. Chem. Soc.,* 78, 2583, 1956.

161. **Brown, H.C., Krishnamurthy, S., and Yoon, N.M.,** 9-Borabicyclo[3.3.1]nonane in tetrahydrofuran as a new selective reducing agent in organic synthesis, *J. Org. Chem.,* 41, 1778, 1976.

162. **Brown, H.C., Bigley, D.B., Arora, S.K., and Yoon, N.M.,** Reaction of disiamylborane in tetrahydrofuran with selected organic compounds containing representative functional groups, *J. Am. Chem. Soc.,* 92, 7161, 1970.

163. **Yoon, N.M. and Gyoung, Y.S.,** Reaction of disobutylaluminum hydride with selected organic compounds containing representative functional groups, *J. Org. Chem.,* 50, 2443, 1985.

164. **Kozuka, S., Akasaka, T., and Furumai, S.,** Cleavage of semipolar linkages by a free radical reactions of amine *N*-oxides and betaines with tri-*n*-butyltin hydride, *Chem. Ind.,* 452, 1974.

165. **Brindle, J.R., Liard, J.L., and Bérubé, N.,** Reactions of NaBH₂S₃ with some typical organic nitrogen compounds, *Can. J. Chem.,* 54, 871, 1976.

166. **Zandersons, A., Lusis, V., Muceniece, D., and Duburs, G.,** Synthesis of *N*-unsubstituted and *N*-methyl substituted 4-aryl-2,6-dimethyl-1,2-dihydropyridine-3,5-dicarbonitriles, *Khim. Geterotsikl. Soedin.,* 81, 1987.

167. **Kawazoe, Y. and Tachibana, M.,** Reduction of 4-substituted quinoline *N*-oxides with sodium borohydride, *Chem. Pharm. Bull.,* 13, 1103, 1965.

168. **Brown, H.C., Cha, J.S., Nazer, B., Kim, S.C., Krishnamurthy, S., and Brown, C.A.,** Potassium trisopropoxyborohydride as a selective reducing agent in organic synthesis, *J. Org. Chem.,* 49, 885, 1984.

169. **Brown, H.C., Mathew, C.P., Pyun, C., Son, J.C., and Yoon, N.M.,** Reaction of lithium 9-boratabicyclo[3.3.1]nonane with selected organic compounds, *J. Org. Chem.,* 49, 3091, 1984.

170. **Wagner, W.R. and Rastetter, W.H.,** Requisite cation complexation for the reduction of 2,6-pyrido-18-crown-6-*N*-oxide and an analogue by potassium tri-*sec*-butylborohydride, *J. Org. Chem.,* 48, 294, 1983.

171. **Corbett, J.F. and Holt, P.F.,** The bromination of benzo[*c*]cinnoline and the preparation and ultraviolet absorption spectra of reference compounds, *J. Chem. Soc.,* 5029, 1961.

172. **Traber, W., Hubmann, M., and Karrer, P.,** Reduction products of *N*-methyl-*p*-phenanthroline methiodide, isoquinoline *N*-oxide, and phenanthridine *N*-oxide, *Helv. Chim. Acta.,* 34, 265, 1960.

173. **Naumann, K., Zon, G., and Mislow, K.,** The use of hexachlorodisilane as a reducing agent. Stereospecific deoxygenation of acyclic phosphine oxides, *J. Am. Chem. Soc.,* 91, 7012, 1969.

174. **Homaidan, F.R. and Issidorides, C.H.,** Deoxygenation of 2,3-disubstituted quinoxaline 1,4-dioxides, *Heterocycles,* 16, 411, 1981.

175. **Hortmann, A.G., Koo, J.Y., and Yu, C.C.,** The utility of hexachlorodisilane for the deoxygenation of nitrones, 2*H*-imidazole 1-oxides, 5*H*-pyrazole 1-oxides and related *N*-hydroxycompounds, *J. Org. Chem.,* 43, 2289, 1978.

176. **Vorbrüggen, H. and Krolikiewicz, K.,** A simple reduction of aromatic heterocyclic *N*-oxides with hexamethyldisilane: reaction with hexamethyldisilane and fluoride ion, *Tetrahedron Lett.,* 5337, 1983.

177. **Hwu, J.R. and Wetzel, J.,** Mechanistic studies in the deoxygenation of pyridine 1-oxide: new 1,2-elimination, *J. Org. Chem.,* 50, 400, 1985.

178. **Jousseaume, B. and Chanson, E.,** Mild selective deoxygenation of amine oxides by tin-tin bonded derivatives, *Synthesis,* 55, 1987.

179. **Moore, R.E. and Furst, A.,** Reductions with hydrazine hydrate catalyzed by Raney nickel. III. Effect of the catalyst on the reduction of 2,2'-dinitrobiphenyl, *J. Org. Chem.,* 23, 1505, 1958.

180. **Di Nunno, L. and Florio, S.,** Deoxygenation of benzofurazan *N*-oxide with sodium azide in carboxylic acids, *Chim. Ind. (Milan),* 57, 243, 1975.

181. **Kurbatova, A.S., Kurbatov, Y.V., Avezov, M.R., Otoshchenko, O.S., and Sadykov, A.S.,** Deoxidation of pyridine and quinoline *N*-oxides with lower alcohols, *Tr. Samarkand. Gos Univ.,* 167, 38, 1969; *Chem. Abs.,* 75, 35663n, 1971.

182. **Okano, T., Fujiwara, K., Konishi, H., and Kiji, J.,** Reduction of oxidized nitrogen compounds with carbon monoxide and water catalyzed by tetrapyridine(1,5-cyclooctadiene) ruthenium bis (tetraphenylborate), *Bull. Chem. Soc. Jpn.,* 55, 1975, 1982.

183. **Wenkert, D. and Woodward, R.B.,** Studies of 2,2'-bipyridyl *N,N'*-dioxides, *J. Org. Chem.,* 48, 283, 1983.

184. **Maffei, S. and Bettinetti, G.F.,** Nitration of phenazine *N*-oxide, *Ann. Chim. (Rome),* 45, 1031, 1955.

185. **Kanoktanaporn, S. and MacBride, J.A.H.,** Preparation of biphenylene and octachlorobiphenylene by vacuum pyrolysis of benzo[c]cinnolines; pyrolysis of benzo[c]cinnoline *N*-oxide, *J. Chem. Res. (S),* 203, 1980.

186. **Antkoviak, W.Z. and Gessner, W.P.,** Synthesis of 2-(2'-hydroxyphenyl)pyridine *N*-oxide and its thermal decomposition as a model reaction of orellanine deoxidation, *Tetrahedron Lett.,* 4045, 1984.

187. **Ito, H.,** Reaction mechanism of oxidation of quinoline, *Chem. Pharm. Bull.,* 12, 345, 1964.

188. **Roberts, S.M. and Suschitzky, H.,** Oxidation of pentachloropyridine and its *N,N*-disubstituted amino-derivatives with peroxyacids, *J. Chem. Soc. (C),* 1537, 1968.

189. **Kröhnke, F. and Schäfer, H.,** Preparation of 4-nitropyridine and its molecular compounds with phenols, *Chem. Ber.,* 95, 1098, 1962.

190. **van Ammers, M. and den Hertog, H.J.,** On the sulfonation of pyridine *N*-oxide, *Recl. Trav. Chim. Pays-Bas,* 78, 586, 1959.

191. **Akropdzhanyan, K.K., Bumber, A.A., Klimov, E.S., and Okhlbystin, O.Y.,** Anion radicals of 2,4,6-triphenylpyridine *N*-oxide, *Khim. Geterotsikl. Soedin,* 1267, 1985.

192. **Kazakova, V.M., Sokol, O.G., Dvoryantseva, G.G., Musatova, I.S., and Elina, A.S.,** EPR spectra of anion radicals formed in the electrochemical reduction of some derivatives of quinoxaline 1,4-dioxide, *Khim. Geterotsikl. Soedin.,* 376, 1980.

193. **Andruzzi, R., Trazza, A., Greci, L., and Marchetti, L.,** Reduction mechanism and electron spin resonance study of cyanoquinoline and their *N*-oxides in DMF, *J. Electroanal. Chem. Interfacial Electrochem.,* 113, 127, 1980.

194. **Andruzzi, R., Trazza, A., Greci, L., and Marchetti, L.,** Reduction mechanism and electron spin resonance study of 2-phenyl 4-substituted quinoline 1-oxides in DMF, *J. Electroanal. Chem. Interfacial Electrochem.,* 108, 59, 1980.

195. **Horner, L. and Röder, H.,** Hydrogenation cleavage of cyclic quaternary ammonium salts and some compounds with bromine at the bridgeheads. Nuclear hydrogenation of pyridinium salts and deoxygenation of some 1-oxides, *Chem. Ber.,* 101, 4179, 1968.

196. **McGinn, F.A. and Brown, G.B.,** Purine *N*-oxides. IX. Polarographic studies, *J. Am. Chem. Soc.,* 82, 3193, 1960.

197. **Matschiner, H. and Tanneberg, H.,** Electrochemical reduction of 2-cyanobenzimidazole 3-oxide, *Z. Chem.,* 20, 263, 1980.

198. **Murray, K.,** Reduction of nicotinamide *N*-oxide by xanthine oxidase, *Methods Enzymol.,* 18B, 210, 1971.

199. **Tada, M.,** Methabolism of 4-nitroquinoline 1-oxide and related compounds, *Carcinog.-Comp. Surv.,* 65, 1981.

200. **Ochiai, E., Kobayashi, G., and Hasegawa, J.,** Synthesis of quinine and dihydroquinine *N*-oxides, *Yakugaku Zasshi,* 67, 101, 1947.

201. **Ochiai, E., Okamoto, T., and Kobayashi, Y.,** Synthesis of 2'-hydroxy derivatives of quinoline base, *Yakugaku Zasshi,* 68, 109, 1948.

202. **Taylor, E.C. and Boyer, N.E.,** Nicotine 1-oxide, nicotine 1'-oxide, nicotine 1,1'-dioxide, *J. Org. Chem.,* 24, 275, 1959.

203. **Johnson, A.W., King, T.J., and Turner, J.R.,** New transformation products of nicotine, *J. Chem. Soc.,* 3230, 1958.

204. **Wieczorek, J.S. and Plazek, E.,** 2-Dimethylaminopyridine N-oxides, *Rec. Trav. Chim. Pays-Bas,* 84, 785, 1965.

205. **Dirlam, J.P. and McFarland, J.W.,** Selective monodeoxygenation of certain quinoxaline 1,4-dioxides with trimethylphosphite, *J. Org. Chem.,* 42, 1360, 1977.

206. **Habib, M.S. and Rees, C.W.,** The oxidation of 3-hydroxyquinoxaline-2-carboxyanilide and its N-methyl derivatives, *J. Chem. Soc.,* 3386, 1960.

207. **Otomasu, H., Takahashi, H., and Yoshida, K.,** Substituent effects on N-oxidation and deoxygenation of aromatic N-heterocyclic compounds. I. β-phenazine derivatives, *Yakugaku Zasshi,* 84, 1080, 1964.

208. **Patcher, I.J. and Kloetzel, M.C.,** Structure of benzo[a]phenazine oxides and syntheses of 1,6-dimethoxyphenazine and 1,6-dichlorophenazine, *J. Am. Chem. Soc.,* 73, 4958, 1951.

209. **Kawazoe, Y., Ohnishi, M., and Yoshioka, Y.,** Deuteration of pyridines and pyridazines, *Chem. Pharm. Bull.,* 12, 1384, 1964.

210. **Bean, G.P., Brignell, P.J., Johnson, C.D., Katritzky, A.R., Ridgewell, B.J., Tarhan, H.O., and White, A.M.,** Acid-catalyzed hydrogen exchange at the α-, β- and γ-positions of substituted pyridine 1-oxides, *J. Chem. Soc. (B),* 1222, 1967.

211. **Katritzky, A.R., Ridgewell, B.J., and White, A.M.,** Acid-catalyzed hydrogen exchange of pyridine 1-oxides, *Chem. Ind.,* 1576, 1964.

212. **El-Anani, A., Banger, J., Bianchi, G., Clementi, S., Johnson, C.D., and Katritzky, A.R.,** Standardization of acid-catalyzed hydrogen exchange rates, *J. Chem. Soc., Perkin Trans. 2,* 1065, 1973.

213. **Bressel, U., Katritzky, A.R., and Lea, J.R.,** Acid-catalyzed hydrogen exchange of quinoline, isoquinoline, and their N-oxides, *J. Chem. Soc. (B),* 4, 1971.

214. **Kawazoe, Y. and Ohnishi, M.,** Electrophilic deuteration of quinoline and its 1-oxide, *Chem. Pharm. Bull.,* 15, 826, 1967.

215. **Uehara, N. and Kawazoe, Y.,** Tritium-labeling of carcinogenic 4-nitroquinoline 1-oxide and related compound, *Chem. Pharm. Bull.,* 18, 203, 1970.

216. **Pohjala, E.K.,** Preparation and cyclization of deuterium labeled 3-(2-pyridyl)methylene-2,4-pentadienones and ethyl 3-oxo-2-(2-pyridyl)methylenebutanoates, *Finn. Chem. Lett.,* 126, 1980.

217. **Tupitsyn, I.F., Zatsepina, N.N., Belyashova, A.I., and Kane, A.A.,** Acid-catalyzed deuterium exchange in methoxylated quinolines and their N-oxides, *Khim. Geterotsikl. Soedin.,* 636, 1986.

218. **Bellingham, P., Johnson, C.D., and Katritzky, A.R.,** Hydrogen exchange of 3,5-dimethylphenol and heterocyclic analogs, *J. Chem. Soc. (B),* 866, 1968.

219. **Lezina, V.P., Gashev, S.B., Borunov, M.M., Smirnov, L.D., and Dyumaev, K.M.,** PMR study of reactivity and mechanism of hydrogen/deuterium exchange in 5-hydroxypyrimidine 1-oxides, *Izv. Akad. Nauk SSSR, Ser. Khim.,* 2697, 1984.

220. **El-Anani, A., Jones, P.E., and Katritzky, A.R.,** Acid-catalyzed hydrogen-exchange of some 2-aminopyridine derivatives, *J. Chem. Soc. (B),* 2363, 1971.

221. **Ochiai, E., Arima, K., and Ishikawa, M.,** Nitration of pyridine 1-oxide and its homologs, *Yakugaku Zasshi,* 63, 79, 1943.

222. **Gleghorn, J., Moodie, R.B., Schofield, K., and Williamson, M.J.,** The kinetics of nitration of pyridine 1-oxide, 2,6-lutidine 1-oxide, and isoquinoline 2-oxide, *J. Chem. Soc. (B),* 870, 1966.

223. **Johnson, C.D., Katritzky, A.R., Shakir, N., and Viney, M.,** The α-, β-, and γ-nitration of pyridine 1-oxide, *J. Chem. Soc. (B),* 1213, 1967.

224. **Ochiai, E. and Zai-Ren, S.,** Simple syntheses of pyridine and quinoline 1-oxides and their nitro derivatives, *Yakugaku Zasshi,* 65, 73, 1945.

225. **Talik, T. and Talik, Z.,** Preparation of some derivatives of 4-nitropyridine N-oxide, *Roczniki Chem.,* 36, 539, 1962.

226. **Profft, E., Krueger, W., Kuhn, P., and Lietz, W.,** 4-Alkoxypyridine 1-oxides, East German patent, 69, 126, 1969; *Chem. Abstr.,* 72, 90309w, 1970.

227. **Matsumura, E.,** 4-Nitro derivatives of pyridine N-oxides and their homologs, Japanese patent, 68, 28, 455, 1968; *Chem. Abstr.,* 70, 77799v, 1969.

228. **Wright, G.C.,** 2-(2-Acetylhydrazino)-4-nitropyridine 1-oxide, U.S. patent, 4216327, 1980; *Chem. Abstr.,* 93, 239251q, 1980.

229. **Evans, R.F. and Kynaston, W.,** A new method of isolation of 2,3-dimethyl- and 2,4,6-trimethyl-pyridine, *J. Chem. Soc.,* 5556, 1961.

230. **Finger, G.C. and Starr, L.D.,** 2-Fluoropyridines, *J. Am. Chem. Soc.,* 81, 2674, 1959.

231. **Jujo, R.,** Synthesis of 3-bromopyridine derivatives, *Yakugaku Zasshi,* 66, 21, 1946.

232. **Essery, J.M. and Schofield, K.,** Some derivatives of 4-amino- and 4-nitro-pyridine, *J. Chem. Soc.,* 4953, 1960.

233. **Profft, E. and Steinke, W.,** N-oxides substituted picolinic acids, *J. Prakt. Chem.,* [4], 13, 58, 1961.

234. **Ishikawa, M.,** Nitration of methyl homologs of pyridine 1-oxide, *Yakugaku Zasshi,* 65A, 6, 1945.

235. **Tanida, H., Irie, T., and Wakisaka, Y.,** Nitration of 5,6,7,8-tetrahydro-5,8-methanoisoquinoline N-oxide with other aromatic substitutions. Isolation and mutagenicity of an α-nitro pyridine N-oxide, *J. Heterocycl. Chem.,* 23, 177, 1986.

236. **den Hertog, H.J. and van Ammers, M.,** The nitration of 2- and 3-methoxypyridine N-oxide, *Rec. Trav. Chem. Pays Bas,* 74, 1160, 1955.

237. **Nesnow, S. and Heidelberger, C.,** Pyridine nucleosides related to 5-fluorouracil, *J. Heterocycl. Chem.,* 10, 779, 1973.

238. **den Hertog, H.J., van Ammers, M., and Schukking, S.,** Nitration of 3-bromo-5-methoxy- and 3,5-dimethoxypyridine N-oxide, *Rec. Trav. Chim. Pays Bas,* 74, 1171, 1955.

239. **den Hertog, H.J., Henkens, C.H., and Dilz, K.,** Nitration of 3,5-dibromo- and 3,5-diethoxy-pyridine N-oxide, *Rec. Trav. Chim. Pays-Bas,* 72, 296, 1953.

240. **Dehmlow, E.V. and Schulz, H.J.,** A note on the chemical reactivity of hydroxylated pyridines, *J. Chem. Res. (S),* 364, 1987.

241. **Brissel, E.R. and Swansinger, R.W.,** Nitration of some 3,5-disubstituted pyridine N-oxides, *J. Heterocycl. Chem.,* 24, 59, 1987.

242. **Dyumaev, K.M. and Lokhov, R.E.,** Mechanism of nitration of 2-substituted 3-hydroxypyridine N-oxides, *Khim. Geterotsikl. Soedin.,* 7, 1003, 1971.

243. **Undheim, K., Nordal, V., and Borka, L.,** N-oxides of 3-hydroxypyridines, *Acta Chem. Scand.,* 23, 2075, 1969.

244. **Ochiai, E. and Futaki, K.,** Nitration and bromination of 4-hydroxypyridine 1-oxide, *Yakugaku Zasshi,* 72, 274, 1952.

245. **Lott, W.A. and Shaw, E.,** Various antibacterial heterocyclic hydroxamic acids, *J. Am. Chem. Soc.,* 71, 70, 1949.

246. **Wieczorek, J.S. and Plazek, E.,** Nitration of N^1-oxide of 2–dimethylaminopyridine, *Rec. Trav. Chim. Pays Bas,* 83, 249, 1964.

247. **Burton, A.G., Frampton, R.D., Johnson, C.D., and Katritzky, A.R.,** Preparation and kinetic nitration of 2-, 3-, and 4-(dimethylamino)pyridines and their 1-oxides in sulfuric acid, *J. Chem. Soc., Perkin Trans. 2,* 1940, 1972.

248. **Talik, T. and Talik, Z.,** Rearrangement of nitraminopyridine N-oxides, *Pr. Nauk. Akad. Ekon. im. Oskara Langego Wroklawiu,* 397, 141, 1987; *Chem. Abst.,* 109, 190207n, 1988.

249. **Prostakov, N.S., Soldatenko, A.T., Krapivko, A.P., Formichev, A.A., Mikaya, A.I., and Ustenko, A.A.,** Nitration of alkyl-substituted 4-aryl(benzoyl) pyridines, *Zh. Org. Khim.,* 18, 1106, 1982.

250. **Katritzky, A.R. and Kingsland, M.,** The mononitration of 2-phenylpyridine and its N-oxide, *J. Chem. Soc. (B),* 862, 1968.

251. **Kurbatov, Y.V., Abdullaev, S.V., Bazhenova, N.S., Otroshenko, O.S., and Sadykov, A.S.,** Nitration of some dipyridyl derivatives containing electron-donor groups, *Tezisy Vses. Soveshch. Khim. Nitrosoedinenii,* 5, 38, 1974; *Chem. Abstr.,* 86, 189668p, 1977.

252. **Kurbatov, Y.V., Tsukervanik, V.S., Otroshchenko, O.S., and Sadykov, A.S.,** Behaviour of 4,4′-bipyridyl *N,N′*–dioxide in nitration reactions, *Nauch. Tr. Samarkand. Univ.,* 167, 65, 1969; *Chem. Abstr.,* 74, 141475y, 1971.

253. **Ochiai, E. and Okamoto, T.,** Nitration of quinoline 1-oxide, *Yakugaku Zasshi,* 70, 384, 1950.

254. **Ochiai, E. and Satake, K.,** Nitration of quinaldine-1-oxide, *Yakugaku Zasshi,* 71, 1078, 1951.

255. **Kawazoe, Y. and Tachibana, M.,** Syntheses of some derivatives of 4-nitro- and 4-hydroxyarninoquinoline 1-oxide, *Chem. Pharm. Bull.,* 15, 1, 1967.

256. **Hamana, M. and Nagayoshi, T.,** Nitration of 6-substituted quinoline 1-oxides, *Chem. Pharm. Bull.,* 14, 319, 1966.

257. **Ochiai, E. and Ikehara, M.,** Nitration of isoquinoline 2-oxides, *Yakugaku Zasshi,* 73, 666, 1953.

258. **Hamana, M. and Saito, H.,** Nitration of 1-cyanoisoquinoline 2-oxide and isoquinoline 2-oxide, *Heterocycles,* 8, 403, 1977.

259. **Hunt, R.R., McOmie, J.F.W., and Sayer, E.R.,** Pyrimidine, 4,6-dimethylpyrimidine, and their 1-oxides, *J. Chem. Soc.,* 525, 1959.

260. **Koelsch, C.F. and Gumprecht, W.H.,** Some diazine-N-oxides, *J. Org. Chem.,* 23, 1603, 1958.

261. **Itai, T. and Natsume, S.,** Nitration of pyridazine 1-oxide, *Chem. Pharm. Bull.,* 11, 83, 1963.

262. **Nakagome, T.,** Nitration of 3,4-dimethylpyridazine N-oxide derivatives, *Chem. Pharm. Bull.,* 11, 726, 1963.

263. **Suzuki, I. and Sueyoshi, S.,** Electrophilic reactions of 3,6-dimethylpyridazine 1,2-dioxide, *Yakugaku Zasshi,* 93, 59, 1973.

264. a. **Yanai, M., Kinoshita, T., and Takeda, S.,** Nucleophilic reaction of 3-ethoxy-4,6-dinitropyridazine 1-oxide, *Chem. Pharm. Bull.,* 19, 2181, 1971; b. **Igeta, H., Tsuchiya, T., Kanakima, M., Sekiya, T., Kumoki, Y., Nakai, T., and Nojina, H.,** Halogenation and nitration of 3-hydroxypyridazine 1-oxides, *Chem. Pharm. Bull.,* 17, 756, 1969.

265. **Suzuki, I., Nakashima, T., and Nagasawa, N.,** Nitration of cinnoline 2-oxide, *Chem. Pharm. Bull.,* 14, 816, 1966.

266. a. **Gleghorn, J.T., Moodie, R.B., Qureshi, E.A., and Schofield, K.,** Kinetics of nitration of 2-cinnoline 2-oxide and quinoline 1-oxide in sulfuric acid. Mechanism of nitration of N-heteroaromatic oxides, with special references to 2,6-lutidine 1-oxide, *J. Chem. Soc. (B),* 316, 1968; b. **Taylor, R.,** Electrophilic aromatic reactivities *via* pyrolysis of 1-arylacetates. Part X. Pyridine N-oxide, *J. Chem. Soc. Perkin Trans. 2,* 277, 1975.

267. **Barton, J.W. and Cockett, M.A.,** Benzo[c]cinnolines. Nitration of benzo[c]cinnoline and its N-oxide, *J. Chem. Soc.,* 2454, 1962.

268. **Corbett, J.F., Holt, P.F., and Vickery, M.L.,** The nitration of benzo[c]cinnoline, its oxide and some methylderivatives, *J. Chem. Soc.,* 4384, 1962.

269. **Silk, J.A.,** Quinoxaline N-oxides. V. Further bz-substituted derivatives, *J. Chem. Soc.,* 2058, 1956.

270. **Otomasu, H. and Nakajima, S.,** Nitration of quinoxalines, *Chem. Pharm. Bull.,* 6, 566, 1958.

271. **Otomasu, H.,** Nitration of phenazine and its derivatives, *Pharm. Bull.,* 2, 383, 1954.

272. **Klemm, L.H., Barnish, I.T., and Zell, R.,** Substitution products derived from thieno[2,3-*b*]pyridine 7-oxide, *J. Heterocycl. Chem.,* 7, 81, 1970.

273. **Berezovskii, V.M., Kirillova, N.I., Glebova, G.D., and Litvak, Z.I.,** Nitration of the 5-N-oxides of methyl-substituted allo- and isoalloxazines, *Zh. Org. Khim.,* 15, 1313, 1979.

274. **Ferguson, I.J., Grimmett, M.R., and Schofield, K.,** Nitration of 1-methylpyrazole 2-oxide, *Tetrahedron Lett.,* 2771, 1972.

275. a. **Ferguson, I.J., Schofield, K., Barnett, J.W., and Grimmett, M.R.,** Nitration of 1,4,5-trimethylimidazole 3-oxide and 1-methylpyrazole 2-oxide, and some reactions of the products, *J. Chem. Soc., Perkin Trans. 1,* 672, 1977; b. **Begtrup, M. and Jonsson, G.,** 3-Substituted 1,2,3-triazole 1-oxides. Preparation and reactions, *Acta Chem. Scand.,* 41B, 724, 1987.

276. **Noland, W.E., Rush, K.R., and Smith L.R.,** Nitration of 2-phenylindole, *J. Org. Chem.,* 31, 65, 1966.

277. **Ghosh, P.B. and Whitehouse, M.W.,** Preparation and in vitro pharmacological activity of some 2,1,3-benzoxadiazoles (benzofurazans) and their N-oxides (benzofuroxans), *J. Med. Chem.,* 11, 305, 1968.

278. **Norris, W.P., Chafin, A., Spear, R.J., and Read, R.W.,** Synthesis and thermal rearrangement of 5-chloro-4,6-dinitrobenzofuroxan, *Heterocycles,* 22, 271, 1984.

279. **Kulibabina, T.N., Kuzilina, E.V., Malinina, L.A., and Pevzner, M.S.,** Nitration of benzotriazole N-oxide, *Khim. Geterotsikl. Soedin.,* 1570, 1976.

280. **Amiet, R.G. and Johns, R.B.,** N-oxidation and subsequent nitration of dibenzodiazepinone, *Aust. J. Chem.,* 20, 723, 1967.

281. **Szmant, H.H. and Chow, Y.L.,** Derivatives of dibenzo[*b,f*] [1,4,5]thiadiazepine. VI. Nitro and amino compounds, *J. Org. Chem.,* 36, 2889, 1971.

282. **Ochiai, H.,** *Aromatic Amine Oxides,* Elsevier, Amsterdam, 1967, 224.

283. **Ochiai, E. and Kaneko, C.,** Nitration of pyridine 1-oxide with acyl nitrates, and the properties of 3,5-dinitropyridine 1-oxide, *Chem. Pharm. Bull.,* 8, 28, 1960.

284. **Nakadate, M., Takano, Y., Hirayama, T., Sakaizawa, S., Hirano, T., Okamoto, K., Hirao, K., Kawamura, T., and Kimura, M.,** Preparation of 5-nitronicotinic acid and its related compounds, *Chem. Pharm. Bull.,* 13, 113, 1965.

285. **Ochiai, E. and Kaneko, C.,** A new nitration of quinoline N-oxide, *Pharm. Bull.,* 5, 56, 1957.

286. **Ochiai, E. and Kaneko, C.,** A new nitration of quinoline N-oxide, 2, *Chem. Pharm. Bull.,* 7, 191, 1959.

287. **Ochiai, E. and Kaneko, C.,** A new nitration of quinoline 1-oxide, 3, *Chem. Pharm. Bull.,* 7, 267, 1959.

288. **Ochiai, E. and Ohta, A.,** A new nitration of quinoline 1-oxide by means of metal nitrates, *Chem. Pharm. Bull.,* 10, 349, 1962.

289. **Osman, A.N. and Khalifa, M.,** A novel approach to the interpretation of the properties and behavior of heterocyclic N-oxides, *Egypt. J. Chem.,* 27, 705, 1985; *Chem. Abstr.,* 104, 185738d, 1986.

290. **Ochiai, E. and Kaneko, C.,** Nitration of 6-methyl- and 3-bromo quinoline N-oxide, *Chem. Pharm. Bull.,* 7, 195, 1959.

291. **Hamana, M. and Yatabe, M.,** Reaction of quinoline N-oxide with sodium nitrite in the presence of tosyl chloride, *Yakugaku Zasshi,* 94, 566, 1974.

292. **Ochiai, E. and Kaneko, C.,** A new nitration of quinoline 1-oxide, 4, *Chem. Pharm. Bull.,* 8, 284, 1960.

293. **Ochiai, E. and Tanida, H.,** Nitration of quinaldine N-oxide by benzoylnitrate, *Pharm. Bull.,* 5, 313, 1957.

294. **Tanida, H.,** Nitration of quinoline homolog N-oxides, *Yakugaku Zasshi,* 78, 1079, 1958.

295. **Itai, T. and Natsume, S.,** Nitration of pyridazine 1-oxide, 2, *Chem. Pharm. Bull.,* 11, 342, 1963.

296. **Itai, T. and Natsume, S.,** Reaction of 3-substituted pyridazine 1-oxide with benzoyl nitrate, *Chem. Pharm. Bull.,* 12, 228, 1964.

297. **van Ammers, M., den Hertog, H.J., and Haase, B.,** Bromination of pyridine N-oxide in fuming sulfuric acid, *Tetrahedron,* 18, 227, 1962.

298. **Saito, H. and Hamana, M.,** γ-Bromination of quinoline and pyridine N-oxides, *Heterocycles,* 12, 475, 1979.

299. **Paudler, W.W. and Jovanovic, M.V.,** Bromination of some pyridine and diazine N-oxides, *J. Org. Chem.,* 48, 1064, 1983.

300. **Hayashi, E.,** Polarization in heterocyclic rings with aromatic character. LXXXV, *Yakugaku Zasshi,* 71, 213, 1951.

301. **Ochiai, E. and Okamoto, T.,** Bromination of quinoline 1-oxide, *Yakugaku Zasshi,* 67, 87, 1947.

302. **Okusa, G. and Osada, M.,** Bromination of 5-pyridazinol 1-oxide derivatives, *Yakugaku Zasshi,* 88, 479, 1968.

303. a. **Radel, R.J., Atwood, J.L., and Paudler, W.W.,** Bromination of some 1,2,4-triazine 2-oxides, *J. Org. Chem.,* 43, 2514 (1978); b. **Kamiya, S., Miyahara, M., Sueyoshi, S., Suzuki, I., and Odashima, S.,** Synthesis of antitumor pyridazine *N*-oxides having (2-chloroethyl)nitrosoureidomethyl and bis(2-chloroethyl)aminomethyl groups, *Chem. Pharm. Bull.,* 26, 3884, 1978.

304. **Ramaiah, K. and Srinivasan, V.R.,** Some substitution reactions of 8-hydroxyquinoline 1-oxide, *Indian J. Chem.,* 6, 635, 1968.

305. a. **Jovanovic, M.V.,** Iodination of some diazines and diazine *N*-oxides, *Heterocycles,* 22, 1195, 1984; b. **Lewicka, K. and Plazek, E.,** Substitution reactions of 3-hydroxypyridine *N*-oxide. III. Iodination, *Rocz. Chem.,* 40, 1875, 1966.

306. **Abdullaev, S.V., Eshmuradova, M., and Kadyrova, M.,** Study of the iodination of pyridine, quinoline, 2,3'-bipyridyl, and their *N*-oxides, *Mater. Resp. Nauchno-Tekh. Konf. Molodykh. Uch. Pererab. Nefti Neftekhim.,* 3rd, 2, 19, 1976; *Chem. Abstr.,* 90, 38758z, 1979.

307. **Ghosh, P.B. and Whitehouse, M.W.,** Potential antileukemic and immunosuppressive drugs. II. Further studies with benz-2,1,3-oxadiazoles (benzofurazans) and their *N*-oxides (benzofuroxans), *J. Med. Chem.,* 12, 505, 1969.

308. **Hamana, M. and Yamazaki, M.,** Bromination of pyridine and quinoline 1-oxides, *Chem. Pharm. Bull.,* 9, 414, 1961.

309. **Yamazaki, M., Chono, Y., Noda, K., and Hamana, M.,** Bromination of aromatic *N*-oxides in the presence of acetic anhydride, *Yakugaku Zasshi,* 85, 62, 1965.

310. **Sedova, V.F., Lisitsyn, A.S., and Mamaev, V.P.,** Bromination of substituted 4-phenyl-1-pyrimidine oxides, *Khim. Geterotsikl. Soedin.,* 1392, 1978.

311. **van Dahm, R.A. and Paudler, W.W.,** Bromination reactions of 1,5- and 1,8-naphthyridine 1oxides, *J. Org. Chem.,* 40, 3068, 1975.

312. **Dyumaev, K.M. and Lokhov, R.E.,** Sulfonation of pyridine, 3-hydroxypyridine, and their *N*-oxides, *Khim. Geterotsikl. Soedin,* 813, 1973.

313. **van Ammers, M. and den Hertog, H.J.,** The mercuration of pyridine 1-oxide. II, *Rec. Trav. Chim.,* 81, 124, 1962.

314. **Ikehawa, N., Hoshino, O., and Honma, Y.,** Mercuration of quinoline 1-oxide, *Chem. Pharm. Bull.,* 17, 906, 1969.

315. **Mosher, H.S. and Welch, F.J.,** The sulfonation of pyridine *N*-oxide, *J. Am. Chem. Soc.,* 77, 2902, 1955.

316. **Saiova, F.M. and Lapteva, T.Y.,** Benzylation of pyridine *N*-oxide in the presence of aluminum chloride, *Zh. Org. Khim.,* 11, 1349, 1975.

317. a. **Bain, B.M. and Saxton, J.E.,** The reaction of nicotinic acid 1-oxide and 3-picoline 1-oxide with acetic anhydride, *J. Chem. Soc.,* 5216, 1961; b. **Nagano, H., Hamana, M., and Nawata, Y.,** Reinvestigation of the reaction of nicotinic acid 1-oxide with acetic anhydride, *Heterocycles,* 26, 1263, 1987; c. **Nagano, H., Nawata, Y., and Hamana, M.,** The mechanism of reaction of nicotinic acid 1-oxide with acetic anhydride, *Chem. Pharm. Bull.,* 35, 4068, 1987; d. **Nagano, H., Nawata, Y., and Hamana, M.,** Reactions of nicotinic acid 1-oxide with propionic, phenylacetic and benzoic anhydrides, *Chem. Pharm. Bull.,* 35, 4093, 1987.

318. **Prostakov, N.S., Banerjee, S.K., and Rodes, L.G.,** 4-Phenyl-2,5-diaroylpyridines and 4-phenyl-2,5-diaminopyridine, *Khim. Geterotsikl. Soedin,* 275, 1968.

319. **Kalatzis, E. and Papadopoulos, P.,** Kinetics of nitrosation of 2- and 4-methylaminopyridine and their 1-oxide derivatives, *J. Chem. Soc. Perkin Trans. 2,* 239, 1981.

320. **Kalatzis, E. and Papadopoulos, P.,** Kinetics of the nitrosation of secondary and diazotization of primary β-aminopyridines, *J. Chem. Soc., Perkin Trans. 2,* 248, 1981.

321. **Sliwa, W.,** Synthesis of 3,5-diamino-2-phenylazopyridine *N*-oxide, *Acta Pol. Pharm.,* 24, 259, 1967.

322. **Badger, G.M. and Buttery, R.G.,** A study of intramolecular hydrogen bonding in 8-hydroxyquinoline, *J. Chem. Soc.,* 614, 1956.

323. **Smirnov, L.D., Zharavlev, V.S., Lezina, V.P., and Dyumaev, K.M.,** Electrophilic reactions of 5-benzyl-3-hydroxypyridine 1-oxide, *Isv. Akad. Nauk SSSR Ser. Khim.,* 2106, 1974.

324. **Smirnov, L.D., Zharavlev, V.S., Merzon, E.E., Zaitsev, B.E., Lezina, V.P., and Dyumaev, K.M.,** Electrophilic reactions of 2-benzyl-3-hydroxypyridine and its *N*-oxide, *Khim. Geterosikl. Soedin.,* 518, 1973.

325. **Smirnov, L.D., Kuzmin, V.I., Zharskii, A.E., Lezina, V.P., and Dyumaev, K.M.,** Synthesis and study of some electrophilic reactions of 2-phenyl-3-hydroxypyridine *N*-oxides, *Isv. Akad. Nauk. SSSR, Ser. Khim.,* 861, 1973.

326. **Dyumaev, K.M. and Lokhov, R.E.,** Aminomethylation in a series of 3-hydroxypyridine *N*-oxides, *Zh. Org. Khim.,* 8, 416, 1972.

327. **Okusa, G., Kamiya, S., and Itai, T.,** The *C*-alkylaminomethylation of 3-pyridazinol 1-oxide derivatives, *Chem. Pharm. Bull.,* 15, 1172, 1967.

328. **Okusa, G. and Kamiya, S.,** The Mannich reaction of 5-pyridazinol 1-oxide, *Chem. Pharm. Bull.,* 16, 142, 1968.

329. **Gashev, S.B., Sedova, V.F., Smirnov, L.D., and Mamaev, V.P.,** Study of some electrophilic reactions of 4-phenyl-5-hydroxypyridine and its 1-oxide, *Khim. Geterotsikl. Soedin.,* 1256, 1983.

330. **Golitsova, L.V., Lezina, V.P., Kuzimin, V.I., and Smirnov, L.D.,** Study of electrophilic reactions of 3-hydroxyquinoline 1-oxide, *Isv. Akad. Nauk, SSSR, Ser. Khim.,* 1660, 1984.

331. **Freifelder, M. and Stone, G.R.,** Reductions with ruthenium. II., *J. Org. Chem.,* 26, 3805, 1961.

332. **Leonard, F. and Wajngurt, A.,** Synthesis of some sulfur-containing derivatives of pyridine 1-oxide and quinoline 1-oxide, *J. Org. Chem.,* 21, 1077, 1956.

333. **Kato, T. and Hayashi, H.,** Reaction of 4-nitro-2-picoline 1-oxide with acetyl chloride, *Yakugaku Zasshi,* 83, 352, 1963.

334. **den Hertog, H.J. and Combe, W.P.,** Preparation of 4-substituted derivatives of pyridine *N*-oxide and pyridine, *Rec. Trav. Chim. Pays-Bas,* 70, 581, 1951.

335. **Gardner, J.N. and Katritzky, A.R.,** The tautomerism of 2- and 4-amino and -hydroxy-pyridine 1-oxide, *J. Chem. Soc.,* 4375, 1957.

336. **Hamana, M. and Yamazaki, M.,** Synthesis of 2-substituted pyridine 1-oxide, *Yakugaku Zasshi,* 81, 612, 1961.

337. **Adams, R. and Reifschneider, W.,** A synthesis of 6-hydroxy-4-quinolizinones, *J. Am. Chem. Soc.,* 81, 2537, 1959.

338. **Mautner, H.G., Chu, S.H., and Lee, M.L.,** 2-Selenopyridine and related compounds, *J. Org. Chem.,* 27, 3671, 1962.

339. **Evans, R.F. and Brown, H.C.,** The sulfonic acids derived from pyridine and 2,6-lutidine and the corresponding *N*-oxides, *J. Org. Chem.,* 27, 1329, 1962.

340. **Abramovitch, R.A. and Cue, B.W.,** Ring contraction of 2-azidoquinoline and 2-azidoquinoxaline 1-oxides, *Heterocycles,* 1, 227, 1973.

341. **Furia, T.E. and Steinberg, D.H.,** 2-Mercaptoquinoline *N*-oxide salts in antidandruff shampoos, *Ger Offen* 2, 035, 171; *Chem. Abst.* 74, 91155w, 1971.

342. **Okamoto, T.,** Reactivity of the nitro group of nitroquinoline 1-oxides, *Yakugaku Zasshi,* 71, 297, 1951.

343. **Hamada, Y.,** Quinoline derivatives. V., *Yakugaku Zasshi,* 80, 1573, 1960.

344. **Buchmann, G.,** Quinoline *N*-oxide derivatives and their biological action, *Chem. Tech.,* 9, 388, 1957.

345. **Itai, I. and Kamiya, S.,** 4-Azidoquinoline and 4-azidopyridine derivatives, *Chem. Pharm. Bull.,* 9, 87, 1961.

346. **Ochiai, E., Ohta, A., and Nomura, H.,** 4-Hydroxylaminoquinoline *N*-oxide, *Chem. Pharm. Bull.,* 5, 310, 1957.

347. **Saneyoshi, M. and Ikehara, M.,** Synthesis and properties of several 4-alkyl or arylsulfonylquinoline 1-oxides, *Chem. Pharm. Bull.,* 16, 1390, 1968.

348. a. **Stefanska, B. and Ledochowski, A.,** Derivatives of 1- and 3-nitro-9-aminoacridine 10-oxides, *Rocz. Chem.,* 42, 1973, 1968; b. **Stefanska, B. and Ledochowski, A.,** N^9-derivatives of 1-nitro-9-amino-,1,9-diaminoacridine and N^{12}-derivatives of 12-aminobenz[a]acridine, *Rocz. Chem.,* 42, 1535, 1968.

349. **Hanna, P.E.,** 3-Amino-5-phenyl-1-(2-pyridyl)pyrrolidines, *J. Heterocycl. Chem.,* 10, 747, 1973.

350. **Jovanovic, M.V.,** Synthesis and reactivity of 2-fluoropyrazine 1-oxide, *Heterocycles,* 22, 1105, 1984.

351. **Itai, T. and Kamiya, S.,** Synthesis of 4- and 5-azidopyridazine 1-oxide, *Chem. Pharm. Bull.,* 11, 1059, 1963.

352. **Katritzky, A.R., Millet, G.H., Noor, R.H.M., and Yates, F.S.,** Reactions of heteroaromatic *N*-oxides with pyridine and diazoles, *J. Org. Chem.,* 43, 3957, 1978.

353. a. **Abushanab, E.,** Nucleophilic displacement of sulfinyl and sulfonyl groups in acid media. A novel method for the preparation of 2-haloquinoline 1-4-dioxides, *J. Org. Chem.,* 38, 3105, 1973; b. **Titkova, R.M., Elina, A.S., and Kostyuchenko, N.P.,** *N*-oxides of 2-amino and 2-hydroxy-1,5-naphthyridine, *Khim. Geterotsikl. Soedin.,* 1237, 1972 .

354. a. **Giner-Sorolla, A.,** Derivatives from 6-chloropurine 3-oxide, *J. Heterocycl. Chem.,* 8, 651, 1971; b. **Scheinfeld, I., Parham, J.C., and Brown, G.B.,** Synthesis of 2-amino-6-chloro- and 2,6-diaminopurine 3-oxides, *J. Heterocycl. Chem.,* 19, 1231, 1982.

355. **Muth, C.W., Darlak, R.S., and Patton, J.C.,** Reaction of substituted pyridine and their *N*-oxides with ethanolic sodium hydroxide and various alkoxydes, *J. Heterocycl. Chem .,* 9, 1003, 1972 .

356. **Hajos, G., Messmer, A., Nesmelyi, A., and Parkanyi, L.,** Synthesis and structural study of azidonaphto-*as*-triazines, *J. Org. Chem.,* 49, 3199, 1984.

357. **Johnson, R.M.,** Kinetics of the reactions between sodium ethoxide and 2- and 4-bromo-, 4-chloro-, 2-, 3-, and 4-nitro, 4-chloro-3,5-dimethyl- and 3,5-dimethyl-4-nitro-pyridine 1-oxide in anhydrous ethanol, *J. Chem. Soc. (B),* 1058, 1966.

358. **Johnson, R.M.,** Kinetics of the reactions between piperidine and 2- and 4-bromo-, 4-chloro-, and 2- and 4-nitropyridine 1-oxide in anhydrous ethanol, *J. Chem. Soc. (B),* 1062, 1966 .

359. **Coppens, G., Declerck, F., Gillet, C., and Nasielski, J.,** Reactivity of 2- and 4-chloropyridines, chloropyridine oxides and chloronitrobenzenes with piperidine in methanol, *Bull. Soc. Chim. Belg.,* 72, 572, 1963.

360. **Abramovitch, R.A., Helmer, F., and Liveris, M.,** Kinetics of reactions between some halogeno-pyridines and -picolines and their *N*-oxides with methoxide ion in methanol and in dimethyl sulfoxide, *J. Chem. Soc. (B),* 492, 1968.

361. **Kato, T., Hayashi, H., and Anzai, T.,** Reactivities of 4-chloropyridine derivatives and their 1-oxides, *Chem. Pharm. Bull.,* 15, 1343, 1967.

362. **Das Gracas Carvalho Cito, M., Dantas Lopes, J.A., Miller, J., and Moran, P.J.S.,** Marked changes in relative nucleophilic strength in comparing S_N reactions of some heterocyclic and homocyclic aromatic systems, *J. Chem. Res. (S),* 184, 1983.

363. **Okamoto, T., Hayatsu, H., and Baba, Y.,** Kinetics of the reaction of 4-haloquinoline 1-oxides and related compounds with piperidine, *Chem. Pharm. Bull.,* 8, 892, 1962.

364. **Delarge, J.,** Synthesis of some mercaptopyridinecarboxylic acids, *J. Pharm. Belg.,* 22, 257, 1967.

365. **Barnes, J.H., Hartley, F.R., and Jones, C.E.L.,** The preparation of 4- and 6-chloro-2-chloromethylpyridines, *Tetrahedron,* 38, 3277, 1982.

366. **Itai, T.,** Reaction of acid chlorides with 4-substituted pyridine and quinoline 1-oxides, *Yakugaku Zasshi,* 65, 70, 1945.

367. **Ballestreros, P., Claramunt, R.M., and Elguero, J.,** Study of the catalytic properties of tris-(3,6-dioxaheptyl)amine (TDA-l) in heteroaromatic nucleophilic substitution of chloropyridines and their *N*-oxides, *Tetrahedron,* 43, 2557, 1987.

368. a. **Ciurla, H. and Talik, Z.,** Some reactions of 3-halo-4-nitropicoline *N*-oxides, *Pol. J. Chem.,* 59, 1089, 1986; b. **Johnson, R.M. and Rees, C.W.,** Light-catalyzed reaction between 4-nitropyridine 1-oxide and piperidine in ethanol, *J. Chem. Soc. (B),* 15, 1967.

369. **Joiner, K.A. and King, F.D.,** Synthesis of sterically hindered 4-(dialkylamino)pyridines, *Tetrahedron Lett.,* 3733, 1987.

370. **Katritzky, A.R., Abdallah, M., Cutler, A.T., Dennis, N., Parton, S.K., Rahimi-Rastgoo, S., Sabongi, G.J., Zamora, H.J.S., and Wuerthwein, E.U.,** 1-(1-Oxido-4-pyridyl)-3-oxidopyridinium, 2-(5-nitro-2-pyridyl)-4-oxidoquinolinium, and 2[trans-3-(4-chlorophenyl)-3-oxoprop-2-enyl]-4-oxidoisoquinolinium, *J. Chem. Res. (S),* 249, 1980.

371. **Hamana, M., and Kumadaki, S.,** Rearrangements of *S*-(4-quinolyl)cysteine 1-oxide and 4-(2-aminoethylthio)quinoline 1-oxide, *Yakugaku Zasshi,* 88, 665, 1968.

372. **Okano, T. and Kano, T.,** *S*-(1-Oxido-4-quinolyl)-DL-methionine salt, a product from the reaction of 4-nitroquinoline 1-oxide and DL-methionine, *Chem. Pharm. Bull.,* 19, 1293, 1971.

373. a. **Ban-Oganowska, H.,** Kinetics of the reaction between 3-halo-2,6-dimethylpyridine *N*-oxides and methoxy ion in methanol and dimethyl sulfoxide, *Pol. J. Chem.,* 56, 275, 1982; b. **Talik, T. and Talik, Z.,** Synthesis of 3-pyridylmethyl sulfide and its oxidation products, *Pr. Nauk. Akad. Ekon. in Oskara Langego Wroclawiu,* 191, 91, 1982; *Chem. Abst.,* 98, 143244s, 1983.

374. **Bellas, M. and Suschitzky, H.,** Fluoropyridine and fluoroquinoline *N*-oxides, *J. Chem. Soc.,* 4007, 1963.

375. **Klein, B., O'Donnel, E., and Auerback, J.,** The amination of chloropyrazines and chloropyrazine *N*-oxides, *J. Org. Chem.,* 32, 2 412, 1967.

376. **Sedova, V.F., Krivopalov, V.P., and Manaev, V.P.,** Photochemical conversion of 5-azido-4-phenylpyrimidine to 5*H*-pyrimido[5,4-*b*]indole, *Khim. Geterotsikl. Soedin,* 986, 1979.

377. a. **Itai, T. and Kamiya, S.,** Syntheses and reactions of 3- and 6-azidopyridazine 1-oxide, *Chem. Pharm. Bull.,* 11, 348, 1963; b. **Yanai, M. and Kinoshita, T.,** Transesterification in alkoxypyridazine *N*-oxides, *Yakugaku Zasshi,* 87, 114, 1967; c. **Nair, M.D. and Mehta, S.R.,** Reaction of isoquinoline *N*-oxides with amines, *Indian J. Chem.,* 5, 224, 1967.

378. a. **Martens, R.J. and den Hertog, H.J.,** *N*-oxides of 2,3-and 3,4-pyridyne, *Rec. Trav. Chim. Pays-Bas,* 86, 655, 1967; b. **Peereboom, R. and dan Hertog, H.J.,** Mechanism of reactions of 3-bromopyridine 1-oxide and derivatives with potassium amide in liquid ammonia, *Rec. Trav. Chim. Pays Bas,* 93, 281, 1974; c. **Kato, T. and Niitsuma, T.,** Reactions of chloropicoline 1-oxides in liquid ammonia in the presence of potassium amide, *Heterocycles,* 1, 233, 1973. d. **Kato, T. and Niitsuma, T.,** Amination of chloro-2,6-lutidines and their *N*-oxides via a hetarine mechanism, *Chem. Pharm. Bull.,* 13, 963, 1965.

379. **Moshchitskii, S.D., Zalasskii, G.A., and Pavlenko, A.F.,** Reaction of pentachloropyridine *N*-oxide with potassium hydrosulfide, *Khim. Geterotsikl. Soedin.,* 96, 1974.

380. **Roberts, S.M. and Suschitzky, H.,** Oxidation of 2- and 4-(*N*-alkylamino)tetrachloropyridines and nucleophilic substitution of tetrachloronitropyridines, *J. Chem. Soc. (C),* 2844, 1968.

381. **Mack, A.G., Suschitzky, H., and Wakefield, B.J.,** Synthesis of the bromo- and iodo-tetrachloropyridines, *J. Chem. Soc., Perkin Trans. 1,* 1472, 1979.

382. **Bratt, J. and Suschitzky, H.,** Reactions of polyhalopyridines and their *N*-oxides with benzenethiols, with nitrite, and with trialkylphosphite, and of pentachloropyridine *N*-oxide with magnesium, *J. Chem. Soc., Perkin Trans. 1,* 1689, 1973.

383. **Ager, E., Iddon, B., and Suschitzky, H.,** Reactions of tetrachloro-4-(methylsulfonyl)pyridine and related compounds, *J. Chem. Soc., Perkin Trans. 1,* 133, 1972.

384. **Licht, H.H. and Ritter, H.,** 2,4,6-Trinitropyridine and related compounds, synthesis and characterization, *Propellants, Explos. Pyrotechn.,* 13, 25, 1988.

385. **Tortorella, V., Macieci, F., and Poma, G.,** Reactions of 2-halonitropyridine *N*-oxide with substituted benzoic acids, *Farmaco, Ed. Sci.,* 23, 236, 1968.

386. **Moerkved, E.H. and Cronyn, M.W.,** Synthesis of some derivatives of 4-(dialkylamino)-2-aminopyridine, *J. Prakt. Chem.,* 322, 343, 1980.

387. **Wright, G.C.,** 2-(2-Acetylhydrazino)-4-nitropyridine 1-oxide, U.S. Patent 4216327, 1980; *Chem. Abst.,* 93, 239251q, 1980.

388. **Talik, Z.,** Some reactions of 2-halo-4-nitropyridine *N*-oxides, *Rocz. Chim.,* 35, 475, 1961.

389. **Talik, Z.,** Substitution reactions of halogens and the nitro group in 2-halo-4-nitropyridine *N*-oxides, *Bull. Acad. Polon. Sci., Ser. Sci. Chim.,* 9, 561, 1961.

390. **Yamazaki, M., Noda, K., Onoyama, J., and Hamana, M.,** Synthesis of 2-halo-4-nitroquinoline 1-oxides and their reactions with acylating agents, *Yakugaku Zasshi,* 88, 656, 1968.

391. **Talik, T. and Talik, Z.,** *N*-oxides of fluoronitropyridines. *Bull. Acad. Pol. Sci., Ser. Sci. Chim.,* 16, 1, 1968.

392. **Araki, M., Saneyoshi, M., Harada, H., and Kawazoe, Y.,** Synthesis and reaction of 3-fluoro-4-nitroquinoline 1-oxide, *Chem. Pharm. Bull.,* 16, 1742, 1968.

393. **Ban-Oganowska, H. and Talik, T.,** Reactivities of 3-fluoro-, 3-chloro-, 3-bromo-, and 3-iodo-4-nitro-2,6-dimethylpyridine *N*-oxides, *Pol. J. Chem.,* 54, 1041, 1980.

394. **Yanai, M., Kinoshita, T., Takeda, S., and Sadaki, H.,** Reaction of 3-alkoxy-4-nitro (and 4,6-dinitro)pyridazine 1-oxides with amines, *Chem. Pharm. Bull.,* 20, 166, 1972.

395. **Novinson, T., Robins, R.K., and O'Brien, D.E.,** The synthesis and reactions of certain pyridazine 1-oxides, *J. Heterocycl. Chem.,* 10, 835, 1973.

396. **Pollak, A., Stanovnik, B., and Tisler, M.,** Oxidation products of some simple and bicyclic pyridazines, *J. Org. Chem.,* 35, 2478, 1970.

397. **Sako, S.,** The reactivity of chlorine atoms in 3- and 6-positions of 3,6-dichloropyridazine 1-oxide, *Chem. Pharm. Bull.,* 10, 956, 1962.

398. **Ochiai, M., Okada, T., Morimoto, A., and Kawakita, K.,** Nucleophilic displacement reactions of 3,6-dichlorophyridazine 1-oxide with sulfur nucleophiles, *J. Chem. Soc., Perkin Trans. 1,* 1988, 1976.

399. **Klinge, D.E. and van der Plas, H.C.,** ¹H-NMR investigations on σ-adduct formation of pyridazine, of pyridazine 1-oxide and some of its derivatives with ammonia. A new substitution mechanism, *Rec. Trav. Chim. Pays-Bas,* 94, 233, 1975.

400. **Sakamoto, T. and van der Plas, H.C.,** Reaction of 3,6-disubstituted 4-nitropyridazine 1-oxides with methanolic ammonia and liquid ammonia, *J. Heterocycl. Chem.,* 14, 790, 1977.

401. **Tamura, Y., Fujita, M., Chen, L.C., Kiyokawa, H., Ueno, K., and Kita, Y.,** Novel and facile synthetic routes to 3-amino-5-hydroxypyridine derivatives, *Heterocycles,* 15, 871, 1981.

402. a. **Caldwell, S.R. and Martin, G.E.,** The synthesis and ¹³C-NMR spectroscopy of 2-azaphenoxathiin, *J. Heterocycl. Chem.,* 17, 989, 1980; b. **Lam, W.W., Martin, G.E., Lynch, V.M., Simonsen, S.H., Lindsay, C.M., and Smith, K.,** The synthesis of 2-azathianthrene and selected analogs, *J. Heterocycl. Chem.,* 23, 785, 1986.

403. **Smith, K., Matthews, I., and Martin, G.E.,** Preparation and characterization of the 1,6-, 1,7-, 1,8-, and 1,9-diazaphenoxaselenines: an unexpected divergence between closely related sulfur and selenium systems, *J. Chem. Soc., Perkin Trans. 1,* 2839, 1987.

404. **Lindsay, C.M., Smith, K., and Martin, G.E.,** The synthesis of various diazathianthrenes and the discrimination of isomeric structures using carbon-13 NMR and lanthanide-induced shift data, *J. Heterocycl. Chem.,* 24, 1357, 1987.

405. **Lindsay, C.M., Smith, K., and Martin, G.E.,** Synthesis and carbon-13 NMR of 1,8-diazaphenoxathiin, *J. Heterocycl. Chem.,* 24, 211, 1987.

406. **El-Shafei, A.K. and El-Sayed, A.M.,** Studies on the nucleophilic reactivity of sulfur and nitrogen in 2-mercapto-4(3*H*)quinazolinone under solid-liquid phase-transfer catalysis conditions, *Rev. Roum. Chim.,* 33, 291, 1988.

407. **Gasco, A., Mortarini, V., Ruà, G., and Serafino, A.,** Methylnitrofuroxans, its structure and behaviour toward nucleophilic substitution, *J. Heterocycl. Chem.,* 10, 587, 1973.

408. **Calvino, R., Gasco, A., Menziani, E., and Serafino, A.,** Chloromethylfuroxans, *J. Heterocycl. Chem.,* 20, 783, 1983.

409. **Takahashi, S. and Kano, H.,** The reactivity of ethyl 1-methyl-2-benzimidazolecarboxylate 3-oxide and related compounds, *Chem. Pharm. Bull.,* 16, 527, 1968.

410. **Wagner, K. and Oehlmann, L.,** Synthesis and reactivity of 2-cyanobenzothiazole *N*-oxides and 2-benzothiazolecarbonitriles, *Chem. Ber.,* 109, 611, 1976.

411. **Jawdosiuk, M., Makosza, M., Molinowska, E., and Wilczynksi, W.,** Aromatic nucleophilic substitution of chlorine and nitro group in 4-chloro- and 4-nitropyridine *N*-oxides by carbanions as derived from 2-phenylalkanenitriles, *Pol. J. Chem.,* 53, 617, 1979.

412. **Takahashi, M. and Tanabe, R.,** Derivation of 9-methyl-1,2,3,4,6,7,12,12b-octahydroindolo[2,3-*a*]quinolizine from 6-methylquinoline, *Chem. Pharm. Bull.,* 15, 793, 1967.

413. **Hamana, M., Sato, F., Kimura, Y., Nishikawa, M., and Noda, H.,** Reaction of 4-chloroquinoline 1-oxide with carbanions, *Heterocycles,* 11, 371, 1978.

414. **Yamanaka, H., Ogawa, S., and Konno, S.,** Nucleophilic substitution of 4-chloropyrimidines and related compounds with carbanions, *Chem. Pharm. Bull.,* 29, 98, 1981.

415. **Matsumura, E. and Ariga, M.,** Reaction of nucleophilic reagents at the β-position of 3-bromo-4-nitropyridine *N*-oxide, *Bull. Chem. Soc. Jpn.,* 46, 3144, 1973 and 50, 237, 1977.

416. a. **Molina, P., Arques, A., Garcia, M.L., and Vinader, M.V.,** Reaction of 4-methylthioquinazoline 3-oxides with active methylene compounds: synthesis of the novel isoxazolo[2,3-*c*]quinazoline ring system, *Synth. Commun.,* 17, 1449, 1987; b. **Sato, F., Yamashita, M., Nishikawa, M., Noda, H., Saeki, S., Hamana, M.,** New alkylation of quinoline *N*-oxides, *Fukusakan Kagaku Toronkai Koen Yoshishu, 12th,* 136, 1979; *Chem. Abstr.,* 93, 71501, 1980.

417. **Suschitzky, H., Wakefield, B.J., and Whitten, J.P.,** Reactions of enamines with polyhalogenopyridines and their *N*-oxides, *J. Chem. Soc., Perkin Trans. 1,* 2709, 1980.

418. **Hartman, G.D., Hartman, R.D., and Cochran, D.W.,** Nucleophilic aromatic substitution by 3-amino-2-butenoates, *J. Org. Chem.,* 48, 4119, 1983.

419. **Binns, F. and Suschitzky, H.,** Grignard reactions on pentachloropyridine 1-oxide, *J. Chem. Soc. (C),* 1223, 1971.

420. **Kozhevnikova, A.N., Shvartsberg, M.S., and Kotlyarevkii, I.L.,** Acetylene condensation in pyridine and pyridine *N*-oxide series, *Izv. Akad. Nauk. SSSR, Ser. Khim.,* 1168, 1973.

421. **Ohta, A., Inoue, A., Ohtsuka, K., and Watanabe, T.,** Reactions of chloropyrazine *N*-oxides with trimethylaluminum, *Heterocycles,* 23, 133, 1985.

422. **Ohta, A., Ohta, M., Igarashi, Y., Saeki, K., Yuasa, K., and Mori, T.,** Ethylation of pyrazines using alkylmetals such as triethylaluminum, diethylzinc, and triethylborane, *Heterocycles,* 26, 2449, 1987.

423. **Akita, Y., Noguki, T., Sugimoto, M., and Ohta, A.,** Cross-coupling reaction of chloropyrazines with ethylenes, *J. Heterocycl. Chem.,* 23, 1481, 1986.

424. **Akito, Y., Inoue, A., and Ohta, A.,** Cross-coupling reaction of chloropyrazines with acetylenes, *Chem. Pharm. Bull.,* 34, 1447, 1986.

425. **Ohta, A., Ohta, M., and Watanabe, T.,** Coupling reaction of chloropyrazines and their *N*-oxides with tetraphenyltin, *Heterocycles,* 24, 785, 1986.

426. a. **Abe, Y., Ohsawa, A., Arai, H., and Igeta, H.,** Alkynylation of halopyridazines and their *N*-oxides using palladium-phosphine complex, *Heterocycles,* 9, 1397, 1978; b. **Ohsawa, A., Abe, Y., and Igeta, H.,** Alkynylation of halopyridazines and their *N*-oxides, *Chem. Pharm. Bull.,* 28, 3488, 1980.

427. **Akita, Y. and Ohta, A.,** Dechlorination of some chloropyrazines and their *N*-oxides, *Heterocycles,* 16, 1325, 1981.

428. **Zoltewicz, J.A., Kauffman, G.M., and Smith, C.L.,** Mechanism of base-catalyzed hydrogen exchange for *N*-methylpyridinium ion and pyridine *N*-oxide. Relative positional rates, *J. Am. Chem. Soc.,* 90, 5939, 1968.

429. **Zoltewicz, J.A. and Helmick, L.S.,** Base-catalyzed hydrogen-deuterium exchange of *N*-substituted pyridinium ions. Inductive effects and internal return, *J. Am. Chem. Soc.,* 92, 7547, 1970.

430. **Abramowitch, R.A., Singer, G.M., and Vimetha, A.R.,** Base-catalyzed deprotonation in pyridine *N*-oxides and pyridinium salts, *Chem. Commun.,* 55, 1967.

431. **Krueger, S.A. and Paudler, W.W.,** Hydrogen-deuterium exchanges in pyridine *N*-oxide, *J. Org. Chem.,* 37, 4188, 1972.

432. **Paudler, W.W. and Humphrey, S.A.,** Basicities and H-D exchange of pyrazine *N*-oxides, *J. Org. Chem.,* 35, 3467, 1970.

433. **Tupitsyn, I.F., Zatsepina, N.N., Kirova, A.V., and Kaputsin, Y.M.,** Activation parameters of base-catalyzable deuterium exchange reaction of nitrogen heterocycles and substituted benzenes, *Reakts. Sposobnost. Org. Soedin.,* 5, 636, 1968; *Chem. Abst.,* 70, 76943a, 1969; Effect of the solvent composition on the rate of hydrogen exchange of nitrogen heterocycles, *ibid.,* 5, 806, 1968; *Chem. Abst.,* 70, 76944b, 1969.

434. **Tupitsyn, I.F., Zatsepina, N.N., and Kirova, A.V.,** Quantum-chemical study of deuterium exchange of some *N*-oxides, *Reakts. Sposobnost. Org. Soedin.,* 5, 626, 1968; *Chem. Abstr.,* 70, 76942z, 1969.

435. **Zoltewicz, J.A. and Kauffman, G.M.,** Hydrogen-deuterium exchange in some halopyridine *N*-oxides: relative positional reactivities, *Tetrahedron Lett.,* 337, 1967.

436. **Grochowski, E., Krowicki, K., and Ryzinski, E.,** d₅-Pyridine, Polish patent 123059, 1984; *Chem. Abst.,* 101, 90775n, 1984.

437. **Damico, R.A.,** Pyridine *N*-oxide carbanion salts and their derivatives, U.S. patent 3773770, 1973; *Chem. Abst.,* 80, 47851v, 1973.

438. **Abramowitch, R.A., Saha, M., Smith, E.M., and Coutts, R.T.,** A new convenient alkylation and acylation of pyridine *N*-oxides, *J. Am. Chem. Soc.,* 89, 1537, 1967.

439. **Damico, R.A.,** Pyridine *N*-oxide carbanion salts, U.S. patent 3590035, 1971; *Chem. Abst.,* 75, 63616z, 1971.

440. **Abramovitch, R.A. and Knaus, E.E.,** Synthesis of 1-hydroxy-2-pyridinethiones, 1-hydroxy-2-pyridones and halopyridine 1-oxides: reactions of lithiopyridine 1-oxides, *J. Heterocycl. Chem.,* 6, 989, 1969.

441. **Abramovitch, R.A. and Knaus, E.E.,** Direct thionation and aminoalkylation of pyridine 1-oxides and related reactions, *J. Heterocycl. Chem.,* 12, 683, 1975.

442. **Abramovitch, R.A., Campbell, J., Knaus, E.E., and Silhankova, A.,** Halogenation of pyridine 1-oxides, *J. Heterocycl. Chem.,* 9, 1367, 1972.

443. **Abramovitch, R.A., Coutts, R.T., and Smith, E.M.,** Direct acylation of pyridine 1-oxides, *J. Org. Chem.,* 37, 3584, 1972.

444. **Abramovitch, R.A., Smith, E.M., Knaus, E.E., and Saha, M.,** Direct alkylation of pyridine 1-oxides, *J. Org. Chem.,* 37, 1690, 1972.

445. **Taylor, S.L., Lee, D.Y., and Martin, J.C.,** Direct, regiospecific 2-lithiation of pyridines and pyridine 1-oxides with in situ electrophilic trapping, *J. Org. Chem.,* 48, 4156, 1983.

446. a. **Hamana, M., Iwasaki, G., and Saeki, S.,** Nucleophilic substitution of 4-chloroquinoline 1-oxide and related compounds by means of hydride elimination, *Heterocycles,* 17, 177, 1982; b. **Hayashi, E. and Oishi, E.,** The reaction of 1-phenylphthalazine 3-oxide with acetophenone, *Yakugaku Zasshi,* 87, 940, 1967.

447. a. **Kobayashi, Y., Kumadaki, I., Sato, H., Yokoo, C., and Mimura, T.,** Methylation of heteroaromatic *N*-oxides, *Chem. Pharm. Bull.,* 21, 2066, 1973; b. **Hamada, Y. and Takeuchi, I.,** Reactions of benzo[*f* or *h*]quinolines and their *N*-oxides with methylsulfinyl carbanion, *J. Org. Chem.,* 42, 4209, 1977.

448. **Hamana, M., Fujimura, Y., and Haradahira, T.,** Reactions of quinoline and 4-chloroquinoline 1-oxide with phenylacetonitrile, chloromethylsulfone and methyl thiomethyl-*p*-tolylsulfone, *Heterocycles,* 25, 229, 1987.

449. **Blumenthal, J.H.,** Ethynylpyridine *N*-oxides, and ethynylquinoline *N*-oxides, U.S. patent 2874162, 1959; *Chem. Abst.,* 53, 12311b, 1959.

450. **Okamoto, T. and Takahashi, H.,** Reactions of nitroquinoline 1-oxides with potassium cyanide, *Chem. Pharm. Bull.,* 19, 1809, 1971.

451. **Richter, H.J. and Rustad, N.E.,** The reaction between 4-nitroquinoline 1-oxide and diethyl sodiomalonate, *J. Org. Chem.,* 29, 3381, 1964.

452. **Yamamori, T., Noda, H., and Hamana, M.,** Reactions of 4-nitroquinoline 1-oxide with enamines of isobutyraldehyde, *Tetrahedron,* 31, 945, 1975.

453. **Kobayashi, Y., Kumadaki, I., and Sato, H.,** Oxidative cyanation of heteroaromatic *N*-oxides, *J. Org. Chem.,* 37, 3588, 1972.

454. **Kobayashi, Y., Kumadaki, I., Hirose, Y., and Hanzawa, Y.,** Syntheses and reactions of (trifluoromethyl)indoles, *J. Org. Chem.,* 39, 1836, 1974.

455. **Kobayashi, Y., Kumadaki, I., and Sato, H.,** Oxidative cyanation of heteroaromatic amine oxides, *Chem. Pharm. Bull.,* 18, 861, 1970.

456. **Miura, K., Kasai, T., and Ueda, T.,** Chemical conversion of adenosine to guanosine, *Chem. Pharm. Bull.,* 23, 464, 1975.

457. **Tondys, H. and van der Plas, H.C.,** Amination of 4-nitropyridazine 1-oxides by liquid ammonia-potassium parmanganate, *J. Heterocycl. Chem.,* 23, 621, 1986.

458. **Rykowski, A. and van der Plas, H.C.,** Liquid ammonia-potassium permanganate, a useful reagent in the Chichibabin amination of 1,2,4-triazines, *Synthesis,* 884, 1985.

459. **Hamana, M. and Yamazaki, M.,** Alkaline ferricyanide oxidation of some aromatic *N*-oxides, *Chem. Pharm. Bull.,* 10, 51, 1962.

460. **Ochiai, E. and Ohta, A.,** Reaction of quinoline *N*-oxides with lead tetraacetate, *Sci. Papers Ist. Phys. Chem. Res., (Tokyo),* 56, 290, 1962; *Chem. Pharm. Bull.,* 10, 1260, 1962.

461. **Ohta, A.,** Reaction of the *N*-oxides of the quinoline series with lead tetraacetate. 2., *Chem. Pharm. Bull.,* 11, 1568, 1963.

462. **Coutts, R.T., Hindmarsh, K.W., and Myers, G.E.,** Quinoline *N*-oxides and hydroxamic acids with anti-bacterial properties, *Can. J. Chem.,* 48, 2392, 1970.

463. **Gutteridge, N.J.A. and McGillan, F.J.,** Oxidation of 1-pyrroline 1-oxide and the corresponding 1-hydroxy-2-pyrrolidone, *J. Chem. Soc. (C),* 641, 1970.

464. **Elsworth, J.F. and Lamchen, M.,** The oxidation of cyclic nitrones by iron (III) chloride, *J. Chem. Soc. (C),* 1477, 1966.

465. **Elsworth, J.F. and Lamchen, M.,** Kinetic and mechanistic studies on the oxidation of cyclic nitrones and cyclic hydroxylamines by iron(III)chloride, *J. S. Afr. Chem. Inst.,* 25, 1, 1972.

466. **Pennings, M.L.M. and Reinhoudt, D.N.,** Chemistry of 4-membered cyclic nitrones (2,3-dihydrozete 1-oxides): a novel one-step syntheses of *N*-acetoxy β-lactams, *Tetrahedron Lett.,* 1153, 1981.

467. **Alderson, G.W., Black, D.S.C., Clark, V.M., and Todd, A.,** Oxidative decarboxylation of 1-hydroxypyrrolid-ine-2-carboxylic acids and oxidation of some Δ¹-pyrroline 1-oxides by hypobromite, *J. Chem. Soc. Perkin Trans. 1,* 1955, 1976.

468. a. **Moriconi, E.J., Creegan, F.J., Donovan, C.K., and Spano, F.A.,** Ring expansion of 2-alkyl-1-indanones to isocarbostyril derivatives, *J. Org. Chem.,* 28, 2215, 1963; b. **Moriconi, E.J. and Spano, F.A.,** Heteropolar ozonization of aza-aromatics and their *N*-oxides, *J. Am. Chem. Soc.,* 86, 38, 1964.

469. **Picot, A., Milliet, P., Cherest, M., and Lusinchi, X.,** Action of the superoxide anion of *N*-methylbenzopyridine fluorosulfonate and benzopyridine *N*-oxides, *Tetrahedron Lett.,* 3811, 1977.

470. **Natsume, J. and Itai, T.,** Autoxidation of quinoline 1-oxide, *Chem. Pharm. Bull.,* 14, 557, 1966.

471. **Bellas, M. and Suschitzky, M.,** Fluoroisoquinoline *N*-oxides, *J. Chem. Soc.,* 4561, 1964.

472. **Endo, H., Tada, M., and Katagiri, K.,** Synthesis of 1- and 2-piperidinophenazine derivatives, *Bull. Chem. Soc. Jpn.,* 42, 502, 1969.

473. **Norris, W.P. and Osmundsen, J.,** 4,6-Dinitrobenzofuroxan, I. Covalent hydration, *J. Org. Chem.,* 30, 2407, 1965.

474. **Brown, N.E. and Keyes, R.T.,** Structure of salts of 4,6-dinitrobenzofuroxan, *J. Org. Chem.,* 30, 2452, 1965.

475. **Boulton, A.J. and Clifford, D.P.,** The potassium salt of 4,6-dinitrobenzofuroxan, and 3,4-dimethyl-4-(3,4-dimethyl-5-isoxazole)isoxazolin-5-one, *J. Chem. Soc.,* 5414, 1965.

476. **Terrier, F., Chatrousse, A.P., and Millot, F.,** Concurrent methoxide ion attack at the 5- and 7-carbons of 4-nitrobenzofurazan and 4-nitrobenzofuroxan, *J. Org. Chem.,* 45, 2666, 1980.

477. **Moir, M.E. and Norris, A.R.,** Kinetic studies of the reactions of 4-nitrobenzofuroxan with cyanide ion and isopropoxide ion in isopropanol, *Can. J. Chem.,* 58, 1691, 1980.

478. **Norris, W.P., Spear, R.J., and Read, R.W.,** Explosive Meisenheimer complexes formed by addition of nucleophilic reagents to 4,6-dinitrobenzofurazan 1-oxide, *Aust. J. Chem.,* 36, 297, 1983.

479. **Terrier, F., Sorkhabi, H.A., Millot, F., Halle, J.C., and Schaal, R.,** Water attack at an sp^2 hybridized-carbon: the hydroxyl σ-complex of 4,6-dinitrobenzofuroxan in water Me$_2$SO-tetramethylammonium chloride mixtures, *Can. J. Chem.,* 58, 1155, 1980.

480. **Terrier, F., Chatrousse, A.P., Sondais, Y., and Hlaibi, M.,** Methanol attack on highly electrophilic 4,6-dinitrobenzofurazan and 4,6-dinitrobenzofuroxan, *J. Org. Chem.,* 49, 4176, 1984.

481. **Read, R.W. and Norris, W.P.,** The nucleophilic substitution reactions of 5- and 7-chloro-4,6-dinitrobenzofurazan 1-oxide by aromatic amines, *Aust. J. Chem.,* 38, 435, 1985.

482. **Sharmin, G.P. and Mukharlamov, R.I.,** Study of the reactivity of 5-chloro-4,6-dinitrobenzofuroxan, *Zh. Org. Khim.,* 19, 2358, 1983.

483. **Buncel, E., Chaqui-Offermanns, N., Moir, R.Y., and Norris, A.R.,** Competitive demethylation and σ-complex formation in reaction of 4,6-dinitro-7-methoxybenzofuroxan with nucleophiles, *Can. J. Chem.,* 57, 494, 1979.

484. **Read, R.W., Spear, R.J., and Norris, W.P.,** Meisenheimer complex formation between 4,6-dinitrobenzofurazan 1-oxide and primary, secondary, and tertiary aromatic amines, *Aust. J. Chem.,* 37, 985, 1984.

485. **Buncel, E., Renfrow, R.A., and Strauss, M.J.,** Reactivity-selectivity relationships in reactions of ambident nucleophiles with the superelectrophiles 4,6-dinitrobenzofuroxan and 4,6-dinitro-2-(2′,4′,6′-trinitrophenyl)benzotriazole 1-oxide, *J. Org. Chem.,* 52, 488, 1987.

486. **Strauss, M.J., Renfrow, R.A., and Buncel, E.,** Ambident aniline reactivity in Meisenheimer complex formation, *J. Am. Chem. Soc.,* 105, 2473, 1983.

487. **Buncel, E., Dust, J.M., Park, K.T., Renfrow, R.A., and Strauss, M.J.,** Reactions of ambident nucleophiles with nitroaromatic electrophiles and super-electrophiles, *Adv. Chem. Ser.,* 215, 369, 1987.

488. **Buncel, E. and Park, K.T.,** σ-Complex formation and nucleophilic displacement in reactions of nitroaromatic electrophiles and super-electrophiles with ambident nucleophiles, *Stud. Org. Chem.,* 31, 247, 1987.

489. **Terrier, F., Simonnin, M.P., Pouet, M.J., and Strauss, M.J.,** Reactivity of carbon acids toward 4,6-dinitrobenzofuroxan. Studies of keto-enol equilibrium and diastereoisomerism in carbon-bonded anionic σ-complexes, *J. Org. Chem.,* 46, 3537, 1981.

490. **Terrier, F., Hallé, J.C., Pouet, M.J., and Simonnin, M.P.,** The proton sponge as nucleophile, *J. Org. Chem.,* 51, 409, 1986.

491. **Hallé, J.C., Pouet, M.J., Simonnin, M.P., and Terrier, F.,** Synthesis of some novel 7-substituted 4,6-dinitrobenzofuroxans and -benzofurazans, *Tetrahedron Lett.,* 1307, 1985.

492. **MacCormack, P., Hallé, J.C., Pouet, M.J., and Terrier, F.,** Unusual structure in Meisenheimer complex formation from the highly electrophilic 4,6-dinitrobenzofuroxan, *J. Org. Chem.,* 53, 4407, 1988.

493. **Hallé, J.C., Simonnin, M.P., Pouet, M.J., and Terrier, F.,** The π-excessive five-membered ring heteroaromatics acting as carbon nucleophiles towards an electrophilic aromatic, *Tetrahedron Lett.,* 2255, 1983.

494. **Terrier, F., Hallé, J.C., Simonnin, M.P., and Pouet, M.J.,** Ring vs. side-chain reactivity of 2,5-dimethyl five-membered ring heterocycles toward electron-deficient aromatics, *J. Org. Chem.,* 49, 4363, 1984.

495. **Tennant, G. and Yacomeni, C.W.,** Nucleophilic addition reactions of [1,2,5]-oxadiazolo[3,4-*d*]pyrimidine 1-oxides (pyrimidofuroxans), *J. Chem. Soc., Chem. Commun.,* 60, 1982.

496. **Renfrow, R.A., Strauss, M.J., Cohen, S., and Buncel, E.,** The chemical reactivity and molecular structure of 4,6-dinitro-2-(2,4,6-trimethylphenyl)benzotriazole 1-oxide, *Aust. J. Chem.,* 36, 1843, 1983.

497. a. **Newkome, G.R. and Paudler, W.W.,** *Contemporary Heterocyclic Chemistry,* Wiley, New York, 1982, p 262; b. **Buncel E., Renfrow, R.A., and Strauss, M.J.,** The ambident reactivity of aromatic amines in their reactions with N-2-(2,4,6-trinitrophenyl)-4,6-dinitrobenzotriazole 1-oxide, *Can. J. Chem.,* 61, 1690, 1983.

498. **Meisenheimer, J.,** About pyridine, quinoline and isoquinoline N-oxides, *Ber. Dtsch. Chem. Ges.,* 59, 1848, 1926.

499. **Nasielski, J., Heilporn, S., Nasielski-Hinkens, R., and Gerts-Evrard, F.,** An unexpected course of the Meisenheimer reaction: aryl phosphates in the reaction of phosphoryl chloride with 2,3-diphenylquinoxaline 1-oxide, *Tetrahedron,* 43, 4329, 1987.

500. **Abramovitch, R.A. and Smith, E.M.,** Pyridine 1-oxide, in *Pyridine and Its Derivatives,* Vol. 2, Abramovitch, R.A., ed., Interscience, New York, 1974, p 112.

501. **Duty, R.C. and Lyons, G.,** Reactions of nitrosobenzene, *o*-nitrosotoluene and pyridine *N*-oxide with phosphorus pentachloride, *J. Org. Chem.,* 37, 4119, 1972.

502. **Palat, K., Celadnik, M., Novacek, L., and Polster, M.,** Hydrazides and hydroxamic acid derived from 2-alkoxy- and 2,6-dialkoxynicotinic acids, *Ceskoslov. Farm.,* 6, 369, 1957.

503. **Suzuki, Y.,** Syntheses of 4-substituted pyridines and their homologs, *Yakugaku Zasshi,* 81, 1204, 1961.

504. **Lyle, R.E., Bristol, J.A., Kane, M.J., and Portlock, D.A.,** Synthesis of an analog of camptothecin by a general method, *J. Org. Chem.,* 38, 3268, 1973.

505. **Rokach, J. and Girard, Y.,** An unexpected and efficient synthesis of 3-chloro-4-cyanopyridine, *J. Heterocycl. Chem.,* 15, 683, 1978.

506. **Büchi, G., Manning, R.E., and Hochstein, F.A.,** Structure and synthesis of flavocarpine, *J. Am. Chem. Soc.,* 84, 3393, 1962.

507. **Kato, T.,** Reaction of α-picoline 1-oxide with phosphoryl chloride, *Yakugaku Zasshi,* 75, 1239, 1955.

508. **Kato, T.,** Reaction of 2,6-lutidine 1-oxide with phosphoryl chloride, *Yakugaku Zasshi,* 75, 1236, 1955.

509. **Gilman, H. and Edward, J.T.,** Preparation of antimalarials from pyridine, *Can. J. Chem.,* 31, 457, 1953.

510. **Hamana, M. and Yamazaki, M.,** Reaction of 2-substituted pyridine 1-oxides with acetic anhydride, *Yakugaku Zasshi,* 81, 574, 1961.

511. **Kametani, T., Nemoto, H., Takeda, H., and Takano, S.,** Synthetic approach to comptothecin, *Tetrahedron,* 26, 5753, 1970.

512. a. **Taylor, A.C. and Crovetti, A.J.,** Synthesis of some nicotinic acid derivatives, *J. Org. Chem.,* 19, 1633, 1954; b. **van Euler, H., Hasselquist, H., and Heidenberger, O.,** About *N*-oxides, *Chem. Ber.,* 92, 2266, 1959.

513. **Said, A.,** 2-chloronicotinic acid amide, *Ger. Offen* 2713346, 1977, *Chem. Abst.,* 87, 201334g, 1977.

514. **Koncewitz, J. and Skrowaczewska, Z.,** Transformation of 4-nitro-2,5-lutidine *N*-oxide, *Rocz. Chem.,* 42, 1873, 1968,

515. **Celadnik, M., Novacek, L., and Palat, K.,** Nucleus-halogenated derivatives of nicotinic acid, *Chem. Zvesti,* 21, 109, 1967.

516. **den Hertog, H.J., Maas, J., Kolder, C.R., and Combé, W.P.,** The action of acid chlorides on 4-chloro and 4-hydroxypyridine *N*-oxide and some of their derivatives, *Rec. Trav. Chim. Pays Bas,* 74, 59, 1955.

517. **den Hertog, H.J. and Hoogzand, C.,** Reactivity of 3,5-disubstituted derivatives of pyridine *N*-oxide towards sulphuryl chloride, *Rec. Trav. Chim. Pays Bas,* 76, 261, 1957.

518. **Constable, E.C. and Seddon, K.R.,** A novel rearrangement of 2,2′-pyridine *N,N′*-dioxides, *Tetrahedron,* 39, 291, 1983.

519. **Newkome, G.R., Hager, D.C., and Kiefer, G.E.,** Synthesis of halogenated terpyridines and incorporation of the terpyridine nucleus into a polyethereal macrocycle, *J. Org. Chem.,* 51, 850, 1986.

520. **Hull, R., MacBride, J.A.H., and Wright, P.M.,** Electrophilic and nucleophilic reactions and basicities of 1,6-, 1,8-, and 2,7-diazabiphenylenes, *J. Chem. Res. (S),* 328, 1984.

521. **Bachman, G.B. and Cooper, D.E.,** Quinoline derivatives form 2- and 4-chloroquinolines, *J. Org. Chem.,* 9, 302, 1944.

522. **Colonna, M.,** Action of sulfuryl chloride on quinoline *N*-oxide, *Boll. Sci. Fac. Chim. Ind. Bologna,* 2, 86, 1941.

523. **Lyle, R.E., Portlock, D.E., Kano, M.J., and Bristol, J.A.,** Benzylic halogenation of methylquinolines, *J. Org. Chem.,* 37, 3967, 1972.

524. **Hamana, M. and Shimizu, K.,** Reaction of 2-phenyl-, 4-phenyl-, and 2,4-diphenylquinoline *N*-oxides with acylating agents, *Yakugaku Zasshi,* 86, 59, 1966.

525. **Gouley, R.W., Moersch, G.W., and Mosher, H.S.,** 4,8-Diaminoquinoline and derivatives, *J. Am. Chem. Soc.,* 69, 303, 1947.

526. **Brown, E.V. and Plasz, A.C.,** The Meisenheimer reaction in the isoquinoline series, *J. Heterocycl. Chem.,* 8, 303, 1971.

527. **Kliegl, A. and Fehrle, A.,** About *N*-oxyacridon and "acridol", *Ber. Dtsch. Chem. Ges.,* 47, 1629, 1914.

528. **Mamalis, P. and Petrow, V.,** Some heterocyclic *N*-oxides, *J. Chem. Soc.,* 705, 1950.

529. **Clark, B.A.H. and Parrick, J.,** Preparation and properties of 1,7-diazaindene 7-oxide and 6,7,8,9-tetrahydro-γ-carboline 2-oxide, *J. Chem. Soc., Perkin Trans. 1,* 2270, 1974.

530. **Itoh, T., Sugawara, T., and Mizuno, Y.,** A novel synthesis of 1-deazaadenosine, *Heterocycles,* 17, 305, 1982.

531. a. **Pokorny, D.J. and Paudler, W.W.,** The Meisenheimer reaction of the 1,X-naphthyridine 1-oxides, *J. Org. Chem.,* 37, 3101, 1972; b. **Paudler, W.W. and Pokorny, D.J.,** Meisenheimer reaction of 1,5- and 1,6-naphthyridine 1-oxldes, *J. Org. Chem.,* 36, 1720, 1971.

532. a. **Pokorny, D.J. and Paudler, W.W.,** Meisenheimer reaction of 1,6-naphthyridine 1,6-dioxide, *J. Heterocycl. Chem.,* 9, 1151, 1972; b. **Koboyashi, Y., Kumadaki, I., and Sato, H.,** Synthesis and some reactions of 1,6-naphthyridine *N*-oxides, *Chem. Pharm. Bull.,* 17, 1045, 1969.

533. **Newkome, G.R. and Garbis, S.J.,** Authenticity of 2,6-dichloro-1,5-naphthyridine, *J. Heterocycl. Chem.,* 15, 685, 1978.

534. **Kloc, K. and Mlochowski, J.,** Syntheses of 7-substituted 1,8 phenanthrolines and their *N*-oxides, *Pol. J. Chem.,* 54, 917, 1980.

535. a. **van den Haak, H.J.W., Bouw, J.P., and van der Plas, H.C.,** Ring transformation of 8-bromo-1,7-phenanthroline in reaction with potassium amide/liquid ammonia, *J. Heterocycl. Chem.,* 20, 447, 1983; b. **Hamada, Y. and Takeuchi, I.,** Chemical reactivity of 4,6-phenantholine, *Chem. Pharm. Bull.,* 24, 2769, 1976.

536. **Ogata, M.,** Synthesis of cyanopyridazines, *Chem. Pharm. Bull.,* 11, 1522, 1963.

537. **Igeta, H.,** 3-Methoxy- and 3-hydroxy-pyridazine 1-oxides, *Chem. Pharm. Bull.,* 7, 938, 1959.

538. **Sako, S.,** 4- or 5-Chloro-3,6-dimethylpyridazine 1-oxide, *Chem. Pharm. Bull.,* 11, 337, 1963.

539. **Igeta, H.,** Reaction of 3,6-dimethoxypyridazine 1-oxide with phosphoryl chloride, *Chem. Pharm. Bull.,* 8, 368, 1960.

540. **Tikhonov, A.Y., Sedova, V.F., Valodarski, L.B., and Mamaev, V.P.,** Interaction of substituted 4-phenylpyrimidine 1,3-dioxides with phosphorus oxychloride, *Khim. Geterotsikl. Soedin,* 526, 1981.

541. **Klein, B., Hetman, N.E., and O'Donnel, M.E.,** The action of phosphoryl chloride on pyrazine *N*-oxides, *J. Org. Chem.,* 28, 1682, 1963.

542. **Bernardi, L., Palamidessi, G., Leone, A., and Larini, G.,** Pyrazine *N*-oxides and their reaction with phosphorous oxychloride, *Gazz. Chim. It.,* 91, 1431, 1961.

543. a. **Sato, N. and Kobayashi, M.,** A novel formation of 2-aminopyrazine from the reaction of pyrazine 1-oxide with phosporyl chloride, *Bull. Chem. Soc. Jpn.,* 57, 3015, 1984; b. **Okada, S., Kosasayama, A., Konno, T., and Uchimaru, F.,** Synthesis, reactions, and spectra of pyrazine *N*-oxide derivatives, *Chem. Pharm. Bull.,* 19, 1344, 1971.

544. a. **Ohta, A., Watanabe, T., Akita, Y., Yoshida, M., Toda, S., Akamatsu, T., Ohno, H., and Suzuki, A.,** Some reactions of mono substituted pyrazine monoxides, *J. Heterocycl. Chem.,* 19, 1061, 1982; b. **Inoue, M., Abe, R., Tamamura, H., Ohta, M., Asami, K., Kitani, H., Kamei, H, Nakamura, Y., Watanabe, T., and Ohta, A.,** Reaction of 2,5-diisopropyl- and 2,5-di-sec-butylpyrazine 1-oxide derivatives with phosphorylchloride and acetic anhydride, *J. Heterocycl. Chem.,* 22, 1291, 1985.

545. **Ohta, A., Masano, S., Iwakura, S., Tamura, A., Watahiki, H., Tsutsui, M., Akita, Y., Watanabe, T., and Kurihara, T.,** Syntheses and reactions of some 2,3-disubstituted pyrazine monoxides, *J. Heterocycl. Chem.,* 19, 465, 1982.

546. **Sato, N.,** Substitution effect on reaction of pyrazine *N*-oxides with phosphoryl chloride, *J. Chem. Res. (S),* 318, 1984.

547. **Palamidessi, G., Vigevani, A., and Zarini, F.,** Reaction of phosphorus oxychloride with pyrazinecarboxamide 4-oxide, *J. Heterocylc. Chem.,* 11, 607, 1974.

548. **Sato, N.,** Chlorination of 1-hydroxy-2(1*H*)-pyrazinones with phosphoryl chloride. Formation of 2,5-dichloro-3-phenylpyrazine from 1-hydroxy-3-phenyl-2(1*H*)-pyrazinone, *J. Heterocycl. Chem.,* 23, 149, 1986.

549. **Cheeseman, G.W.H. and Godwin, R.A.,** 2,6-Dihydroxy-3,5-diphenylpyrazine and related compounds, *J. Chem. Soc. (C),* 2977, 1971.

550. a. **Taylor, E.C., Abdulla, K.F., Tanaka, K., and Jacobi, P.A.,** Syntheses of xanthopterin and isoxanthopterin, *J. Org. Chem.,* 40, 2341, 1975; b. **Taylor, E.C. and Reiter, L.A.,** Unequivocal total synthesis of deoxyurothione, *J. Org. Chem.,* 47, 528, 1982.

551. **Neilsen, J.B., Broadbent, H.S., and Hennen, W.J.,** Unequivocal syntheses of 6-methyl and 6-phenylisoxanthopterin, *J. Heterocycl. Chem.,* 24, 1621, 1987.

552. **Ohta, A. and Fujii, S.,** Synthesis of DL-aspergillic acid and DL-deoxyaspergillic acid, *Chem. Pharm. Bull.,* 17, 851, 1969.

553. **Ohta, A., Masano, S., Tsutsui, M., Yamamoto, F., Suzuki, S., Makita, H., Tamamura, H., and Akita, Y.,** Syntheses of some 2-hydroxypyrazine 1-oxides, *J. Heterocycl. Chem.,* 18, 555, 1981.

554. **Ohta, A., Akita, Y., and Nakano, Y.,** Conversion of 2,5-diphenyl- and 2,5-dibenzylpyrazines to 2,5-diketopiperazines, *Chem. Pharm. Bull.,* 27, 2980, 1979.

555. **Ohta, A., Akita, Y., and Hara, M.,** Syntheses and reactions of some 2,5-disubstituted pyrazine monoxides, *Chem. Pharm. Bull.,* 27, 2027, 1979.

556. **Landquist, J.K.,** The oxidation of quinoxaline and its *Bz*-substituted derivatives, *J. Chem. Soc.,* 2816, 1953.

557. **Ahmad, Y., Habib, M.S., Qureshi, M.I., and Farooqi, M.A.,** Reaction of quinoxaline 1,4-dioxide and some of its derivatives with acetyl chloride, *J. Org. Chem.,* 38, 2176, 1973.

558. **Ahmad, Y., Habib, M.S., Ziauddin, and Bashir, N.,** An unusual chlorine substitution during the reaction of 3-hydroxy-2-phenylquinoxaline 1-oxide with acetyl chloride, *Bull. Chem. Soc. Jpn.,* 38, 1654, 1965.

559. **Nasielski-Hinkens, R., Vande Vyver, E., and Nasielski, J.,** Regioselectivity in the reaction of nitroquinoxaline *N*-oxides with phosporyl chloride, *Bull. Soc. Chim. Belg.,* 95, 663, 1986.

560. a. **Postovskii, I.Y. and Abramova, E.I.,** Synthesis of some *N*-oxides of phenazine derivatives, *Zh. Obshch. Khim.,* 24, 485, 1954; b. **Hayashi, E. and Oishi, E.,** *N*-oxidation of 1-phenylphthalazine and chemical properties of 1-phenylphthalazine 3-oxide, *Yakugaku Zasshi,* 86, 576, 1966; c. **Higashino, T.,** Reaction of 4-alkoxyquinazoline 1-oxide with nucleophilic reagents, *Yakugaku Zasshi,* 79, 699, 1959.

561. **Kawashima, H. and Kumashiro, I.,** Reaction of hypoxanthine 1-*N*-oxide and 2′,3′,5′-tri-*O*-acetylinosine l-*N*-oxide with phosphoryl chloride, *Bull. Chem. Soc. Jpn.,* 40, 639, 1967.

562. **Steppan, H., Hammier, J., Bauer, R., Gottlieb, R., and Pfleiderer, W.,** Synthesis and properties of 6- and 7-amino-1,3-dimethylumazines, *Liebigs Ann. Chem.,* 2135, 1982.

563. **Goya, P. and Pfleiderer, W.,** Synthesis, properties, and reactivity of lumazine 5-oxide ribosides, *Chem. Ber.,* 114, 707, 1981.

564. **Nishigaki, S., Ichiba, M., and Senga, K.,** Syntheses of azolopyrimido[5,4-*e*]-*as*-triazines and azolopyrimido[4,5-*c*]-pyridazines related to fervenulin, *J. Org. Chem.,* 48, 1628, 1983.

565. **Yoneda, F., Shinozuke, K., Hiromatsu, K., Matsushita, R., Sakume, Y., and Hamana, M.,** A new synthetic approach to 8-chloroflavins and their conversion into 8-(substituted-amino)flavins, *Chem. Pharm. Bull.,* 28, 3576, 1980.

566. a. **Katada, M.,** Reaction between pyridine 1-oxide and acid anhydrides, *Yakugaku Zasshi,* 67, 51, 1947; b. **Tanida, H.,** Nucleophilic substitution on the β-position of quinoline 1-oxide, *Yakugaku Zasshi,* 78, 1083, 1958.

567. **Traynelis, V.J.,** Rearrangement of *O*-acylated heterocyclic *N*-oxides, in *Mechanism of Molecular Migrations,* Vol. 2, Thyagarjan, B.S., Ed., Interscience, New York, 1969, p. 1.

568. **Markgraf, J.H., Brown, H.B., Mohr, S.C., and Peterson, R.G.,** The rearrangement of pyridine *N*-oxide with acetic anhydride: kinetics and mechanism, *J. Am. Chem. Soc.,* 85, 958, 1963.

569. **Oae, S. and Kozuka, S.,** The mechanism of the reaction of pyridine *N*-oxide with acetic anhydride, *Tetrahedron,* 21, 1971, 1965.

570. **Cohen, T. and Deets, G.L.,** The reaction of pyridine *N*-oxide with acetic anhydride in anisole and in benzonitrile, *J. Org. Chem.,* 37, 55, 1972.

571. **Klaebe, A. and Lattes, A.,** Chromatographic study of the reaction between acetic anhydride and pyridine *N*-oxide, *J. Chromatog.,* 27, 502, 1967.

572. a. **Moran, D.B., Morton, G.O., and Albright, J.D.,** Synthesis of (pyridinyl)-1,2,4-triazolo[4,3-*a*]pyridines, *J. Heterocycl. Chem.,* 23, 1071, 1986; b. **Koenig, T.W and Wieczorek, J.,** Pyridine *N*-oxide trichloroacetic anhydride reaction, *J. Org. Chem.,* 35, 508, 1970.

573. **Bublitz, D.E.,** Nucleophilic displacement of polyhalogenated heterocyclic *N*-methoxy methylsulfates, *J. Heterocycl. Chem.,* 9, 471, 1972.

574. **Boekelheide, V. and Lehn, W.L.,** The rearrangement of substituted pyridine *N*-oxides with acetic anhydride, *J. Org. Chem.,* 26, 428, 1961.

575. **Nagano, H. and Hamana, M.,** Reactions of 2-acylpyridine 1-oxides with acetic anhydride, *Heterocycles,* 26, 3249, 1987.

576. **Bain, B.M., and Saxton, J.E.,** Rearrangement of nicotinic acid *N*-oxide and 3-picoline *N*-oxide, *Chem. Ind.,* 402, 1960.

577. **Oae, S. and Kozuka, S.,** The mechanism of the reaction of 3-picoline *N*-oxide with acetic anhydride, *Tetrahedron,* 20, 269, 1964.

578. **Cava, M.P. and Weinstein, B.,** The rearrangement of some simple 3-halopyridine-*N*-oxides, *J. Org. Chem.,* 23, 1616, 1958.

579. **Carrol, F.I., Berrang, B.D., and Linn, C.P.,** Synthesis of naphthyridinone derivatives as potential antimalarials, *J. Heterocycl. Chem.,* 18, 941, 1981.

580. **Sakata, Y., Adachi, K., Akahori, Y., and Hayashi, E.,** On 2-(trimethylsilyl)pyridine 1-oxide, *Yakugaku Zasshi,* 87, 1374, 1967.

581. **Geiger, R.E., Lalonde, M., Stoller, H., and Schleich, K.,** Cobalt — catalyzed cycloaddition of alkynes and nitriles to pyridines: a new route to pyridoxine, *Helv. Chim. Acta,* 67, 1274, 1984.

582. a. **Ochiai, E. and Okamoto, T.,** Reaction of acetic anhydride on quinoline 1-oxide, *Yakugaku Zasshi,* 68, 88, 1948; b. **Montanari, F. and Risaliti, A.,** Aromatic *N*-oxides: reaction with acetic anhydride, *Gazz. Chim. It.,* 83, 278, 1953.

583. **Daeniker, H.U. and Druey, J.,** Rearrangement of heterocyclic *N*-oxides with trifluoroacetic anhidride, *Helv. Chim. Acta,* 41, 2148, 1958.

584. **Hamana, M. and Funagoshi, K.,** Reaction of quinoline 1-oxide with benzoyl chloride, *Yakugaku Zasshi,* 80, 1031, 1960.

585. **Yoshikawa, T.,** Synthesis of pyrido[2,3-*g*]quinoline derivatives, *Yakugaku Zasshi,* 81, 1601, 1961.

586. **Ochiai, E. and Ikehara, M.,** A rearrangement of isoquinoline *N*-oxide by tosyl chloride, *Pharm. Bull.,* 3, 454, 1955.

587. **Robinson, M.L. and Robinson, B.L.,** The rearrangement of isoquinoline *N*-oxides, *J. Org. Chem.,* 21, 1337, 1956.

588. **Dorgan, R.J.J., Parrick, J., and Hardy, C.R.,** Some reactions of 1-benzyl-1*H*-pyrazolo[3,4-*b*]pyridine and its 7-oxide, *J. Chem. Soc. Perkin Trans. 1,* 938, 1980.

589. **Igeta, H.,** 3-Methoxy- and 3-hydroxypyridazine 1-oxide, *Chem. Pharm. Bull.,* 7, 938, 1959.

590. **Yanai, M., Kinoshita, T., Takeda, S., Sadaki, H., and Watanabe, H.,** Synthesis of 4,5-diaminopyridazine derivatives, *Chem. Pharm. Bull.,* 18, 1680, 1970.

591. **Bredereck, M., Gompper, R., and Herlinger, H.,** Preparation, properties and reactions of pyrimidine, *Chem. Ber.,* 91, 2832, 1958.

592. **Elina, A.S.,** Quinoxaline *N*-oxides. XV. Mono- and di-*N*-oxides in the reactions of oxidation and interaction with acetic anhydride, *Khim. Geterotsikl. Soedin.,* 545, 1968.

593. **Ahmed, Y., Qureshi, M.I., Habib, M.S., and Farooqi, M.A.,** The reactions of quinoxaline 1,4-dioxides with acetic anydride, *Bull. Chem. Soc. Jpn.,* 60, 1145, 1987.

594. a. **Parham, J.C., Templeton, M.A., and Teller, M.N.,** Redox and rearrangement of 1,7-dimethylguanine 3-oxide with anhydrides, *J. Org. Chem.,* 43, 2325, 1978; b. **Woelcke, U., Pfleiderer, W., Delia, T.J., and Brown, G.B.,** Rearrangement of purine 3-*N*-oxides on acylation and methylation, *J. Org. Chem.,* 34, 981, 1969.

595. a. **Taylor, E.C. and Jacobi, P.A.,** Facile synthesis of xanthopterin, *J. Am. Chem. Soc.,* 95, 4455, 1973; b. **Perez-Rubalcaba, A. and Pfleiderer, W.,** Reactivity of lumazine 5- and 8-oxides and 5,8-dioxides, *Liebigs Ann. Chem.,* 852, 1983.

596. **Parnell, E.W.,** 1-Methylpyrazole 2-oxide, *Tetrahedron Lett.,* 3941, 1970.

597. a. **Matsuda, T., Honjo, B., Yamazaki, M., and Goto, Y.,** Reaction of 2,5-diarylthiazole 3-oxides with acetic anhydride, *Chem. Pharm. Bull.,* 25, 3270, 1977; b. **Volodarskii, L.B. and Kobrin, V.S.,** Rearrangement of 2,4,4-trimethyl-5-phenyl-4*H*-imidazole 1-oxide to 2,4,4-trimethyl-1-phenyl-2-imidazolin-5-one, *Khim. Getero-sikl Soedin,* 1423, 1976; c. **Doepp, D. and nour-el-Din, A.M.,** Indolenine oxides. VI. Catalysis of an isomerization by ethylenetetracarbonile and oxiranetetracarbonile, *Chem. Ber.,* 111, 3952, 1978.

598. a. **Castagnoli, N. and Sadee, W.,** Mechanism of the reaction of a 1,4-benzodiazepine *N*-oxide with acetic anhydride, *J. Med. Chim.,* 15, 1076, 1972. b. **Gatta, F., Del Giudice, M.R., Di Simone, L., Settimj, G.,** Synthesis of 2-amino-3-hydroxy- and of 2,3-diamino-3*H*-1,4-benzodiazepines, *J. Heterocycl. Chem.,* 17, 865, 1980; c. **Schlager, L.H.,** 3-Di-*n*-propylacetoxybenzodiazepin-2-ones, *Austrian Pat.* 358051, 1980, *Chem. Abst.* 94, 139854r, 1981.

599. **Barton, D.H.R., Gutteridge, N.J.A., Hesse, R.H., and Pechet, M.M.,** Some reactions of 2, 4, 4-trimethyl-1-pyrroline 1-oxide, *J. Org. Chem.,* 34, 1473, 1969.

600. **Gutteridge, N.J.A. and Dales, J.R.M.,** Schotten-Bauman benzoylation of 2,4,4-trimethyl-1-pyrroline 1-oxide: reinvestigation, *J. Chem. Soc. (C),* 122, 1971.

601. **Popp, F.D.,** Reissert compounds, *Adv. Heterocycl. Chem.,* 9, 1, 1968.

602. **Reissert, A.,** The introduction of a benzoyl group in cyclic tertiary bases, *Ber. Dtsch. Chem. Ges.,* 38, 1603, 1905.

603. **Henze, M.,** Benzoylation of quinoline *N*-oxide, *Ber. Dtsch Chem. Ges.,* 69, 1566, 1936.

604. **Kaneko, C.,** A modified Reissert reaction in the quinoline 1-oxide series, *Chem. Pharm. Bull.,* 8, 286, 1960.

605. **Ochiai, E. and Nakayano, I.,** Reissert reaction of quinoline and pyridine 1-oxides, *Yakugaku Zasshi,* 65A, 7, 1945.

606. **Nakayane, I.,** Syntheses of 4-substituted quinaldamides, *Yakugaku Zasshi,* 70, 355, 1950.

607. **Georgian, V., Harrisson, R.J., and Skaletzky, L.L.,** The synthesis of 5-oxoperhydroisoquinoline, *J. Org. Chem.,* 27, 4571, 1962.

608. **Hamana, M. and Shimizu, K.,** Reactions of 2-phenyl-, 4-phenyl-, and 2,4-diphenylquinoline *N*-oxides with acylacing agents, *Yakugaku Zasshi,* 86, 59, 1966.

609. **Hamana, M., Takeo, S., and Noda, M.,** Some reactions of 2-phenylquinoline 1-oxide and its derivatives, *Chem. Pharm. Bull.,* 25, 1256, 1977.

610. **Hayashi, E. and Miyashita, A.,** The reaction of 2,4-disubstituted quinoline 1-oxides with benzoylchloride and potassium cyanide, *Yakugaku Zasshi,* 97, 211, 1977.

611. **Hayashi, E. and Miyashita, A.,** The reaction of 2-substituted and 2,3-disubstituted quinoline 1-oxides with aroyl chloride and potassium cyanide, *Yakugaku Zasshi,* 96, 968, 1976.

612. **Hayashi, E. and Miyashita, A.,** On the reaction of 1-substituted and 1,4-disubstituted isoquinoline 2-oxides with aroyl chloride and potassium cyanide, *Yakugaku Zasshi,* 97, 1334, 1977.

613. **Colonna, M. and Fatutta, S.,** Nucleophilic reactions of angular benzoquinoline *N*-oxides, *Gazz. Chim. It.,* 83, 622, 1953.

614. **Kobayashi, Y., Kumadaki, I., and Sato, H.,** Cyanation of 1,6-naphthyridine *N*-oxides, *Chem. Pharm. Bull.,* 17, 2614, 1969.

615. **Hamada, Y., Morishita, K., and Hirota, M.,** The Reissert reaction of 4,6-phenanthroline, *Chem. Pharm. Bull.,* 26, 350, 1978.

616. **Kloc, K. and Mlochowski, J.,** Reactivity of phenanthrolines. IV. Reissert reaction of *N*-oxides, *Rocz. Chem.,* 49, 1621, 1975.

617. **Klemm, L.H. and Muchiri, D.R.,** The Reissert-Henze reaction as a route to a simple *C*-substituents alpha to the heteronitrogen atom, *J. Heterocycl. Chem.,* 20, 213, 1983.

618. **Hayashi, E., Higashino, T., and Shimada, N.,** The reaction of 1-phenyl-1*H*-pyrazolo[3,4-*d*]pyridine 5-oxide with methanosulfonyl chloride and potassium cyanide, *Yakugaku Zasshi,* 98, 234, 1978.

619. **Kobayashi, Y. and Kumadaki, I.,** Preparation and reaction of *N*-oxides of trifluoromethylated pyridines, *Chem. Pharm. Bull.,* 17, 510, 1969.

620. **Ochiai, E. and Yamanaka, H.,** Amine oxides of the pyrimidine series, *Pharm. Bull.,* 3, 173, 1955.

621. **Yamanaka, H.,** 4-Alkoxy-6-methylpyrimidine *N*-oxides, *Chem. Pharm. Bull.,* 6, 633, 1958.

622. **Brown, R., Joseph, M., Leigh, T., and Swain, M.L.,** Synthesis and reactions of 7,8-dihydro-8-methylpterin and 9-methylguanine 7-oxides, *J. Chem. Soc. Perkin Trans. 1,* 1003, 1977.

623. **Corey, E.J., Borror, A.L., and Foglia, T.,** Transformation in the 1,10-phenanthroline series, *J. Org. Chem.,* 30, 288, 1965.

624. **Tani, H.,** Reactions of *N*-alkoxypyridinium derivatives, *Yakugaku Zasshi,* 80, 1418, 1960; 81, 141, 1961; 81, 182, 1961.

625. **Ferles, M. and Jankovsky, M.,** Reaction of 1-alkoxypyridinium and 1-alkoxy-3-alkylpyridinium halides with potassium cyanide, *Coll. Czech. Chem. Commun.,* 33, 3848, 1968.

626. **Jankovsky, M. and Ferles, M.,** On the reaction of 1-alkoxy-3-isobutylpyridinium salts with potassium cyanide, *Coll. Czech. Chem. Commun.,* 35, 2797, 1970.

627. **Matsumura, E., Ariga, M., and Ohfuji, T.,** Reissert-Kaufmann-type reaction of 4-nitropyridine *N*-oxide and of its homologs, *Bull. Chem. Soc. Jpn.,* 43, 3210, 1970.

628. **Okamoto, J. and Tani, H.,** Synthesis of 2- and 4-cyanopyridines, *Chem. Pharm. Bull.,* 7, 130, 1959.

629. **Villani, F.J., Daniels, P.J.L., Ellis, C.A., Mann, T.A., and Wang, K.C.,** Derivatives of 10,11-dihydro-5*H*-dibenzo[*a,d*]cycloheptene and related compounds. III. Azaketones, *J. Heterocycl. Chem.,* 8, 73, 1971.

630. **Warawa, E.J.,** Synthesis of 3-alkyl-3,6-dicyanopyridines by a unique rearrangement. Preparation of fusaric acid analog, *J. Org. Chem.,* 40, 2092, 1975.

631. **Rao, A.V.R., Chavan, S.P., and Sivadasan, L.,** Synthesis of lavendamycin, *Tetrahedron,* 42, 5065, 1986.

632. **Nakagome, T., Castle, R.N., and Murakami, H.,** Synthesis of pyrimido[4,5-*c*]pyridazines and pyrimido[5,4-*c*]pyridazines, *J. Heterocycl. Chem.,* 5, 523, 1968.

633. **Takahashi, S. and Kano, H.,** Reaction of 3-methoxy-1-methyl- and 1,2-dimethylbenzimidazolium iodide with various nucleophiles, *Chem. Pharm. Bull.,* 14, 375, 1966.

634. **Fife, W.K.,** Regioselective cyanation of pyridine 1-oxides with trimethylsilane carbonitrile: a modified Reissert-Henze reactions, *J. Org. Chem.,* 48, 1375, 1983.

635. **Vorbrüggen, H. and Krolikiewicz, K.,** A simple one-step conversion of aromatic heterocyclic *N*-oxides to α-cyano aromatic *N*-heterocycles, *Synthesis,* 316, 1983.

636. **Harusawa, S., Hamada, Y., and Shioiri, T.,** Reaction of diethyl phosphorocyanidate (DEPC) with aromatic amine oxides. A modified Reissert-Henze reaction, *Heterocycles,* 15, 981, 1981.

637. **Fife, W.K.,** Regioselective cyanation of 3-substituted pyridine 1-oxides, *Heterocycles,* 22, 93, 1984.

638. **Sakamoto, T., Kaneda, S., Nishimuira, S., and Yamanaka, H.,** Site-selectivity in the cyanation of 3-substituted pyridine 1-oxides with trimethylsilanecarbonitrile, *Chem. Pharm. Bull.,* 33, 565, 1985.

639. **Hilpert, H.,** Synthesis of 3-(2-carboxy-4-pyridyl)- and 3-(6-carboxy-3-pyridyl)-DL-alanine, *Helv. Chim. Acta,* 70, 1307, 1987.

640. **Yamanaka, H., Nishimura, S., Kaneda, S., and Sakamoto, T.,** Synthesis of pyrimidinecarbonitriles by reaction of pyrimidine *N*-oxides with trimethylsilyl cyanide, *Synthesis,* 681, 1984.

641. **Yamanaka, H., Sakamoto, T., Nishimura, S., and Sagi, M.,** Site-selectivity in the reaction of 5-substituted and 4,5-disubstituted pyrimidine *N*-oxides with trimethylsilyl cyanide, *Chem. Pharm. Bull.,* 35, 3119, 1987.

642. **Gribble, G.W., Barden, T.C., and Johnson, D.A.,** Synthesis of sempervirine, *Tetrahedron,* 44, 3195, 1988.

643. a. **Hayashi, E. and Shimada, N.,** The reaction of quinolinecarbonitrile *N*-oxides with cyanide ion in dimethyl sulfoxide, *Yakugaku Zasshi,* 97, 123, 1977; b. **Hayashi, E., Makino, H., and Higashino, T.,** Reaction of 4-isoquinolinecarbonitrile and its 2-oxide with cyanide ions in dimethyl sulfoxide, *Yakugaku Zasshi,* 94, 1041, 1974; c. **Pietra, S. and Casiraghi, G.,** Nucleophilic reactions with 2-nitrophenazine l0-oxide. II., *Gazz. Chim. It.,* 97, 1817, 1967.

644. a. **Koyama, T., Namba, T., Hirota, T., Ohimori, S., and Yamato, M.,** Reactions of pyridine *N*-oxides with formamide, *Chem. Pharm. Bull.,* 25, 964, 1977; b. **Hirota, T., Namba, T., and Sasaki, K.,** Reaction of pyridine *N*-oxides and pyrazine di *N*-oxides with formamide, *J. Heterocycl. Chem.,* 24, 949, 1987; c. **Hirota, T., Namba, T., and Sasaki, K.,** Reaction of pyridine and quinoline *N*-oxides with *N*-methylformamide, *Chem. Pharm. Bull.,* 34, 3431, 1986; d. **Sliwa, H. and Raharimanana, C.,** Evidence for a PARC-ANRO mechanism in heterocyclic ring conversion of functionalized *N*-alkoxypyridinium salts, *Tetrahedron Lett.,* 349, 1986; e. **Sliwa, H. and Ouattara, L.,** A new access to 2-benzoylquinoline and 1-benzoylisoquinoline using a novel mode of base-induced decomposition of *N*-alkoxypyridinium salts, *Heterocycles,* 26, 3065, 1987.

645. a. **Pietra, S. and Argentini, M.,** Reactions of sulfonylpyrazines, *Ann. Chim. (Rome),* 61, 290, 1971; b. **Pietra, S. and Casiraghi, G.,** Nucleophilic reactions on 2-nitrophenazine l0-oxide. I and III., *Gazz. Chim. Ital.,* 96, 1630, 1966 and 97, 1826, 1967; c. **Pietra, S., Casiraghi, G., and Selva, A.,** Nucleophilic reactions on 2-nitrophenazine l0-oxide. IV. Reaction with ketones, *Ann. Chim. (Rome),* 58, 1380, 1968.

646. **Hamana, M.,** Some advances in the reaction of aromatic *N*-oxides, *J. Heterocycl. Chem.,* 9, S51, 1972.

647. a. **Bruni, P. and Guerra, G.,** Enolizable cyclic ketones. I. Reaction with activated heteroaromatic *N*-oxides, *Ann. Chim. (Rome),* 57, 688, 1967; b. **Hamana, M. and Yamazaki, M.,** Reactions of quinoline 1-oxide with acetyl cyanide, *Chem. Pharm. Bull.,* 11, 411, 1963.

648. **Okamoto, T. and Takayama, H.,** The reaction of *N*-methoxyquinolinium salts with ketones, *Chem. Pharm. Bull.,* 11, 514, 1963.

649. **Hamana, M. and Yamazaki, M.,** Reactions of quinoline 1-oxides with compounds containing reactive hydrogens in the presence of acetic anhydride, *Chem. Pharm. Bull.,* 11, 415, 1963.

650. **Yousif, M.M., Saeki, S., and Hamana, M.,** Reactions of aromatic N-oxides with 1,3-cyclohexanediones in the presence of acetic anhydride, *Chem. Pharm. Bull.*, 30, 2326, 1982.

651. **Douglass, J.E. and Fortner, H.D.,** Reactions of quinoline 1-oxide with 1,3-diketones, *J. Heterocycl. Chem.*, 10, 115, 1973.

652. **Iwao, M. and Kuraishi, T.,** The facile synthesis of some N-heteroarylacetic esters, *J. Heterocycl. Chem.*, 15, 1425, 1978.

653. **Pollak, A., Stanovnik, B., Tisler, M., and Venetic-Fortuna, J.,** Reactions of some halogenated heterocycles and N-oxides with reactive methylene compounds, *Monatsh. Chem.*, 106, 473, 1975.

654. **Stein, M.L., Manna, F., and Lombardi, C.C.,** The reaction of 3-hydroxypyridine N-oxide with active hydrogen compounds and the synthesis of 3-substituted 2-aminofuro[3.2-b]pyridines, *J. Heterocycl. Chem.*, 15, 1411, 1978.

655. **Sedova, V.F. and Mamaev, V.P.,** Some transformations of pyrimidine di-N-oxides, *Khim. Geterotsikl. Soedin.*, 827, 1979.

656. **Zakhs, E.R., El'tsov, A.V., and Lyashenko, E.V.,** Synthesis and properties of nitroarylmethyl derivatives of quinoline, *Zh. Org. Khim.*, 14, 1992, 1978.

657. **Yamazaki, M., Noda, K., Honjo, N., and Hamana, M.,** Reactions of aromatic N-oxides with N-acylmethylpyridinium salts in the presence of acylating agents, *Chem Pharm. Bull.*, 21, 712, 1973.

658. **Yamazaki, M., Noda, K., and Hamana, M.,** Reactions of quinoline 1-oxide with N-acylmethylpyridinium salts in the presence of acylating agents, *Chem. Pharm. Bull.*, 18, 901, 1970.

659. **Ognyanov, V., Haimova, M., and Mollov, N.,** Synthesis of 8H-dibenzo[a,g]quinolizin-8-ones from homophtalic anhydrides and 1-chloroisoquinolines or isoquinoline N-oxides. A new synthesis of caseadine, *Heterocycles*, 19, 1069, 1982.

660. **Uff, B.C., Al-Kolla, A., Adamali, K.E., and Harutunian, V.,** Formation of cyanohydrin carbonates of aromatic aldehydes and aryl heteroaryl ketones, *Synth. Commun.*, 8, 163, 1978.

661. **Endo, T., Saeki, S., and Hamana, M.,** Reactions of aromatic N-oxides with O-benzoyl aromatic aldehyde cyanohydrins in acetic anhydride, *Heterocycles*, 3, 19, 1975.

662. **Yousif, M.M., Saeki, S., and Hamana, M.,** Reactions of aromatic N-oxides with 2-substituted 2-oxazolin-5-ones in the presence of acetic anhydride, *J. Heterocycl. Chem.*, 17, 1029, 1980.

663. **Yousif, M.M., Saeki, S., and Hamana, M.,** Reactions of aromatic N-oxides with 3-aryl rhodanines in the presence of acetic anhydride, *J. Heterocycl. Chem.*, 17, 305, 1980.

664. a. **Henning, H.G., Gelbin, A., and Schoeder, H.,** Alternative routes to the synthesis of 2-furanoyl- and 2-pyrrolonyl-substituted quinolines, *Pharmazie.*, 43, 45, 1988; b. **Hamana, M. and Kumadaki, I.,** Reactions of aromatic N-oxides with oxindoles in the presence of acylating agents, *Chem. Pharm. Bull.*, 18, 1822, 1970.

665. a. **Sone, M., Tominaga, Y., Natsuki, R., Matsuda, Y., and Kobayashi, G.,** Syntheses and reactions of 4-methylthio-2-buten-4-olide derivatives, *Chem. Pharm. Bull.*, 22, 617, 1974; b. **Baty J.D., Jones, G., and Moore, C.,** 9-Azasteroids. III. The synthesis of some 2-cyclopentylquinolines as models for rings A, B, and D, *J. Org. Chem.*, 34, 3295, 1969.

666. **Klemm, L.H., Muchiri, D.R., and Louris, J.N.,** Direct introduction of C-substituents *gamma* to the heteronitrogen atom in the thieno[2,3-b]pyridine system, *J. Heterocycl. Chem.*, 21, 1135, 1984.

667. **Klemm, L.H., Lu J.J., Greene, D.S., and Boisvert, W.,** Synthesis, tautomerism, and reactions of quinolino and thienopyridine systems which bear a 1-carboethoxy-1-cyanomethyl substituent in the pyridine ring. Part 2, *J. Heterocycl. Chem.*, 24, 1467, 1987.

668. **Saeki, S., Honda, H., Kaku, Y., Funakoshi, K., and Hamana, M.,** A novel formation of cyclopropa[c]quinolines from some 2-substituted quinoline N-oxides, *Heterocycles*, 7, 801, 1977.

669. **Stanovnik, B., Drofenik, I., and Tisler, M.,** Transformations of heterocyclic N-oxides into derivatives of α-heteroaryl substituted α-aminoacids and dipeptides, *Heterocycles*, 26, 1805, 1987.

670. **Hamana, M., Fujimura, Y., and Nawata, Y.,** Reactions of quinoline 1-oxide with cyanoacetic acid derivatives bearing leaving groups, *Heterocycles*, 25, 235, 1987.

671. **Yousif, M.M., Saeki, S., and Hamana, M.,** Reactions of aromatic N-oxides with Meldrum's acid in the presence of acetic anhydride, *Chem. Pharm. Bull.*, 30, 1680, 1982.

672. **Yousif, M.M., Saeki, S., and Hamana, M.,** Reactions of quinoline N-oxide with barbituric acid in the presence of acetic anhydride, *Heterocycles*, 15, 1083, 1981.

673. **Funakoshi, K., Inada, H., and Hamana, M.,** Formation of 1-isoquinoliniomethylides by the reaction of isoquinoline 2-oxide with cyanoacetic acid and benzoylacetonitrile in the presence of acetic anhydride, *Chem. Pharm. Bull.*, 32, 4731, 1984.

674. **Dorneanu, M., Carp. E., and Zugravescu, I.,** o-Diazabicyclic derivative of benzo[c]cinnoline and 3,8-dimethylbenzo[c]cinnoline, *An. Stiint. Univ. Iasi. Sec. Ic*, 19, 223, 1973; *Chem. Abst.*, 80, 133371v, 1974.

675. **Douglass, J.E. and Koottungal, D.,** 2,3-Annelation on quinoline and pyridine 1-oxides, *J. Org. Chem.*, 37, 2913, 1972.

676. **Douglass, J.E. and Hunt, D.A.,** Synthesis of quinolizinones by the condensation of ylidenemalononitriles with quinoline 1-oxide, *J. Org. Chem.*, 42, 3974, 1977.

677. **Kato, T., Yamanaka, H., Sakamoto, T., and Shiraishi, T.,** Reaction of diketene with quinoline 1-oxide. Supplement, *Chem Pharm. Bull.,* 22, 1206, 1974.

678. a. **Yamanaka, H., Sakamoto, T., and Shiraishi, T.,** Reaction of isoquinoline 2-oxide with diketene, *Chem. Pharm. Bull.,* 29, 1044, 1981; b. **Kato, T., Chiba, T., and Daneshtalab, M.,** Reactions of diketene with acridine *N*-oxide, *Heterocycles,* 2, 315, 1974.

679. a. **Hamana, M. and Noda, H.,** Reaction of aromatic *N*-oxides with enamine of cyclohexanone in the presence of acylating agents, *Chem. Pharm. Bull.,* 133, 912, 1965; b. *ibid.,* 14, 762, 1966; c. *ibid.,* 15, 474, 1967.

680. **Nakanishi, M., Yatabe, M., and Hamana, M.,** Reactions of pyridine *N*-oxides with enamines of *N*-substituted 4-piperidones in the presence of an acylating agent, *Heterocycles,* 3, 287, 1975.

681. **Hamana, N. and Noda, H.,** Reaction of aromatic *N*-oxides with 1-morpholinoisobutene in the presence of acylating agents, *Yakugaku Zasshi,* 89, 641, 1969.

682. **Colonna, M. and Poloni, M.,** Reaction of electron-rich systems with activated azomethines, *Gazz. Chim. Ital.,* 115, 187, 1985.

683. **Saeki, S., Yamashita, A., Matsukura, Y., and Hamana, M.,** Reactions of pyridine and quinoline *N*-oxides with 1(10)-dehydroquinolizidine, *Chem. Pharm. Bull.,* 22, 2341, 1974.

684. **Saeki, S., Yamashita, A., Morinaka, Y., and Hamana, M.,** Reactions of substituted dehydroquinolizidines with pyridine *N*-oxides, *Yakugaku Zasshi,* 96, 456, 1976.

685. **Colonna, M., Bruni, P., and Guerra, A.M.,** Indoles, pyrroles and pyrazoles in the reaction with heteroaromatic *N*-oxides, *Gazz. Chim. Ital.,* 96, 1410, 1966.

686. **Hamana, M. and Kumadaki, I.,** Reaction of aromatic *N*-oxides with indoles in the presence of acylating agents, *Chem. Pharm. Bull.,* 15, 363, 1967.

687. **Hamana, M. and Kumadaki, I.,** Reactions of aromatic *N*-oxides with indoles in the presence of acylating agents, *Chem. Pharm. Bull.,* 18, 1742, 1970.

688. **Nagayoshi, J., Saeki, S., and Hamana, M.,** Some reactions of 1-hydroxy-2-phenyl indole, *Heterocycles,* 6, 1666, 1977.

689. a. **Matoba, K., Shibata, M., and Yamazaki, T.,** Synthesis of the 9,13-diazasteroid system, *Chem. Pharm. Bull.,* 30, 1718, 1982; b. **Yamaneka, H., Egawa, H., and Sakamoto, T.,** Condensation of quinoline and isoquinoline *N*-oxides with isoxazoles, *Chem. Pharm. Bull.,* 26, 2759, 1978.

690. **Hamana, M. and Hoshino, O.,** Reactions of quinoline 1-oxide with amines in the presence of acylating agents, *Yakugaku Zasshi,* 84, 35, 1964.

691. **Hamana, M. and Noda, H.,** Reactions of aromatic *N*-oxides with antipyrine in the presence of acylating agents, *Chem. Pharm. Bull.,* 15, 1380, 1967.

692. **Hamana, M., Noda, H., Narimatsu, K., and Ueda, I.,** Reaction of *N*-alkoxyquinolinium salts with enamines of ketones, *Chem. Pharm. Bull.,* 23, 2918, 1975.

693. **Hamana, M. and Noda, H.,** Reactions of aromatic *N*-oxides with enol ethers in the presence of benzoyl chloride, *Chem. Pharm. Bull.,* 18, 26, 1970.

694. a. **Cervinka, O.,** Reactions of Grignard derivatives with quaternary salts of pyridine *N*-oxide and homologues, *Coll. Czech. Chem. Commun.,* 27, 567, 1962; b. **Webb, T.R.,** Regioselective synthesis of 2-substituted pyridines via Grignard addition to 1-(alkoxycarboxy)pyridinium salts, *Tetrahedron Lett.,* 3191, 1985; c. **Al-Arnaout, A., Courtois, G., and Miginiac, L.,** Regioselective organometallic synthesis of pyridines, 4-picolines, and 3,5-lutidines substituted in the 2-position by an unsaturated and/or functional group, *J. Organomet. Chem.,* 333, 139, 1987.

695. a. **Vorbrueggen, H. and Krolikiewicz, K.,** Conversion of heterocyclic *N*-oxides into α-alkylated heterocycles. Trimethylsilanol as leaving group, *Tetrahedron Lett.,* 889, 1983; b. **Kudo, T., Nose, A., and Hamana, M.,** Reaction of trialkylborane with aromatic amine *N*-oxide, *Yakugaku Zasshi,* 96, 521, 1975.

696. a. **Furukawa S., Kinoshita, T., and Watanabe, M.,** Reaction of dimethoxo- and dimethylsulfonium benzolmethylide with aromatic amine *N*-oxides, *Yakugaku Zasshi,* 93, 1064, 1973; b. **Watanabe, M., Kodera, M., Kinoshita, T., and Furukawa, S.,** Reaction of dimethylsulfonium acetylmethoxycarbonylmethylide and dimethylsulfonium diacetylmethylide with quinoline 1-oxide, *Chem. Pharm. Bull.,* 23, 2598, 1975; c. **Kinoshita, T., Onoue, T., Watanabe, M., and Furukawa, S.,** The reaction of dimethylsulfonium acetylcarbamoylmethylide with quinoline 1-oxide, *Chem. Pharm. Bull.,* 28, 795, 1980.

697. **Ektova, L.V., Shishkina, R.P., and Fokin, E.P.,** Reaction of 2-methylcaramidinone *N*-oxide with amines, *Izv. Sib. Otd. Akad. Nauk SSSR, Ser. Khim. Nauk,* 105, 1971, *Chem. Abst.,* 77, 61772q, 1972.

698. **Solekhova, M.A., Kurbatov, Y.V., Otroschenko, O.S., and Sadykov, A.S.,** Reaction of *N*-oxides of pyridine, bipyridine and quinoline with ammonia and ammonium salts, *Khim. Geterotsikl. Soedin.,* 229, 1976.

699. a. **Kurbatov, Y.V. and Solekhova, M.A.,** Reaction of pyridine *N*-oxide with aniline and *p*-toluensulfonyl chloride in alkaline medium, *Zh. Org. Khim.,* 19, 663, 1983; b. **Kurbatov, Y.V. and Solekhova, M.A.,** Amination of pyridine *N*-oxide with aniline and *p*-anisidine, and tosylation of the products, *Khim. Geterotsikl. Soedin.,* 936, 1986.

700. **Tanida, H.,** New dimethylaminomethylation of *N*-oxides of quinoline series, *Yakugaku Zasshi,* 78, 608, 1958.

701. **Kurbatov, Y.V. and Solekhova, M.A.,** High-temperature benzoylarnination of quinoline *N*-oxide, *Zh. Org. Khim.*, 17, 1121, 1981.

702. **Hamana, M. and Funakoshi, K.,** Reaction of quinoline 1-oxide with acyl chloride in the presence of pyridine, *Yakugaku Zasshi*, 82, 512, 1962.

703. **Hamana, M. and Funakoshi, K.,** Reaction of pyridine 1-oxides derivatives with tosyl chloride in the presence of pyridine, *Yakugaku Zasshi*, 84, 23, 1964.

704. **Hamana, M. and Funakoshi, K.,** Reactions of quinoline 1-oxides with tosyl chloride in the presence of pyridine, *Yakugaku Zasshi*, 84, 28, 1964.

705. **Elina, A.S.,** Reaction of quinoxaline 1-oxides with benzenesulfonyl chloride and benzoyl chloride, *Khim. Geterotsikl. Soedin.*, 724, 1967.

706. **Wölcke, U., Birdsall, N.J.M., and Brown, G.B.,** Preparation of 8-substituted xanthines and guanines by nucleophile displacement of a 3-substituent, *Tetrahedron Lett.*, 785, 1969.

707. **Abramovitch, R.A. and Singer, G.M.,** Direct acylamination of pyridine 1-oxides, *J. Org. Chem.*, 39, 1795, 1974; b. **Parham, W.E. and Sloan, K.B.,** The reaction of heterocycle *N*-oxides with benzimidoyl chloride, *Tetrahedron Lett.*, 1947, 1971.

708. **Abramovitch, R.A., Pilski, J., Konitz, A, and Tomasik, P.,** Direct acylamination of pyridine 1-oxide with *N*-phenylareneimidoyl chlorides and fluorides, *J. Org. Chem.*, 48, 4391, 1983.

709. a. **Abramovitch, R.A., Rogers, R.B., and Singer, G.M.,** Direct acylamination of quinoline, isoquinoline, benzimidazole, pyridazine and pyrimidine 1-oxides. Novel 1,5-sigmatropic shift, *J. Org. Chem.*, 40, 41, 1975; b. **Abramovitch, R.A. and Rogers, R.B.,** Direct acylamination of 3-substituted pyridine 1-oxides. Directive effect of the substituent, *J. Org. Chem.*, 39, 1802, 1974.

710. a. **Abramovitch, R.A. and You-Xiong, W.,** Regiospecific amination of 3-substituted pyridines using imidoyl chloride functionalized polystyrene, *Heterocycles*, 26, 2065, 1987; b. **Abramovitch, R.A. and Bailey, T.D.,** Direct side chain amination of pyridine 1-oxides. New rearrangement, *J. Heterocycl. Chem.*, 12, 1079, 1975.

711. **Wachi, K. and Terada, A.,** Synthesis of primary 2-aminopyridines via the reaction of imidoyl chlorides of 1,3-benzoxazines with pyridine *N*-oxides, *Chem. Pharm. Bull.*, 28, 465, 1980.

712. a. **Wachi, K. and Terada, A.,** New rearrangement modes in the reaction of 4-chloro-2,2-dimethyl-2*H*-1,3-benzoxazine with substituted pyridine *N*-oxides, *Chem. Pharm. Bull.*, 28, 3020, 1980; b. **Saczewski, F.,** 2-Chloro-4,5-dihydroimidazole. I. Reactions with some heteroaromatic *N*-oxides, cyclic nitrones and aldoximes, *Synthesis*, 170, 1984.

713. **Ramirez, F. and von Oswalden, P.W.,** The action of heterocyclic *N*-oxides on 2-bromopyridine, *J. Am. Chem. Soc.*, 81, 156, 1959.

714. **Kajihara, S.,** Reactions of aromatic heterocyclic *N*-oxides with 4-bromopyridine and 4-bromoquinoline, *Nippon Kagaku Zasshi*, 85, 672, 1964.

715. **Kajihara, S.,** Reactions of isoquinoline 2-oxide with heterocyclic halides, *Nippon Kagaku Zasshi*, 86, 93, 1965.

716. a. **Kajihara, S.,** Substituent effects on the reaction of pyridine 1-oxides with 2-bromopyridine, *Nippon Kagaku Zasshi*, 86, 839, 1965; b. **Kajihara, S.,** The mechanism of the reactions of aromatic heterocyclic *N*-oxides with heterocyclic halides, *Nippon Kagaku Zasshi*, 86, 1060, 1965.

717. **Deegan, A. and Rose, F. L.,** Reaction of pyridine *N*-oxides with 3,6-dichloropyridazine, *J. Chem. Soc. (C)*, 2756, 1971.

718. **de Villiers, P.A. and den Hertog, H.J.,** The reaction of pyridine *N*-oxide with 2-, 3-, and 4-pyridyl *p*-toluensulphonate, *Rec. Trav. Chim. Pays Bas*, 76, 647, 1957.

719. **Hamana, M. and Kumadaki, S.,** Reaction of quinoline *N*-oxide with cyanogen bromide, *Chem. Pharm. Bull.*, 21, 800, 1973.

720. **Hamana, M. and Kumadaki, S.,** Reaction of quinoline 1-oxide derivatives with cyanogen bromide, *Chem. Pharm. Bull.*, 22, 1506, 1974.

721. **Hamana, M. and Kumadaki, S.,** Reactions of isoquinoline and pyridine *N*-oxides derivatives with cyanogen bromide, *Yakugaku Zasshi*, 95, 87, 1975.

722. **Hamana, M. and Kumadaki, S.,** Reaction of quinoline *N*-oxides with potassium cyanate and tosyl chloride in ethanol, *Chem. Pharm. Bull.*, 23, 2284, 1975.

723. **Nishiyama, K. and Miyata, I.,** Reaction of trimethylsilyl azide with C=N-O bond, *Bull. Chem. Soc. Jpn.*, 58, 2419, 1985.

724. **Reddy, K.S., Iyengar, D.S., and Bhalerao, U.T.,** The reaction of pyridine *N*-oxide and its benzo analogs with arenesulfonyl azides: novel synthesis of tetrazoloazines, *Chem. Lett.*, 1745, 1983.

725. **Bauer, L. and Prachayasittikul, S.,** Deoxidative substitutions of pyridine 1-oxides by thiols, *Heterocycles*, 24, 161, 1986.

726. **Bauer, L., Dickerhofe, T.E., and Tserng, K.Y.,** Deoxidative substitution of pyridine *N*-oxide by mercaptans in the presence of carbamyl, carbonyl, and sulfamyl chlorides, *J. Heterocycl. Chem.*, 12, 797, 1975.

727. **Mikrut, B.A., Khullar, K.K., Chan, P.Y.P., Kokosa, J.M., Bauer, L., and Egan, R.S.,** (1-Adamantanethio)pyridines and tetrahydropyridines from the reaction of 1-adamantanethiol with pyridine 1-oxide in acetic anhydride, *J. Heterocycl. Chem.*, 11, 713, 1974.

728. **Hershenson, F.M., and Bauer, L.,** Steric and electronic effects influencing the deoxidative substitution of pyridine *N*-oxides by mercaptans in acetic anhydride, *J. Org. Chem.,* 34, 655, 1969.

729. **Mikrut, B.A., Hershenson, F.M., King, K.F., Bauer, L., and Egan, R.S.,** The effect of triethylamine on the deoxidative substitution of pyridine *N*-oxides by mercaptans, *J. Org. Chem.,* 36, 3749, 1971.

730. **Prachayasittikul, S. and Bauer, L.,** The deoxidative substitution reactions of nicotinamide and nicotinic acid *N*-oxides by 1-adamantanethiol in acetic anhydride, *J. Heterocycl. Chem.,* 22, 771, 1985.

731. **Prachayasittikul, S., Kokosa, J.M., Bauer, L., Fesik, S.W., and Egan, R.S.,** Tetrahydropyridines from 3-picoline 1-oxide and *tert*-butyl and 1-adamantylmercaptan in acetic anhydride. Structural elucidation by long range 2D J(C-H) resolved NMR spectroscopy, *J. Org. Chem.,* 50, 997, 1985.

732. **Kokosa, J.M., Chu, I., Bauer, L., and Egan, K.S.,** Tetrahydropyridines and furans from the reaction of 4-*tert*-butylpyridine 1-oxide with *tert*-butyl and 1-adamantyl mercaptan, *J. Heterocycl. Chem.,* 13, 861, 1976.

733. a. **Kokosa, J.M., Chu, I., Bauer, L., and Egan, R.S.,** The structure of the tetrahydropyridines from 3,5-lutidine 1-oxide and mercaptans in acetic anhydride, *J. Heterocycl. Chem.,* 15, 785, 1978; b. **Kokosa, J.M., Bauer, L., and Egan, R.S.,** Tetrahydropyridinediols and related aldehydes from the reaction of pyridine *N*-oxide with mercaptans, *J. Heterocycl. Chem.,* 13, 321, 1976.

734. **Kokosa, J.M., Bauer, L., and Egan, R.S.,** Revised structures of some tetrahydropyridines isolated from the reaction of pyridine *N*-oxides with mercaptans and acid anhydrides, *J. Org. Chem.,* 40, 3196, 1975.

735. **Bauer, L. and Dickerhofe, T.E.,** The substitution of quinoline 1-oxide by mercaptans, *J. Org. Chem.,* 31, 939, 1966.

736. **Gogte, V.N., Sathe, R.N., and Tilak, B.D.,** Synthesis of thiazolo[3,2-*a*]quinoline and thiazolo[2,3-*a*]isoquinoline derivatives, *Indian J. Chem.,* 11, 1115, 1973.

737. **Hayashi, E. and Shimada, N,** The reaction of pyridine 1-oxide with sulfonic acid chloride and potassium cyanide, *Yakugaku Zasshi,* 98, 95, 1978.

738. **Hayashi, E. and Shimada, N.,** The reaction of quinoline *N*-oxides with sulfonic acid chloride and potassium cyanide, *Yakugaku Zasshi,* 97, 627, 1977.

739. **Hayashi, E. and Shimada, N.,** The reaction of 2,4-disubstituted quinoline 1-oxides with sulfonic acid chloride and potassium cyanide, *Yakugaku Zasshi,* 97, 641, 1977.

740. **Hayashi, E. and Shimada, N.,** On the reaction of isoquinoline 2-oxides with sulfonic acid chloride and potassium cyanide, *Yakugaku Zasshi,* 97, 1345, 1977.

741. **Markgraf, J.H., Ahn, M.K., Carson, C.G., and Lee, G.A.,** Reaction of acridine *N*-oxide with acetyl sulfide, *J. Org. Chem.,* 35, 3983, 1970.

742. **Harpp, D.N. and MacDonald, J.G.,** Synthesis of thioamides from aldonitrones utilizing thiocarbonyl transfer reagents, *Tetrahedron Lett.,* 4927, 1983.

743. **Kreher, R. and Morgenstern, H.,** 5*H*-Dibenz[*c,e*]azepines with sulfur-containing molecular functions, *Z. Chem.,* 22, 258, 1982.

744. **Hamana, M. and Muraoka, K.,** Reaction of quinoline *N*-oxide with *N,N*-diethylcarbamyl chloride, *Heterocycles,* 1, 241, 1973.

745. a. **Redmore, D.,** Pyridine phosphonic acids, U.S. patent 3,810,907, 1974; *Chem. Abst.,* 81, 37646p, 1974; b. **Redmore D.,** Quinoline and isoquinoline phosphonates, U.S. patent 3,888,627, 1975; *Chem. Abst.,* 83, 97570k, 1975; c. **Redmore, D.,** Phenanthridine phosphonic compounds, U.S. patent 3,888,626, 1975; *Chem. Abst.,* 83, 97569s, 1975.

746. **Conary, G.S., Russel, A.A., Paine, R.T., Hall, J.H., and Ryan, R.R.,** Synthesis and coordination chemistry of 2-(diisopropoxyphosphino)pyridine *N,P*-dioxide, *Inorg. Chem.,* 27, 3242, 1988.

747. **Pennings, M.L.M. and Reinhoudt, D.N.,** Oxidation of *N*-hydroxyazetines: a novel synthesis of *N*-acetoxy β-lactams and four-membered cyclic nitrones, *Tetrahedron Lett.,* 1003, 1982.

748. **Black, D.S.C., Blackman, N.A., and Johnstone, L.M.,** Reaction of Grignard reagents with 2-oxo-1-pyrroline 1-oxides and synthesis of 2*H*-pyrrole 1-oxides, *Aust. J. Chem.,* 32, 2025, 1975.

749. **Black, D.S.C., Clark, V.M., Odell, B.G., and Todd, A.,** Synthesis and some reactions of β-oxo-Δ¹-pyrroline 1-oxides, *J. Chem. Soc. Perkin Trans. 1,* 1944, 1976.

750. **Alazard, J.P. and Lusinchi, X.,** Reactions of methylmagnesium iodide with various conanine-type heterocyclic nitrones. Air oxidation of the resulting hydroxylamines, *Bull. Soc. Chim. Fr.,* 40, 1814, 1973.

751. **Nedenskov, P. and Mandrup, M.,** Arylation in the 5-position of 1,4-benzodiazepine 4-oxide, *Acta Chem. Scand.,* B31, 701, 1977.

752. **Ruggli, P., Hegedüs, B., and Caspar, E.,** On some addition products of isatogen, *Helv. Chim. Acta,* 22, 411, 1939.

753. **Berti, C., Colonna, M., Greci, L., and Marchetti, L.,** Stable nitroxide radicals from phenylisatogen and arylimino-derivatives with organometallic compounds, *Tetrahedron,* 31, 1745, 1975.

754. **Schiess, P., Monnier, C., Ringele, P., and Sendi, E.,** Grignard addition to pyridine *N*-oxide, *Helv. Chim. Acta,* 57, 1676, 1974.

755. **Schiess, P. and Ringele, P.,** Grignard reaction with pyridine 1-oxide, *Tetrahedron Lett.,* 311, 1972.

756. **Kato, T., Yamanaka, H., Adachi, T., and Hiranuma, H.,** The ring opening reaction of 1-hydroxy-2-phenyl-1,2-dihydropyridine and related compounds, *J. Org. Chem.,* 32, 3788, 1967.

757. **van Bergen, T.J. and Kellog, R.M.,** Reaction of aryl Grignard reagents with pyridine 1-oxide, *J. Org. Chem.,* 36, 1705, 1971.
758. **Ochiai, E. and Arima, K.,** Reaction of Grignard reagent on pyridine *N*-oxide, *Yakugaku Zasshi,* 69, 51, 1949.
759. **Kato, T. and Yamanaka, H.,** Reaction of pyridine and quinoline *N*-oxides with phenylmagnesium bromide, *J. Org. Chem.,* 30, 910, 1965.
760. **Fritzsche, U. and Huenig, S.,** Ring opening of pyridine *N*-oxide by sodium acetylide, *Justus Liebigs Ann. Chem.,* 1407, 1974.
761. **Crombie, L., Kerton, N.A., and Pattenden, G.,** The synthesis of terminal enynes by Grignard addition to pyridazine 1-oxide, *J. Chem. Soc., Perkin Trans. 1,* 2136, 1979.
762. **Okusa, G., Kumagai, M., and Itai, T.,** Reaction of pyridazine 1-oxide with Grignard reagent, *Chem. Pharm. Bull.,* 17, 2502, 1979.
763. **Giraudi, E. and Teisseire, P.,** New synthesis of naturally-occurring 1,3,5-undecatrienes, *Tetrahedron Lett.,* 24, 489, 1983.
764. **Garito, A.F.,** Diacetylenes having liquid crystal phases, European patent 137679 Al, 1985; *Chem. Abstr.* 103, 170331w, 1985.
765. **Oishi, E., Endo, T., Asahina, Y., and Hayashi, E.,** Chemical properties of 7-methoxy-1-phenyl-1*H*-pyrazolo[3,4-*d*]pyridazine 5-oxide, *Yakugaku Zasshi,* 105, 129, 1985.
766. **Igeta, H., Tsuhiya, T., Nakai, T., Okusa, G., Kumagai, M., Miyoshi, J., and Itai, T.,** Reaction of cinnoline 2-oxides with phenyl magnesium bromides, *Chem. Pharm. Bull.,* 18, 1497, 1970.
767. **Igeta, H., Nakai, T., and Tsuchiya, T.,** Reaction of 5,6-benzo-1,2,4-triazine 1-oxides with Grignard reagents, *J. Chem. Soc., Chem. Commun.,* 622, 1973.
768. a. **Hayashi, E., Oishi, E., Tezuka, T., and Ema, K.,** *N*-Oxidation of 1-methyl- and 1-benzylphthalazine, *Yakugaku Zasshi,* 88, 1333, 1968; b. **Hayashi, E. and Oishi, E.,** *N*-oxidation of 1-phenylphthalazine and chemical properties of 1-phenylphthalazine 3-oxide, *Yakugaku Zasshi,* 86, 576, 1966.
769. **Colonna, M., Greci, L., and Poloni, M.,** *Ipso* attack in the reaction between 2-methoxy and 2-cyanoquinoline *N*-oxide and phenylmagnesium bromide, *J. Heterocycl. Chem.,* 17, 293, 1980.
770. **Colonna, M. and Risaliti, A.,** Aromatic *N*-oxides. Action of organo-magnesium derivatives, *Gazz. Chim. Ital.,* 83, 58, 1953.
771. **Teranishi, M., Suzuki, K., Kase, H., Kitamura, S., Shuto, K., and Omori, K.,** Quinoline *N*-oxide derivatives of pharmaceutical compositions, European patent 177764, 1986; *Chem. Abst.* 105, 97343m, 1986.
772. a. **Hayashi, E. and Miura, Y.,** On *N*-oxidation of 2-alkylquinoxalines, *Yakugaku Zasshi,* 87, 643, 1967; b. **Hayashi, E. and Iijima, C.,** Reaction of 2-pheylquinoxaline 4-oxide with Grignard reagent, *Yakugaku Zasshi,* 86, 571, 1966.
773. **Black, D.S.C., Clark, V.H., Thakur, R.S., and Todd, A.,** A new synthesis of 2-cyano-Δ¹-pyrroline 1-oxides and their behavior towards a Grignard reagent, *J. Chem. Soc., Perkin Trans. 1,* 1951, 1976.
774. **Colonna, M., Greci, L., and Poloni, M.,** Reaction between 2-cyano- and 4-cyanobenzoquinoline *N*-oxides and the Grignard reagent, *J. Heterocycl. Chem.,* 17, 1473, 1980.
775. a. **Berti, C., Colonna, M., Greci, L., and Marchetti, L.,** Stable nitroxide radicals from 2-substituted quinoline *N*-oxides with organometallic compounds, *Tetrahedron,* 32, 2147, 1976; b. **Berti, C., Colonna, M., Greci, L., and Marchetti, L.,** Stable nitroxide radicals from acridine *N*-oxides with Grignard reagents, *Gazz. Chim. Ital.,* 108, 659, 1978.
776. a. **Azev, Y.A., Mudretsova, I.I., Goleneva, A.F., and Aleksandrova, G.A.,** Synthesis and some properties of 1,3-dimethyl-5-nitrosouracil hydrazides, *Khim.-Farm. Zh.,* 21, 1446, 1987; b. **Yamanaka, H., Nitsuma, S., Sakai, M., and Sakamoto, T.,** Ring transformation of 4-alkoxypyrimidine 1-oxides by reaction with diketene, *Chem. Pharm. Bull.,* 36, 168, 1988.
777. a. **Higashino, T., Suzuki, K., and Hayashi, E.,** Transformation of quinazoline 3-oxide into quinoline derivatives, *Chem. Pharm. Bull.,* 23, 746, 1975; b. **Higashino, T., Nagano, Y., and Hayashi, E.,** Transformation of quinazoline 3-oxide into quinoline derivatives, *Chem. Pharm. Bull.,* 21, 1943, 1973.
778. **Higashino, T., Suzuki, K., and Hayashi, E.,** Transformation of pyrido[2,3-*d*]pyrimidine 3-oxide into 1,8-naphthyridines, *Chem. Pharm. Bull.,* 23, 2939, 1975.
779. **Takeuchi, I., Ozawa, I., Hamada, Y., Masuda, H., and Hirota, M.,** 3-Methylsulfonyl-3*H*-benz[*d*]azepine, *Chem. Lett.,* 519, 1976.
780. **Peereboom, R. and van del Plas, H.C.,** Influence of leaving group mobility on reactions of 4-X-6-methyl (or phenyl)-pyrimidine 1-oxides with liquid ammonia and with potassium amide in liquid ammonia, *Rec. Trav. Chim. Pays-Bas,* 93, 284, 1974.
781. **Peereboom, R. and van der Plas, H.C.,** Nitrogen-15 study on amination and isoxazole formation in reactions of 6-substituted 4-chloro- and 4-phenoxypyrimidine 1-oxides with liquid ammonia and with potassium amide in liquid ammonia, *Rec. Trav. Chim. Pays-Bas,* 93, 277, 1974.
782. **van der Plas, H.C., Vollering, M.C., Jongejan, H., and Zuurdeeg, B.,** Conversion of pyrimidines into isoxazoles, *Rec. Trav. Chim. Pays-Bas,* 93, 225, 1974.
783. **Higashino, T.,** Quinazoline 3-oxide, *Chem. Pharm. Bull.,* 9, 635, 1961.

784. a. **Vercek, B., Leban, I., Stanovnik, B., and Tisler, M.,** 1,2,4-Oxadiazolylpyridines and pyrido[2,3-*d*]pyrimidine 3-oxides, *J. Org. Chem.,* 44, 1695, 1979; b. **Higashino, T. and Hayashi, E.,** Pyrido[2,3-*d*]pyrimidine 3-oxide, *Chem. Pharm. Bull.,* 21, 2643, 1973.

785. **Kocevar, M., Stanovnik, B., and Tisler, M.,** Ring transformations of some 4-aminopteridine 3-oxides and derivatives, *Tetrahedron,* 39, 823, 1983.

786. **Suehiro, H., Kikugawa, K., Ichino, M., and Nakamura, T.,** 2-Thioadenosine, Japan Kokai 51/52198, 1976; *Chem. Abst.,* 85, 124311b, 1976.

787. **Fujii, T., Saito, T., Kizu, K., Hayashibara, H., Kumazawa, Y., and Nakajima, S.,** New synthetic route to 2-deuterioadenines substituted or unsubstituted at the 9-position, *Heterocycles,* 24, 2449, 1986.

788. **Meisenheimer, J., Senn, O., and Zimmermann, P.,** About the oxime of *o*-aminobenzo- and acetophenone, *Ber. Dtsch. Chem. Ges.,* 60, 1736, 1927.

789. a. **Sasaki, T. and Minamoto, K.,** Structural studies on the oxidation products of 3-amino-5-phenyl-*as*-triazine with organic peracids, *J. Org. Chem.,* 31, 3917, 1966; b. **Fusco, R. and Bianchetti, G.,** Alkaline demolition in the *as*-triazine series, *Gazz. Chim. Ital.,* 90, 1113, 1960.

790. **Hayashi, E. and Miura, Y.,** On syntheses of 2-alkyl- and 2-alkoxybenzimidazole 3-oxides from 2-alkyl- and 2-alkoxyquinoxaline 4-oxides, *Yakugaku Zasshi,* 87, 648, 1967.

791. **Armarego, W.L.F.,** Covalent hydration in quinazoline 3-oxides, *J. Chem. Soc.,* 5030, 1962.

792. a. **Kato, T., Yamanaka, H., and Yasuda, N.,** Hydrolysis of pyrimidine *N*-oxides to give isoxazole derivatives, *J. Org. Chem.,* 32, 3593, 1967; b. **Sakamoto, T., Niitsuma, S., Mizugaki, M., and Yamanaka, H.,** On the structural determination of pyrimidine *N*-oxides, *Heterocycles,* 8, 257, 1977

793. **Lewis, A.F. and Townsend, L.B.,** Synthesis of the novel *C*-nucleoside 5-amino-3-(β-D-ribofuranosyl)pyrazolo[4,3-*d*]pyrimidin-7-one, a guanosine analogue related to the nucleoside antibiotic formycin B, *J. Am. Chem. Soc.,* 104, 1073, 1982.

794. **Azev, Y.A., Mudretsova, I.I., Pidemskii, E.L., Goloneva, A.F., and Alexandrova, G.A.,** Products of the reactions of indoles with fervenulin and fervenulin-3-one 4-oxides, *Khim.-Farm. Zh.,* 20, 1228, 1986.

795. **Sliwa, H. and Tartar, A.,** An access to 4-isoxazolones starting from pyridine *N*-oxides, *Tetrahedron Lett.,* 311, 1977.

796. **Sliwa, H. and Tartar, A.,** Competitive elimination during the basic decomposition of *N*-alkoxypyridinium salts, *Tetrahedron,* 33, 3111, 1977.

797. **Brown, G.B., Levin, G., Murphy, S., Sele, A., Reilly, H.C., Tarnowski, G.S., Schmid, F.A., Teller, M.N., and Stock, C.C.,** Synthesis of 6-mercaptopurine 3-*N*-oxide — its chemotherapeutic possibilities, *J. Med. Chem.,* 8, 190, 1965.

798. **Ohsawa, A., Itoh, T., and Igeta, H.,** Flash vacuum pyrolysis of pyridazine *N*-oxides, *Heterocycles,* 26, 2677, 1987.

799. **Flynn, A.P.,** Imidazole *N*-oxide hazard, *Chem. Br.,* 20, 30, 1984.

800. **Champman, J.A., Crosby, J., Cummings, C.A., Rennie, R.A.C., and Paton, R.M.,** Furazan *N*-oxides: a convenient source of both nitrile oxides and isocyanates, *J. Chem. Soc., Chem. Commun.,* 240, 1976.

801. **Prokudin, V.G. and Nazin, G.M.,** Gas-phase cyclodecomposition of furazan and its *N*-oxide, *Izv. Akad. Nauk. SSSR, Ser. Khim.,* 221, 1987.

802. **Prokudin, V.G., Nazin, G.M., and Ovchinnikov, I.V.,** Thermal decomposition of isomeric 3(4)-nitro-4(3)-phenylfuroxans in the gas phase, *Izv. Akad. Nauk. SSSR, Ser. Khim.,* 2841, 1987.

803. **Prokudin, V.G., Nazin, G.M., and Manelis, G.B.,** Thermal degradation mechanism of furazans and furoxans, *Dokl. Akad. Nauk SSSR,* 255, 917, 1980.

804. **Oyumi, Y. and Brill, T.B.,** High-rate thermolysis of benzofuroxans and 3,4-dimethylfuroxan, *Combust. Flame,* 65, 313, 1986.

805. **Mitchell, W.R. and Paton, R.M.,** Isolation of nitrile oxides from the thermal fragmentation of furazan *N*-oxides, *Tetrahedron Lett.,* 2443, 1979.

806. **Whitney, R.A. and Nicholas, E.S.,** Furoxans as nitrile oxide precursors: cycloaddition reactions of bis(benzenesufonyl)furoxan, *Tetrahedron Lett.,* 3371, 1981.

807. **Shimizu, T., Hayashi, Y., Taniguchi, T., and Teramura, K.,** Reaction of 3,4-disubstituted 1,2,5-oxadiazole 2-oxides with dipolarophiles. Substituent and solvent effect on the reaction courses, *Tetrahedron,* 41, 727, 1985.

808. **Curran, D.P. and Fenk, C.J.,** Thermolysis of bis[2-[(trimethylsilyl)oxy]propyl]furoxan (TOP-furoxan). The first practical method for intermolecular cycloaddition of an *in situ* generated nitrile oxide witn 1,2-di- and trisubstituted olefins, *J. Am. Chem. Soc.,* 107, 6023, 1985.

809. **Barnes, J.F., Paton, R.M., Ashcroft, P.L., Bradbury, R., Crosby, J., Joyce, C.J., Holmes, D.R., and Milner, J.A.,** Sulfur dioxide mediated conversion of strained furazan *N*-oxides into diisocyanates, *J. Chem. Soc., Chem. Commun.,* 113, 1978.

810. **Ashcroft, P.L., Barnes, J.F., Barron, K., Bradbury, R., Crosby, J., Joyce, C.J., Harding, M.M., Holmes, D.R., Milner, J.A., and Paton, R.M.,** Synthesis of di-isocyanates from strained norbonane furazan *N*-oxides, *J. Chem. Soc., Perkin Trans. 1,* 601, 1986.

811. a. **Crosby, J., Rennie, R.A.C., and Paton, R.M.,** Organic isocyanates, *Ger Offen* 2555830, 1976; *Chem. Abst.,* 85, 124937k, 1976; b. **Altaf-ur-Rahman and Boulton, A.J.,** A new ring opening of strained furoxans, *Chem. Commun.,* 73, 1968; c. **Boulton, A.J. and Mathur, S.S.,** Acenaphtho[1,2-*d*]furazan, *J. Org. Chem.,* 38, 1054, 1973; d. **Katzman, S.M. and Moffat, J.,** Preparation of nitriles from 1,2,5-oxadiazoles by reduction with triphenylphosphite, *J. Org. Chem.,* 37, 1842, 1972; e. **Bulacinski, A.B., Scriven, E.F.V., and Suschitzky, H.,** Reaction of benzofuroxan with *p*-anisylazide. Trapping of the *o*-dinitroso intermediate, *Tetrahedron Lett.,* 3577, 1975.

812. **Ley, K. and Seng, F.,** Syntheses using benzofuroxans, *Synthesis,* 415, 1975.

813. **Haddadin, M.J. and Issidorides, C.H.,** Application of benzofurazan oxide to the synthesis of heteroaromatic N-oxides, *Heterocycles,* 4, 767, 1976.

814. **Heyns, K., Behse, E., and Francke, W.,** Synthesis of 2,3-dialkypyrazines, *Chem. Ber.,* 114, 240, 1981.

815. **Heyns, K., Behse, E., and Francke, W.,** Reaction of benzofuroxan with ketones, *Chem. Ber.,* 114, 246, 1981.

816. **Haddadin, M.J., Agopian, G., and Issidorides, C.H.,** Synthesis and photolysis of some substituted quinoxaline di-N-oxides, *J. Org. Chem.,* 36, 514, 1971.

817. **McFarland, J.W.,** 2,3-Dihydroquinoxaline 1,4-dioxides as intermediates in the reaction between benzofurazan 1-oxide an enamines, *J. Org. Chem.,* 36, 1842, 1971.

818. **Marchetti, L. and Tosi, G.,** Reactions of benzofuroxan and benzofurazan with isomerizable azomethines, *Ann. Chim. (Rome),* 57, 1414, 1967.

819. **Konwar, D., Boruah, R.C., and Sandhu, J.J.,** A base-catalyzed reaction of arylidenemalononitrile with 2,1-benzisoxazole, *Heterocycles,* 23, 2557, 1985.

820. **El-Haj, M.J.A., Dominy, B.W., Johnston, J.D., Haddadin, M.J., and Issidorides, C.H.,** New route to phenazine 5,10-dioxides and related compounds, *J. Org. Chem.,* 37, 589, 1972.

821. **Nazer, M.Z., Issidorides, C.H., and Haddadin, M.J.,** Reactions of benzofurazan oxide with amines. I. Reaction with diethylamine, *Tetrahedron,* 35, 681, 1979.

822. **Haddadin, M.J. and Issidorides, C.H.,** Enamines with isobenzofuroxan: a novel synthesis of quinoxaline di-N-oxides, *Tetrahedron Lett.,* 3253, 1965.

823. **Stuart, K.L.,** Furazans, *Heterocycles,* 3, 651, 1975.

824. a. **Zamet, J.J., Haddadin, M.J., and Issidorides, C.H.,** Reaction of benzofurazan-3(2*H*)-ones, and a new synthesis of benzofuro[2,3-*b*]quinoxalines, *J. Chem. Soc., Perkin Trans. 1,* 1687, 1974; b. **Monge, A., Llamas, A., and Pascual, M.A.,** New synthesis of quinoxaline N,N'-dioxide. Part II, *An. Quim.,* 73, 1208, 1977.

825. **Mufarrij, N.A., Haddadin, M.J., Issidorides, C.H., McFarland, J.M., and Johnston, J.D.,** Reactions of benzofurazan 1-oxides with enamines, *J. Chem. Soc., Perkin Trans. 1,* 965, 1972.

826. **El-Abadelah, M.M., Sabri, S.S., Nazer, M.Z., and Zaater, M.F.,** Synthesis and chiroptical properties of some N-(3-methyl-2-quinoxaloyl) L-amino acids and their dioxides, *Tetrahedron,* 32, 2931, 1976.

827. **Haddadin, M.J., Taha, M.U., Jarrar, A.A., and Issidorides, C.H.,** Reaction of benzofurazan oxide with unsymmetrical 1,3-diketones; steric and polar effects, *Tetrahedron,* 32, 719, 1976.

828. a. **Duerckheimer, W.,** Reactions of benzofuroxans with esters of β-keto acids, *Justus Liebigs Ann. Chem.,* 756, 145, 1972; b. **Sabri, S.S., El-Abadelah, M.M., and Al-Bitar, B.A.,** Synthesis and spectroscopic studies on some new substituted 2-quinoxalinecarboxamides and their N-oxides, *Heterocycles,* 26, 699, 1987.

829. a. **Seng, F. and Ley, K.,** 2-Aminocarbonyl-1-hydroxybenzimidazole 3-oxides, *Synthesis,* 606, 1972; b. **Takabatake, T. and Hasegawa, M.,** A new synthesis of 2,3-disubstituted quinoxaline 1,4-dioxides catalyzed by molecular sieves, *J. Heterocycl. Chem.,* 24, 529, 1987.

830. **Seng, F. and Ley, K.,** New synthesis of lumichrome, *Angew. Chem. Int. Ed. Engl.,* 11, 1010, 1972.

831. **Ley, K., Seng, F., Eholzer, U., Nast, R., and Schubart, R.,** New synthesis of quinoxaline and phenazine di-N-oxides, *Angew. Chem., Int. Ed. Engl.,* 8, 596, 1969.

832. **Seng, F. and Ley, K.,** Simple synthesis of 3-amino-1,2,4-benzotriazine 1,4-dioxide, *Angew. Chem., Int. Ed. Engl.,* 11, 1009, 1972.

833. **Issidorides, C.H., Atfah, M.A., Sabounji, J.J., Sidani, A.R., and Haddadin, M.J.,** Application of 1,2-diketones in the synthesis of phenazine oxides, *Tetrahedron,* 34, 217, 1978.

834. **Kluge, A.F., Maddox, M.L., and Lewis, G.S.,** Formation of quinoxaline monoxides from reaction of benzofurazan oxide with enones and carbon-13 NMR correlations of quinoxaline N-oxides, *J. Org. Chem.,* 45, 1909, 1980.

835. **Lewis, G.S. and Kluge, A.F.,** Benzofurazan oxide chemistry: a novel reaction for the synthesis of quinoxaline monoxides, *Tetrahedron Lett.,* 2491, 1977.

836. **Borah, H.N., Devi, P., Sandhu, J.S., and Baruah, J.N.,** Reaction of benzofuroxans with dienamines; synthesis of a novel class of quinoxaline N,N-dioxide enamines, *Tetrahedron,* 40, 1617, 1984.

837. **Monge, A., Llamas, A., and Pascual, M.A.,** A new synthesis of quinoxaline N,N'-dioxides, *An. Quim.,* 73, 912, 1977.

838. **Sakamoto, M., Shibano, M., and Tomimatsu, Y.,** Cycloaddition reactions of furazan derivatives with anethole, *Yakugaku Zasshi,* 93, 1643, 1973.

839. a. **Ley, K., Seng, F., and Heitzer, H.,** Phenazine-N,N'-dioxide phosphonium betaines, *Synthesis,* 258, 1970; b. **Argyropoulos, N.G., Gallos, J.K., and Nicolaides, D.N.,** Reactions of furoxans with phosphor ylides, *Tetrahedron,* 42, 3631, 1986.

840. **Binder, D., Noe, C.R., Nussbaumer, J., and Prager, B.C.,** Position selective synthesis of pyrido[2,3-*b*]pyrazine-1,4-dioxides, *Monatsh. Chem.,* 111, 407, 1980.

841. a. **von Dobeneck, H., Weil, E., Brummer, E., Deubel, H., and Wolkenstein, D.,** Annelated nitrogen-containing 5,5-ring systems from 3-pyrrolin-2-ones, *Justis Liebigs Ann. Chem.,* 1424, 1978; b. **Vrettou, M.S., Gallos, J.K., and Nicolaides, D.N.,** Reactions of furoxano[3,4-*b*]quinoxaline with alkynes and alkenes. Synthesis of pyrazino[2,3-*b*]quinoxaline 1,4-dioxides, *J. Heterocycl. Chem.,* 25, 813, 1988.

842. **Seng, F. and Ley, K.,** A simple synthesis of 1-hydroxybenzimidazole-2-carboxylic acid, *Synthesis,* 703, 1975.

843. **Latham, D.W.S., Meth-Cohn, O., Suschitzky, H., and Herbert, J.A.L.,** Conversion of benzofurazan *N*-oxides into 2*H*-benzimidazoles and some unusual reaction of 2*H*-benzimidazoles, *J. Chem. Soc., Perkin Trans. 1,* 470, 1977.

844. **Abu El-Haj, M.J.,** Novel synthesis of 1-hydroxy-1*H*-benzimidazole 3-oxides and 2,2-dialkyl-2*H*-benzimidazole 1,3-dioxides, *J. Org. Chem.,* 37, 2519, 1972.

845. **Claypool, D.P., Sidani, A.R., and Flanagan, K.J.,** Benzofurazan oxide. Reaction with sulfur enolate anion, *J. Org. Chem.,* 37, 2372, 1972.

846. **Boulton, A.J., Gripper Gray, A.C., and Katritzky, A.R.,** Alkylation of benzofuroxans with formation of 1-hydroxybenzimidazole 3-oxides, *J. Chem. Soc. (B),* 911, 1967.

847. a.**Borah, H.N., Boruah, R.C., and Sandhu, J.S.,** A novel ring transformation of benzofuroxans to 1-hydroxy-2-arylbenzimidazole 3-*N*-oxide with nitrones, *Heterocycles,* 23, 1625, 1985; b. **Schmidt, J. and Zinner, G.,** 1,3-Dihydroxyureas, part II, *Arch. Pharm.,* 312, 1019, 1979.

848. **Noland, W.E. and Jones, D.A.,** Novel ring expansion and nitrogen insertion reactions of isatogens, *J. Org. Chem.,* 27, 341, 1962.

849. a. **Gasco, A. and Boulton, A.J.,** Furoxans and benzofuroxans, *Adv. Heterocycl. Chem.,* 29, 252, 1981; b. **Burakevich, J.V., Butler, R.S., and Volpp, G.P.,** Phenylfurazan oxide chemistry, *J. Org. Chem.,* 37, 593, 1972.

850. a. **Boulton, A.J., Coe, D.E., and Tsoungas, P.G.,** The isoxazoline rearrangement of furoxans, *Gazz. Chim. Ital.,* 111, 167, 1981; b. **Calvino, R., Ferrarotti, B., Gasco, A., Serafino, A., and Pelizzetti, E.,** Studies on 3-phenylfuroxan, *Gazz. Chim. Ital.,* 113, 811, 1983.

851. **Eremeev, A.V., Adrianov, V.G., and Piskunova, I.P.,** Reaction of dibenzoylfuroxan with primary amines and methylhydrazine, *Khim. Geterotsikl. Soedin.,* 616, 1978.

852. **Rakitin, O.A., Godonkova, T.I., Strelenko, Y.A., and Khmel'hitski, L.I.,** Unusual reaction of chloronitrofuroxan with ammonia, *Isv. Akad. Nauk. SSSR, Ser. Khim.,* 2398, 1986.

853. **Bertelson, R.C., Glanz, K.D., and McQuain, D.B.,** The reactions of 3,4-diacylfuroxans with phenylhydrazine and with aniline, *J. Heterocycl. Chem.,* 6, 317, 1969.

854. **Latham, D.W.S., Meth-Cohn, O., and Suschitzky, H.,** Reaction of benzofurazan *N*-oxide with secondary aliphatic amines; preparation of *N,N*-dialkyl-*N'*-(*o*-nitrophenyl)hydrazines, *J. Chem. Soc., Perkin Trans. 1,* 2216, 1976.

855. **Belton, J.G. and McElhinney, R.S.,** Nucleophilic attack on 4,7-disubstituted benzofurazans and their *N*-oxides: synthesis of tetrazolo[1,5-*a*]azepines, *J. Chem. Soc., Perkin Trans. 1,* 145, 1988.

856. **Tennant, G. and Wallace, G.M.,** Unprecedented aminative ring opening reactions of [1,2,5]oxadiazolo[3,4-*d*]pyrimidines 1-oxides (pyrimidofuroxans), *J. Chem. Soc. Chem. Commun.,* 267, 1982.

857. **Bailey, A.S., Peach, J.M., Prout, C.K., and Cameron, T.S.,** Triphenylphosphine-benzotrifuroxan reaction, *J. Chem. Soc. (C),* 2277, 1969.

858. **Yoneda, F., Tachibana, T., Tanoue, J., Yano, T., and Sakuma, Y.,** A conversion of [1,2,5]oxadiazolo[3,4-*d*]pyrimidine 1-oxides into purines, *Heterocycles,* 15, 341, 1981.

859. **Dornow, A., Fust, K.J., and Jordan, H.D.,** On further reductive cleavage of C-C bonds, *Chem. Ber.,* 90, 2124, 1957.

860. a. **Angeli, A.,** On the configuration of some glyoximes, *Gazz. Chim. Ital.,* 46 (II), 300, 1916; b. **Ponzio, G.,** The peroxide of α-phenylglyoxime, *Gazz. Chim. Ital.,* 66, 119, 1936.

861. **Ponzio, G. and Longo, G.,** About dioximes. LXXI., *Gazz. Chim. Ital.,* 60, 893, 1930.

862. **Meloy, C.R. and Shirley, D.A.,** Arylethanolamines from diaroylfurazan oxides, *J. Org. Chem.,* 32, 1255, 1967.

863. **Thompson, C.D. and Foley, R.T.,** Electrochemical reduction of benzofuroxans. I. Aqueous solutions, *J. Electrochem. Soc.,* 119, 177, 1972.

864. **Bigiavi, D.,** On glyoximes and peroxides, *Gazz. Chim. Ital.,* 51, (II), 324, 1921.

865. **Boulton, A.J., Ghosh, P.B., and Katritzky, A.R.,** Rearrangement of 4-arylazo- and 4-nitroso-benzofuroxans: new syntheses of the benzotriazole and benzofuroxan ring systems. *J. Chem. Soc. (B),* 1004, 1966.

866. **Boulton, A.J., Ghosh, P.B., and Katritzky, A.R.,** Rearrangement of 4-acyl- and 4-imido-alkyl-benzofuroxans: new syntheses of the anthranil and indazole ring system. *J. Chem. Soc. (B),* 1011, 1966.

867. **Ghosh, P.B.,** Preparation and study of some 5- and 7-substituted 4-nitrobenzofurazans and their *N*-oxides. Retro-Boulton-Katritzky rearrangement, *J. Chem. Soc. (B),* 334, 1968.

868. **Balasubrahmanyam, S.N., Radhakrishna, A.S., Boulton, A.J., and Thoe, K.W.,** Interconversion of anthranils, benzofurazan oxides, and indazoles, *J. Org. Chem.,* 42, 897, 1977.

869. a. **Hobin, T.P.,** Some aminodinitro derivatives of benzofurazan and benzofurazanoxide, *Tetrahedron,* 24, 6145, 1968; b. **Boulton, A.J., Frank, F.J., and Huckstep, M.R.,** The rearrangement of a furoxan oxime, *Gazz. Chim. Ital.,* 112, 181, 1982.

870. a. **Freeman, J.P. and Kassner, J.A.,** Base-catalyzed conversion of 2,5-dicarbomethoxy-3,4-diazacyclopentadienone 3,4-dioxide to 3,5-dicarbomethoxy-4-hydroxyisoxazole, *J. Org. Chem.,* 48, 2441, 1983; b. **Unterhalt, B. and Pindur, U.,** Fragmentation reactions of 4-oxo-4*H*-pyrazole 1,2-dioxides, *Arch. Pharm.,* 312, 353, 1979.

871. a. **Argo, C.B., Robertson, I.R., and Sharp, J.T.,** The preparation of 3*H*-1,2-diazepine 2-oxides and their rearrangement to give 3-alkenyl-3*H*-pyrazole 2-oxides, *J. Chem. Soc., Perkin Trans. 1,* 2611, 1984; b. **Gilman, N.W., Blount, J.F., and Sternbach, L.H.,** Base-catalyzed rearrangement of 2-dimethylamino-5-phenyl-7-chloro-3*H*-1,4-benzodiazepine 4-oxide, *J. Org. Chem.,* 37, 3201, 1972; c. **Meguro, K., Natsugari, H., Tawada, H., and Kuwada, Y.,** Reactions of 2-amino-3*H*-1,4-benzodiazepines with primary amines and hydroxylamines, *Chem. Pharm. Bull.,* 21, 2366, 1973; d. **Ning, R.Y., Fryer, R.I., and Sluboski, B.C.,** 1-Hydroxy-1,3-dihydro-2*H*-1,4-benzodiazepin-2-one, a hydroxamic acid via an amidine *N*-oxide, *J. Org. Chem.,* 42, 3301, 1977; e. **Pennings, M.L., Reinhoudt, D.N., Harkema, S., and Van Hummel, G.J.,** Extension of the Woodward-Hoffmann rules to heterocyclic systems: stereospecific thermal isomerization of 1-azacyclobutene 1-oxides, *J. Am. Chem. Soc.,* 102, 7570, 1980; f. **van Eijk, P.J.S.S., Reinhoudt, D.N., Harkema, S., and Visser, R.,** Chemistry of four membered cyclic nitrones. Reaction with a nonnucleophilic base, stereoselective ring opening of *in situ* generated 1,2-dihydroazetes and structure elucidation of the resulting α,β-unsaturated oximes by X-ray analysis, *Rec. Trav. Chim. Pays-Bas,* 105, 103, 1986.

872. **Huisgen, R., Seidl, H., and Brüning, I.,** Kinetics and mechanism of the addition of nitrones to unsaturated compounds, *Chem. Ber.,* 102, 1102, 1969.

873. a. **Confalone, P.N. and Huie, E.M.,** The [3+2]nitrone-olefin cycloaddition reaction, *Org. React.,* 36, 1, 1988; b. **Torsell, K.B.G.,** *Nitrile Oxides, Nitrones and Nitronates in Organic Synthesis,* VCH Pub., Weinheim, 1988.

874. **Mailey, E.A. and Ocone, L.R.,** Fluoroalkylpyridines. A novel rearrangement, *J. Org. Chem.,* 33, 3343, 1968.

875. **Matsuoka, T., Harano, K., and Hisano, T.,** 1,3-Dipolar cycloaddition of aromatic *N*-oxides to *N*-phenylmaleimides, *Chem. Pharm. Bull.,* 31, 2948, 1983.

876. **Hisano, T., Harano, K., Matsuoka, T., Yamada, H., and Kurihara, M.,** Stereoselective exo-cycloaddition of 3,5-lutidine *N*-oxide with *N*-substituted maleimides and a frontier molecular orbital and mechanistic study, *Chem. Pharm. Bull.,* 35, 1049, 1987.

877. **Hamana, M., Funakoshi, K., and Kuchino, Y.,** Reaction of quinoline *N*-oxide with acrylonitrile, *Chem. Pharm. Bull.,* 22, 1806, 1974.

878. **Hamana, M., Funakoshi, K., Shigyo, H., and Kuchino, Y.,** Reactions of quinoline and isoquinoline *N*-oxides and propiolates and methacrylonitrile, *Chem. Pharm. Bull.,* 23, 346, 1975.

879. **Funakoshi, K., Kuchino, Y., Shigyo, H., Sonoda, H., and Hamana, M.,** Re-examination of the reaction of quinoline 1-oxide with acetic anhydride and the related reactions, *Chem. Pharm. Bull.,* 24, 2356, 1976.

880. **Huisgen, R., Seidl, H., and Wulff, J.,** Reactions of aromatic amine oxides with carboxylic esters of the ethylene and acetylene series, *Chem. Ber.,* 102, 915, 1969.

881. **Wittig, G. and Steinhoff, G.,** Azatryptycene, *Liebigs Ann. Chem.,* 676, 21, 1964.

882. **Kabzinska, K. and Wrobel, J.T.,** Thiophene dioxide derivatives. IV. 1,3-Dipolar cycloadditions, *Bull. Acad. Pol. Sci., Ser. Sci. Chim.,* 22, 843, 1974.

883. **Adler, J., Boehnisch, V., and Neunhoeffer, H.,** Reactions of 1,2,4-triazine *N*-oxides with dienophiles, *Chem. Ber.,* 111, 240, 1978.

884. **Huisgen, R. and Gambra, F.P.,** 1,3-Dipolar cycloadditions of aromatic azoxy compounds to strained cyclohexenes, *Tetrahedron Lett.,* 55, 1982.

885. **Haddadin, M.J., Amalu, S.J., and Freeman, J.P.,** Self-condensation of 3*H*-pyrrol-3-one 1-oxides, *J. Org. Chem.,* 49, 2824, 1984.

886. **Doepp, D., Krueger, C., Makedakis, G., and Nour-el-Din, A.M.,** Novel polycyclic linearly conjugated cyclohexadienimines from rearrangement of unstable tetrahydroisoxazolo[2,3-*a*]indoles, *Chem. Ber.,* 118, 568, 1985.

887. **Freeman, J.P. and Hoare, M.J.,** Cycloaddition reactions of 3,4-diazacyclopentadienone oxides with olefins and acetylenedicarboxylic ester, *J. Org. Chem.,* 36, 19, 1971.

888. **Aversa, M.C., Giannetto, P., Ferlazzo, A., and Romeo, G.,** Tetrahydroisoxazolo[2,3-*d*][1,4]benzodiazepinone ring system: synthesis, stereochemistry and conformation, *J. Chem. Soc., Perkin Trans. 1,* 2701, 1982.

889. **Abramovitch, R.A. and Shinkai, I.,** Aromatic substitution via new rearrangements of heteroaromatic *N*-oxides, *Acc. Chem. Res.,* 9, 192, 1976.

890. **Abramovitch, R.A. and Shinkai, I.,** β-Alkylation of pyridine and quinoline 1-oxides with acetylenes. Mechanism. *J. Chem. Soc., Chem. Commun.,* 569, 1973.

891. **Abramovitch, R.A., Grins, G., Rogers, R.B., Atwood, J.L., Williams, M.D., and Crider, C.,** Novel β-alkylation of pyridine and quinoline 1-oxides, *J. Org. Chem.,* 37, 3383, 1972.

892. **Murthi, G.S.S., Gangopadhyay, S.K., and Murthi, M.,** Synthesis of methyl 2-methylpyridine 3-acetate and methyl 2,6-dimethylpyridine 3-acetate, *Indian J. Chem., Sec. B,* 18B, 274, 1979.

893. **Abramovitch, R.A. and Shinkai, I.,** Consecutive 3,5-shifts in pyridine 1-oxide rearrangements. One-step furopyridine synthesis, *J. Am. Chem. Soc.,* 97, 3227, 1975.

894. **Abramovitch, R.A., Kishore, D., Konieczny, M., and Danter, Z.,** Reaction of pyridine 1-oxide with methyl propyolate: a pyridooxepin and other novel products, *Heterocycles,* 25, 13, 1987.

895. **Abramovitch, K.A., Deeb, A., Kishore, D., Mpango, G.B.W., and Shinkai, I.,** The reaction of 4-chloropyridine 1-oxide with activated acetylenes. A convenient one-step synthesis of furo[3,2-c]pyridines, *Gazz. Chim. Ital.,* 118, 167, 1988.

896. **Abramovitch, R.A. and Shinkai, I.,** Reaction of pyridine 1-oxide with benzyne. β-hydroxyarylation of pyridines via ($\sigma^{2s} + \pi^{2a} + \pi^{4s}$) rearrangements, *J. Am. Chem. Soc.,* 96, 5265, 1974.

897. **Morita, N. and Miller, S.I.,** α- and β-rearrangement products, benzoylpyridyltriphenylphosphonium methylides and phenylethynylpyridines, from pyridine N-oxides and phenylethyltriphenylphosphonium bromide, *J. Org. Chem.,* 42, 4245, 1977.

898. **Krebs, A., Colberg, H., Hoepfner, U., Kimling, H., and Odenthal, J.,** Addition of sulfur and pyridine N-oxide to seven membered cycloalkynes, *Heterocycles,* 12, 1153, 1979.

899. **Ishiguro, Y., Funakoshi, K., Saeki, S., Hamana, M., and Ueda, I.,** The reaction of 4-methoxyquinoline 1-oxide with dimethyl acetylene dicarboxylate, *Heterocycles,* 14, 179, 1980.

900. **Canonne, P., Lemay, G., and Abramovitch, R.A.,** The reaction of 4-chloroquinoline 1-oxide with activated acetylenes: furo[3,2-c]quinolines, *Heterocycles,* 9, 1217, 1978.

901. **Kobayashi, Y., Kumadaki, I., and Fujino, S.,** 1,3-Dipolar cycloaddition reaction of aromatic N-oxide with hexafluorobutyne-2, *Heterocycles,* 7, 871, 1977.

902. **Ishiguro, Y., Funagoshi, K., Saeki, S., Hamana, M., Ueda, I., and Kawano, S.,** The reaction of 2-substituted quinoline 1-oxides with dimethyl acetylenedicarboxylate: formation of 1-benzazepine derivatives, *Heterocycles,* 20, 1545, 1983.

903. **Haddadin, M.J. and Atfah, M.A.,** Some addition reactions of 2-substituted quinoxaline 1,4-dioxides, *J. Org. Chem.,* 47, 1772, 1982.

904. **Kaupp, G., Voss, H., and Frey, H.,** Dipyrrolo[1,2-a:2',1'-c]quinoxalines: a new heterocyclene system, *Angew. Chem., Int. Ed. Engl.,* 26, 1280, 1987.

905. **Acheson, R.M. and Selby, I.A.,** Rotational isomerism of 6-aralkyl-5(oxidovinyl)phenanthridiniums, *J. Chem. Soc. Perkin Trans. 1,* 423, 1974.

906. **Acheson, R.M., Ansell, P.J., and Murray, J.R.,** Synthesis and reactions of diethyl ethynylphosphonate and tetraethyl ethynylphosphonate, *J. Chem. Res. (S),* 378, 1986.

907. **Oishi, E., Yamamoto, H., and Hayashi, E.,** Reaction of 1-substituted phthalazine 3-oxides with acetylenic dienophiles, *Yakugaku Zasshi,* 101, 1042, 1981.

908. **Huisgen, R. and Palacios, F.G.,** Aromatic azoxy compounds and strained cycloalkenes, *Chem. Ber.,* 115, 2242, 1982.

909. **Challand, S.R., Rees, C.W., and Storr, R.C.,** Benzo[c]cinnoline N-oxides as 1,3-dipoles, *J. Chem. Soc., Chem. Commun.,* 837, 1973.

910. **Igeta, H., Arai, H., Hasegawa, H., and Tsuchiya, T.,** Reaction of pyridazine N-oxides and pyridazinium ylides with benzyne, *Chem. Pharm. Bull.,* 23, 2791, 1975.

911. **Hasegawa, H., Arai, H., and Igeta, H.,** The reaction of substituted N-acetyliminopyridazinium ylides with benzyne, *Chem. Pharm. Bull.,* 25, 192, 1977.

912. **Oishi, E., Nakamura, K., and Hayashi, E.,** Chemical properties of 5-phenylpyrido[2,3-d]pyridazine 7-oxides, *Yakugaku Zasshi,* 103, 631, 1983.

913. **Freeman, J.P., Kassner, J.A., and Grabiak, R.C.,** Conversion of 1,3,4-oxadiazin-6-one 4-oxides to substituted butenolides, *J. Org. Chem.,* 40, 3402, 1975.

914. **Freeman, J.P. and Grabiak, R.C.,** Reaction of benzyne with 1,3,4-oxadiazin-6-one 4-oxides and related compounds, *J. Org. Chem.,* 41, 2531, 1976.

915. **Ichiba, M., Kanazawa, H., Tamura, Z., and Senga, K.,** A new ring contraction of pteridine 5-oxides to purines and a 7-deazapurine by the 1,3-dipolar cycloaddition reaction, *Heterocycles,* 23, 2317, 1985.

916. **Senga, K., Ichiba, M., and Nishigaki, S.,** The 1,3-dipolar cycloaddition reaction of fervenulin 4-oxides with dimethyl acetylenedicarboxylate, a novel synthesis of pyrrolo[3,2-d]pyrimidines (9-deazapurines), *Heterocycles,* 9, 793, 1978.

917. **Kanazawa, H., Ichiba, M., Shimizu, N., Tamura, Z., and Senga, K.,** Further studies on the ring transformation of pyrimido[5,4-e]-as-triazine 4-oxides to pyrrolo[3,2-d]pyrimidines involving 1,3-dipolar cycloaddition reaction, *J. Org. Chem.,* 50, 2413, 1985.

918. **Senga, K., Ichiba, M., and Nishigaki, S.,** New synthesis of pyrrolo[3,2-d]pyrimidines (9-deazapurines) by the 1,3-dipolar cycloaddition reaction of fervenulin 4-oxides with acetylenic esters, *J. Org. Chem.,* 44, 3830, 1979.

919. **Stauss, U., Haerter, H.P., Neuenshwander, M., and Shindler, O.,** 1,3-Dipolar cycloaddition of acetylenedicarboxylic acid esters to 2-rnethyl-4-phenylquinazoline 3-oxides, *Helv. Chim. Acta,* 55, 771, 1972.

920. **Clark, B.A.J., Evans, T.J., and Simmonds, R.G.,** Preparation and reactions of 2,2-diaryl-2*H*-imidazole 1-oxides, *J. Chem. Soc., Perkin Trans. 1*, 1803, 1975.

921. **Kanazawa, H., Tamura, Z., and Senga, K.,** A new ring transformation of thiazolo[4,5-*g*]quinazoline 3-oxides into [1,4]thiazino[3,2-*g*]quinazolines by the 1,3-dipolar cycloaddition reaction, *Chem. Pharm. Bull.*, 34, 1384, 1986.

922. **Kanazawa, H., Ichiba, M., Tamura, Z., Senga, K., Kawai, K., and Otomasu, H.,** 1,3-Dipolar cycloaddition reactions of thiazolo[5,4-*d*]pyrimidine 1-oxides with acetylenic esters involving new ring transformations of the thiazole nucleus, *Chem. Pharm. Bull.*, 35, 35, 1987.

923. **Butler, R.N., Cunningham, D., Marren, E.G., and McArdle, P.,** 2,5-Dihydro-1,2,3-triazine derivatives, from attempted 1,3-dipolar cycloaddition of 1,2,3-triazole *N*-oxides with dimethyl acetylenedicarboxylate: a new ring expansion, *J. Chem. Soc., Chem. Commun.*, 706, 1987.

924. **Entenmann, G.,** Thiazolo[3,2-*a*]pyrimidones from 2-aminothiazole 3-oxides and esters of acetylenedicarboxylic acid, *Z. Naturforsch, B*, 31b, 251, 1976.

925. **Freeman, J.P., Duthie, E.G., O'Hare, M.J., and Hansen, J.F.,** Novel ring expansion of a diazacyclopentadienone dioxide, *J. Org. Chem.*, 37, 2756, 1972.

926. **Noland, W.E. and Modler, R.F.,** Novel ring expansion and carbon insertion reactions of isatogens, *J. Am. Chem. Soc.*, 86, 2086, 1964.

927. **Senga, K., Ichiba, M., Kanazawa, H., and Nishigaki, S.,** 1,3-Dipolar cycloaddition of a thiazolo[5,4-*d*]pyrimidine 1-oxide to dimethyl acetylenedicarboxylate. New ring transformation to a pyrrolo[3,2-*d*]pyrimidine via a pyrimido[4,5-*b*] [1,4]thiazine, *J. Chem. Soc., Chem. Commun.*, 278, 1981.

928. **Capozzi, G., Ottana, R., Romeo, G., Sindona, G., Uccella, N., and Valle, G.,** Cycloaddition of benzodiazepine nitrones to alkynes, synthesis and x-ray analysis of some tricyclic quinoxalines, *J. Chem. Res. (S)*, 234, 1986.

929. **Capozzi, G., Liguori, A., Ottana, R., Romeo, G., Russo, N., and Uccella, N.,** Regioselectivity of 1,4-benzodiazepinic nitrones in 1,3-dipolar cycloadditions, *J. Chem. Res. (S)*, 96, 1985.

930. **Hisano, T., Matsuoka, T., Ichikawa, M., and Hamana, M.,** Reexamination of the reaction of pyridine 1-oxide with phenyl isocyanate. The isolation of 2,3-dihydropyridine intermediate and the study of the reaction course, *Heterocycles*, 14, 19, 1980.

931. **Hisano, T., Ichikawa, M., Matsuoka, T., Hagiwara, H., Muraoka, K., Komori, T., Harano, K., Ida, Y., and Christensen, A.T.,** Cycloadditions of substituted phenylisocyanates to 3,5-dimethyl and 3,5-dibromopyridine *N*-oxides and x-ray crystal structures of the isomeric cycloadducts, *Chem. Pharm. Bull.*, 27, 2261, 1979.

932. **Hisano, T., Matsuoka, T., and Ichikawa, M.,** Reaction of β-alkylpyridine *N*-oxides with phenylisocyanate, *Chem. Pharm. Bull.*, 24, 533, 1976.

933. **Abramovitch, R.A., Shinkai, I., and van Dahm, R.,** Reaction of pyridine 1-oxides with isocyanates: structure of the intermediates. Rationalization of rearrangements of six-membered heteroaromatic *N*-oxide derivatives, *J. Heterocycl. Chem.*, 13, 171, 1976.

934. **Hisano, T., Matsuoka, T., and Ichikawa, M.,** Reaction of 3,5-dibromopyridine *N*-oxide with phenylisocyanate, *Heterocycles*, 2, 163, 1974.

935. **Hisano, T., Yoshikawa, S., and Muraoka, K.,** Reaction products of 3-picoline 1-oxide with phenylisocyanate, *Chem. Pharm. Bull.*, 22, 1611, 1974.

936. **Hisano, T., Matsuoka, T., Tsutsumi, K., Muraoka, K., and Ichikawa, M.,** Factors affecting the 1,3-cycloaddition of pyridine 1-oxide with phenylisocyanates, *Chem. Pharm. Bull.*, 29, 3706, 1981.

937. **Hisano, T., Matsuoka, T., Fukunaga, K., and Ichikawa, M.,** 1,3-Cycloaddition of 2-substituted pyridine *N*-oxides with phenylisocyanates, *Chem. Pharm. Bull.*, 30, 3776, 1982.

938. **Hisano, T., Matsuoka, T., Ichikawa, M., Muraoka, K., Komori, T., Harano, K., Ida, Y., and Christensen, A.T.,** X-ray crystal structure analysis of isomeric cycloadducts of 3-picoline *N*-oxide with *p*-chlorophenylisocyanate, *Org. Prep. Proced. Int.*, 10, 300, 1978.

939. **Matsuoka, T., Shinada, M., Suematsu, F., Harano, K., and Hisano, T.,** Effect of aromaticity on 1,3-dipolar cycloaddition reactivity of substituted pyridine *N*-oxides and preparation of oxazolo[4,5-*b*]pyridine derivatives, *Chem. Pharm. Bull.*, 32, 2077, 1984.

940. **Matsuoka, T., Harano, K., Kubo, H., and Hisano, T.,** Formation of 1:2 cycloadduct of 2-alkylpyridine *N*-oxides and phenylisocyanates and a frontier molecular orbital study, *Chem. Pharm. Bull.*, 34, 572, 1986.

941. **Harano, K., Suematsu, F., Matsuoka, T., and Hisano, T.,** Further studies on the 1,3-dipolar cycloaddition reaction of pyridine *N*-oxides with phenylisocyanates, *Chem. Pharm. Bull.*, 32, 543, 1984.

942. **Hisano, T., Harano, K., Fukuoka, R., Matsuoka, T., Muraoka, K., and Shinohara, I.,** 1,3-Dipolar cycloaddition reaction of pyridine *N*-oxides with tosyl isocyanate and one-pot synthesis of 2-oxooxazolo[4,5-*b*]pyridine derivatives, *Chem. Pharm. Bull.*, 34, 1485, 1986.

943. **Harano, K., Kondo, R., Murase, M., Matsuoka, T., and Hisano, T.,** Role of charge-transfer complexes in 1,3-dipolar cycloaddition of pyridine *N*-oxide to phenylisocyanate, *Chem. Pharm. Bull.*, 34, 966, 1986.

944. **Huisgen, R.,** 1,3-Dipolar cycloadditions. Past and future, *Angew. Chem. Int. Ed. Eng.*, 2, 565, 1963.

945. **Seidl, H., Huisgen, R., and Grashey, R.,** Some reactions of nitrones and heteroaromatic amine oxides with phenylisocyanate and phenylisothiocyanate, *Chem. Ber.*, 102, 926, 1969.

946. **Hayashi, E.,** Reaction of phenanthridine 5-oxide with phenylisocyanate, *Yakugaku Zasshi,* 81, 1030, 1961

947. **Hamana, M., Noda, H., and Aoyama, M.,** Reaction of 3-substituted quinoline *N*-oxides with phenylisocyanate, *Heterocycles,* 2, 167, 1974.

948. **Yamanaka, H., Niitsuma, S., and Sakamoto, T.,** Reaction of 4-alkoxypyrimidine 1-oxides with phenylisocyanate and phenylisothiocyanate, *Chem. Pharm. Bull.,* 27, 2642, 1979.

949. **Onaka, T.,** Reaction of *N*-sulfinyl-*p*-toluensulfonamide with aromatic amine oxides, *Itsuu Kenkyusho Nempo,* 29, 1968; *Chem. Abst.,* 72, 90231q, 1968.

950. **Takahashi, S. and Kano, H.,** 1,3-Dipolar cycloaddition reaction with 1-methylbenzimidazole 3-oxide, *Tetrahedron Lett.,* 1687, 1963.

951. **Wagner, K., Ley, K., and Ochlmann, L.,** Synthesis of ethyl 1-(arylimino) (S^{IV}) benzothiazole-2-carboxylate, *Chem. Ber.,* 107, 414, 1974.

952. **Honjo, N., Niiya, T., and Goto, Y.,** Reactions of 2,4- and 2,5-disubstituted thiazole *N*-oxides with aryl isocyanates, *Chem. Pharm. Bull.,* 30, 1722, 1982.

953. **Goto, Y. and Honjo, N.,** Reaction of 4-phenyloxazole 3-oxides with arylisocyanates to give imidazolino[4,5-*d*] isoxazolid-2-one derivatives, *Chem. Pharm. Bull.,* 26, 3798, 1978.

954. **Katagiri, N., Niwa, R., Furuya, Y., and Kato, T.,** Reactions of dichloroketene with aromatic amine *N*-oxides, *Chem. Pharm. Bull.,* 31, 1833, 1983.

955. **Freeman, J.P. and Grabiak, R.C.,** Synthesis of 3-dialkylaminocyclopentadienones, *J. Org. Chem.,* 41, 1887, 1976.

956. **Wittig, G. and Steinhoff, G.,** Benzo-azatriptycen, *Chem. Ber.,* 95, 203, 1962.

957. a. **Riediker, M. and Graf, W.,** Isolated olefinic double bonds as 2π components in [8+2]cycloadditions, *Angew. Chem. Int. Ed. Engl.,* 26, 481, 1981; b. **Wada, K., Funakoshi, K., Saeki, S., and Hamana, M.,** Oxidative 1,4-dipolar cycloaddition of quinaldine *N*-oxide with dimethyl acetylenedicarboxylate, *Heterocycles,* 24, 1095, 1986.

958. **Brown, R.F.C., Subrahmanyan, L., and Whittle, C.P.,** Reactions of α-oxo nitrones at nitrone carbon, *Aust. J. Chem.,* 20, 339, 1967.

959. **Iwamura, M. and Inamoto, N.,** Novel formation of nitroxide radicals by radical addition to nitrones, *Bull. Chem. Soc. Jpn.,* 40, 703, 1967.

960. **Janzen, E.G.,** Spin trapping, *Acc. Chem. Res.,* 4, 31, 1971.

961. **Marriot, P.R., Perkins, M.J., and Griller, D.,** Spin trapping for hydroxyl in water: a kinetic evaluation of two popular traps, *Can. J. Chem.,* 58, 803, 1980.

962. **Harbour, J.R. and Hair, M.L.,** Spin trapping of the CO_2- radical in aqueous medium, *Can. J. Chem.,* 57, 1150, 1979.

963. **Bullock, A.T., Gavin, D.L., and Ingram, M.D.,** Electron spin resonance detection of spin-trapped radicals formed during the glow-discharge electrolysis of aqueous solutions, *J. Chem. Soc., Faraday Trans. 1,* 76, 648, 1980.

964. **Finkelstein, E., Rosen, G.M., and Ranckman, E.J.,** Kinetics of the reaction of superoxide and hydroxyl radicals with nitrones, *J. Am. Chem. Soc.,* 102, 4994, 1980.

965. **Kremers, W. and Singh, A.,** Electron spin resonance study of spin-trapped azide radicals in aqueous solution, *Can. J. Chem.,* 58, 1592, 1980.

966. **Bullock, A.T., Gavin, D.L., and Ingram, M.D.,** 5,5-Dimethylpyrroline *N*-oxide: a trap for the unwary, *J. Chem. Soc., Chem. Commun.,* 327, 1980.

967. **Brown, J.K., Coldrick, P.J., and Forbes, E.J.,** 4-Nitrosopyridine 1-oxide and its derivatives. A new range of spin-traps, *J. Chem. Soc., Chem. Commun.,* 770, 1982.

968. **Bertini, F., Caronna, T., and Cecere, M.,** *tert*-alkylation of pyridine *N*-oxide, *Org. Prep. Proced. Int.,* 5, 109, 1973.

969. **Abramovitch, R.A. and Koleoso, O.A.,** Homolytic arylation of aromatic compounds with benzenediazonium tetrafluoborate and pyridine in homogeneous solution, *J. Chem. Soc. (B),* 1292, 1968.

970. **Elofson, R.M., Gadallah, F.F., and Schulz, K.F.,** Homolytic arylation of pyridine and pyridine *N*-oxide and the effect of localization energy and temperature on arylation patterns, *J. Org. Chem.,* 36, 1526, 1971.

971. **Vernin, G., Lebreton, M.A., Don, H.J.M., Metzger, J., and Vernin, G.,** Pseudo Gomberg reaction. Decomposition of pyridinyl 2-aminothiazoles in protic and aprotic media. New route to thiazolylpyridines, *Tetrahedron,* 30, 4171, 1974.

972. **Endo, T., Saeki, S., and Hamana, M.,** Substitution of aromatic *N*-oxides with radicals produced from azo compounds, *Chem. Pharm. Bull.,* 29, 3105, 1981.

973. **Kurbatova, A.S., Kurbatov, Y.V., and Dmitrieva, N.M.,** Diarylhydroxymethylation of pyridine *N*-oxide by anion radicals of some aromatic ketones, *Zh. Org. Khim.,* 16, 648, 1980.

974. **Kurbatova, A.S., Kurbatov, Y.V., and Niyazova, D.A.,** Reaction of benzophenone ketyls with pyridine *N*-oxide, *Zh. Org. Khim.,* 15, 2004, 1979.

975. **Kurbatova, A.S., Kurbatov, Y.V., Niyazova, D.A., and Atayan, P.S.,** Diphenylhydroxymethylation of 2-methylpyridine *N*-oxide by ketyls and dianions of benzophenone, *Zh. Org. Khim.,* 20, 187, 1984.

976. **Kurbatova, A.S., Kurbatov, Y.V., and Niyazova, D.A.,** Reduction of 4-chloropyridine *N*-oxide by metal ketyls and dianions of benzophenone, *Zh. Org. Khim.,* 22, 679, 1986.

977. **Natsume, M., Kumadaki, S., and Tanabe, R.,** Free radical arylation of aromatic amine *N*-oxides, *Itsuu Kenkyusho Nempo,* 25, 1971; *Chem. Abst.,* 77, 61765a, 1971.

978. **Natsume, M., Natsume, S., and Itai, T.,** Five-step synthesis of *N*-alkylmorphinan, *Itsuu Kenkyusho Nempo,* 9, 1968; *Chem. Abst.,* 71, 113119e, 1968.

979. **Natsume, M. and Tanabe, R.,** Syntheses of model intermediates for isoquinoline and indole alkaloids, *Istuu Kenkyusho Nempo,* 21, 1968; *Chem. Abst.,* 71, 124736h, 1968.

980. **Dziembowska, T. and Szafran, M.,** Homolytic amidation of 4-substituted quinoline *N*-oxides, *Rocz. Chem.,* 48, 2293, 1974.

981. **Spence, G.G., Taylor, E.C., and Buchardt, O.,** The photochemical reactions of azoxy compounds, nitrones and aromatic amine *N*-oxides, *Chem. Res.,* 70, 231, 1970.

982. **Albini, A. and Alpegiani, M.,** The photochemistry of the *N*-oxide function, *Chem. Res.,* 84, 43, 1984.

983. **Bellamy, F. and Streith, J.,** The photochemistry of aromatic *N*-oxides. A critical review, *Heterocycles,* 4, 1931, 1976.

984. a. **Yamada, S., Ishikawa, K., and Kaneko, C.,** Photochemistry of acridine 10-oxide, *Chem. Pharm. Bull.,* 23, 2818, 1973; b. **Nastasi, M. and Streith, J.,** Photochemical rearrangements involving three-membered rings, in *Rearrangements in Ground and Excited States,* Vol. 3, de Mayo, P., Ed., Academic Press, New York, 1980, p. 445.

985. **Kaneko, C., Yamada, S., and Yokoe, I.,** Evidence for the formation of oxaziridine during the light irradiation of aromatic amine oxides, *Tetrahedron Lett.,* 2333, 1970.

986. **Tokumura, K., Itoh, M., and Kaneko, C.,** A possible oxaziridine intermediate in the photorearrangement reaction of 6-cyanophenanthridine 5-oxide, *Tetrahedron Lett.,* 2027, 1979.

987. **Tokumura, K., Goto, H., Kashibara, H., Kaneko, C., and Itoh, M.,** Formation and reaction of oxaziridine intermediate in the photochemical reaction of 6-cyanophenanthridine 5-oxide at low temperature, *J. Am. Chem. Soc.,* 102, 5643, 1980.

988. **Kawata, H., Kikuchi, K., and Kokubun, H.,** Studies of the photoreactions of heterocyclic *N,N*-dioxides: identification of the oxaziridine intermediate of quinoxaline 1,4-dioxide, *J. Photochem.,* 21, 343, 1983.

989. **Tomer, K.B., Harrit, N., Rosenthal, I., Buchardt, O., Kumler, P.L., and Creed, D.,** Photochemical behaviour of aromatic 1,2-diazine *N*-oxides, *J. Am. Chem. Soc.,* 95, 7402, 1973.

990. **Lohse, C.,** Primary photoprocesses in isoquinoline *N*-oxides, *J. Chem. Soc. Perkin Trans. 2,* 229, 1972.

991. **Lohse, C., Hagedorn, L., Albini, A., and Fasani, E.,** Photochemistry of pyridine *N*-oxide, *Tetrahedron,* 44, 2591, 1988.

992. **Kaneko, C., Yokoe, I., and Ishikawa, M.,** Photochemical reaction of 2-cyanoquinoline 1-oxides with amines: a new approach to *N*-aminocarbostyrils, *Tetrahedron Lett.,* 5237, 1967.

993. **Albini, A., Fasani, E., and Buchardt, O.,** Radicaloid intermediates in the photochemistry of 6-cyanophenanthridine *N*-oxide, *Tetrahedron Lett.,* 4849, 1982.

994. **Albini, A., Fasani, E., and Frattini, V.,** Medium and substituent effect on the photochemistry of phenanthridine *N*-oxides. Is an intermediate of diradical character involved in the photorearrangement of heterocyclic *N*-oxides?, *J. Chem. Soc., Perkin Trans. 2,* 235, 1988.

995. **Christidis, T.C. and Heinenken, F.W.,** Electron spin resonance and ENDOR study of the photochemical decomposition of substituted quinoxaline bis-*N*-oxides, *J. Chem. Soc., Faraday Trans. 1,* 84, 3263, 1988.

996. **Shukun, L. and Hanqing, W.,** Radical mechanism in the photoreaction of organic *N*-oxides: some paradiazine *N,N*-dioxides, *Heterocycles,* 24, 659, 1986.

997. a. **Hata, N.,** The photochemical magnetic field effect of isoquinoline *N*-oxides, *Bull. Chem. Soc. Jpn.,* 58, 1088, 1985; b. **Hata, N.,** Photochemical magnetic-field effects of isoquinoline *N*-oxide in various alcohols, *Bull. Chem. Soc. Jpn.,* 59, 2723, 1986.

998. a. **Hata, N.,** The effect of external magnetic field on the photochemical reaction of isoquinoline *N*-oxide, *Chem. Lett.,* 547, 1976; b. **Hata, N.,** The effect of external magnetic field on the photochemical reaction of isoquinoline *N*-oxide and 2-cyanoquinoline *N*-oxide, *Chem. Lett.,* 1359, 1978; c. **Hata, N., Ono, Y., and Nakagawa, F.,** The effect of external magnetic field on the photochemical reaction of isoquinoline *N*-oxide in various alcohols, *Chem. Lett.,* 603, 1979.

999. **Bellamy, F. and Streith, J.,** Effects of copper salts on the photochemistry of monosubstituted pyridine *N*-oxides, *J. Chem. Res. (S),* 18, 1979.

1000. a. **Hata, N.,** Photochemical isomerization of quinoline *N*-oxides. Effect of inorganic anions and reactive intermediates, *Nippon Kagaku Kaishi,* 44, 1984; b. **Aloisi, G.G. and Favaro, G.,** Photorearrangement of quinoline 1-oxides: relevance of ground and excited state basicity and effect of heavy atom quenchers, *J. Chem. Soc., Perkin Trans. 2,* 456, 1976.

1001. **Scholz, M.,** MO-calculations of the electronic spectrum of pyridine *N*-oxide. A critical analysis, *J. Prakt. Chem.,* 324, 85, 1982.

1002. **Streith, J. and Sigwalt, C.,** Contraction of aromatic heterocycles by photochemical means, *Tetrahedron Lett.,* 1347, 1966.

1003. **Streith, J. and Sigwalt, C.,** Photochemistry of pyridine N-oxides, *Bull. Soc. Chim. Fr.,* 1157, 1970.

1004. **Bellamy, F., Streith, J., and Fritz, H.,** Photochemical behavior of monosubstituted pyridine N-oxides in water solution, *Nouv. J. Chim.,* 3, 115, 1979.

1005. **Finsen, L., Becher, J., Buchardt, O., and Koganty, R.R.,** N-substituted 5-amino-2,4-pentadienals, their oximes, and 5-amino-2,4-pentadienenitriles, *Acta Chem. Scand.,* B34, 513, 1980.

1006. **Buchardt, O., Christensen, J.J., Nielsen, P.E., Koganty, R.R., Finsen, L., Lohse, C., and Becher, J.,** Photochemical ring-opening of pyridine N-oxide to 5-oxo-2-pentenenitrile and/or 5-oxo-3-pentenenitrile. A reassignment of structure, *Acta Chem. Scand.,* B34, 31, 1980.

1007. **Becher, J., Finsen, L., Winchelmann, I., Koganty, R.R., and Buchardt, O.,** Photochemical and thermal preparation of 5-amino-2,4-pentadienenitriles, *Tetrahedron,* 37, 789, 1981.

1008. **Albini, A., Fasani, E., and Lohse, C.,** Photochemistry of pyridine N-oxide. Trapping of an intermediate, *Heterocycles,* 27, 113, 1988.

1009. **Tsuchiya, T., Arai, H., and Igeta, H.,** Photolysis of pyridazine N-oxides; formation of cyclopropenyl ketones, *J. Chem. Soc. Chem. Commun.,* 550, 1972.

1010. **Tsuchiya, T., Arai, H., and Igeta, H.,** Formation of cyclopropenyl ketones and furans from pyridazine N-oxides by irradiation, *Tetrahedron,* 29, 2747, 1973.

1011. **Tsuchiya, T., Arai, H., Tonami, T., and Igeta, H.,** Photolysis of polyphenylpyridazine N-oxides, *Chem. Pharm. Bull.,* 20, 300, 1972.

1012. **Streith, J., Leibovici, C., and Martz, P.,** Photochemistry of pyrimidine N-oxide, *Bull. Soc. Chim. Fr.,* 38, 4152, 1971.

1013. **Bellamy, F., Martz, P., and Streith, J.,** Pyrimidine mono-N-oxide photochemistry, *Tetrahedron Lett.,* 3189, 1974.

1014. **Ishikawa, M., Kaneko, C., Yokoe, I., and Yamada, S.,** Photolysis of 2,6-dicyanopyridine 1-oxides, *Tetrahedron,* 25, 295, 1969.

1015. **Deeleman, R.A.F. and van der Plas, H.C.,** On the photochemistry of pyrimidine N-oxides, *Rec. Trav. Chim. Pays-Bas,* 92, 317, 1973.

1016. **Roeterdink, F., van der Plas, H.C., and Kandijs, A.,** Photochemistry of pyrimidine N-oxides, *Rec. Trav. Chim. Pays-Bas,* 94, 16, 1975.

1017. **Buchardt, O., Pedersen, C.L., and Harrit, N.,** Light-induced ring expansion of pyridine N-oxides, *J. Org. Chem.,* 37, 3592, 1972.

1018. **Mukai, T. and Sukawa, H.,** Synthesis of 2-phenyl-1,3-oxazepine by the irradiation of 4-phenyl-2,3-oxabicyclo[3.2.0]hepta-3,6-diene, *Tetrahedron Lett.,* 1835, 1973.

1019. **Tezuka, T., Seshimoto, O., and Mukai, T.,** Thermal rearrangement of 2-phenyl-1,3-oxazepine into 1-formyl-2-phenylpyrrole, *Tetrahedron Lett.,* 1067, 1975.

1020. **Streith, J., Danner, B., and Sigwalt, C.,** Photochemistry of pyridine oxide: atomic oxygen transfer in solution: a new synthesis of phenol, *Chem. Commun.,* 979, 1967.

1021. **Neunhoeffer, H. and Böhnisch, V.,** Reactions of 1,2,4-triazine 4-oxides, *Liebigs Ann. Chem.,* 153, 1976.

1022. **Ikekawa, N., Honma, Y., and Kenkyusho, R.,** Photochemical reactions of pyrazine-N-oxides, *Tetrahedron Lett.,* 1197, 1967.

1023. **Kawata, H., Niizuma, S., and Kokubun, H.,** Photoreaction of pyridazine N-dioxide in aqueous solution, *J. Photochem.,* 13, 261, 1980.

1024. **Ager, E., Chivers, G.E., and Suschitzky, H.,** Photochemistry of pentachloropyridine and some derivatives, *J. Chem. Soc., Perkin Trans. 1,* 1125, 1973.

1025. **Ohsawa, A., Arai, H., Igeta, H., Akimoto, T., and Tsuji, A.,** The photoisomerization of pyridazine 1,2-dioxides: formation of 3a,6a-dihydroisoxazolo[5,4-d]isoxazole, *Tetrahedron,* 35, 1267, 1979.

1026. **Albini, A., Bettinetti, G.F., and Minoli, G.,** On 1,3-benzoxazepine and 3,1-benzoxazepine, *Tetrahedron Lett.,* 3761, 1979.

1027. **Albini, A., Bettinetti, G.F., and Minoli, G.,** 1,3-Oxazepines via photoisomerization of heterocycle N-oxides, *Org. Synth.,* 61, 98, 1983.

1028. **Streith, J., Darrah, H.K., and Weil, M.,** Photochemical ring-contraction of quinoline 1-oxide hydrate, *Tetrahedron Lett.,* 5555, 1966.

1029. **Buchardt, O.,** The photolysis of 2-phenyl substituted quinoline N-oxides. A tentative assignment of 4,5-benz-1,3-oxazepine structures to some of the products, *Tetrahedron Lett.,* 6221, 1966.

1030. **Buchardt, O., Kumler, P.L., and Lohse, C.,** Photolysis of phenylquinoline N-oxide in solution. Solvent influence on the products distribution, *Acta Chem. Scand.,* 23, 2149, 1969.

1031. **Hata, N. and Ono, I.,** The primary process of the photochemical isomerization of azanaphthalene N-oxides, *Bull. Chem. Soc. Jpn.,* 49, 1794, 1976.

1032. **Buchardt, O., Jensen, B., and Larsen, I.K.,** The formation of benz[d]-1,3-oxazepines in the photolysis of quinoline N-oxides in solution, *Acta Chem. Scand.,* 21, 1841, 1967.

1033. **Kaneko, C., Yamada, S., and Ishikawa, M.,** Irradiation of N-oxides of α-cyanoazanaphthalenes in aprotic solvents, *Tetrahedron Lett.,* 2145, 1966.

1034. **Kaneko, C. and Yamada, S.,** Isomerization of 1aH-oxazirino[2,3-*a*]-quinoline-1a-carbonitrile and its substituted derivatives to the corresponding 3-hydroxyquinoline derivatives, *Chem. Pharm. Bull.,* 15, 663, 1967.

1035. **Albini, A., Fasani, E., and Maggi Dacrema, L.,** Photochemistry of methoxy substituted quinoline and isoquinoline N-oxides, *J. Chem. Soc., Perkin Trans. 1,* 2738, 1980.

1036. **Kaneko, C., Hayashi, S., and Kobayashi, Y.,** Photolysis of 2-(trifluoromethyl)quinoline 1-oxides and 1-(trifluoromethyl)-isoquinoline 2-oxide, *Chem. Pharm. Bull.,* 22, 2147, 1974.

1037. **Yokoe, I., Ishikawa, M., and Kaneko, C.,** Photolysis of 2-styrylquinoline 1-oxide, *Iyo Kizai Kenkyusho Hokohu, Tokyo Ika Skuka Daigaku,* 6, 18, 1972; *Chem. Abst.,* 81, 105233d, 1974.

1038. **Irvine, R.W., Summers, J.C., and Taylor, W.C.,** Photochemistry of dimethyl quinoline-3,4-dicarboxylate N-oxides, *Aust. J. Chem.,* 36, 1419, 1983.

1039. **Kaneko, C., Fujii, H., Kawai, S., Yamamoto, A., Hashiba, K., Kimata, T., Hayashi, R., and Somei, M.,** A novel synthesis of substituted indoles by photochemical ring contraction of 3,1-benzoxazepines, *Chem. Pharm. Bull.,* 28, 1157, 1980.

1040. **Kaneko, C., Yamamoto, A., and Hashiba, M.,** Ring contraction reactions of methyl quinoline 1-oxide 5-carboxylates via the corresponding benz[*d*]-1,3-oxazepines. A facile synthesis of methyl indole-4-carboxylate and its derivatives, *Chem. Pharm. Bull.,* 27, 946, 1979.

1041. **Kaneko, C. and Kitamura, R.,** Photochemical ring-contraction reactions of benz[*d*]-3,1-oxazepines having no substituent at the 5-position to indole-3-carboxaldehydes, *Heterocycles,* 6, 111, 1977.

1042. **Kaneko, G. and Kitamura, R.,** Photochemical ring contraction reactions of quinoline 1-oxides having a carboxylic acid function in the 4 position to indole-3-carboxaldehydes via the corresponding oxazepines, *Heterocyles,* 6, 117, 1977.

1043. **Buchardt, O., Lohse, C., Duffield, A.M., and Djerassi, C.,** The photolysis of 1-phenyl and 1-cyano substituted isoquinoline N-oxides to benz[*f*]-1,3-oxazepines, *Tetrahedron Lett.,* 2741, 1967.

1044. **Simonsen, O., Lohse, C., and Buchardt, O.,** The photolysis of 1-phenyl and 1-cyano substituted isoquinoline N-oxides to benz[*f*][1,3]oxazepines, *Acta Chem. Scand.,* 24, 268, 1970.

1045. **Brenmer, J.B. and Wiryachitra, P.,** The photochemistry of papaverine N-oxide, *Aust. J. Chem.,* 26, 437, 1973.

1046. **Kaneko, C., Okuda, W., Karasawa, Y., and Somei, M.,** A photochemical synthesis of 4-hydroxyindole, *Chem. Lett.,* 547, 1980.

1047. **Albini, A., Colombi, R., and Minoli, G.,** Photochemistry of quinoxaline 1-oxide and some of its derivatives, *J. Chem. Soc., Perkin Trans. 1,* 924, 1978.

1048. **Kaneko, C., Yamada, S., Yokoe, I., and Ishikawa, M.,** Structure of the stable photo-products derived from quinoline 1-oxides and quinoxaline 1-oxides, *Tetrahedron Lett.,* 1873, 1967.

1049. **Kaneko, C., Yokoe, I., Yamada, S., and Ishikawa, M.,** Photochemical synthesis of 1aH-oxazirino[2,3-a]quinoxaline derivatives and their thermal reactions, *Chem. Pharm. Bull.,* 14, 1316, 1966.

1050. **Field, G.F. and Sternbach, L.H.,** Quinazoline and 1,4-benzodiazepines. XLII. Photochemistry of some N-oxides, *J. Org. Chem.,* 33, 4438, 1968.

1051. a. **Horspool, W.M., Kerskshaw, J.R., Murray, A.W., and Stevenson, G.M.,** Photolysis of 4-substituted 1,2,3-benzotriazine 3-N-oxides, *J. Am. Chem. Soc.,* 95, 2390, 1973; b. **Horspool, W.M., Kerkshaw, J.R., and Murray, A.W.,** Photolysis of 4-methylcinnoline 1- and 2-N-oxides, *J. Chem. Soc., Chem. Commun.,* 345, 1973.

1052. **Burrel, R.A., Cox, J.M., and Savins, E.G.,** Quinoxaline precursors of fungitoxic benzimidazolylcarbamates: syntheses and photochemically-induced transformations, *J. Chem. Soc., Perkin Trans. 1,* 2707, 1973.

1053. **Buchardt, O., Kumler, P.L., and Lohse, C.,** The liquid phase photolysis of quinoline N-oxides unsubstituted in the 2-position, *Acta Chem. Scand.,* 23, 159, 1969.

1054. **Buchardt, O., Becher, J., and Lohse, C.,** Photorearrangement of methyl substituted quinoline-N-oxides in acqueous solution and in potassium bromide discs, *Acta Chem. Scand.,* 20, 2467, 1966.

1055. **Ishikawa, M., Yamada, S., Hotta, H., and Kaneko, C.,** Photochemistry of the N-oxides of azanaphthalenes and their substituted derivatives, *Chem. Pharm. Bull.,* 14, 1102, 1966.

1056. **Buchardt, O., Tomer, K.B., and Madsen, V.,** A photochemical 1,2-deuterium shift. Irradiation of quinoline N-oxide and quinoline N-oxide-2-d$_1$, *Tetrahedron Lett.,* 1311, 1971.

1057. **Ishikawa, M., Yamada, S., and Kaneko, C.,** Photochemical rearrangement of quinaldine 1-oxide, *Chem. Pharm. Bull.,* 13, 747, 1965.

1058. **Jarrar, A.A. and Fataftah, Z.A.,** Photolysis of some quinoxaline-1,4-dioxides, *Tetrahedron,* 33, 2127, 1977.

1059. **Kobayashi, Y., Kumadaki, I., and Sato, H.,** Photolysis of 1,6-naphthyridine 1,6-dioxide and its theoretical consideration, *Tetrahedron Lett.,* 2337, 1970.

1060. **Haddadin, M.J., Agapian, G., and Issidorides, C.H.,** Synthesis and photolysis of some substituted quinoxaline di-N-oxides, *J. Org. Chem.,* 36, 514, 1971.

1061. **Jarrar, A.A., Halawi, S.S., and Haddadin, M.J.,** Photolysis of some quinoxaline 1,4-dioxides, *Heterocycles,* 4, 1077, 1976.

1062. **Albini, A., Bettinetti, G.F., and Minoli, G.,** Photochemistry of some azaphenanthrene N-oxides, *J. Chem. Soc., Perkin Trans. 1,* 1159, 1980.

1063. **Taylor, E.C. and Spence, G.G.,** Further studies on the photochemical reactions of 6-substituted phenanthridine 5-oxides, *Chem. Commun.,* 1037, 1968.

1064. **Kaneko, C., Hayashi, R., Yamamori, M., Tokumura, K., and Itoh, M.,** Photochemical reactions of 6-cyanophenanthridine 5-oxide, *Chem. Pharm. Bull.,* 26, 2508, 1978.

1065. **Taylor, E.C. and Spence, G.G.,** Group migration in the photolysis of 6-substituted phenanthridine 5-oxides, *Chem. Commun.,* 767, 1966.

1066. **Clarcke, S.I. and Prager, R.H.,** Synthetic studies on 5-(3,4-dimethoxyphenyl)-5,6-dihydrophenanthridin-6-ol, an analog of perlonine, *Aust. J. Chem.,* 35, 1645, 1982.

1067. **Ridley, A.B. and Taylor, W.C.,** Synthesis of 6-(3,4-dimethoxyphenyl)-4-oxo-3,4-dihydrobenzo[c][2,7]naphthyridin-6-ium chloride (perlonine chloride), *Aust. J. Chem.,* 40, 631, 1987.

1068. **Prager, R., Duong, T., and Clarke, S.,** The synthesis of dehydroperlonine, *Heterocycles,* 18, 237, 1982.

1069. **Yamada, S., Ishikawa, M., and Kaneko, C.,** Photolysis of 2,7-dimethylacridine 10-oxide, *Tetrahedron Lett.,* 971, 1972.

1070. **Albini, A., Bettinetti, G.F., and Pietra, S.,** Photoreactions of phenazine 5-oxide, *Tetrahedron Lett.,* 3657, 1972.

1071. **Kawata, H., Niizuma, S., and Kokubun, H.,** Photochromism of phenazine N-dioxides in organic solvents, *J. Photochem.,* 9, 463, 1978.

1072. **Yamada, S. and Kaneko, C.,** Photochemistry of acridine 10-oxides: syntheses and reaction of dibenz[c,f]-1,2-oxazepines, *Tetrahedron,* 35, 1273, 1979.

1073. **Yamada, S., Ishikawa, M., and Kaneko, C.,** Identification of the presence of unstable intermediate in the photolyses of acridine 10-oxides, *Tetrahedron Lett.,* 977, 1972.

1074. **Mantsch, H., Zanker, V., and Prell, G.,** 11-Alkoxy-5.11-dihydro-dibenz[b,e]-1,4-oxazepine, product of photoreduction of acridine N-oxide in alcohols, *Liebigs Ann. Chem.,* 723, 95, 1969.

1075. **Albini, A., Bettinetti, G.F., and Pietra, S.,** Photoisomerization of substituted phenazine N-oxides, *Gazz. Chim. Ital.,* 105, 15, 1975.

1076. **Albini, A., Barinotti, A., Bettinetti, G.F., and Pietra, S.,** The effect of substituents on the photoisomerization of phenazine 5-oxide, *J. Chem. Soc., Perkin Trans. 2,* 238, 1977.

1077. **Kawata, H., Niizuma, S., Kumagai, T., and Kokubun, H.,** Photoisomerization of 1-methoxyphenazine N,N-dioxide in protic and aprotic solvents, *Chem. Lett.,* 767, 1985.

1078. **Kaneko, C., Yamada, S., and Ishikawa, M.,** Synthesis of dibenzo[c,g]-2,5-diaza-1,6-oxido[l0]annulene and 12-methyldibenz[c,g]-2-aza-1,6-oxido-[10]annulene, *Tetrahedron Lett.,* 2329, 1970.

1079. **Albini, A., Barinotti, A., Bettinetti, G.F., and Pietra, S.,** The photoisomerization of benzo[a]phenazine 7-oxide, *Gazz. Chim. Ital.,* 106, 871, 1976.

1080. **Brown, G.B., Levin, G., and Murphy, S.,** Purine N-oxides. XII. Photochemical changes induced by ultraviolet radiation, *Biochemistry,* 3, 880, 1964.

1081. **Cramer, F. and Schlingloff, G.,** Photolysis of adenine l-N-oxide, *Tetrahedron Lett.,* 3201, 1964.

1082. **Lam, F.L. and Parham, J.C.,** Photochemistry of 1-hydroxy- and 1-methoxyhypoxanthines, *J. Org. Chem.,* 38, 2397, 1973.

1083. **Lam, F.L. and Parham, J.C.,** Photochemistry of purine 3-oxides in hydroxylic solvents, *Tetrahedron,* 38, 2371, 1982.

1084. **Lam, F.L., Brown, G.B., and Parham, J.C.,** Photoisomerization of 1-hydroxy to 3-hydroxyxanthine. Photochemistry of related 1-hydroxypurines, *J. Org. Chem.,* 39, 1391, 1974.

1085. **Lam, F.L. and Parham, J.C.,** The photoreactions of 6-methyl- and 6,9-dimethylpurine 1-oxides, *J. Am. Chem. Soc.,* 97, 2839, 1975.

1086. **Lam, F.L. and Lee, T.C.,** Photochemistry of some pteridine N-oxides, *J. Org. Chem.,* 43, 167, 1978.

1087. **Bose, S.N., Kumar, S., Davies, J.M., Sethi, S.K., and McCloskey, J.A.,** Conversion of formycin into the fluorescent isoguanosine analogue 7-amino-3-(β-D-ribofuranosyl)-lH-pyrazolo[4,3-d]pyrimidin-5(4H)-one, *J. Chem. Soc., Perkin Trans. 1,* 2421, 1984.

1088. **Timpe, H.J. and Becher, H.G.O.,** The photochemistry of isoelectronic 1,2,4-triazole 4-N-oxide and 4-imino-1,2,4-triazolium-ylide, *J. Prakt. Chem.,* 314, 324, 1972.

1089. **Fielden, R., Meth-Cohn, O., and Suschitzky, H.,** Photorearrangements of benzimidazole N-oxides and related systems, *J. Chem. Soc., Perkin Trans. 1,* 702, 1973.

1090. **Ogata, M., Matsumoto, H., Takahashi, S., and Kano, H.,** Photolysis of l-benzyl-2-ethylbenzimidazole 3-oxide, *Chem. Pharm. Bull.,* 18, 964, 1970.

1091. **Woolhonse, A.D.,** Photoisomerization of tetraaryl-lH-imidazole 3-oxides, *Aust. J. Chem.,* 32, 2059, 1979.

1092. **Haddadin, M.J., Hawi, A.A., and Nazer, M.Z.,** Photolysis of some 1-hydroxybenzimidazole 3-oxides, *Tetrahedron Lett.,* 4581, 1978.

1093. **Pedersen, C.L., Lohse, C., and Poliakoff, M.,** Photolysis of benzo[c]-1,2,5-thiadiazole 2-oxide. Spectroscopic evidence for the reversible formation of 2-thionitrosonitrobenzene, *Acta. Chem. Scand.,* B32, 625, 1978.

1094. **Pedersen, C.L.,** Photolysis of benzo[c]1,2,5-selenadiazole 2-oxide. Spectroscopic evidence for the formation of 2-selenonitronitroso benzene and 2-nitroso N-selenoseleninylaniline, *Tetrahedron Lett.,* 745, 1979.

1095. **Calzaferri, G., Gleiter, R., Knauer, K.H., Martin, H.D., and Schmidt, E.,** Photochromism of 4-substituted benzofuroxans, *Angew. Chem. Int. Ed. Engl.,* 13, 86, 1974.

1096. **Bryce, M.R., Reynolds, C.D., Hanson, P., and Vernon, J.M.,** Formation and rearrangement of adducts from benzyne and substituted 2,1,3-benzoselenadiazoles, *J. Chem. Soc., Perkin Trans. 1,* 607, 1981.

1097. **Braun, H.P., Zeller, K.P., and Meier, H.,** Photolysis of 1,2,3-thiadiazole 2-oxide, *Liebigs Ann. Chem.,* 1257, 1975.

1098. **Servé M.P., Feld, W.N., Seybold, P.C., and Steppel, R.N.,** Synthesis of 1-methyl-1,2,3-benzotriazole 2-oxide, *J. Heterocycl. Chem.,* 12, 811, 1975.

1099. **Albini, A., Fasani, E., and Frattini, V.,** Photochemistry of 2- and 4-benzoylpyridine N-oxides, *J. Photochem.,* 37, 355, 1987.

1100. **Lin, S.K. and Feng, L.B.,** An ESR study of the oxygen transfer from excited heteroaromatic N-oxides to triethylamine to form diethyl nitroxide, *Chem. Phys. Lett.,* 128, 319, 1986.

1101. **Lin, S.K.,** Radical mechanism in the photoreaction of organic N-oxides: oxygen transfer from photoexcited paradiazine di-N-oxides to amines, *J. Photochem.,* 37, 363, 1987.

1102. **Seki, H., Takematsu, A., and Arai, S.,** Photoinduced electron transfer from amino acids and proteins to 4-nitroquinoline 1-oxide in aqueous solutions, *J. Phys. Chem.,* 91, 176, 1987.

1103. **Hata, N., Ono, I., and Kawasaki, M.,** Photoinduced deoxygenation reaction of heterocyclic N-oxides, *Chem. Lett.,* 25, 1975.

1104. **Kaneko, C., Yamamori, M., Yamamoto, A., and Hayashi, R.,** Irradiation of aromatic amine oxides in dichloromethane in the presence of triphenylphosphine: a facile deoxygenation procedure of aromatic amine N-oxides, *Tetrahedron Lett.,* 2799, 1978.

1105. **Pietra, S., Bettinetti, G.F., Albini, A., Fasani, E., and Oberti, R.,** Photochemical reactions of nitrophenazine 10-oxides with amines, *J. Chem. Soc., Perkin Trans. 2,* 185, 1978.

1106. **Kaneko, C., Yamamoto, A., and Gomi, M.,** A facile method for the preparation of 4-nitropyridine and -quinoline derivatives: reduction of aromatic amine N-oxides with triphenylphosphite under irradiation, *Heterocycles,* 12, 227, 1979.

1107. **Boekeheide, V. and Linn, W.J.,** A novel synthesis of pyridyl carbinols and aldehydes, *J. Am. Chem. Soc.,* 76, 1286, 1954.

1108. **McKillop, A. and Bhagrath, M.K.,** The Katada reaction: a study of experimental conditions, *Heterocycles,* 23, 1697, 1985.

1109. **Konno, K., Hashimoto, K., Shirahama, H., and Matsumoto, T.,** Improved procedures for preparation of 2-pyridones and 2-(hydroxymethyl)pyridines from pyridine N-oxides, *Heterocycles,* 24, 2169, 1986.

1110. **Wieczorek, J.S. and Wojtowski, R.,** Reaction of quinaldine N-oxide and trichloroacetylchloride, *Bull. Acad. Pol. Sci., Ser. Sci. Chim.,* 20, 309, 1972.

1111. **Kato, T., Hamaguchi, F., and Oiwa, T.,** Reaction of 2-picoline 1-oxide with ketene, *Yakugaku Zasshi,* 78, 422, 1958.

1112. **Kato, T., Goto, Y., and Yamamoto, Y.,** Reaction of methylpyridine 1-oxide and methylquinoline 1-oxide with ketene, *Yakugaku Zasshi,* 82, 1649, 1962.

1113. **Oal, S., Tamagaki, S., Negoro, T., Ogino, K., and Kozuka, S.,** Kinetic studies on the reactions of 2- and 4-alkyl-substituted heteroaromatic N-oxides with acetic anhydride, *Tetrahedron Lett.,* 917, 1968.

1114. **Kozuka, S., Tamagaki, S., Negoro, T., and Oae, S.,** Uneven distribution of ^{18}O in the resulting esters formed in the reaction of 2-picoline, 2,6-lutidine and quinaldine N-oxide with acetic anhydride, *Tetrahedron Lett.,* 923, 1968.

1115. **Bodalski, R. and Katritzky, A.R.,** The mechanism of the acetic anhydride rearrangement of 2-alkylpyridine 1-oxides, *J. Chem. Soc. (B),* 831, 1968.

1116. **Berg-Nielsen, K. and Skattebol, L.,** Evidence for a ionic mechanism for the reaction of 2-cyclopropylpyridine N-oxide with acetic anhydride, *Acta Chem. Scand.,* B29, 985, 1975.

1117. **Cohen, T. and Deets, G.L.,** Trapping of pycolyl cations in the reaction of 2- and 4-picoline N-oxide with acetic anhydride, *J. Am. Chem. Soc.,* 94, 932, 1972.

1118. **Furukawa S.,** Mechanism of rearrangement of picoline 1-oxide derivatives with acetic anhydride, *Yakugaku Zasshi,* 59, 487, 1959.

1119. **Cohen, T. and Fager, J.H.,** The reaction of 2- and 4-picoline N-oxide with phenylacetic anhydride, *J. Am. Chem. Soc.,* 87, 5701, 1965.

1120. **Jenkins, J.A. and Cohen, T.,** Deuterium isotope effects and the influence of solvent in the redox and rearrangement reactions of 2-picoline N-oxide and phenylacetic anhydride, *J. Org. Chem.,* 40, 3566, 1975.

1121. **Iwamura, H., Iwamura, M., Nishida, T., and Miura, I.,** CIDNP in the reaction of 2-picoline N-oxide with acetic anhydride, *Tetrahedron Lett.,* 3117, 1970.

1122. **Kato, T., Kitagawa, T., Shibata, T., and Nakai, K.,** Reaction of hydroxy derivatives of 2,6-lutidine 1-oxide with acetic anhydride, *Yakugaku Zasshi,* 82, 1647, 1962.

1123. **Iwamura, H., Iwamura, M., Nishida, T., and Sato, S.,** The mechanism of the reaction of 4-picoline N-oxide with acetic anhydride studied by nuclear magnetic resonance spectroscopy, *J. Am. Chem. Soc.,* 92, 7474, 1970.

1124. a. **Traynelis, V.J. and Martello, R.F.,** The mechanism of the reaction of 4-picoline *N*-oxide with acetic anhydride, *J. Am. Chem. Soc.,* 82, 2744, 1960; b. **Traynelis, V.J., Yamauchi, K., and Kimball, J.P.,** Nature of the rearrangement step in the reaction of 4-alkylpyridine *N*-oxides and acid anhydrides, *J. Am. Chem. Soc.,* 96, 7289, 1974.

1125. **Furukawa, S.,** Preparation of several derivatives of 2-pyridinealdehyde, *Yakugaku Zasshi,* 77, 11, 1957.

1126. **Nishimoto, N. and Nakashima, T.,** Synthesis of 2-acyl-4-alkylpyridines, *Yakugaku Zasshi,* 81, 88, 1961.

1127. **Hardegger, E. and Nikles, E.,** Preparation of fusarinic acid from 2,5-1utidine, *Helv. Chim. Acta,* 40, 2428, 1957.

1128. **Ginsburg, S. and Wilson, I.B.,** Oximes of the pyridine series, *J. Am. Chem. Soc.,* 79, 481, 1957.

1129. **Kobayashi, G., Furukawa, S., and Kawada, Y.,** Reaction of 2,6-lutidine 1-oxide with acetic anhydride, *Yakugaku Zasshi,* 74, 790, 1954.

1130. **Popp, F.D. and McEwen, W.E.,** Approaches to the synthesis of emetine from Reissert compounds, *J. Am. Chem. Soc.,* 80, 1181, 1958.

1131. a. **Endo, M. and Nakashima, T.,** Syntheses of 4-substituted compounds of picoline acid and 5-ethylpicolinic acid, *Yakugaku Zasshi,* 80, 875, 1960; b. **Schnekenburger, J.,** Benz[1,2-*b*]indolizines as secondary products in rearrangement of 2-substituted pyridine *N*-oxides, *Arch. Pharm.,* 306, 360, 1973.

1132. **Michalski, J., Piechucki, C., and Zajac, H.,** Reaction of 1,2-di(2-pyridyl)ethane *N,N'*-dioxide with acetic anhydride, *Bull. Acad. Pol. Sci., Ser. Sci. Chim.,* 14, 505, 1966.

1133. **Achremowicz, L. and Syper, L.,** Reactions of 3-nitropicoline *N*-oxides with acetic anhydride, *Rocz. Chem.,* 46, 409, 1972.

1134. **Sliwa, W. and Skrowaczewska, Z.,** 6-Methyl-3-aminopyridine-2-carboxaldehyde by rearrangement of 3-acetamido-2,6-lutidine *N*-oxide, *Rocz. Chem.,* 44, 1941, 1970.

1135. **Murakami, Y. and Sunamoto, J.,** Reaction of 6-methylpicolinic acid *N*-oxide with acetic anhydride, *Bull. Chem. Soc. Jpn.,* 42, 3350, 1969.

1136. a. **Kobayashi, G., Furukawa, S., Akimoto, Y., and Hoshi, T.,** Reaction of lepidine 1-oxide and quinoline 1-oxide with acetic anhydride, *Yakugaku Zasshi,* 74, 791, 1954; b. **Muth, C.W., Darlak, R.S., DeMatte M.L., and Chovanec, G.F.,** Some rate studies and their mechanistic implications for the reaction of 2-methylquinoline 1-oxide and acetic anhydride, *J. Org. Chem.,* 33, 2762, 1968.

1137 **Furukawa, S.,** Reaction of 2,4-lutidine 1-oxide and 2,4-dimethylquinoline 1-oxide with acetic anhydride, *Pharm. Bull.,* 3, 413, 1955.

1138. **Tamagaki, S., Ogino, K., Kozuka, S., and Oae, S.,** Reaction of lepidine, quinaldine and 1-methylisoquinoline *N*-oxides with acetic anhydride, *Tetrahedron,* 26, 4675, 1970 .

1139. **Meghani, P., Street, J.D., and Joule, J.A.,** A synthetic approach to benzo[1,2-*b*:4,3-*b'*]dipyrroles from isoquinolines, *J. Chem. Soc., Chem. Commun.,* 1406, 1987.

1140. a. **Sueyoshi, S. and Suzuki, I.,** Nucleophilic substitution of 3,6-dimethylpyridazine 1,2-dioxide, *Yakugaku Zasshi,* 95, 1327, 1975; b. **Ogata, M. and Kano, H.,** 3-Methylpyridazine *N*-oxides, *Chem. Pharm. Bull.,* 11, 29, 1963; c. **Tikhonov, A.Y., Volodarskii, L.B., Vakolova, O.A., and Podgarnaya, M. I.,** Reaction of 2,4,6-trialkylpyrimidine 1,3-dioxides with electrophilic reagents, *Khim. Geterotsikl. Soedin,* 110, 1981; d. **Sakamoto, T., Yoshizawa, H., Kaneda, S., and Yamanaka, H.,** Substituent effect on the reaction of 4-substituted 2,6-dimethylpyrimidine 1-oxides with acetic anhydride, *Chem. Pharm. Bull.,* 32, 728, 1984.

1141. a. **Gumprecht, W.H., Beukelman, T.E., and Paju, R.,** Identification and separation of the isomeric 2-methylpyrazine mono *N*-oxides, *J. Org. Chem.,* 29, 2477, 1964; b. **Klein, B., Berkowitz, J., and Hetman, N.E.,** The rearrangement of pyrazine-*N*-oxides, *J. Org. Chem.,* 26, 126, 1961.

1142. **Elina, A.S., Musatova, I.S., and Syrova, G.P.,** Synthesis and properties of 2- and 2,3-substituted pyrazine *N*-oxide, *Khim. Geterotiskl. Soedin,* 1275, 1972.

1143. **Elina, A.S.,** Oxidation-reduction reactions of *N*-oxides of 2, 3-dimethylquinoxaline, *Zhur. Obshei Khim.,* 31, 1018, 1961; b. **Haddadin, M.J. and Issidorides, C.H.,** Quinoxaline mono-*N*-oxide derivatives: a facile elimination, *Tetrahedron Lett.,* 4609, 1968.

1144. **Haddadin, M.J. and Salameh, A.S.,** Reactions of 1,2,3,4-tetrahydrophenazine di-*N*-oxide with acetic anhydride, *J. Org. Chem.,* 33, 2127, 1968.

1145. a. **Stevens, M.A., Giner-Sorolla, A., Smith, H.W., and Brown, G.B.,** Purine *N*-oxides. X. The effect of some substituents on stability and reactivity, *J. Org. Chem.,* 27, 567, 1962; b. **Sutherland, D.K. and Brown, G.B.,** Reactions of 3-acetoxy-8-methylxanthine, *J. Org. Chem.,* 38, 1291, 1973; c. **Zondler, H., Forrest, H.S., and Lagowski, J.M.,** Synthesis of 6-hydroxymethyl-1,3-dimethyllumazine by rearrangement of the corresponding 6-methyllumazine 5-oxide, *J. Heterocycl. Chem.,* 4, 124, 1967.

1146. **Ochiai, E., Takahashi, M., and Tanabe, R.,** Effect of acetic anhydride on 5,6,7,8-tetrahydroquinaldine 1-oxide, *Chem. Pharm. Bull.,* 15, 1385, 1967; b. **Ochiai, E. and Takahashi, M.,** Reaction of 3-methyl-5,6,7,8-tetrahydroisoquinoline 2-oxide with acetic anhydride, *Itsuu Kenkyusho Nempo,* 15, 1971; *Chem. Abst.,* 77, 48172z, 1972.

1147. **Epsztajn, J., Hahn, W.E., and Bieniek, A.,** Reaction of 2-methylcycloalkeno[*b*]pyridine *N*-oxides with acetic acid anhydride, *Rocz. Chem.,* 50, 1681, 1976.

1148. a. **Klimov, G.A. and Tilichenko, M.N.,** Synthesis based on the 9-chloro-*sym*-octahydroacridine *N*-oxide, *Khim. Geterostikl. Soedin.*, 572, 1969; b. **Robinson, M.M.,** The preparation of 1,5-pyrindine, *J. Am. Chem. Soc.*, 80, 6254, 1958.

1149. **Klimov, G.A., Tilichenko, M.N., and Karaulov, Y.S.,** Synthesis based on *syn*-octahydro-4-acridinol *N*-oxide, *Khim. Geterotsikl. Soedin*, 297, 1969.

1150. **Parham, W.E. and Olson, P.E.,** Stereochemistry of reactions of heterocyclic *N*-oxides with acetic anhydride, acetyl chloride, and *p*-toluenesulfonyl chloride, *J. Org. Chem.*, 39, 2916, 1974.

1151. **Parham, W.E., Sloan, K.B., Reddy, K.R., and Olson, P.E.,** Reaction of aromatic amine oxides with acid halides, sulfonyl halides, and phosphorous oxychloride. Stereochemical configuration of substituents in the 1-position of 12,13-benzo-16-chloro[10](2,4)-pyridinophanes, *J. Org. Chem.*, 38, 927, 1973.

1152. **Parham, W.E., Davenport, R.W., and Biasotti, J.B.,** 1.3-Bridged aromatic systems. VII. Quinolines, *J. Org. Chem.*, 35, 3775, 1970.

1153. **Eish, J.J., Gopal, H., and Kuo, C.T.,** Synthesis and tautomeric character of cyclopenta[*c*]quinoline (benzo[*c*][2]pyrindine), *J. Org. Chem.*, 43, 2190, 1978.

1154. **Epsztajn, J. and Bieniek, A.,** A general study of the reaction of lithium alkyls with pyridine ketones, *J. Chem. Soc., Perkin Trans. 1*, 213, 1985.

1155. a. **Vitolo, M.J. and Marquez, V.E.,** Cis and trans isomers of 3,9-dimethyl-4b,5,6,10b,11,12-hexahydroquino[8,7-*h*]quinoline, *J. Org. Chem.*, 42, 2187, 1977; b. **Hahn, W.E. and Tomczyk, D.,** Reactions of 2,3-quinolinocycloparaffins, *Rocz. Chem.*, 46, 837, 1972.

1156. **Kende, A.S., Veits, J.E., Lorah, D.P., and Ebetino, F.H.,** Anomalous condensation of a β-ketoenamine with a tetronic acid, *Tetrahedron Lett.*, 2423, 1984.

1157. **Giam, C.S. and Ambrozich, D.,** A new approach to the synthesls of 1,6- and 1,7-naphthyridines, *J. Chem. Soc, Chem. Commun.*, 265, 1984.

1158. **Neunhoeffer, H. and Metz, H.J.,** Synthesis of benzo[*f*]cyclopent[*c*]lisoquinolines and benzo[*n*]cyclopent[*c*]isoquinolines (12-azasteroids), *Liebigs Ann. Chem.*, 1476, 1983.

1159. **Kessar, S.V., Sobti, A.K., and Joshi, G.S.,** Synthesis of 7-hydroxy-2-methoxy-7,8, 9, 10-tetrahydrobenzo[*i*]phenanthridine, *J. Chem. Soc. (C)*, 259, 1971.

1160. **Chang, J.C., El-Sheikh, M., Harmon, A., Avasthi, K., and Cook, J.M.,** Synthesis of 1,6-diazaphenalene, *J. Org. Chem .*, 46, 4188, 1981.

1161. **Walker, G.N., Engle, A.R., and Kempton, R.J.,** Novel syntheses of 1,4-benzodiazepines, isoindolo[2,1-*d*][1,4]benzodiazepines, isoindolo[1,2-*a*][2]benzazepines, and indolo[2,3-*d*][2]benzazepine, based on the use of the Strecker reaction, *J. Org. Chem.*, 37, 3755, 1972.

1162. a. **Jones, G., Jones, R.K., and Robinson, M.J.,** Synthesis of 4*H*-cyclohepta[*b*]thiophen-4-ones, 4*H*-cyclohepta[*b*]furan-4-one, and 9*H*-cyclohepta[*b*]pyridin-9-one, *J. Chem. Soc., Perkin Trans. 1*, 968, 1973; b. **O'Leary, M.H. and Payne, J.R.,** Improved synthesis of 3-hydroxypyridine-4-carboxyaldehyde, *J. Med. Chem.*, 14, 773, 1971.

1163. **Boyle, F.T. and Jones, R.A.Y.,** Deoxygenation and side-chain substitution reactions of 1-methoxypyrazole 2-oxides, *J. Chem. Soc. Perkin Trans. 1*, 170, 1973.

1164. **Ciba-Geigy, A.-G.,** 2-Phenyl-2*H*-1,2,3-triazoles, *Jpn. Kokai* 56/133272, 1981; *Chem. Abst.*, 96, 104252g, 1982.

1165. **Anderson, H.J., Barnes, D.J., and Khan, Z.M.,** The rearrangment of 2,4-dimethylthiazole 3-oxide with acetic anhydride, *Can. J. Chem.*, 42, 2375, 1964.

1166. a. **Hass, W. and Koenig, W.A.,** Synthesis of *N*-terminal aminoacids of nikkomycins I, J, X, and Z, *Liebigs Ann. Chem.*, 1615, 1982; b. **Brana, M.F. and Lopez Rodriguez, M.L.,** Synthesis and reactivity of *N*-[(α-acetoxy)-4-pyridylmethyl]-3,5-dimethylbenzamide, *J. Heterocycl. Chem.*, 18, 869, 1981.

1167. **Ikeda, K., Tsuchida, K., Monma, T., and Mizuno, Y.,** Reaction of 2-picolyl 1-oxide derivatives with acetic anhydride, *J. Heterocycl. Chem.*, 11, 321, 1974.

1168. a. **Elina, A.S., Tsyrul'nikova, L.G., and Syrova, G.P.,** Redox reactions in a series of α-hydroxymethylquinoxaline *N*-oxides, *Khim. Geterotikl. Soedin*, 149, 1969; b. **Eholzer, U., Heitzer, H., Seng, F., and Ley, K.,** Redox reactions with 2-chloromethylquinoxaline di-*N*-oxide, *Synthesis*, 296, 1974; c. **Anderson, R.C. and Fleming, R.H.,** Pyrrolo[3,4-*b*]quinoxalines. Stable analogs of benz[*f*]isoindole, *Tetrahedron Lett.*, 1581, 1969.

1169. **Muth, C.W., Patton, J.C., Bhattacharya, B., Giberson, D.L., and Ferguson, C.A.,** Base-catalyzed reactions of α-hydroxyalkylazaaromatic *N*-oxides, *J. Heterocycl. Chem.*, 9, 1299, 1972.

1170. **Elina, A.S., Musatova, I.S., Titkova, R.M., Dubinskii, R.A., and Goizman, M.S.,** Redox transformation of quinoxaline-2-aldehyde di *N*-oxide hydrate and diethyl acetal in an alkaline medium, *Khim. Geterotsikl. Soedin*, 1106, 1980.

1171. a. **Traynelis, V.J. and Yamauchi, K.,** Reactions of 4-alkylpyridine 1-oxides with dimethyl sulfoxide, *Tetrahedron Lett.*, 3619, 1969; b. **Sammes, P.G., Serra-Errante, G., and Tinker, A.C.,** Intramolecular aromatic hydroxylation via irradiation of pyridine *N*-oxides, *J. Chem. Soc., Perkin Trans. 1*, 853, 1978; c. **Carlson, R.M. and Heinis, L.J.,** Intramolecular rearrangements of tertiary amine oxides, *J. Org. Chem.*, 44, 2530, 1979.

1172. **Koenig, T.W. and Wieczorek, J.S.,** Reaction of trichloroacetyl chloride with 2-picoline *N*-oxide and pyridylcarbinols, *J. Org. Chem.,* 33, 1530, 1968.

1173. **Matsumura, E.,** The reaction between 2-picoline 1-oxide and *p*-toluensulfonyl chloride, *Nippon Kagaku Zasshi,* 74, 363, 1953.

1174. **Matsumura, E., Hirooka, T., and Imagawa, K.,** The reaction of *N*-oxides of pyridine and thiazole homologs with *p*-tosylchloride, *Nippon Kagaku Zasshi,* 82, 616, 1961.

1175. **Matsumura, E., Nashima, T., and Ishibashi, F.,** The synthesis of 1-methyl-2-pyridones from 2-chloromethylpyridines, *Bull. Chem. Soc. Jpn.,* 43, 3540, 1970.

1176. **Tanida, H.,** Synthesis of chloromethylquinoline from quinaldine 1-oxide and lepidine 1-oxide, *Yakugaku Zasshi,* 78, 611, 1958.

1177. a. **Tsai, C.Y. and Sha, C.K.,** 2*H*-Pyrrolo[3,4-*b*]pyridine and 2*H*-pyrrolo[3,4-*c*]pyridine: synthesis of the parent ring system and the Diels-Alder reaction, *Tetrahedron Lett.,* 1419, 1987; b. **Hayashi, E. and Shimada, N.,** *p*-Tolylsulfonylation of the side-chain carbon in 2-methylquinoline 1-oxide and related compounds, *Yakugaku Zasshi,* 98, 1503, 1978; c. **Schichiri, K., Funakoshi, K., Saeki, S., and Hamana, M.,** Reactions of quinoline 1-oxide derivatives with tosyl chloride in the presence of triethylamine, *Chem. Pharm. Bull.,* 28, 493, 1980.

1178. **Taylor, E.C. and Cheeseman, G.W.H.,** Synthesis and properties of pyrrolo[1,2-*a*]quinoxalines, *J. Am. Chem. Soc.,* 86, 1830, 1964.

1179. **Cohen, S., Thom, E., and Bendich, A.,** The preparation and properties of 6-halomethylpurines, *J. Org. Chem.,* 27, 3545, 1962.

1180. **Bauer, L. and Gardella, L.A.,** The reaction of 1-alkoxypicolinium salts with mercaptide and thiophenoxide ions, *J. Org. Chem.,* 28, 1323, 1963.

1181. **Sakamoto, T., Yoshizawa, H., Kaneda, S., Hama, Y., and Yamanaka, H.,** Reaction of 4-substituted 2,6-dimethylpyrimidine 1-oxides with phosphoryl chloride, *Chem. Pharm. Bull.,* 31, 4533, 1983.

1182. **Brana, M.F., Lopez Rodriguez, M.L., Garrido, J., and Roldan, C.M.,** Synthesis of *N*,*N*′-diacyl-1,2-di-(4-pyridyl)ethylenediamines, *J. Heterocycl. Chem.,* 18, 1305, 1981.

1183. **Brana, M.F., Castellano, J.M., and Lopez Rodriguez, M.L.,** Improvement in the synthesis of *N*,*N*′-diacyl-1,2-di-(4-pyridyl)ethylenediamines, *J. Heterocycl. Chem.,* 20, 1723, 1983.

1184. **Ohsawa, A., Kawaguchi, T., and Igeta, H.,** Flash vacuum pyrolysis of 2-benzylpyridine *N*-oxides. Synthesis of methylpyrido[1,2-*a*]indoles, *Chem. Lett.,* 1737, 1981.

1185. **Oshawa, A., Kawaguchi, T., and Igeta, H.,** Flash vacuum pyrolisis of substituted pyridine *N*-oxides and its application to syntheses of heterocyclic compounds, *J. Org. Chem.,* 47, 3497, 1982.

1186. **Brana, M.F., Lopez Rodriguez, M.L., Rodriguez, C., and Garrido-Pertierra, A.,** Reaction of *N*-(4-pyridylmethyl)benzamide *N*-oxides with ethyl cyanoacetate in the presence of acetic anhydride, *J. Heterocycl. Chem.,* 23, 1019, 1986.

1187. **Brana, M.F., Castellano, J.M., Redondo, M.C., and Yunta, M.J.R.,** Reaction of 4-(acylaminomethyl)pyridine *N*-oxides with 5-monosubstituted barbituric acid in the presence of acetic anhydride, *J. Heterocycl. Chem.,* 24, 833, 1987.

1188. **Brana, M.F., Castellano, J.M., Redondo, M.C., and Esplugues, J.,** Reaction of 4-acylaminomethylpyridine *N*-oxides with phenylbutazone in the presence of acetic anhydride, *J. Heterocycl. Chem.,* 24, 741, 1987.

1189. **Hamana, M. and Funakoshi, K.,** Reaction of pyridine 1-oxides with tosyl chloride in the presence of pyridine, *Yakugaku Zasshi,* 82, 516, 1962.

1190. **Brana, M.F. and Yunta, J.R.,** Structural analogues of picobenzide. I. Synthesis and reactivity of alkylthioethers, *An. Quim., Ser. C,* 78, 210, 1982.

1191. **Abramovitch, R.A., Abramovitch, D.A., and Tomasik, P.,** Mechanism of direct side-chain acylamination and aminoarylation of 2- and 4-picoline 1-oxides, *J. Chem. Soc., Chem. Commun.,* 561, 1981; *ibid.,* 956, 1979.

1192. **Chilton, W.S. and Bultler, A.K.,** Preparation of aldehyde derivatives from picoline *N*-oxides, *J. Org. Chem.,* 32, 1270, 1967.

1193. **Mizuno, Y. and Kobayashi, J.,** Novel reaction involving 2-formylpyridine *N*-oxides and cyclohexyl isocyanide, *J. Chem. Soc., Chem. Commun.,* 308, 1975.

1194. **Gallagher, A.I., Lalinsky, B.A., and Cuper, C.M.,** Base-catalyzed deuterium exchange in pyridine *N*-oxides, *J. Org. Chem.,* 35, 1175, 1970.

1195. **Zatsepina, N.N., Tupitsin, K.F., Dushina, V.P., Kapustin, Y.M., and Kaminskii, Y.L.,** Basic deuterium exchange in substituted methylnaphthalenes, methylquinolines, and their *N*-oxides, *Reakts. Sposobnost Org. Soedin.,* 9, 745, 1972.

1196. **Kawazoe, Y., Ohnishi, M., and Yoshioka, Y.,** Base-catalyzed hydrogen exchange of α-hydrogens in alkyl groups substituted on some nitrogen-containing heteroaromatic rings, *Chem. Pharm. Bull.,* 15, 1225, 1967.

1197. **Kawazoe, Y., Yoshioka, Y., Yamada, M., and Igeta, H.,** Base-catalyzed hydrogen exchange of methyl hydrogens of methylpyridazines and their *N*-oxides, *Chem. Pharm. Bull.,* 15, 2000, 1967.

1198. **Goto, Y., Niiya, T., Hojo, N., Sakamoto, T., Yoshizawa, H., Yamanaka, H., and Kubota, T.,** Molecular orbital study of the reactivity of active alkyl groups of pyridine and pyrimidine derivatives, *Chem. Pharm. Bull.,* 30, 1126, 1982.

1199.   **Kaiser, E.M., Thomas, W.R., Synos, T.E., McClure, J.R., Mansour, T.S., Garlich, J.R., and Chastain, J.E.,** Regiointegrity of carbanions derived by selective metalation of dimethylpyridines and -quinolines, *J. Organomet. Chem.,* 213, 405, 1981.

1200.   **Kaiser, E.M.,** Lateral metalation of methylated nitrogenous heterocycles, *Tetrahedron,* 39, 2055, 1983.

1201.   **Kato, T. and Niitsuma, T.,** Arylation of methylpyridine and its derivatives, *Yakugaku Zasshi,* 94, 766, 1974.

1202.   **Cervinka, O., Fabryova, A., and Josef, J.,** Reaction of 2-picoline 1-oxide with benzyl chloride and potassium, *Z. Chem.,* 18, 173, 1978.

1203.   **Ohsawa, A., Uezu, T., and Igeta, H.,** Nucleophilic reaction of lithiated methylpyridazines, *Yakugaku Zasshi,* 100, 774, 1980.

1204.   **Brzezinski, B.,** Synthesis of compounds with heteroconjugated intramolecular hydrogen bonds with strong proton polarizability, *Pol. J. Chem.,* 57, 249, 1983.

1205.   **Parker, E.D. and Furst, A.,** A new method for the preparation of dialkylaminostyril derivatives of pyridine and quinoline and their *N*-oxides, *J. Org. Chem.,* 23, 201, 1958.

1206.   **Katritzky, A.R. and Monro, A.M.,** Per-acid oxidation of some conjugated pyridines, *J. Chem. Soc.,* 150, 1958.

1207.   **Pentimalli, L.,** Preparation and peracid oxidation of 2-(*p*-dimethylamino)-styrylpyridine, *Tetrahedron,* 14, 151, 1961.

1208.   **Jerchel, D. and Heck, H.E.,** Condensation of methylpyridines with benzaldehyde, *Liebigs Ann. Chem.,* 613, 171, 1958.

1209.   **Somani, R. and Swamy, R.V.,** Use of *tert*-butoxide in condensation reaction of 2- and 4-picoline *N*-oxides with benzaldehydes, *Indian J. Chem.,* 16B, 927, 1978.

1210.   **Itai, T., Sako, S., and Okusa, G.,** Reaction of 3,6-dimethylpyridazine 1-oxide and methylpyridazine 1-oxides with benzaldehydes, *Chem. Pharm. Bull.,* 11, 1146, 1963.

1211.   **Igeta, H., Tsuchiya, T., Kaneko, C., and Suzuki, S.,** Synthesis of 1,2-bis(3′-pyridazinyl)ethylenes, -ethanes, and their di-*N*-oxides, *Chem. Pharm. Bull.,* 21, 125, 1973.

1212.   **Plevachuk, N.E. and Baranov, S.N.,** Electrophilic substitution in 2,3′-dimethylquinoxaline and its *N*-oxides, *Khim. Geterotsikl. Soedin,* 729, 1968.

1213.   **Musatova, I.S., Elina, A.S., and Padeiskaya, E.N.,** Reactions of methylquinoxaline di-*N*-oxides with benzene and furan aldehydes, *Khim-Farm. Zh.,* 16, 1063, 1982.

1214.   **Elina, A.S. and Tsyrulnikova, L.G.,** β-Acetoxyethylderivatives of quinoxaline and their *N*-oxides, *Khim. Geterosikl. Soedin.,* 432, 1966.

1215.   **Miyano, S. and Abe, N.,** A new synthesis of 1,2-dipyrydiylethylene and related compounds, *Chem. Pharm. Bull.,* 15, 511, 1967.

1216.   **Yamanaka, H., Ogawa, S., and Konno, S.,** Reaction of active methyl groups on pyrimidine *N*-oxides, *Chem. Pharm. Bull.,* 28, 1526, 1980.

1217.   a. **Adams, R. and Miyano, S.,** Condensation reactions of picoline 1-oxides, *J. Am. Chem. Soc.,* 76, 3168, 1954; b. **Ozegowski, W., Wunderwald, M., and Krebs, D.,** β-(2-Quinolyl-1-oxide)-DL-alanine, *J. Prakt. Chem.,* [4], 9, 54, 1959; c. **Sakamoto, T., Yoshizawa, H., Yamanaka, H., Goto, Y., Niiya, T., and Honjo, N.,** Site-selective effect of *N*-oxide function on six membered *N*-heteroaromatics, *Heterocycles,* 17, 73, 1982.

1218.   **Connor, D.T., Yong, P.A., and von Strandtmann, M.,** Preparation of pyridine *N*-oxide substituted chromones, chromanones, coumarins, quinolones, dihydroquinolones and cinnolones, *J. Heterocycl. Chem.,* 14, 143, 1977.

1219.   **Bredereck, H., Simchen, G., and Wahl, R.,** Reaction of active methyl group of substituted toluenes and heterocycles with aminal-*tert*-butylester to form enamines, *Chem. Ber.,* 101, 4048, 1968.

1220.   a. **Kreutzberger, A. and Abel, D.,** Novel formation of the indolizine ring system, *Arch. Pharm.,* 302, 701, 1969; b. **Landquist, J.K. and Silk J.A.,** Derivatives of *Py*-hydroxyalkyl-, -aminoalkyl- and -carboxyquinoxalines, *J. Chem. Soc.,* 2052, 1956.

1221.   **Kato, T. and Goto, Y.,** The reaction of picolines and their *N*-oxides with amyl nitrite, *Chem. Pharm. Bull.,* 11, 461, 1963.

1222.   **Schnekenburger, J.,** Syn- and anti-2(and 4-)pyridinecarboxaldehyde 1-oxide oximes, *Arch. Pharm.,* 302, 494, 1969.

1223.   **Kato, T., Goto, Y., and Kondo, M.,** Reaction of quinaldine, lepidine and their *N*-oxides with amyl nitrite, *Yakugaku Zasshi,* 84, 290, 1964.

1224.   **Ogata, M.,** Reaction of methylpyridazine *N*-oxides with amyl nitrite, *Chem. Pharm. Bull.,* 11, 1517, 1963.

1225.   a. **Ogata, M.,** The reaction of methyl-substituted 4-nitropyridazine 1-oxides with acetyl chloride, *Chem. Pharm. Bull.,* 11, 1511, 1963; b. **Feuer, H. and Lawrence, J.P.,** Nitration of alkyl substituted heterocyclic compounds, *J. Org. Chem.,* 37, 3662, 1972.

1226.   a. **Cislak, F.E.,** 2-Cyanopyridines, U.S. patent 2989534, 1961; *Chem. Abst.* 56, 3466e, 1962; b. **Hamana, M., Umezawa, B., Gotoh, Y., and Noda, K.,** King reaction of 2-picoline 1-oxide, *Chem. Pharm. Bull.,* 8, 692, 1960.

1227.   **Mischina, T.M. and Efros, L.S.,** Reactivity of methylderivatives of *N*-oxides of aromatic heterocycles, *Zh. Obshch. Khim.,* 32, 2217, 1962.

1228.   **Bedford, G.R., Katritzky, A.R., and Wuest, H.M.,** *N*-oxides and related compounds. Part XXIII. Some pyridoxine analogs, *J. Chem. Soc.,* 4600, 1963.

1229.   **Colonna, M.,** Reaction of lepidine *N*-oxide with nitroso derivatives, *Gazz. Chim. Ital.,* 90, 1197, 1960.

1230. **Elina, A.S., Tsyrul'nikova, L.C., and Musatova, I.S.,** Reaction of di-*N*-oxides of α-methylderivatives of quinoxaline and pyrazine with *p*-nitrosodimethylaniline, *Khim-Farm. Zh.,* 1, 10, 1967.

1231. a. **Hamana, M., Umezawa, B., and Nakashima, S.,** Reactions of *N*-(*p*-dimethylaminophenyl)nitrones having pyridine *N*-oxide, quinoline or its *N*-oxide as α-substituents, *Chem. Pharm. Bull.,* 10, 961, 1962; b. **Sakamoto, T., Sakaai, T., and Yamanaka, H,** Site-selective oxidation of dimethylpyrimidines with selenium dioxide to pyrimidine-monoaldehydes, *Chem. Pharm. Bull.,* 29, 2485, 1981; c. **Achremowicz, L.,** Oxidative demethylation of some methylnitropyridine 1-oxides, *Tetrahedron Lett.,* 2433, 1980; d. **Suszko, J. and Szafran, M.,** Selective oxidation of 2,6-lutidine *N*-oxide and its derivatives, *Roczn. Chem.,* 38, 1795, 1964; e. **Szafran, M. and Brzezinski, B.,** Permanganate oxidation of 2,6-, 2,4-, and 2,5-lutidine *N*-oxide, *Rocz. Chem.,* 44, 1951, 1970; f. **Sugimoto, N., Kugita, H., and Tanaka, T.,** Oxidation of alkylpyridines, *Yakugaku Zasshi,* 76, 1308, 1956; g. **Granzow, A. and Wilson, A.,** Oxidation of 3-picoline 1-oxide and 4-nitro-3-picoline-1-oxide, *J. Org. Chem.,* 37, 3063, 1972; h. **Sagae, H., Fujihira, M., Lund, H., and Osa, T.,** Oxidation of methylpyridines and methylpyridine *N*-oxides with electrogenated superoxide ion, *Heterocycles,* 13, 321, 1979.

1232. a. **Zhuravlev, V.S., Smirnev, L.D., Lezina, V.P., Stolyarova, L.G., and Dyumaev, K.M.,** Sulfonation of 5-benzyl-3-hydroxypyridine and *N*-oxide, *Isv. Akad Nauk SSSR, Ser Khim.,* 2374, 1974; b. **Zhuravlev, V.S., Smirnov, L.D., Gugunava, M.A., Lezina, V.P., Zaitzev, B.F., and Dyumaev, K.,** Electrophilic reactions of 6-hydroxy(methoxy)-2-benzyl -3-hydroxypyridines, *Izv. Akad. Nauk SSSR, Ser. Khim.,* 206, 1973.

1233. **Smirnov, L.D., Zhuravlev, V.S., Lezina, V.P., and Dyumaev, K.M.,** Nitration of 2- and 5-benzyl-3-hydroxypyridines and their *N*-oxides, *Khim. Geterotsikl. Soedin.,* 1094, 1974.

1234. **Calvino, R., Ferrarotti, B., Gasco, A., and Serafino, A.,** Nitrophenyl derivatives of the furazan and furoxan ring system, *J. Heterocycl. Chem.,* 20, 1419, 1983.

1235. **Kresze, G. and Bathelt, H.,** Diels-Alder reactions of nitro substituted benzofuroxans, *Tetrahedron,* 29, 1043, 1973.

1236. **Boruah, R.C., Devi, P., and Sandhu, J.S.,** Reaction of diazomethane with 6-nitro-2,l-benzisoxazole and nitrobenzofurazan oxides, *J. Heterocycl. Chem.,* 16, 1555, 1979.

1237. **Jarvis, B.B. and Marien, B.A.,** Reactions of triarylphosphines with halomethylpyridylphenyl sulfones, *J. Org. Chem.,* 42, 2676, 1977.

1238. **Elina, A.S.,** Haloquinoxaline *N*-oxides in nucleophilic substitution reactions, *Khim. Geterotsikl. Soedin.,* 940, 1967.

1239. **Tanida, H., Irie, T., and Hayashi, Y.,** Neighboring group participation by pyridine rings and *N*-oxides. Synthesis and solvolysis of 5,8-dihydro-5,8-methanoisoquinoline derivatives, *J. Org. Chem.,* 49, 2527, 1984.

1240. **Sullivan, P.T., Kester, M., and Norton, S.J.,** Synthesis and study of pyridylalanine *N*-oxides, *J. Med. Chem.,* 11, 1172, 1968.

1241. **Majewski, P., Bodalski, R., and Michalski, J.,** Synthesis and reactions of 2-chloromethylpyridine 1-oxide. Route to compounds containing the 1-oxido-2-pyridylmethyl group, *Synthesis,* 140, 1971.

1242. **Nishikawa, M., Saeki, S., Hamana, M., and Noda, H.,** Reactions of 2-chloromethylquinoline derivatives with 2-nitropropane, *Chem. Pharm. Bull.,* 28, 2436, 1980.

1243. **Wuensch, K.H., Krumpholz, I., Perez-Zayas, J., Tapanes-Peraza, R., and Schulz, G.,** Reaction of 6-chloro-2-chloromethyl-4-phenylquinazoline 3-oxide with some nucleophilic reagents, *Z. Chem.,* 10, 113, 1970.

1244. **Eiden, F. and Dusemund, J.,** Reactions of 2-halomethyl-4-phenylquinazoline derivatives with nucleophilic reagents, *Arch. Pharm.,* 304, 729, 1971.

1245. **Broadbent, H.S., Anderson, R.C., and Kuchar, M.C.J.,** Attack of *S*-vs *N*-nucleophiles on 2-halomethylquinazoline 3-oxides, *J. Heterocycl. Chem.,* 14, 289, 1977.

1246. **Archer, G.A. and Sternbach, L.H.,** The chemistry of benzodiazepines, *Chem. Rev.,* 68, 747, 1968.

1247. a. **Sternbach, L.H. and Reeder, E.,** The rearrangement of 6-chloro-2-chloromethyl-4-phenylquinazoline 3-oxide into 2-amino derivatives of 7-chloro-5-phenyl-3*H*-1,3-benzodiazepine 4-oxide, *J. Org. Chem.,* 26, 1111, 1961; b. **Bell, S.C., Gochman, G., and Childress, S.J.,** Some analogs of chlordiazepoxide, *J. Med. Pharm. Chem.,* 5, 63, 1962.

1248. **Stempel, A., Reeder, E., and Sternbach, L.H.,** Mechanism of ring enlargement of quinazoline 3-oxides with alkali to 1,4-benzodiazepin-2-one 4-oxides, *J. Org. Chem.,* 30, 4267, 1965.

1249. **Derieg, M.E., Fryer, R.I., and Sternbach, L.H.,** The reaction of hydrazines with 6-chloro-2-chloromethyl-4-phenylquinazoline 3-oxide, *J. Chem. Soc. (C),* 1103, 1968.

1250. **Field, G.F., Zally, W.J., and Sternbach, L.H.,** Novel ring enlargements of 1,2-dihydroquinazoline 3-oxides, *Tetrahedron Lett.,* 2609, 1966.

1251. **Field, G.F., Zally, W.J., and Sternbach, L.H.,** Three tautomeric forms of the benzodiazepine ring system, *J. Am. Chem. Soc.,* 89, 332, 1967.

1252. **Vejdelek, Z., Rajsner, M., Svatek, E., Holubeck, J., and Protiva, M.,** Synthesis of 7-chloro-5-(6,7,8,9-tetrahydro-5*H*-benzocyclohepten-2-yl)-2-(methylamino)-3*H*-1,4-benzodiazepine 4-oxide and of some related compounds, *Collect. Czech. Chem. Commun.,* 44, 3604, 1979.

1253. **Shenoy, U.D.,** 1,5-Benzodiazocine derivatives, Ger. Offen 2525094, 1976; *Chem. Abst.,* 86, 106683r, 1977.

1254. **Mizuno, Y. and Endo, T.,** A novel type of neighboring group participation involving pyridine *N*-oxides in acylation and phosphorylation. 1. *J. Org. Chem.,* 43, 684, 1978.

1255. **Benderly, A., Fuller, G.B., Knaus, E.E., and Redda, K.,** Synthesis and reactions of 2-epoxyethylpyridine 1-oxide with nitrogen, oxygen and sulfur nucleophiles, *Can. J. Chem.,* 56, 2673, 1978.

1256. **Okano, V., Toma, H.E., and do Amaral, L.,** Hydration constants of pyridinecarboxyladehyde N-oxides, *J. Org. Chem.,* 46, 1018, 1981.

1257. **Queguiner, G., Salaun Boulx, M., and Pastour, P.,** Hydration of diformylpyridines and their N-oxides, *Bull. Soc. Chim. Fr.,* 38, 3690, 1970.

1258. **Okano, V., Bastos, M.P., and Do Amaral, L.,** Kinetics and mechanism for pyridine N-oxide carboxaldehyde phenylhydrazone formation, *J. Am. Chem. Soc.,* 102, 4155, 1980.

1259. **Hamana, M., Umezawa, B., and Goto, Y.,** Some reactions of N(p-dimethylaminophenyl)-α(1-oxidopyridyl)nitrone.2., *Yakugaku Zasshi,* 80, 1519, 1960.

1260. **Bodalski, R., Pietrusiewicz, M., Majewski, P., and Michalski, J.,** Diethyl 2- and 4-pyridylmethylphosphonate 1-oxides and their use in the synthesis of 2- and 4-alkenylpyridine 1-oxides, *Coll. Czech. Chem. Commun.,* 36, 4079, 1971.

1261. **Falkner, P.R. and Harnson, D.,** The kinetics of alkaline hydrolysis of 2-, 3-, and 4-ethoxycarbonylpyridines and their 1-oxides, *J. Chem. Soc.,* 1171, 1960.

1262. **Kobayashi, Y., Kumadaki, I., and Taguchi, S.,** Alcoholysis of (trifluoromethyl) quinolines, *Chem. Pharm. Bull.,* 19, 624, 1971.

1263. **Ochiai, E. and Katada, M.,** Reduction of 4-nitropyridine 1-oxide, *Yakugaku Zasshi,* 63, 186, 1943.

1264. **Thunus, L. and Delarge, J.,** Synthesis of 3- and 4-pyridinesulfonic acids, *J. Pharm. Belg.,* 21, 485, 1966.

1265. **Okano, T. and Tsuji, K.,** Conversion of 4-nitroquinoline 1-oxide into 4-hydroxyaminoquinoline 1-oxide in the presence of ascorbic acid, *Yakugaku Zasshi,* 89, 67, 1969.

1266. **Kato, T. and Hamaguchi, F.,** Reduction of 4-nitro-2,6-lutidine 1-oxide and 4-nitro-3-picoline 1-oxide, *Pharm. Bull.,* 4, 174, 1956.

1267. **Berson, J.A. and Cohen, T.,** A synthesis of 4-carbomethoxy-5-ethyl-2-methylpyridine, *J. Org. Chem.,* 20, 1461, 1955.

1268. **Katritzky, A.R. and Simmons, P.,** The basicities of (amino- and nitro-phenyl)-pyridines and pyridine 1-oxides, *J. Chem. Soc.,* 1511, 1960.

1269. **Itai, T. and Sako, S.,** 3-Substituted 6-chloropyridazine 1-oxides, *Chem. Pharm. Bull.,* 10, 989, 1962.

1270. **Shaw, E.,** The tautomerism of the hydroxypyridine-N-oxides, *J. Am. Chem. Soc.,* 71, 67, 1949.

1271. **Mitsui, S., Sakai, T., and Saito, H.,** Selectivity of catalysts in hydrogenation of pyridine derivatives, *Nippon Kagaku Zasshi,* 86, 409, 1965.

1272. **Ferrier, B.M. and Campbell, N.,** Some pyridine derivatives, *Chem. Ind.,* 1089, 1958.

1273. **Kubota, S. and Akita, T.,** Reduction of 4-nitropyridine 1-oxide and 4-nitro-2-picoline 1-oxide with hydrazine, *Yakugaku Zasshi,* 78, 248, 1958.

1274. a. **Ochiai, E. and Mitarashi, H.,** 4-Hydroxyaminopyridine N-oxide from 4-nitropyridine N-oxide, *Chem. Pharm. Bull.,* 11, 1084, 1963; b. **Morita, I., Iigo, M., Fukino, H., Sakai, K., and Yamane, Y.,** Effect of molybdenum on reduction of 4-nitroquinoline 1-oxide, *Yakugaku Zasshi,* 95, 954, 1975.

1275. **Okabayashi, T. and Yoshimoto, A.,** Reduction of 4-nitroquinoline 1-oxide by microorganisms, *Chem. Pharm. Bull.,* 10, 1229, 1962.

1276. **Kataoka, N., Imamura, A., Kawazoe, Y., Chihara, G., and Nagota, C.,** Structure of the free radical produced from carcinogenic 4-hydroxyaminoquinoline l'-oxide, *Bull. Chem. Soc. Jpn.,* 40, 62, 1967.

1277. **Makhova, N.N., Ovchimikov, I.V., Khasanov, B.N., and Khmel'intski, L.I.,** Synthesis of isomeric 3(4)-nitro-4(3)-phenylfuroxans, *Izv. Akad. Nouk SSSR, Sci. Khim.,* 646, 1982.

1278. **Calvino, R., Gasco, A., Serafino, A., and Viterbo, A.,** 3-Nitro-4-phenylfuroxan: reaction with sodium methoxide and x-ray structural analysis, *J. Chem. Soc., Perkin Trans. 2,* 1240, 1981.

1279. **Maurer, M., Orth, W., and Fickert, W.,** Bis(2-pyridyl)disulfide, Ger. Offen. 2904548, 1980; *Chem. Abst.* 93, 239253s, 1980.

1280. **Sega, T., Pollak, A., Stanovnik, B., and Tisler, M.,** Oxidative transformations of pyridazinyl sulfides, *J. Org. Chem.,* 38, 3307, 1973.

1281. **Gardner, J.N. and Katritzky, A.R.,** The tautomerism of 2- and 4-amino and -hydroxy-pyridine 1-oxide, *J. Chem. Soc.,* 4375, 1957.

1282. **Bojarska-Dahlig, H.N.,** On the synthesis of derivatives of N-(4-pyridonyl)-oxyacetic acid, *Rec. Trav. Chim. Pays-Bas,* 78, 981, 1959.

1283. **Ochiai, E., Teshigawara, T., Oda, K., and Naito, T.,** Syntheses of γ-substituted pyridine derivatives, 2., *Yakugaku Zasshi,* 65, 5/6A, 1, 1945.

1284. **Ochiai, E. and Hayashi, E.,** 4-Hydroxypyridine 1-oxide, *Yakugaku Zasshi,* 67, 151, 1947.

1285. **Newbold, G.T. and Spring, F.S.,** Cyclic hydroxamic acids observed from pyridine and quinoline, *J. Chem. Soc.,* 1864, 1948.

1286. **Ochiai, E. and Hayashi, E.,** 4-Hydroxyquinoline 1-oxide, *Yakugaku Zasshi,* 67, 154, 1947.

1287. **Nakagome, T.,** 3-Methoxy-6-pyridazinol 1-oxide, *Yakugaku Zasshi,* 82, 244, 1962.

1288. **Nakagome, T.,** Structure of 6-methyl-3-pyridazinol 1-oxide, *Yakugaku Zasshi,* 82, 1206, 1962.

1289. **Deady, L.W. and Stanborough, M.S.,** Kinetics and mechanism of the reaction of pyridinamine 1-oxides with acetylating agents in water and aprotic solvents, *Aust. J. Chem.,* 35, 1841, 1982.

1290. **Katritzky, A.R.,** The structure and reactivity of 2-aminopyridine 1-oxide, *J. Chem. Soc.,* 191, 1957.

1291. **Jones, R.A. and Katritzky, A.R.,** The tautomerism of mercapto- and acylamino-pyridine 1-oxides, *J. Chem. Soc.,* 2937, 1960.

1292. **Sedova, V.F., Mustafina, T.Y., and Mamaev, V.P.,** Alkylation and acylation of 2-aminopyrimidine *N*-oxides, *Khim. Geterotsikl. Soedin.,* 1515, 1981.

1293. **Hayashi, E., Higashino, T., and Tomisaka, S.,** Methylation of 4-aminoquinazoline 3-oxide with diazomethane, *Yakugaku Zasshi,* 87, 578, 1967.

1294. **Montgomery, J.A. and Thomas, H.J.,** The synthesis of 2-azapurine nucleosides from purine nucleosides accomplished via ring opening followed by reclosure with nitrous acid, in *Nucleic Acid Chemistry,* Vol. 2, Townsend, L.B., Tipson, R.S., Eds., Wiley, New York, 1978, 681.

1295. **Fujii, T., Wu, C.C., Itaya, T., Moro, S., and Saito, T.,** Synthesis of *N*-alkoxyadenosines and their 2′,3′-*O*-isopropylidene derivatives, *Chem. Pharm. Bull.,* 21, 1676, 1973.

1296. **Fujii, T. and Itaya, T.,** Alkylation of adenine 1-oxide and 1-alkoxyadenines: synthesis of 1-alkoxy-, 1-alkoxy-9-alkyl-, and 9-alkyladenines, *Tetrahedron,* 27, 351, 1971.

1297. **McClure, R.E. and Ross, A.,** *S*-Alkoxymethyl- and *S*-alkylmercaptomethyl derivatives of 2-pyridinethiol 1-oxide, *J. Org. Chem.,* 27, 304, 1962.

1298. a. **Rockett, J.,** Esters of 2-mercaptopyridine *N*-oxide, U.S. patent 2922792, 1960; *Chem. Abst.,* 54, 8857b, 1960; b. **Pluijgers, C.W., Berg, J., and Thorn, G.D.,** Synthesis and antifungal activity of some carbohydrate and amino acid derivatives of dimethyldithiocarbonic acid and of pyridine-4-thiol 1-oxide, *Rec. Trav. Chim. Pays Bas,* 88, 241, 1969.

1299. **Gewald, K., Buchwalder, M., and Peukert, M.,** Reaction of 6-amino-2*H*-thiopyran-2-thiones with amines, *J. Prakt. Chem.,* 315, 679, 1973.

1300. **Dinan, F.J. and Tieckelmann, H.,** Rearrangements of alkoxypyridine 1-oxides, *J. Org. Chem.,* 29, 1650, 1964.

1301. **Litster, J.E. and Tieckelmann, H.,** The thermal rearrangements of 2-alkenyloxypyridine 1-oxides, *J. Am. Chem. Soc.,* 90, 4361, 1968.

1302. **Yamazaki, M., Noda, K., Chono, Y., Honjo, N., and Hamana, M.,** Rearrangements of 2-alkoxyquinoline 1-oxides, *Yakugaku Zasshi,* 88, 661, 1968.

1303. **Schoellkopf, U. and Hoppe, I.,** Indications of a suprafacial 1,4-sigmatropic shift in the rearrangement of 2-alkoxypyridine *N*-oxide to *N*-alkoxy-2-pyridone, *Tetrahedron Lett.,* 4527, 1970.

1304. **Schoellkopf, U. and Hoppe, I.,** Evidence for a [1s,4s]-sigmatropic shift in the rearrangements of 2-alkoxypyridine 1-oxides to *N*-alkoxy-2-pyridone, *Justus Liebigs Ann. Chem.,* 765, 153, 1972.

1305. **Gerhart, F. and Wilde, L.,** 2-Alkoxyquinoline *N*-oxide rearrangement, *Tetrahedron Lett.,* 475, 1974.

1306. **Le Noble, W.J. and Daka, M.R.,** Effect of pressure on concerted and stepwise sigmatropic shifts, *J. Am. Chem. Soc.,* 100, 5961, 1978.

1307. **Schoellkopf, U. and Driessler, F.,** Mechanism of the rearrangement of 2-aroxypyridine 1-oxides to *N*-aroxy-2-pyridones, *Justus Liebigs Ann. Chem.,* 1521, 1975.

1308. **Yanai, M. and Kinoshita, T.,** Reaction of 3,6-dimethoxy-4-nitropyridazine 1-oxide with alkylhalides, *Chem. Pharm. Bull.,* 16, 1221, 1968, *Yakugaku Zasshi,* 86, 1124, 1966.

1309. **Tanida, H.,** Rearrangement of 4-benzyloxyquinoline 1-oxide and 4-allyloxyquinoline 1-oxide by boron trifluoride, *Yakugaku Zasshi,* 78, 613, 1958.

1310. **Boyle, F.T. and Jones, R.A.Y.,** Thermal rearrangements of 1-methoxypyrazole 2-oxides, *J. Chem. Soc., Perkin Trans. 1,* 167, 1973.

1311. **Barlos, K., Papaioannou, D., Voliotis, S., Prewo, R., and Bieri, J.H.,** Crystal structure of 3-(*N*ᵅ-tritylmethionyl)benzotriazole 1-oxide, a synthon in peptide synthesis, *J. Org. Chem.,* 50, 696, 1985.

1312. **Nagarajan, S., Wilson, S.R., and Rinehart, K.L.,** Rearrangement of unsaturated acyloxybenzotriazoles, *J. Org. Chem.,* 50, 2174, 1985.

1313. **Paquette, L.A.,** A novel synthesis of cyclic hydroxamic acid esters, *Tetrahedron,* 22, 25, 1966.

1314. **Paquette, L.A.,** Electrophilic additions of acyl and sulfonyl halides to 2-ethoxypyridine 1-oxide. A new class of activated esters and their application to peptide synthesis, *J. Am. Chem. Soc.,* 87, 5186, 1965.

1315. **Wagner, G. and Schmidt, R.,** Synthesis of quinoline glucosides by the Hilbert-Johnson method, *Arch. Pharm.,* 298, 481, 1965.

1316. **Yanai, M. and Yamaguchi, M.,** Reaction of 3,6-dialkoxypyridazine 1-oxide with alkylhalides, haloketones and haloacid esters, and decomposition of reaction products, *Chem. Pharm. Bull.,* 16, 1244, 1968.

1317. **Yanai, M. and Kinoshita, T.,** Reactions of 3,6-dialkoxypyridazine 1-oxide with alkyl halides and acyl halides, *Yakugaku Zasshi,* 86, 314, 1966.

1318. **Ochiai, E. and Naito, T.,** Synthesis of 4-substituted quinoline derivatives. 1., *Yakugaku Zasshi,* 65, 5/6A, 3, 1945.

1319. **Naito, T.,** Synthesis of 4-substituted quinoline derivatives. 2., *Yakugaku Zasshi,* 65, 5/6A, 3, 1945.

1320. **Kalatzis, E. and Mastrokalos, C.,** Kinetics of the diazotization of 2- and 4-aminopyridine 1-oxide, *J. Chem. Soc., Perkin Trans. 2,* 1830, 1977.

1321. **Kalatzis, E. and Mastrokalos, C.,** Kinetics of the diazotization of substituted 2-aminopyridine and 2-aminopyridine 1-oxide, *J. Chem. Soc., Perkin Trans. 2,* 1835, 1977.

1322. **Sako, S.,** Halopyridazine 1-oxides, *Chem. Pharm. Bull.,* 14, 303, 1966.

1323. **Jovanovic, M.V.,** Diazotization of some aminodiazine N-oxides, *Heterocycles,* 20, 2011, 1983.

1324. **Jovanovic, M.V.,** Diazonium coupling reaction of some diazine N-oxides, *Heterocycles,* 22, 1115, 1984.

1325. **Keen, B.T., Radel, R.J., and Paudler, W.W.,** 1,2,4-Triazine 1- and 2-oxides, Reactivities toward some electrophiles and nucleophiles, *J. Org. Chem.,* 42, 3498, 1977.

1326. **Elslager, E.F., Capps, D.B., Kurtz, D.H., Werbel, L.M., and Worth, D.F.,** 5-[4-(2-Diethylaminoethylamino-1-naphthylazo]uracil and related [4-(aminoalkylamino)-1-naphthylazo]heterocyclic compounds, *J. Med. Chem.,* 6, 646, 1963.

1327. **Scriven, E.F.V.,** *Azides and Nitrenes,* Academic Press, Orlando, FL, 1984.

1328. **Itai, T. and Kamiya, S.,** 4-Azidoquinoline and 4-azidopyridine derivatives, *Chem. Pharm. Bull.,* 9, 87, 1961.

1329. **Kamiya, S.,** Reactions of the azido group in quaternary salts of 4-azidoquinoline, 4-azidopyridine, and their 1-oxides, *Chem. Pharm. Bull.,* 10, 669, 1962.

1330. **Kamiya, S.,** Reactions of 4-azidoquinoline 1-oxide, *Chem. Pharm. Bull.,* 10, 471, 1962.

1331. **Kamiya, S.,** Synthesis of 3-, 2-, and 5-azidoquinoline 1-oxide derivatives, *Yakugaku Zasshi,* 81, 1743, 1961.

1332. **Sawanishi, H., Hirai, T., and Tsuchiya, T.,** Photolysis of pyridyl, quinolyl, and isoquinolyl azides in hydrohalic acids, *Heterocycles,* 19, 1043, 1982.

1333. **Sawanishi, H., Hirai, T., and Tsuchiya, T.,** Photolysis of quinolyl and isoquinolyl azides in alcohols containing sulfuric acid, *Heterocycles,* 19, 2071, 1982.

1334. **Sawanishi, H., Hirai, T., and Tsuchiya, T.,** Photolysis and thermolysis of quinolyl and isoquinolyl azides in ethanethiol, *Heterocycles,* 22, 1501, 1984.

1335. **Abramovitch, R.A. and Cue, B.W.,** N-Hydroxypyrroles and related compounds, *J. Org. Chem.,* 38, 173, 1973.

1336. **Abramovitch, R.A. and Cue, B.W.,** Ring contraction of 2-azidopyridine 1-oxides and related compounds. 2-Cyano-1-hydroxypyrroles and -imidazoles, *J. Am. Chem. Soc.,* 98, 1478, 1976.

1337. **Wentrup, C. and Winter, H.W.,** Isolation of diazacycloheptatetraenes from thermal nitrene-nitrene rearrangements, *J. Am. Chem. Soc.,* 102, 6159, 1980.

1338. **Abramovitch, R.A. and Shinkai, I.,** Decomposition of 3-azidopyridazine 2-oxides. Ring opening of the pyridazine ring, *J. Chem. Soc., Chem. Commun.,* 703, 1975.

1339. **Abramovitch, R.A. and Cue, B.W.,** Ring contraction of 2-azidoquinoline and quinoxaline 1-oxides, *J. Org. Chem.,* 45, 5316, 1980.

1340. **Dirlam, J.P., Cue, B.W., and Gombatz, K.J.,** Thermal decomposition of 2-azidoquinoxaline N-oxides, *J. Org. Chem.,* 43, 76, 1978.

1341. **Abramovitch, R.A., Shinkai, I., Cue, B.W., Ragan, F.A., and Atwood, J.L.,** A new ring transformation of 3-halo-2-azidopyridine 1-oxides. A novel synthesis of 1,2-oxazin-6-ones, *J. Heterocycl. Chem.,* 13, 415, 1976.

1342. **Abramovitch, R.A. and Cue, B.W.,** Synthesis of 2H-pyrrole-1-oxides by ring contraction, *Heterocycles,* 2, 297, 1974.

1343. **Tanno, M. and Kamiya, S.,** Syntheses of 4-(3-cyano-1-triazeno)-pyridines and related compounds, *Chem. Pharm. Bull.,* 27, 1824, 1979.

1344. **Kaniya, S. and Tanno, M.,** Synthesis of 4-(3-cyano-1-triazeno)-pyridazine 1-oxides and related compounds, *Chem. Pharm. Bull.,* 28, 529, 1980.

1345. **Sasaki, T., Kanematsu, K., and Murata, M.,** Staudinger reaction of tetrazolopyridines with triphenylphosphine, *Tetrahedron,* 27, 5359, 1971.

1346. **Cresswell, R.M., Maurer, H.K., Strauss, T., and Brown, G.B.,** A total synthesis of a pyrimidine N-oxide, a pteridine 1-N-oxide, and xanthine 3-N-oxide, *J. Org. Chem.,* 30, 408, 1965.

1347. **Taylor, E.C. and Lenard, K.,** An unequivocal synthesis of 6-substituted pteridine 8-oxides, pteridines, and 7,8-dihydropteridines, *J. Am. Chem. Soc.,* 90, 2424, 1968.

1348. **Taylor, E C., Perlman, K.L., Kim, Y.H., Sword, I.P., and Jacobi, P.A.,** Unequivocal route to 2,4-diamino-6-substituted pteridines, *J. Am. Chem. Soc.,* 95, 6413, 1973.

1349. **Taylor, E.C. and Lenard, K.,** An unequivocal synthesis of 6-pteridincarboxaldehyde, *Justus Liebigs Ann. Chem.,* 726, 100, 1969.

1350. **Rao, R.A.,** Preparation of some pyrazolopyridines, *Labdev (A),* 6, 214, 1968; *Chem. Abst.,* 70, 87659p, 1969.

1351. **Cook, P.D. and Castle, R.N.,** Synthesis and reactivity of pyrrolo[2,3-d]pyridazine 5-oxide, *J. Heterocycl. Chem.,* 10, 551, 1973.

1352. **Cook, P.D. and Castle, R.N.,** Synthesis of the novel pyrrolo[3,2-c]pyridazine ring system, *J. Heterocycl. Chem.,* 10, 807, 1973.

1353. **Kregar-Cadez, P., Pollak, A., Stanovnik, B., Tisler, M., and Wechtersbach-Lazetic, B.,** Novel heterocyclic system, pyrido[2,3-c]pyridazine, *J. Heterocycl. Chem.,* 9, 351, 1972.

1354. **Pollak, A., Stanovnik, B., and Tisler, M.,** s-Triazolo[4,3-b]pyridazine 5-oxides, *J. Heterocycl. Chem.,* 5, 513, 1968.

1355. **Matoba, K., Terada, T., Sugiura, M., and Yamazaki, T.,** Reactions of 2-acetyl-3-methylquinoxaline 1,4-dioxide and its derivatives, *Heterocycles,* 26, 55, 1987.

1356. **Edrissi, M., Kooshkabadi, H., and Lalezari, I.,** Formation of 1,6-diazathianthrene by photolysis of 1-hydroxy-2-pyridinethione, *Microchem. J.,* 19, 282, 1974.

1357. **Katritzky, A.R.,** The preparation and properties of pyridino (1′:2′-2:3)-1-oxa-2,4-diazol-5-one, *J. Chem. Soc.,* 2063, 1956.

1358. **Katritzky, A.R.,** Some derivatives of 2-aminopyridine 1-oxide, *J. Chem. Soc.,* 4385, 1957.

1359. **Boyer, J.H., Borgers, R., and Wolford, L.T.,** The azomethine linkage of pyridine in ring-closure isomerizations, *J. Am. Chem. Soc.,* 79, 678, 1957.

1360. **Rousseau, D. and Taurins, A.,** The reaction of 2-aminopyridine 1-oxides with thiophosgene; 2*H*-[1,2,4]Oxadiazolo[2,3-a]pyridin-2-thiones, *Can. J. Chem.,* 55, 3736, 1977.

1361. **Tauda, H.,** Synthesis of 2-aminoquinoline 1-oxide and 1-aminoisoquinoline 2-oxide and their reactivity, *Yakugaku Zasshi,* 79, 1063, 1959.

1362. **Itai, T. and Nakashima, T.,** N-oxidation of 3-aminopyridazine derivatives, *Chem. Pharm. Bull.,* 10, 936, 1962.

1363. **Muller, J.C. and Ramuz, H.,** Oxadiazolopyridine derivatives, Ger. Offen 2804518, 1978; *Chem. Abst.,* 89, 197595a, 1978.

1364. **Muller, J.C., Ramuz, H., Daly, J., and Schoenholzer, P.,** Structure of new 2-oxo-2,8-dihydro [1,2,4]oxadiazolo[2,3-a]pyrimidinecarbamates, *Helv. Chim. Acta,* 65, 1454, 1982.

1365. **Muller, J.C. and Ramuz, H.,** Pharmaceutical oxadiazolotriazine derivatives and intermediate products, European Patent 7643, 1986; *Chem. Abstr.,* 93, 168252z, 1986.

Chapter 5

APPLICATIONS

I. APPLICATIONS OF *N*-OXIDES IN CHEMISTRY

A. AS INTERMEDIATES IN SYNTHESIS

As has been discussed in Chapter 4, heterocycle *N*-oxides undergo both electrophilic and nucleophilic substitution more easily than the parent compounds, and furthermore display a rich reactivity of their own, i.e., deoxidative susbtitution, cycloaddition, and photorearrangement. It is therefore not surprising that these derivatives are often useful intermediates in synthesis. Several examples have been reported during the systematic discussion of reactions, and some representative cases are considered in the present section.

N-oxides are often intermediates for the synthesis of lactams, either as target compounds, e.g., some natural products,[1,2] or in view of further transformations, e.g., for the preparation of azabicyclo[2.2.0] hexenone **1**,[3] a building block for the carbapenem nucleus.

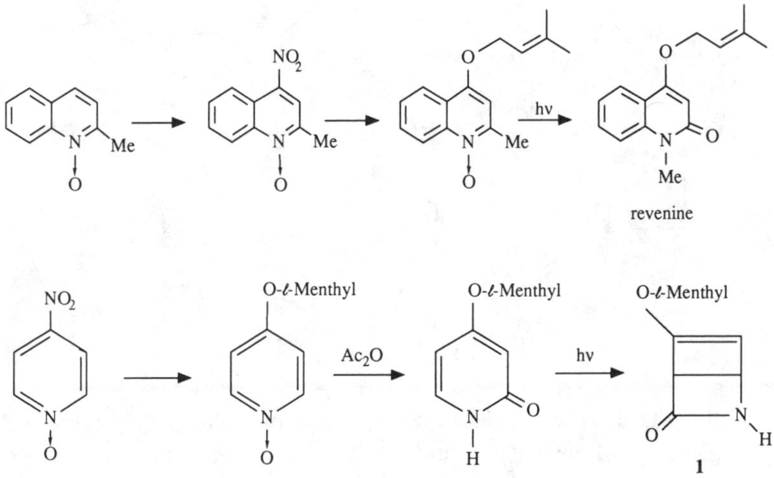

revenine

Synthesis via *N*-oxides is convenient, both in view of the easy thermal or photochemical rearrangement to lactams, and of the possibility of exploiting the easy substitution reactions of the *N*-oxide before the final conversion. With some substituted *N*-oxides, reaction with acetic anhydride leads to β-hydroxyheterocycles rather than to lactams, a pathway that has been used for the synthesis of pyridoxine.[4] Likewise, deoxidative halogenation is a source of haloheterocycles, and thus a starting point for the construction of new rings,[5] a target alternatively reached through deoxidative alkylation, cyanation, etc.[6-9]

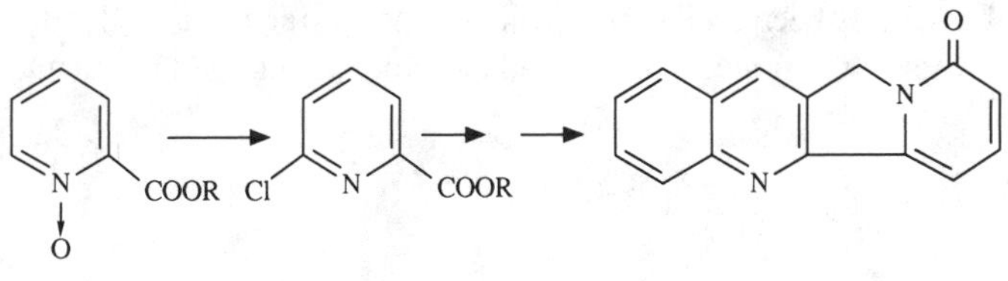

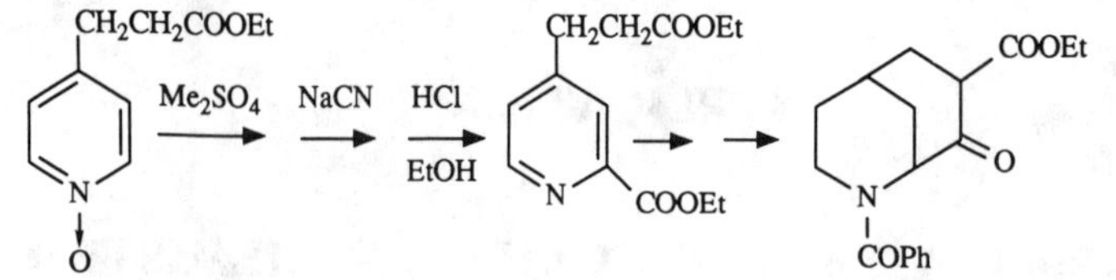

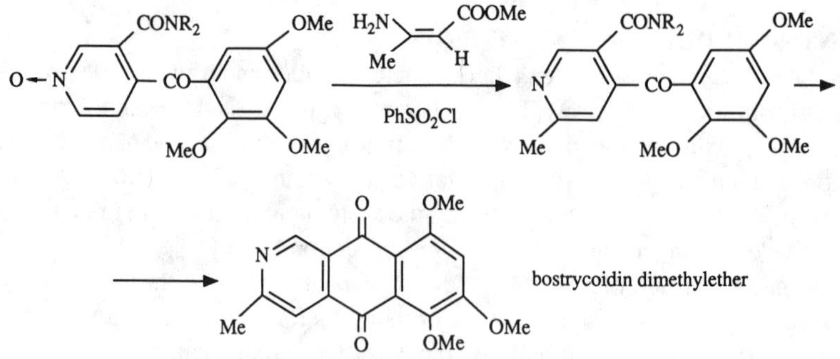

bostrycoidin dimethylether

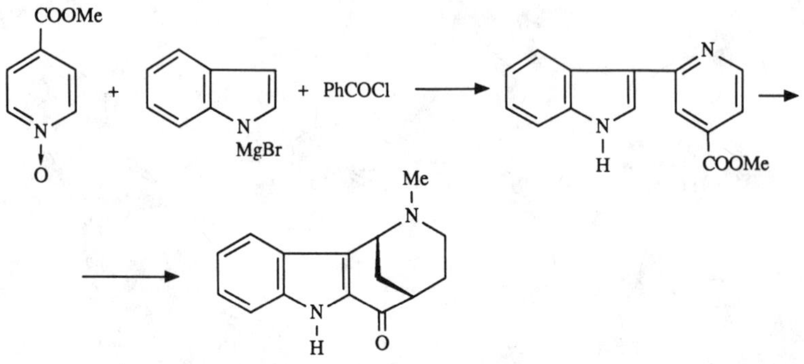

de-ethyldasycarpidone

The alkylation has also been used for the preparation of a variety of functionalized hetero-cycles, e.g., α-heteroaryl-substituted α-amino acids.[10]

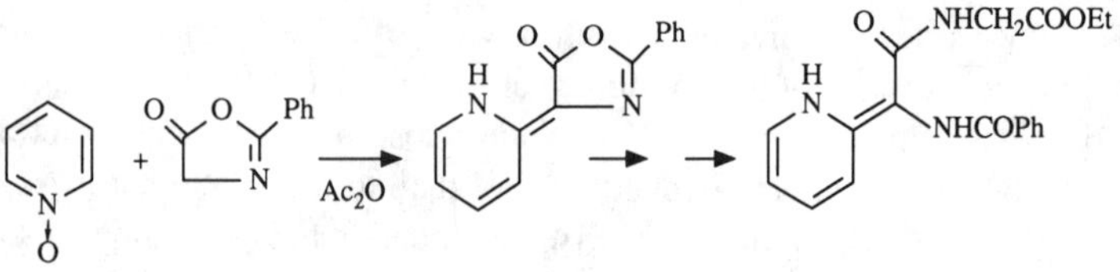

The mild (and regioselective)[11] functionalization of α and γ alkyl groups via deoxidative acetoxylation has largely been used for the synthesis of natural products,[11-14] the introduction of a double bond in a fused saturated ring,[15,16] and the synthesis of ligands and macrocyclics, e.g., the torand **2**.[17-19]

dihydrogarbitanine

2

Many other reactions of the *N*-oxides, ranging from free-radical substitution[20] to nucleophilic ring opening,[21] have been used for the synthesis of natural products. Furthermore, *N*-oxidation modifies the electronic structure of the ring, and thus the reactivity of the substituents. As typical examples, electron withdrawal by the $\geqslant$N$\rightarrow$O group increases the reactivity of the 4-carboxylic group in quinoline derivatives and is sufficient to let it participate into the Curtius reaction;[22] the spin delocalization in the 4-pyridyl 1-oxide ring makes the rearrangement of a 2-substituted methylenecyclopropane much faster,[23] and N^1-oxidation efficiently protects the acid-labile glycosidic bond in the 2-deoxyadenosine series.[24a] On the other hand, α-(2-pyridyl)ethyl acetate is not susceptible to ruthenium tetroxide catalyzed oxidation of the heteroaromatic ring by sodium periodate, while the corresponding *N*-oxide is (probably because it does not complex with intermediate metal species).[24b] Not to be forgotten, some furoxans are clean source of nitrile oxides.[25]

B. AS AUXILIARY AGENTS IN SYNTHESIS

Heterocyclic *N*-oxides can be used as auxiliary agents that are later eliminated in the course of the reaction, yielding the desired product. As an example, the 1-oxido-2-thiopyridine anion is a good leaving group and as such has been used to obtain mild nucleophilic substitution in cephem derivatives.[26] Similarly, 7α-methoxylation of cephalosporins is achieved by treatment of their imidoyl chlorides with 4-methoxypyridine 1-oxide, 1,4-elimination, and addition of methanol.[27]

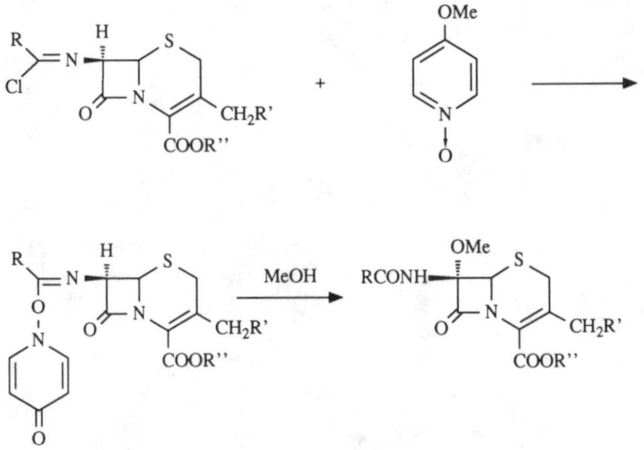

Other examples include a four-component Ugi condensation in the synthesis of amino acid nucleoside derivatives,[28,29] and long-range effects, i.e., in the alkylation of 3-substituted 2-deoxyriboses.[30]

C. AS PROTECTING GROUPS

Heterocyclic derivatives, as well as their *N*-oxides,[31] can be used in the same manner as their carbocyclic analogs, as *N*-protecting groups that are removable by hydrogenolysis. More importantly, one can make recourse to the specific reactions of *N*-oxides, most usually to side chain acetoxylation, and devise new protecting groups that are removable under mild conditions. Thus, the (1-oxido-2-pyridyl)methyl group has been successfully used for the protection of the amino, hydroxy, thiol substituents in heterocycles, as well as ring NH in pyrimidine and purine bases, and hydroxy and phosphate groups in the carbohydrate moiety of nucleosides and nucleotides.[32-36] The protecting group is introduced by reaction with 2-chloromethylpyridine 1-oxide with amino-, hydroxy-, thiol-substituted compounds; by reaction of 2-aminomethylpyrid-ine 1-oxide with sulfonyl derivatives;[35] or, very conveniently, by reaction with (1-oxido-pyrid-

2-yl)diazomethane, which alkylates substrates with $pK_a < 9.8$,[34] and is selectively removed by reaction with acetic anhydride, followed by hydrolysis to yield pyridincarboxyaldehyde and the substrate. This method is effective when the benzyl group gives no satisfactory results, and also has a directing effect on the functionalization.

D. AS OXIDANTS

1. Thermal Reactions

Heterocycle N-oxides are useful in synthesis as oxidants. Various dehydrogenations, oxidative cleavages, and oxygen transfer processes fall under this heading, though the mechanism has not always been elucidated.

Typical examples include dehydrogenation of diphenylethane to stilbene derivatives,[37] aromatization of hydro-aromatics[38] and -heterocyclics,[39-40] as well as cyclic dehydrogenation, e.g., of 2-hydroxychalchones to flavones[41] and of 2-hydroxyphenylimino derivatives to 2-phenylbenzoxazoles (in both cases with pyridine 1-oxide and palladium on charcoal),[42] or of 1-aminoanthraquinone to indanthrone (with pyridine 1-oxide/KOH).[43] Related reactions are the conversion of aromatic alcohols to ketones[46] and oxidative dimerization, e.g., formation of bibenzyl[44] and bipyridyls[45] from the corresponding methyl derivatives.

As discussed in Chapter 4, the reaction with anhydrides leads to N-acyloxy salts, which are themselves useful intermediates, and usually rearrange to α-acyloxy heterocycles in the absence of other nucleophiles. When α-vinyl- and α-aryl-substituted carboxylic anhydrides are used, however, cleavage of the carbon-carbon bond is the faster process and yields carbonyl derivatives with the loss of one carbon unit.[47-52] The mechanism of this reaction has been extensively investigated and appears to involve an intermediate carbocation. As expected from this scheme, α-alkoxy anhydrides similarly cleave or add nucleophiles.[50] Analogous results are obtained by replacing the anhydrides with ketenes or acyl chlorides.[53] The reaction between benzylchloroformiate and pyridine 1-oxide yields benzyl chloride through a related mechanism.[51] Further examples of oxidative bond cleavage have been reported for disulfides[54] and hydrazines[55] (Table 1).

R = Ph , CH$_2$=CH- , ROCH$_2$- , etc.

TABLE 1
Oxidation by N-Oxides

Substrate	Product (% yield)	Conditions	Ref.
n-C$_8$H$_{17}$Br	n-C$_7$H$_{13}$CHO(83)	i) 4-Me$_2$NC$_5$H$_4$NO, MeCN; reflux; ii) DBU[a]	56
PhCH$_2$Br	PhCHO(> 90)	i) 4-Me$_2$NC$_5$H$_4$NO, MeCN, reflux, ii) DBU[a]	56
PhCH$_2$Br	PhCHO(86)	i) Pyridine 1-oxide, MeCN; ii) aq NaOH	59
m-C$_6$H$_4$(CH$_2$Br)$_2$	m-C$_6$H$_4$(CHO)$_2$(60)	i) Pyridine 1-oxide, MeCN; ii) aq NaOH	59
3,5-Dibromo cyclopentene	Cyclopentadienone (low)	i) Pyridine 1-oxide; ii) aq NaOH	62
CH$_3$CHBrCOOEt	CH$_3$COCOOMe(90)	4-Me$_2$NC$_5$H$_4$NO, MeCN, reflux	56
CHBr(COOMe)$_2$	BrC(COOMe)$_2$(62) HOC(COOMe)$_2$	Pyridine 1-oxide, 20°C, 24 h	63
4-MeC$_6$H$_4$COCHBr$_2$	4-MeC$_6$H$_4$COCOOH(95)	Pyridine 1-oxide	65
9,10-Dihydro- anthracene	Anthracene(93)	Quinoline 1-oxide, 210—225°C, 15 min	38
Ph$_2$CHOH	Ph$_2$CO(73)	Pyridine 1-oxide, Raney Ni	46
PhMe	PhCH$_2$OH(40), PhCHO(19) HOC$_6$H$_4$Me (18)[b]	Benz[c,d]indazole 1,2-dioxide	77
1,2,3-Me$_3$C$_6$H$_3$	2,4,6-Me$_3$C$_6$H$_2$OH(23)	6-Methylpyridazine 1-oxide, CH$_2$Cl$_2$, hv	81
Triptophan	3-Indolecarboxylaldehyde(47)	Pyrimido[5,4-g]pteridine N-oxide, hv	84

[a]   DBU, 1,8-Diazabicyclo[5.4.0]undecene.
[b]   Mixture of isomers (2-OH, 11%, 3-OH, 3%, 4-OH, 4%).

The N-alkoxy salts obtained from alkyl halides and N-oxides (see Chapter 4) are decomposed by bases through two main pathways,[56-63] i.e., regeneration of the N-oxide with nucleophile addition to alkyl group and proton abstraction to yield the deoxygenated heterocycle and a carbonyl derivative (the competition of the two processes and its dependence on the hardness or softness of the base has been discussed). The latter sequence corresponds to a mild oxidation of alkyl halides to carbonyls, a conversion that is also obtained with dialkyl sulfoxides and aliphatic amine N-oxides,[64] but the use of heteroaromatic oxides is significantly advantageous in some cases (the most convenient reaction is with 4-dimethylaminopyridine 1-oxide, see Table 1[56]). Pyridine 1-oxide directly converts phenacyl dibromides to glyoxylic acids.[65]

Subsequent reactions of ethoxyacetylene with HgCl$_2$, pyridine 1-oxide, and zinc dust affords a zinc enolate, which, in turn, is reactive with aldehydes.[66]

The formation of carbonyl derivatives by oxidative cleavage of π bonds through reaction with N-oxides has been observed starting from diazo derivatives,[67] azirines,[68] imidoyl chlorides,[69] carbenoid tungsten carbonyl complexes,[70] and keteneimines.[71,72] Oxygen transfer to heteroatoms, e.g., formation of isocyanates from isonitriles,[73] or transfer to phosphorous,[74] sulfur (this appears to be favored in comparison with reaction with an allylic bromide group present in the same molecule),[75] or heterocyclic nitrogen (e.g., oxygen exchange between nicotinic acid 1-oxide and pyridine)[76] has also been observed.

Thermal benzylic oxygenation has been reported in some cases,[77,78] treatment with benz[cd]indazole 1,2-dioxide resulting in a particularly efficient reaction in this respect.[77]

2-Pentanone. *2-bromopentane (38 μl) and 4-dimethylaminopyridine 1-oxide (41 mg) in absolute acetonitrile are refluxed for 40 min. 1,8-Diazabicyclo[5.4.0]undecene (53.7 ml) is added and reflux continued for 16 h, yielding the product quantitatively.[56]*

Methyl 2-oxooctanoate. *Methyl 2-bromooctanoate (167 mg) and 4-dimethylaminopyridine 1-oxide (201 mg) in 2 ml acetonitrile are refluxed for 35 min. HCl (1.1 eqv) in MeCN is added, and the mixture is chromatographed to give the product in a yield of 103 mg (85%).[56]*

2. Photochemical Reactions

Though in most cases heterocyclic *N*-oxides rearrange under irradiation (see Chapter 4), oxygen transfer to various substrates may be an efficient process, and diazine (in particular, pyridazine 1-oxides)[81] and polyazapolycyclic *N*-oxides (e.g., pyrimido[5,4-*g*]pteridine *N*-oxide)[83,87] often give better results in this reaction than, e.g., pyridine or quinoline 1-oxides.

Reported reactions include hydroxylation of alkanes[79,80] and aromatics,[79-83,87d] epoxidation of alkenes,[82,84,85] oxidation of epoxides,[86] dealkylation of alkyamines,[87a,87b] dehydrocyclization of nucleosides to cyclonucleosides,[87c] *S*-oxidation of alkyl sulfides,[82] desulfurization, and deseleniation of phosphorous (V) derivatives.[88-90]

Various mechanisms have been put forward and include the formation of a labile intermediate (e.g., oxaziridine, which has also been proposed as the first step in photorearrangements, see Chapter 4) behaving as a strong oxidizer,[82,91] the liberation of oxygen in an active form ("oxene"),[80,85] a radicalic pathway,[92] and an electron transfer pathway, the last one being well characterized by mechanistic studies, but probably limited to a few polycyclic *N*-oxides.[83,87]

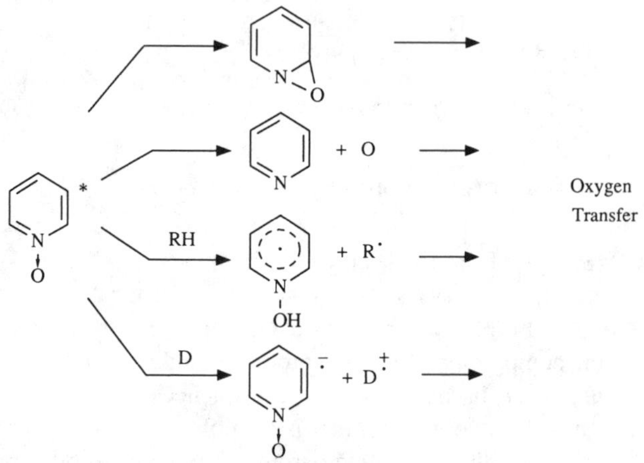

The mild oxidation obtained under these conditions presents some interest, and this is particularly the case for the hydroxylation of aromatics, in that it occurs via arene oxides, as demonstrated using deuterated substrates, in which the label is found in part as exchangeable O-D and in part remains as C-D;[82,89] this is the N.I.H. shift, which is observed also in microsomal oxidation,[93] and photochemical oxidation serves as a model for these biological processes. Electron-transfer photoinduced oxidation by pyrimido[5,4-*g*]pteridine *N*-oxide has also been considered as a model for biological oxidation.[87]

E. AS LIGANDS IN METAL COMPLEXES

Heterocycle *N*-oxides form σ-complexes through the oxygen atom with various metals, though they are not as strongly coordinating as the corresponding bases, which act through the nitrogen atom. The *N*-oxides tend to form complexes with the maximum coordination number of the particular metal, though there are numerous exceptions. Some typical structures are shown in Figure 1 and include examples with the *N*-oxide acting as a bridging ligand.[94-96] Table 2 includes some references to these compounds (Figure 1, Tables 2 and 3).

In view of the oxidizing properties of the *N*-oxides, complexes containing metals in their lowest oxidation number are less numerous, though by no means rare. Indeed, exceptional examples of complexes containing zero-valent metals have been reported, though heteroaromatic *N*-oxides are more generally used (albeit less frequently than their aliphatic counterpart) for the liberation of ligands by decomposition of their metal carbonyl complexes. Thus, e.g.,

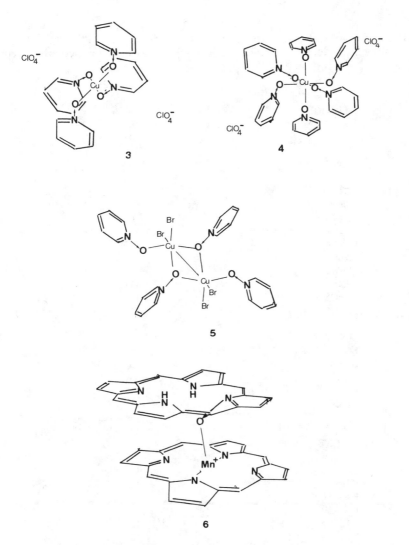

FIGURE 1.   Heterocyclic *N*-oxides as ligands in metal complexes.

germylene carbonyl complexes of chromium and tungsten of formula $X_2GeM(CO)_5$.pyridine 1-oxide have been prepared.[97] In other cases, the *N*-oxide transfers the oxygen atom to the metal, as shown in the preparation of oxo-bridged bimetallic samarium complexes.[98]

$$(C_5Me_5)_2Sm(THF)_2 + \text{pyridine 1-oxide} \rightarrow [(C_5Me_5)_2Sm]_2\,(\mu\text{-O}) + \text{pyridine}$$

The oxidizing properties of the pyridine 1-oxides are apparent in the reaction with palladium and platinum in the presence of organic halides, which results in the oxidation of the metal and yields pyridine complexes.[99]

$$Pt + \text{pyridine 1-oxide} + CHBr_3 \rightarrow Pt(\text{pyridine})_2Br_4$$

The complexes are usually obtained by simply mixing a solution of the *N*-oxide and the metal compounds, under an inert atmosphere or with dehydrating agents when appropriate, or by ligand exchange. The ligand can be easily substitued, e.g., more basic *N*-oxides are replaced by

TABLE 2
Metal Complex of Certain Heterocyclic *N*-Oxides (L = Pyrididine 1-Oxide,
L′-Isoquinoline 2-Oxide, L ″ = Octaethylporphyrin *N*-Oxide, L‴= Octaethylporphyrin)

Complex	Ref.
L_4Li, $NaClO_4$	103
L_6Mg, Ca, Sr, Ba $(ClO_4)_2$	103
$LAlCl_3$	104
L_3AlCl_3	105—106
$L_6Ga(ClO_4)_3$	107
L_3InCl_3	108b
L_2TlBrI_2	108a
$LSnCl_2$	109—110
L_2SnCl_2	109
$L_2Sn(n\text{-Bu})_2(AcO)_2$	111
$L_2Sn(OEt)Cl_3$	112
$L_3Pb(ClO_4)_2$	103
$LPbMe_3Cl$	117
$LSbCl_3$	113—114
L_2SbCl_3	114
L_3SbCl_3	113
$LSbMeCl_4$	115
$LCuCl.H_2O$	116
$LCuCl_2$	96
$LCu(AcO)_2$	96
L_2CuCl_2	96
$LAgClO_4$	117
$LZnSO_4$	118
L_2ZnCl_2	119—120
L_6Zn, $Hg(ClO_4)_2$	120, 125
$LCdCl_2$	122
$L_5Hg_2SiF_6$	123
$LHgCl_2$	119, 124
L_2HgCl_2	125
$L_2Hg(CF_3)_2$	126
$L_6Sc(ClO_4)_3$	127
L_4La, Nd, $Eu(ClO_4)_3$	128
L_8La, Ce, Eu $(ClO_4)_3$	129—131
$LUO_2(MeCOCHCOMe)_2$	132
$L_2UO_2SO_4$	133
L_2Ti, VCl_4	134, 136
$L_5V(ClO_4)_3$	135a
L_6V, $Cr(ClO_4)_3$	136
$LCr(CO)_5GeF_2$	97
L_3CrCl_3	135b
$L_2MoO_2Cl_2$	137
$L_6Mo_2Cl_{12}$	138
$LMnCl_2$	139
$L_6Mn(ClO_4)_2$	136
L″ L‴ Mn OAc	140
L_2MnCl_3	141
$LFeCl_2 \cdot (H_2O)_2$	143
$L_2FeCl_2 \cdot H_2O$	143
L_6Fe, Co, Ni $(ClO_4)_2$	121, 134, 136, 145
L_2FeCl_3	144
$L_6Fe(ClO_4)_3$	134
$L_7Fe(ClO_4)_3$	145
$L_2Fe_2Cl_6$	146
L_3Co, $NiCl_2 \cdot H_2O$	124, 134

TABLE 2 (continued)
Metal Complex of Certain Heterocyclic *N*-Oxides (L = Pyrididine 1-Oxide,
L′-Isoquinoline 2-Oxide, L ″ = Octaethylporphyrin *N*-Oxide, L′″= Octaethylporphyrin)

Complex	Ref.
L_3CoCl_2	147
L' Rh $(C_8H_{12})ClO_4$	142
LRh Cl(CO)$_2$	148
LPtCl$_2$ (C$_2$H$_4$)	149

TABLE 3
Complexes Containing Derivatives of *N*-Oxides as (Potentially)
Chelating Ligands

Ligand (L)	Complexes	Ref.
2-Carboxypyridine l-oxide anion	LNa	150
	L$_2$Ca, Zn, Mn	150, 151
	L$_2$Fe. 2H$_2$O	150
	L$_3$Fe	150, 151
l-Hydroxy-2-pyridone anion	L$_2$Cu, Zn, Mn, Ni	152
	L$_2$Sn(n-Bu)$_2$	153
l-Hydroxy-2-pyridinethione anion	LNa, Ag	154
	L$_2$Ca, Sn, Cu, Zn, Co, Ni, Pd	154
	LTlMe$_2$	155
	L$_2$Sn(n-Bu)$_2$	153
	L$_3$Cr, Fe	154
	L$_2$MoO$_2$, WO$_2$	154
	L$_4$Pt	154
2-Dialkylaminopyridine 1-oxide	L$_2$Cu(BF$_4$)$_2$	156
2-Carboxyamidopyridine 1-oxide	LCuCl	157a
2-Acetylpyridine l-oxide oxime	L$_2$Ni, Co, CuX$_2$	157b
2-Diphenylphosphinopyridine *N,P*-dioxide	LUO$_2$(NO$_3$)$_2$	158
1,10-Phenanthroline 1-oxide	L$_4$CeNO$_3$(ClO$_4$)$_3$.H$_2$O	159

less basic ones in platinum complexes.[100-102] Octaethylporphyrine *N*-oxides react with Mn(II) acetate to yield a complex OEP.Mn(III).OEPH$_2$*N*-oxide (OEP = dianion of octaethylporphyrine) containing a bridged oxygen (see Figure 1).[140]

Various substituted heterocyclic *N*-oxides function as bidentate ligands (see Table 3; the chelating character of the complexation has not been ascertained in every case). These include the anions of picolinic acid 1-oxide,[150,151] of 1-hydroxy-2-pyridone[152,153] and -pyridinethione,[153-155] as well as 2-amino-[156] and -carboxyamidopyridine 1-oxides,[157] diarylphosphinopyridine *N,P*-dioxides,[158] 1,10-phenanthroline 10-oxide,[159] macrocyclics,[160] and others. Bipyridine oxide appears to be a weaker ligand than pyridine 1-oxide.[161] Complexes of type M (pyridine 1-oxide)$_6$X$_2$ (M = Mn, Fe, Co, Ni, Cu, Zn; X = ClO$_4$, BF$_4$, NO$_3$, BrO$_3$, I) have been the subject of extended scrutiny, since the high symmetry of the lattice of the magnetic atoms is suitable for a comparison with theory.[162]

F. AS CATALYSTS

Heterocyclic *N*-oxides have an important role as catalysts or cocatalysts. Their action is due to different mechanisms. These polar molecules intervene in reactions via ionic species, forming adducts or labile complexes, e.g., they act as acyl transfer reagents in the reaction of carboxylic derivatives. In two-phase reactions, *N*-oxides work as phase-transfer catalysts for the selective transport of organic or metallic compounds.

Reported effects include the acceleration of nucleophilic substitution of alkyl[163] and aryl halides,[164] and particularly of reactions involving transfer of acyl,[165-170] sulfonyl,[167,172-177] and phosphoryl groups[170,178,179] (in the second case, it has been shown that the high catalytic activity of N-oxides in comparison with the corresponding bases results from the high thermal stability of intermediates, such as **7**, and their high reactivity towards arylamines).[176] Other important categories of catalytic effects are oxidations[181-184,196,197] and carbon-carbon bond-forming reactions.[185-188] A mixture of triisopropylbenzenesulfonyl chloride and 4-dimethylaminopyridine 1-oxide is an effective condensing agent for the rapid automated synthesis of oligodeoxyribonucleotides on a silica gel support[189] (Table 4).

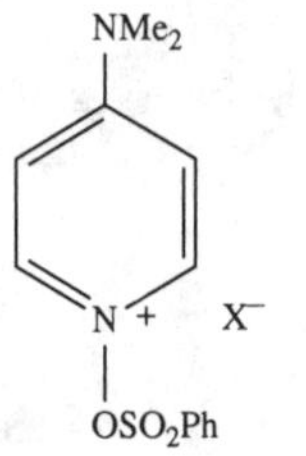

7

TABLE 4
N-Oxides as Catalysts

Substrate and reaction	Catalyst (or cocatalysts)	Ref.
Alkylhalides, nucl. subst. (2-phases)	$2\text{-MeSOC}_5\text{H}_4\text{NO}$, phase transfer	163
Picryl halides, nucl. subst.	Pyridine 1-oxide	164
$RCOOH$ + alcohols	Pyridine 1-oxide	165
$RCOCl + NaHCO_3 \rightarrow (PhCO)_2O$ (2-phases)	Pyridine 1-oxide and polymers	166
$RCOCl + ArOH$	Pyridine 1-oxide - Et_3N	167
$RCOCl + ArNH_2$	Pyridine 1-oxide	168—169
Esters, hydrolysis (2-phases)	$4\text{-}R_2NC_5H_4NO$ and polymers	170
Isocyanates→carbodiimides	Pyridine 1-oxide	171a
Isocyanates→carbamates	Pyridine 1-oxide	171b
$ROC(=S)SMe \rightarrow RSC(=O)SMe$	Pyridine 1-oxide and polymers	172a
$NH_2SO_3H + ROH$	Pyridine 1-oxide-$AcNMe_2$	172b
$ArSO_2X + ArOH$	Pyridine 1-oxide-Et_3N or $4\text{-}R_2NC_5H_4NO$	167, 173
$ArSO_2X + ArNH_2$	Pyridine 1-oxide or $4\text{-}R_2NC_5H_4NO$	174—176
Enols, nitration with $CF_3SO_2NO_3$	Lutidine 1-oxide	177
Phosphoric esters, hydrolysis	Pyridine 1-oxide and polymers	170, 178
Phosphoric esters, preparation	Pyridine 1-oxides	179
$ROH + R_3SiCl$	$4\text{-}NO_2C_5H_4NO$	180
$ArCH_3 + O_2$	Various N-oxides - $CoBr_2$	181
Alkenes + PhIO	Pyridine 1-oxide - metal complexes	182—183
Alkenes + H_2O_2 (2-phases)	Pyridine 1-oxides-Mo, W(VI) complexes	184a
$C_4H_9C\equiv CH + LiAlH_4$	Pyridine 1-oxides	185a
$CO + H_2 \rightarrow (CH_2OH)_2$	Pyridine 1-oxides, $Rh_4(CO)_{12}$, R_3P	185b
$CH_2 = CHOAc + CO + ROH \rightarrow AcOCH_2CH_2COOR$	$4\text{-MeC}_5H_4NO\text{-}Co_2(CO)_8$	186
$CH_2 = CHCN + CO + ROH \rightarrow NCCH_2CH_2COOR$	$4\text{-MeC}_5H_4NO\text{-}CO$	187a
$NaCH(COCR)_2 + R'X$	Pyridine 1-oxide	188a

G. OTHER REACTIONS

The role of N-oxides as ligands in metal complexes has been discussed above. This is related to the several reported applications for the extraction of metals in bichromophoric systems;[190-194] successful applications to the extraction of anions[195a] and of organic molecules[195b] have also been reported. Extraction of oxidizing cations makes N-oxides useful in two-phase oxidation[196, 197] (see Section IF), and, on the other hand, preformed metal-*N*-oxide complexes are in some cases particularly convenient oxidizers.[198]

In contrast to their normal role as oxidants, *N*-oxides have been shown in some cases to function as hydrogen donors.[199] Other uses include such different fields as the determination of *N*-terminal amino acids in proteins and polypeptides,[200-203] use as probes for the determination of the mechanism of biological reactions,[204,205] and incorporation in polymers for the immobilization of biomolecules.[206]

II. COMMERCIAL APPLICATIONS

Many individual *N*-oxides have found application in various fields. Some representative examples are collected in Table 5. Figure 2 shows some of the *N*-oxides that are known by common or commercial names (see also Chapter 1 for *N*-oxides present in nature).

Obviously extensive studies on the toxicity of commonly used *N*-oxides[324,325] (as well as of known carcinogens such as 4-nitroquinoline 1-oxide)[326,327] have also been published (Figure 2, Table 5).

TABLE 5
Some Applications of Heterocyclic *N*-Oxides

N-oxides	Application	Ref.
Pharmaceuticals		
Quinoxaline	Bactericide	207—208
Benzotriazine	Bactericide	209—210
Pyrido[2,3-*b*]quinoxaline	Antitrichomonal, bactericide	211
2-Pyridinethiocarbamate	Antinflammatory	212
Pyridine carboxyamides	Antiphlogistic	213
Imidazolynoxymethylpyridine	Analgesic	214
1,2,4-Triazine	Analgesic	215
Bis(2-pyridine)disulfide	Ophthalmic composition	216
Quinoline (aurachins)	Antibiotic	217
Phenazine (mixin)	Antibiotic	218
Guanine	Antibiotic	219
Pyrimidine (minoxidyl)	Vasodilator	220a
Pyridylcyanoguanidine	Antihypertensive	220b
Quinazoline	Cardiotonic, bronchodilators	221
Thioadenosin	Vasodilator	222
Adenosine	Angine treatment	223
1,2,5-Oxadiazole	Cardiovascular disease	224
3-Phenoxypyridine	Anticonvulsant	225
Pyridinecarboxy acid, amides	Antiulcer	226, 227
Pyridinecarboxylic acid	Antihyperlipemic	228
Quinoline	Antilipolytic	229
Pyrazine (acipimox)	Antilipolytic	230
2-Hydroxypyridine	Iron mobilization	231
Thiazole	Antitumor	232
Benzo-1,4-diazepine	Tranquilizers	233, 234
Aminopyridine	Degenerative brain disease	235
Veterinary, agrochemicals		
Quinoxaline	Antiinfective	236
Pyrazine (emimycin)	Anticoccidiosis	240
l-Ribosylpyrazine	Anticoccidiosis	238
Aminopyridine	Anticoccidiosis	237

TABLE 5 (continued)
Some Applications of Heterocyclic *N*-Oxides

N-oxides	Application	Ref.
Veterinary, agrochemicals (continued)		
3-Pyridinsulfonamide	Anticoccidiosis	239
2,2′-Dithiobispyridine	Bovine mastitis	241
Aminopyridine	Animal feed additive	242
Hydroxypyridine (Cu salts)	Animal feed additive	243
Quinoxaline	Animal feed additive	244—246
Bis(2-pyridine)disulfide	Animal feed additive	247
Pyrimidol[4,5-*b*]quinoxaline	Plant bactericide	248a
Imidazole carboxylic acid	Plant growth regulator	248b, 251
Pyridine	Herbicide	249—255
Pyrimidine	Herbicide	256
Quinoxaline	Herbicide	257
Benzo-1,2,4-triazine	Herbicide	258
Phenazine	Herbicide	259
Tetrahydrobenzotriazole	Herbicide	260
Indazole	Herbicide	263
2-Mercapto (or alkylthio) pyridine	Fungicide	261—262
5-(4-Quinolyl) cysteine	Fungicide	264
2-Acylthiopyridine	Seed sterilant	265
Bis(2-pyridine)disulfide	Fruit abscission enhancement	266
Insecticides, preservatives		
Pyridine, quinoline amides	Insecticide	267
Pyridine	Insecticide	268
Furoxan	Rodenticide	269
Pyrazine	Wood preservative	270
2-Mercaptopyridine	Antiseptic (textiles, paints)	271, 272
Pyridine	Antiseptic (food)	273
Cosmetic		
Aminopyrimidine(minoxidil)	Hair growth, hair lotion	274,275
2-Mercaptopyridine, quinoline	Antidandruff shampoos	276—278
Bis(2-pyridine) disulfide	Antiperspirant, deodorant	279
Quinoxaline	Skin tanning	280
Polymers, coatings		
Pyridine	Grafting maleic anh. onto polyethylene	281
2,2′-Dithiobipyridine	Vulcanization accelerator	282
Benzofuroxan	Crosslinking	283, 284
Pyridine, quinoline	Photocrosslinking	285
Pyridine, quinoline	Hardening epoxy resins	287
Pyridine	Diene polymerization	286
Pyridine	Stereospecific propylene polymerization	288
Pyridine	UV photolitography	289
Benzo[*c*]cinnoline	Sensitizer for photoconductors	290, 291
Benzotriazole	UV adsorbers	292
Pyridine	Light-stable coating	293
2-Mercaptopyridine	Antifouling coating compositions	294, 295
2-Mercaptopyridine polym.	Antimold agent	296
2-Mercaptopyridine	Nylon spinning	297

TABLE 5 (continued)
Some Applications of Heterocyclic *N*-Oxides

N-oxides	Application	Ref.
	Photography	
Pyridine	Hardening photographic films	298
Pyrimidine	Processing photographic material	299
Azopyridine	Oxidant	300
4-Nitropyridine	Silver bleach	301
	Others	
Benzofuroxans	Explosives	302—305
Nitrobenzo[*c*]cinnoline	Explosives	306
3-Methyl-4-nitropyridine	Nonlinear optic	307—311
3-Methyl-4-nitropyridine	Laser harmonic generator	312
Pyridine	Liquid crystals	313
Nitrobenzofuroxan	Liquid crystals	314
Resazurin	Redox indicator (for determination of water toxicity, fish freshness, tooth decay, etc.)	315—317
Pyridine	Metal deposition	318, 319
Benzofuroxan	Dye intermediate	320
Quinoxaline	Dyes	321
Acridine carboxy esters	Chemiluminescence	322
Pyridine	Rust-inhibited gasoline composition	323

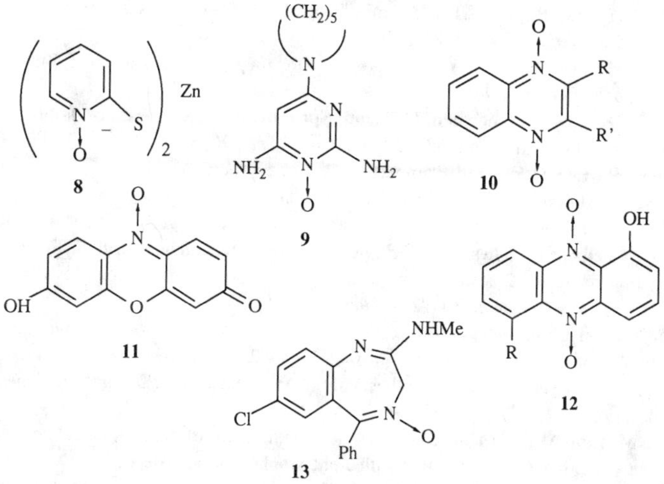

FIGURE 2. Some *N*-oxides known by common or trade names: **8**, Zinc Pirithione (zinc omactine) **9**, Minoxidil, **10** Dioxydine (R,R′=CH₂OH), Carbadox (R=H, R′=CH=NNH-COCH₂CN), Olaquinodox, BayoNox (R=Me,R′= CONHCH₂CH₂OH), **11**, Resazurin **12**, Iodinin (R=OH), Myxin (R=OMe), **13**, Chlordiazepoxide, Librium.

REFERENCES

1. **Konno, K., Hashimoto, K., Ohfune, Y., Shirahama, H., and Matsumoto, T.,** Acromelic acids A and B. Potent neuroexcitatory amminoacids isolated from *Clitocybe acromelalga, J. Am. Chem. Soc.,* 110, 4807, 1988.

2. **Kaneko, C., Naito, T., Hashiba, M., Fujii, H., and Somei, M.,** A novel synthesis of revenine and related alkaloids by means of a photo-rearrangement reaction of 4-alkoxy-2-methylquinoline l-oxides, *Chem. Pharm. Bull.,* 27, 1813, 1979.

3. **Sato, M., Katagiri, N., Muto, M., Haneda, T., and Kaneko, C.,** Practicable synthesis of (lR, 4R-5-L-menthoxy)-2-azabicyclo[2.2.0]hex-5-en-3-one and its derivatives. New building blocks for carbapenem nuclei, *Tetrahedron Lett.,* 6091, 1986.

4. **Geiger, R.E., Lalonde M., Stoller, H., and Schleich, K.,** Cobalt-catalyzed cycloaddition of alkynes and nitriles to pyridines: a new route to pyridoxine (vitamin B_6), *Helv. Chim. Acta,* 67, 1274, 1984.

5. **Kametani, T., Nemoto, H., Takeda, H., and Takano, S.,** Synthetic approach to comptothecin, *Chem. Ind.,* 1323, 1970.

6. **Adachi, J., Nomura, K., Shiraki, K., and Mitsuhashi, K.,** Synthesis of 1,2,3,4,5,6-hexahydro-1,5-methano-2-methylpyrido[2,3-c]azocine, *Chem. Pharm. Bull.,* 22, 658, 1974.

7. **Kametani, T. and Suzuki, T.,** An alternative synthesis of de-ethyldasycarpidone, *J. Chem. Soc. (C),* 1053, 1971.

8. **Kametani, T. and Suzuki, T.,** Total synthesis of (+-)-dasycarpidone and (+-)-3-epidasycarpidone. Formal total syntheses of (+-)-uleine and (+-)-3-epiuleine, *J. Org. Chem.,* 36, 1291, 1971.

9. **Watanabe, M., Shinoda, E., Shimizu, Y., Furukawa, S., Iwao, M., and Kuraishi, T.,** Synthesis of bostrycoidin via directed lithiation of tertiary nicotinamide, *Tetrahedron,* 43, 5281, 1987.

10. **Stanovnik, B., Drofonik, I., and Tisler, M.,** Transformation of heterocyclic *N*-oxides into derivatives of α-heteroaryl substituted α-aminoacids and dipeptides, *Heterocycles,* 26, 1805, 1987.

11. **Iwata, M. and Kuzuhara, H.,** Synthesis of a 6-methylpyridoxine hydrochloride derivative as an electrophilic precursor for pyridoxal-like functionalization, *Bull. Chem. Soc. Jpn.,* 55, 2295, 1982.

12. **Newkome, G.R. and Marston, C.R.,** Nicotinic acid lariat esters: syntheses, complexation, and reduction, *J. Org. Chem., 50, 4238, 1985.*

13. **Weinreb, S.M., Basha, F.Z., Hibino, S., Khatri, N.S., Kim, D., Pye, W.E., and Wu, T.T.,** Total synthesis of the antitumor antibiotic streptonigrin, *J. Am. Chem. Soc.,* 104, 536, 1982.

14. **O'Leary, M.H. and Payne, J.R.,** Vitamin B_6 analogs. Improved synthesis of 3-hydroxypyridine-4-carboxaldheide, *J. Med. Chem.,* 14, 773, 1971.

15. **Giam, C.S. and Ambrozich, D.,** A new approach to the preparation of 1,6- and 1,7-naphthyridines, *J. Chem. Soc., Chem. Commun.,* 265, 1984.

16. **Wenkert, E., Dave, K.C., Gnewuch, C.T., and Sprague, P.W.,** Syntheses of dl-dihydrogambirtannine and aspidosperma-strychnos alkaloid models, *J. Am. Chem. Soc.,* 90, 5251, 1968.

17. **Newkome, G.R., Lee, H.W., and Fronczek, F.R.,** A new route to macrocycles possessing the 2,6-pyridino subunit via bis-1,3-dithianes, *Isr. J. Chem.,* 27, 87, 1986.

18. **Bell, T.W. and Firestone, A.,** Torands: rigid toroidal macrocycles, *J. Am. Chem. Soc.,* 108, 8109, 1986.

19. **Botteghi, C., Chelucci, G., Chessa, G., Delogu, G., Gladiali, S., and Soccolini, F.,** Enantioface-discriminating transfer hydrogenation of 2-acetophenone catalyzed by rhodium (I) complexes with chiral 2-(2'-pyiridyl) pyridines, *J. Organomet. Chem.,* 304, 217, 1986.

20. **Natsume, M., Natsume, S., and Itai, T.,** Free radical reaction on aromatic amine *N*-oxides. I. Five-step synthesis of *N*-alkylmorphinan, *Itsuu Kenkyusho Nempo,* 9, 1968; *Chem. Abst.,* 71, 113119e, 1968.

21. **Giraudi, E. and Teisseire, P.,** New synthesis of naturally-occurring 1,3,5-undecatrienes, *Tetrahedron Lett.,* 489, 1983.

22. **Shioiri, T., Murata, M., and Hamada, Y.,** A new synthesis of α-amino acid and peptide amide of aromatic amines using a modified Curtius reaction with diphenyl phosphorazidate, *Chem. Pharm. Bull.,* 35, 2698, 1987.

23. **Creary, X. and Mehrsheikh-Mohammadi, M.E.,** The pyridine *N*-oxide group. A potent radical stabilizing function, *Tetrahedron Lett.,* 749, 1988.

24. a. **Morin, C.,** Protection of the glycosidic bond in the 2'-deoxyadenosine series, *Tetrahedron Lett.,* 53, 1983;
    b. **Kasai, M. and Ziffer, M.,** Ruthenium tetroxide catalyzed oxidation of aromatic and heteroaromatic rings, *J. Org. Chem.,* 48, 2346, 1983.

25. **Curran, D.P. and Fenk, C.J.,** Thermolysis of bis[2-(trimethylsilyl)oxy]-propyl]furoxan (TOP-furoxan). The first practical method for intermolecular cycloaddition of an in situ generated nitrile oxide with 1,2-di- and trisubstituted olefins, *J. Am. Chem. Soc.,* 107, 6023, 1985.

26. **Ochiai, M., Aki, O., Morimoto, A., Okata, T., and Kaneko, T.,** Nucleophilic substitution of 2-(4-carboxy-7-acylaminoceph-3-em-3-ylmethylthio)pyridine *N*-oxide derivatives in the presence of copper(II) salts, *Tetrahedron Lett.,* 2345, 1972.

27. **Iwamatsu, K., Shudo, K., and Okamoto, T.,** 7-α-Methoxylation of cephalosporins, *Heterocycles,* 20, 5, 1983.

28. **Tsuchida, K., Mizuno, Y., and Ikeda, K.,** Synthesis of 3-(3-amino-3-carboxypropyl)uridine (a modified nucleoside in certain RNAs) by four component condensation, *Heterocycles,* 15, 883, 1981.

29. **Tsuchida, K., Mizuno, Y., and Ikeda, K.,** Synthesis of naturally occurring uridine-α-amino acid derivatives by the application of the Ugi reaction, *Nucleic Acids Symp. Ser.,* 8, 549, 1980.

30. **Narasaka, K., Ichikawa, Y., and Kubota, H.,** Stereoselective preparation of β-*C*-glycosides from 2-deoxyribose utilizing neighboring participation by 3-*O*-methylsulfinylethyl group, *Chem. Lett.,* 2139, 1987.

31. **Bouchaudon, J. and Jolles, G.,** Protecting amino groups with pyridylmethoxycarbonyl groups, especially during peptide synthesis, *Ger. Offen* 2038996, 1971; *Chem. Abst.,* 75, 210172, 1971.

32. **Mizuno, Y., Ikeda, K., Endo, T., and Tsuchida, K.,** Aromatic amine *N*-oxides in syntheses of nucleosides and nucleotides, *Heterocycles,* 7, 1189, 1977.

33. **Mizuno, Y., Limn, W., Tsuida, K., and Ikeda, K.,** A novel protecting group for the synthesis of 7-α-D-pentofuranosylhypoxanthines, *J. Org. Chem.,* 37, 39, 1972.

34. **Endo, T., Ikeda, K., Kawamura, Y., and Mizuno, Y.,** 1-Oxidopyrid-2-yldiazomethane, a water soluble alkylating agent for nucleosides and nucleotides, *J. Chem. Soc. Chem. Commun.,* 673, 1973.

35. **Mizuno, Y., Endo, T., Miyaoka, T., and Ikeda, K.,** Syntheses and deblocking of 1-oxido-2-pyridylmethyl protected nucleosides and nucleotides, *J. Org. Chem.,* 39, 1250, 1974.

36. **Mizuno, Y., Endo, T., Takahashi, A., and Inaki, A.,** Synthesis of (4-substituted 2-picolyl 1-oxide) halides and 2′ and 3′-*O*-(4-substituted-2-picolyl 1-oxide) nucleosides, *Chem. Pharm. Bull.,* 28, 3041, 1980.

37. **Cohen, T., Shaw, C.K., and Jenkins, J.B.,** Conversion of a saturated to an unsaturated acid by pyridine *N*-oxide, *J. Org. Chem.,* 38, 3737, 1973.

38. **Kurbatov, Y.V., Kurbatova, A.S., Otroshchenko, A.S., Sadykov, A.S., and Smiyanova, V.P.,** Homogeneous dehydrogenation of 1,4-dihydronaphthalene and 9,10-dihydroanthracene with pyridine and quinoline *N*-oxides, *Tr. Samarkand Univ.,* 206, 220, 1972; *Chem. Abst.,* 80, 36990j.

39. **Kurbatov, Y.U., Kurbatova, A.S., Zalyalieva, O.V., Otroshchenko, A.S., and Sadykov, A.S.,** Dehydrogenation of anabasine, *Tr. Samarkand Univ.,* 167, 9, 1969; *Chem. Abst.,* 74, 142111p.

40. **Kiroev, G.V., Leont'ev, V.B. Kurbatov, Y.V., Kortavtsev, O.I., Otroschenko, A.S., and Sadikov, A.S.,** Mechanism of the dehydrogenation of Hantsch ester by pyridine *N*-oxide, *Isv. Akad. Nauk SSSR, Ser. Khim.,* 807, 1978.

41. **Haginiwa, J., Higuchi, Y., Kawashima, T., and Shinokawa, H.,** Reactions of 2′-hydroxychalcone derivatives with amine *N*-oxides, *Yakugaku Zasshi,* 96, 195, 1976.

42. **Haginiwa, J., Higuchi, Y., and Ohtsuka, H.,** Reactions of Schiff bases with amine *N*-oxides, *Yakugaku Zasshi,* 96, 209, 1976.

43. **Hardouin, J.C. and Vallette, M.,** Indanthrone vat dyes, *Ger. Offen* 2526197, 1976; *Chem. Abst.,* 85, 34641e, 1976.

44. **Prostakov, N.S., Torres, M., Varlamov, A.V., and Vasiliev, G.A.,** Pyridine *N*-oxide in oxidative dimerization reactions, *Izv. Vyssh. Uchebn. Zaved., Khim. Khim. Tekhnol.,* 24, 650, 1981.

45. **Haginiwa, J., Higuchi, Y., and Maki, I.,** Reactions of 3,3′-ethylenedipyridine derivatives with 2,6-1utidine *N*-oxide, *Yakugaku Zasshi,* 97, 1261, 1977.

46. **Haginiwa, J., Higuchi, Y., and Hirose, M.,** Oxidation of secondary alcohols, *Yakugaku Zasshi,* 97, 459, 1977.

47. **Cohen, T., Song, I.H., Fager, J.H., and Deets, G.L.,** Oxidative decarboxylation of carboxylic acids by pyridine *N*-oxide, *J. Am. Chem. Soc.,* 89, 4968, 1967.

48. **Rüchardt, C. and Krätz, O.,** Mechanism of decarboxylation of acid anhydrides with pyridine *N*-oxide, *Tetrahedron Lett.,* 5915, 1966.

49. **Cohen, T., Deets, G.L., and Jenkins, J.A.,** Oxidation of oxygen-labeled phenylacetic anhydride by pyridine *N*-oxide. Relative nucleophilicities of pyridine and pyridine *N*-oxide, *J. Org. Chem.,* 34, 2550, 1969.

50. **Rüchardt, C., Krätz, O., and Eichler, S.,** Reactions of pyridine 1-oxide with acid anhydrides, *Chem. Ber.,* 102, 3922, 1969.

51. **Koenig, T.W.,** Oxygen transfer reactions of aromatic amine *N*-oxides. Carboxy inversion mechanism, *Tetrahedron Lett.,* 2751, 1967.

52. **Jenkins, J.A. and Cohen, T.,** Deuterium isotope effects and the influence of solvent in the redox and rearrangement reactions of 2-picoline *N*-oxide and phenylacetic anhydride, *J. Org. Chem.,* 40, 3566, 1975.

53. **Koenig, T.W. and Barklow, T.,** Oxygen transfer reactions of amine *N*-oxides. III. Reaction of ketones and acid chlorides, *Tetrahedron,* 25, 4875, 1969.

54. **Ikura, K. and Oae, S.,** The reaction of diaryldisulfides with pyridine *N*-oxides, *Tetrahedron Lett.,* 3791, 1968.

55. **Bruni, P. and Poloni, M.,** Deamination of l,l-disubstituted hydrazines, *Gazz. Chim. Ital.,* 101, 893, 971.

56. **Mukaiyama, S., Inanaga, J., and Yamaguchi, M.,** 4-Dimethylaminopyridine *N*-oxide as an efficient oxidizing agent for alkyl halides, *Bull. Chem. Soc. Jpn.,* 54, 2221, 1981.

57. **Tamaki, Y., Sugie, H., and Noguchi, H.,** Methyl(*Z*)-5-tetradecenoate: sex-attractant pheromone of the soybean beetle, *Anomala rufocuprea* Motschulsky, *Appl. Entomol. Zool.,* 20, 359, 1985.

58. **Sliwa, H., and Tartar, A.,** Isolation and alkaline decomposition of the intermediate pyridinium salts occurring in the pyridine *N*-oxide oxidation of α-halo esters or acids, *J. Org. Chem.,* 41, 160, 1976.

59. **Feely, W., Lehn, W.L., and Boekelheide, V.,** Alkaline decomposition of quaternary salts of amine oxides, *J. Org. Chem.,* 22, 1135, 1957.

60. **Kato, T., Goto, Y., and Yamamoto, Y.,** A new synthetic method for α-oxo aldehydes. Reaction of α-picoline *N*-oxide with α-halo ketones, *Yakugaku Zasshi,* 84, 287, 1964.

61. **Hill, R.K. and Barcza, S.,** The synthesis of 2,2-diphenyl-tetrahydro-3-furonitrile, *J. Org. Chem.,* 27, 317, 1962.

62. **de Puy, C.H. and Zaweski, E.F.,** Cyclopentene-3,5-dione. I. Synthesis and properties, *J. Am. Chem. Soc.,* 81, 4920, 1959.

63. **Titkova, R.M. and Mikhalev, V.A.,** Reaction of pyridine *N*-oxide with esters of bromomalonic acid, *Zh. Org. Khim.,* 1, 1121, 1965.

64. **Franzen, V.,** Octanal, *Org. Synth.,* 47, 96, 19.

65. **Saldabols, N., Cimanis, A., and Hillers, S.,** *N*-oxides in the synthesis of arylglyoxyl acids, *Zh. Org. Khim.,* 10, 1059, 1974.

66. **Mukaiyama, T. and Murakani, M.,** A novel method for the generation of an ester enolate anion from ethoxyacetylene, *Chem. Lett.,* 1129, 1981.

67. **Dyakonov, I.A., Domareva-Mandel'shtam, T.V., and Radul, O.M.,** Reaction of ethyl diazoacetate with pyridine *N*-oxide in the presence of copper stearate, *Zh. Org. Khim.,* 4, 723, 1968.

68. **Padwa, A. and Crosby, K.,** Reaction of 2*H*-azirines with nitrones, *J. Org. Chem.,* 39, 2651, 1974.

69. **Rastetter, W.H., Gadek, T.R., Tane, J.P., and Frost, J.W.,** Oxidations and oxygen transfers effected by a flavin *N*(5)-oxide. A model for flavin-dependent monooxygenases, *J. Am. Chem. Soc.,* 101, 2228, 1979.

70. **Lukehart, C.M. and Zeile, J.W.,** Oxidation of transition metal carbenoid complexes by iodosobenzene, Comparative study, *J. Organomet. Chem.,* 97, 421, 1975.

71. **Barker, M.W. and Sung, H.S.,** Heterocycles from ketenimines. XI. Oxindoles, *J. Heterocycl. Chem.,* 14, 693, 1977.

72. **Barker, M.N. and Sung, H.S.,** Reaction of γ–picoline *N*-oxide with keteneimines in the presence of organic acids: preparation of α-acyloxy amides, *J. Heterocycl. Chem.,* 13, 1351, 1976.

73. **Johnson, H.W. and Krutzsch, H.C.,** Halogen-catalyzed pyridine *N*-oxide oxidation of isonitriles to isocyanates, *J. Org. Chem.,* 32, 1939, 1967.

74. **Stec, W.J., Okruszek, A., and Michalski, J.,** Stereochemistry of oxidation of phosphorus (III) compounds with amine *N*-oxides, *Bull. Acad. Pol. Sci. Ser. Sci. Chim.,* 21, 444, 1973.

75. **Polivka, Z., Licha, I., Taufmann, P., Svatek, E., Holubek, J., and Protiva, M.,** Synthesis of 3-(6,11-dihydrodibenzo[*b,e*]thiepin ll-ylidine)-propanoic acid and related compounds, *Coll. Czech. Chem. Commun.,* 52, 1566, 1987.

76. **Ryzhakov, A.V. and Elaev, N.R.,** Oxygen transfer from nicotinic acid *N*-oxide to pyridine, *Khim. Geterotsikl. Soedin,* 1075, 1987.

77. **Bowman, W.R., Gretton, W.R., Kirby, G.W., and Michael, J.D.,** Thermal oxygen atom transfer from benz[*c d*]indazole 1,2-dioxide to benzenoid derivatives, *J. Chem. Soc., Perkin Trans. 1,* 680, 1976.

78. **Kurbatova, A.S., Kurbatov, Y.V., Otroschenko, O.S., and Sadykov, A.S.,** Pyridine *N*-oxide as an oxidant. Oxidation of 9-anthrone to anthraquinone, *Tr. Samarkand. Univ.,* 206, 143, 1972; *Chem. Abst.,* 80, 36901k, 1974.

79. **Igeta, H., Tsuchiya, T., Yamada, M., and Arai, H.,** Photoinduced oxygenation of hydrocarbons by pyridazine *N*-oxides, *Chem. Pharm. Bull.,* 16, 767, 1968.

80. a. **Strub, H., Strehler, C., and Streith, J.,** Photoinduced nitrene, carbene, and atomic oxygen transfer reactions starting from the corresponding pyridinium *N*-, *C*-, and *O*-ylides, *Chem. Ber.,* 120, 355, 1987; b. **Schneider, H.J. and Sanerbrey, R.,** Photo-oxidation of alkanes with *N*-oxide, *J. Chem. Res. (S),* 14, 1987.

81. **Tsuchiya, T., Arai, H., and Igeta, H.,** Photo-induced oxygenation by pyridazine *N*-oxides. III. Oxygenation of polymethylbenzenes, *Tetrahedron Lett.,* 2213, 1970.

82. **Akhatar, N.M., Boyd, D.R., Neill, J.D., and Jerina, D.M.,** Stereochemical and mechanistic aspects of sulfoxide, epoxide, arene oxide and phenol formation by photochemical oxygen atom transfer from aza-aromatic *N*-oxides, *J. Chem. Soc., Perkin Trans. 1,* 1693, 1980.

83. **Sako, M., Shimada, K., Hirata, K., and Maki, Y.,** Photochemical hydroxylation of benzene derivatives by pyrimido[5,4-*g*]pteridine *N*-oxide, *Tetrahedron Lett.,* 6493, 1985.

84. **Tsuchiya, T., Arai, H., and Igeta, H.,** Photo-induced oxygenation by pyridimine *N*-oxides. II. Formation of epoxides from ethylenic compounds, *Tetrahedron Lett.,* 2747, 1969.

85. **Ogawa, Y., Iwasaki, S., and Okuda, S.,** A study of the transition state in the photooxygenation by aromatic amine *N*-oxides, *Tetrahedron Lett.,* 2277, 1981.

86. **Ito, Y. and Matsuura, T.,** The reaction of epoxides with oxygen-transfer reagents, *J. Chem. Soc., Perkin Trans. 1,* 1871, 1981.

87. a. **Sako, M., Shimada, K., Hirota, K., and Maki, Y.,** Photochemical oxygen atom transfer reaction by heterocyclic *N*-oxides involving a single electron-transfer process: oxidative demethylation of *N,N*-dimethylaniline, *J. Am. Chem. Soc.,* 108, 6039, 1986; b. **Sako, M., Shimada, K., Hirota, K., and Maki, Y.,** Photochemical oxidative C_α-C_β bond cleavage of tryptophan side-chain by pyrimido[5,4-*g*]pteridine *N*-oxide, *Tetrahedron Lett.,* 3877, 1986; c. **Sako, M., Shimada, K., Hirota, K., and Maki, Y.,** Photochemical intramolecular cyclization of purine and pyrimidine nucleosides induced by an electron acceptor, *J. Chem. Soc., Chem. Commun.,* 1704, 1986; d. **Maki, Y., Sako, M., Shimada, K., Murase, T., and Hiroto, K.,** Pyrimido[5,4-*g*]pteridine *N*-oxides as a simple chemical model of hepatic monooxigenase, *Stud. Org. Chem.,* 33, 465, 1988.

88. **Bharaway, R.K. and Dandson, R.S.,** The use of *N*-oxides to photoinduce the oxidative desulfurization and deselenation at pentacovalent phosphorous, *J. Chem. Res. (S),* 406, 1987.

89. **Ogawa, Y., Iwasaki, S., and Okuda, S.,** Photochemical aromatic hydroxylation by aromatic amine *N*-oxides, remarkable solvent effect on NIH-shift, *Tetrahedron Lett.,* 3637, 1981.

90. **Rowley, A.C. and Steedman, J.R.F.,** The mechanism of the conversion of thiophosphoryl compounds into their phosphoryl analogs by photochemically excited 3-methylpyridazine 2-oxide and by 2-methyl-3-*p*-nitrophenyloxaziridine; a comparison, *J. Chem. Soc., Perkin Trans. 2,* 1113, 1983.

91. **Kaneko, C., Yamamori, M., Yamamoto, A., and Hayashi, R.,** Irradiation of aromatic amine oxides in dichloromethane in presence of triphenylphosphine, *Tetrahedron Lett.,* 2799, 1978.

92. **Lin, S.K. and Wang, H.Q.,** Radical mechanism in the photoreaction of organic *N*-oxides: some paradiazine *N,N*-dioxides, *Heterocycles,* 24, 659, 1986.

93. **Daly, J.W., Jerina, D.M., and Witkop, B.,** Arene oxides and NIH shift. Metabolism, toxicity and carcinogenicity of aromatic compounds, *Experientia,* 28, 1129, 1972.

94. **Lee, J.D., Brown, D.S., and Melsom, B.G.A.,** Crystal structure of tetra(pyridine oxide) copper(II) pechlorate, *Acta Crystallogr. Sec. B,* 25, 1378, 1969.

95. **O'Connor, C.J., Sinn, E., and Carlin, R.L.,** Structural and magnetic properties of $[M(C_5H_5NO)_6]L_2$ (M=Cu, Zn; L=ClO$_4^-$, BF$_4^-$), *Inorg. Chem.,* 16, 3314, 1977.

96. **Jotham, R.W., Kettle, S.F.A., and Marks, J.A.,** Binuclear copper(II) complexes with bridging aromatic *N*-oxide groups, *J. Chem. Soc., Dalton Trans.,* 1133, 1972.

97. **Castel, A., Riviere, P., Satge, J., and Ahbala, M.,** Reaction of chromium and tungsten complexes with 1,2-dipolar germylenes, *J. Organomet. Chem.,* 328, 123, 1987.

98. **Evans, W.J., Grate, I.W., Bloom, I., Hunter, W.E., and Atwood, J.L.,** Synthesis and X-ray crystallographic characterization of an oxo-bridged bimetallic organosamarium complex, $[(C_5Me_5)_2Sm]_2$ (μ-O), *J. Am. Chem. Soc.,* 107, 405, 1985.

99. **Makitova, D.D., Letuchii, Y.A., Krasodika, O.N., Roshchupkina, O.S., Lavrentièv, I.P., Atovmyan, L.O., and Khidekel, M.L.,** Composition and structure of crystalline products of the oxidation of rhodium, palladium and platinum metals in the L-halohydrocarbon system, *Koord. Khim.,* 14, 394, 1988.

100. **Orchin, M. and Schmidt, P.J.,** Pyridine *N*-oxide complexes of platinum(II), *Coord. Chem. Rev.,* 3, 345, 1968.

101. **Kaplan, P.D., Schmidt, P., and Orchin, M.,** Ligand lability in platinum(II) complexes, *J. Am. Chem. Soc.,* 89, 4537, 1967.

102. **Weil, T.A., Schmidt, P., Rycheck, M., and Orchin, M.,** Solvent displacement of ligands from carbonyl complexes of platinum(II), *Inorg. Chem.,* 8, 1002, 1969.

103. **Reedijk, J.,** The ligand properties of pyridine *N*-oxide, *Rec. Trav. Chim. Pays-Bas,* 88, 499, 1969.

104. **Brown, D.H., Stewart, D.T., and Jones, D.E.M.,** Spectra and structure of aluminum halide complexes with pyridine derivatives, *Spectrochimica Acta,* Part A, 29, 213, 1973.

105. **Masaguer Fernandez, J.R., Sanchez, A., and Casas, J.S.,** Interaction of aluminum chloride with pyridine oxides, *Acta Cient. Compostelana,* 9, 181, 1972; *Chem. Abst.,* 81, 168946j, 1974.

106. **Masaguer Fernandez, J.R., Sanchez, A., Casas, J.S. and Sordo, J.,** Reaction of aluminum chloride with pyridine oxides, *Acta Cient. Compostelana,* 13, 31, 1976; *Chem. Abst.,* 87, 33014g, 1977.

107. **Hutton, A.T. and Thornton, D.A.,** The infrared spectra of pyridine *N*-oxide complexes of transition metal perchlorates in relation to their structures, *J. Mol. Struct.,* 39, 33, 1977.

108. a. **Brown, D.H. and Stewart, D.T.,** The preparation and far i.r. spectra of some indium trihalide complexes, *J. Inorg. Nucl. Chem.,* 32, 3751, 1970; b. **Bermejo, M.R., Teresa Lage, M., Isabel Fernandez, M., and Castineiras, A.,** Complexes of thallium dibromide iodide with pyridine *N*-oxide and some substituted pyridine N-oxides, *Synth. React. Inorg. Met. Org. Chem.,* 17, 79, 1987.

109. **Morrison, J.S. and Haendler, H.M.,** Some reactions of tin(II) chloride in nonaqueous solution, *J. Inorg. Nucl. Chem.,* 29, 393, 1967.

110. **Kauffman, J.W., Moor, D.H., and Williams, R.J.,** SnCl$_2$-methyl substituted pyridine *N*-oxide adducts, *J. Inorg. Nucl. Chem.,* 39, 1165, 1977.

111. **Graddon, D.P. and Rana, B.A.,** Lewis acidity of organotin carboxylates and isocyanates, *J. Organomet. Chem.,* 136, 19, 1977.

112. **Paul, R. C., Nagpal, V., and Chadha, S.L.,** Addition compounds of alkoxy tin(IV) trichloride with some phosphoryl and amine-oxide ligand, *Inorg. Chim. Acta,* 6, 335, 1972.

113. **Malhotra, K.C., Chawla, G.K., and Chandhry, S.C.,** Nature of antimony trichloride, *J. Indian Chem. Soc.,* 51, 1035, 1974.

114. **Masaguer Fernandez, J.R., Castano, M.V., Casas, J.S., Bermejo, M.R., and Sordo, J.,** Pyridine 2-, 3-, and 4-methylpyridine *N*-oxide complexes with antimony (3+) chloride and antimony (5+) chloride, *Inorg. Chem. Acta,* 19, 139, 1976.

115. **Nishii, N., Hashimota, K., and Okawara, R.,** Preparation and properties of monoorganoantimony tetrachloride adducts, *J. Organomet. Chem.,* 55, 133, 1973.

116. **Kokoszka, G.F., Allen, H.C., and Gordon, G.,** Magnetic and optical spectra of two dimeric copper chloride pyridine *N*-oxide complexes, *J. Chem. Phys.,* 46, 3013, 1967.

117. **Pradhan, G.C. and Ramana Rao, D.V.,** Complexes of pyridine *N*-oxide and -picoline *N*-oxide with Ag(I) nitrate, perchlorate and cyanate, *Ind. J. Chem.,* 15A, 245, 1977.

118. **Ahuia, I.S.,** Pyridine *N*-oxide and 4-methylpyridine *N*-oxide complexes with zinc(II) sulfate, *Ind. J. Chem.,* 9, 173, 1971.

119. **Schmauss, G. and Specker, M.,** Pyridine *N*-oxide complexes of zinc, cadmium, and mercury halides, *Naturwissenschaften,* 54, 258, 1967.

120. **Quagliano, J.V., Fujita, J., Franz, G., Phillips, D.J., Walmsley, J.A., and Tyree, S.Y.,** The donor properties of pyridine *N*-oxide, *J. Am. Chem. Soc.,* 83, 3770, 1961.

121. **Carlin, R.L., O'Connor, C.J., and Bhatia, S.N.,** Magnetic investigation of the electronic structure of hexakis (pyridine *N*-oxide) cobalt(II) perchlorate, *J. Am. Chem. Soc.,* 98, 685, 1976.

122. **Ahuja, I.S. and Restogi, P.,** Complexes of some pyridine *N*-oxides and chloroanilines with cadmium (II) halides, *J. Inorg. Nucl. Chem.,* 32, 2665, 1970.

123. **Potts, R.A. and Allred, A.L.,** Mercury(I) complexes, *Inorg. Chem.,* 5, 1066, 1966.

124. **Powell, H.B., Pappas, A.J., and Osterman, F.A.,** Differential thermal analysis of some pyridine l-oxide co-ordination compounds, *Inorg. Chem.,* 9, 2695, 1970.

125. **Pappas, A.J., Villa, J.F., and Powell, H.B.,** The interaction of mercury(II) salts with pyridine l-oxide, *Inorg. Chem.,* 8, 550, 1969.

126. **Powell, H.B., Maung, M.T., and Lagowski, J.J.,** Some complex compounds of fluoroalkyl- and fluoroaryl-mercurials, *J. Chem. Soc.,* 2484, 1963.

127. **Crawford, N.P. and Melson, G.A.,** Complexes of scandium(II) perchlorate with uni- and bi-dentate oxygen donor ligands, *J. Chem. Soc. (A),* 141, 1970.

128. **Pneumaticakis, G.A.,** Complexes of yttrium and rare earths with pyridine *N*-oxide, *Chem. Ind.,* 770, 1968.

129. **Ramakrishnan, L. and Soundararajan, S.,** Pyridine l-oxide complexes of lanthanide iodides, *Monatsh.,* 106, 625, 1975.

130. **Krishnamurthy, V.N. and Sounderarajan, S.,** Pyridine *N*-oxide complexes of rare earth perchlorates, *Can. J. Chem.,* 45, 189, 1967.

131. **Butter, E. and Seifert, W.,** Quadratic-antiprismatically coordinated pyridine *N*-oxide complexes of europium(II), *Z. Anorg. Allg. Chem.,* 368, 133, 1969.

132. **Haigh, J.M. and Thornton, D.A.,** Ligand substitution effects in uranyl-β-ketoenolates, *J. Mol. Struct.,* 8, 351, 1971.

133. **Ahuja, I.S. and Sing, R.,** Uranyl nitrate and thiocyanate complexes with some pyridine *N*-oxides, *J. Inorg. Nucl. Chem.,* 35, 561, 1973.

134. **McGregor, W.R. and Bridgland, B.E.,** Nitrogen-oxygen stretching frequence for some substituted pyridine *N*-oxide complexes of titanium(IV) and tin(IV)chlorides, *J. Inorg. Nucl. Chem.,* 31, 3325, 1969.

135. a. **Karayannis, N.M., Strocko, M.J., Mikulski, C.M., Bradshaw, E.E., Pytlewski, L.L., and Labes, M.M.,** Vanadium(III)perchlorate interactions with pyridine *N*-oxides, *J. Inorg. Nucl. Chem.,* 32, 3962, 1970; b. **Gutmann, V. and Melcher, G.,** Ligand stabilization and molar extinction for octahedral chromium(III) complexes and their dependence from the ligand properties, *Monatsh.,* 103, 624, 1972.

136. **Nathan, L.C. and Ragsdale, R.O.,** Trends in nitrogen-oxygen stretching frequency of 4-substituted pyridine *N*-oxide coordination compounds, *Inorg. Chem. Acta,* 10, 177, 1974.

137. **Buteher, R.J., Gunz, H.P., Maclagan, R.G.A.R., Kipton, H., Powell, H.K.J., Wilkins, C.J., and Hian, Y.S.,** Infrared spectra and configurations of some molybdenum(VI) dihalide dioxide complexes, *J. Chem. Soc., Dalton Trans.,* 1223, 1975.

138. **Fiedl, R.A., Kepert, D.L., and Taylor, D.,** Acceptor properties and bonding of $(Mo_6Cl_8)^{4+}$ and $(Nb_6Cl_{12})^{2+}$ clusters, *Inorg. Chim. Acta,* 4, 113, 1970.

139. **Brown, D.H., Kenyon, D., and Sharp, D.W.A.,** The thermal decomposition of some pyridine *N*-oxide and substituted pyridine *N*-oxide complexes of manganese(II), cobalt(II), nickel(II) and zinc(II) halides, *J. Chem. Soc. (A),* 1474, 1969.

140. **Arasasingham, R.D., Balch, A.L., Olmstead, M.M., and Renner, M.W.,** Insertion of manganese and cobalt into octaethylporphyrin *N*-oxide. Formation of layered diporphyrin structures joined through M-O-N links, *Inorg. Chem.*, 26, 3562, 1987.

141. **Uson, R., Riera, V., Ciriano, M.A., and Valdemarra, M.,** Pentacoordinate neutral manganese(III) complexes, *Transition Met. Chem.*, 1, 122, 1976.

142. **Christofides, A., Drigas, D., and Tourka, E.,** Synthesis of new cationic rhodium(I) complexes with diolefins and isoquinoline *N*-oxide as ligands, *Inorg. Chim. Acta*, 126, 95, 1987.

143. **Prabhakaran, C.P. and Patel, C.C.,** Antipyrine and pyridine *N*-oxide complexes of iron(II), *J. Inorg. Nucl. Chem.*, 34, 3485, 1972.

144. **Cotton, S.A. and Gibson, J.F.,** Spectroscopic studies on pyridine *N*-oxide complexes of iron(III), *J. Chem. Soc. (A)*, 2105, 1970.

145. **Carlin, R.L.,** Transition metal complexes of pyridine *N*-oxide, *J. Am. Chem. Soc.*, 83, 3773, 1961.

146. **Karayannis, N.M., Cronin, J.T., Mikulski, C.M., Pytlewski, L.L., and Labes, M.M.,** Structures of iron(III) complexes with 4-substituted pyridine *N*-oxides, *J. Inorg. Nucl. Chem.*, 33, 4344, 1971.

147. **Bertrand, J.A. and Plymale, D.L.,** Structure of tris (pyridine *N*-oxide) cobalt(II) chloride and bromide, *Inorg. Chem.*, 3, 775, 1964.

148. **Pribula, A.J. and Drago, R.S.,** Thermodynamic and spectroscopic studies of the reaction of Lewis bases with di-μ-chloro-tetracarbonyldirhodium(I), *J. Am. Chem. Soc.*, 98, 2784, 1976.

149. **Garcia, L., Shupack, S.I., and Orchin, M.,** Preparation and reaction of olefin-pyridine *N*-oxide platinum(II) complexes, *Inorg. Chem.*, 1, 893, 1962.

150. **Boyd, S.A., Kohrman, R.E., and West, D.X.,** Transition metal ion complexes of 2-picolinate *N*-oxide, *J. Inorg. Nucl. Chem.*, 38, 607, 1976.

151. **Lever, A.B.P., Lewis, J., and Nylholm, R.S.,** Metal complexes of picolinic acid *N*-oxide, *J. Chem. Soc.*, 5262, 1962.

152. **Sun, P.J., Fernando, Q., and Freiser, H.,** Formation constants of transition metal complexes of 2-hydroxypyridine 1-oxide and 2-mercaptopyridine 1-oxide, *Anal. Chem.*, 36, 2485, 1964.

153. **Kaufman, C.W.,** Dialkyltin salts of substituted pyridine 1-oxides as fungicides and bactericides, U.S. patent 3705943, 1972; *Chem. Abst.*, 78, 72371e, 1973.

154. **Edrissi, M., Massoumi, A., and Dalzeil, J.A.W.,** Comparative studies of 1-hydroxy-2-pyridinethion and its methyl derivatives as analytical reagents for metal ions, *Microchem. J.*, 16, 538, 1971.

155. **Castano, M.V., Sanchez, A., Casas, J.S., Sordo, J., Brianso, J.L., Piniella, J.F. Solans, X., Germain, G., Debaerdemaeker, T., and Glaser, J.,** Dimethyl thallium(III) complexes with sulfur-nitrogen and sulfur-oxygen ligands; the molecular structure of Me$_2$TlL (L=1-oxidopyridinium-2-thiolato), *Organometallics*, 7, 1897, 1988.

156. a. **West, D.X. and Nipp, C.A.,** Copper(II) complexes of 2-dimethylamino-3-picoline *N*-oxide (3MDM) and 2-diethylamino-3-picoline *N*-oxide (3MDE), *In. Chim. Acta*, 127, 129, 1987; b. **West, D.X. and Daraska, C.A.,** Copper(II) complexes of 2-dialkylaminopyridine *N*-oxide, *Synth. React. Inorg. Met-Org. Chem.*, 17, 431, 1987.

157. **West, D.X., Profilet, R.D., Severns, J.C., and Bunting R.K.,** Metal ion complexes of 2-picolinamide *N*-oxide, *Transition Met. Chem.*, 13, 29, 1988; b. **Landers, A.E. and Phillips, D.J.,** Metal complexes of 2-acetylpyridine *N*-oxide oxime, *Inorg. Chim. Acta*, 59, 41, 1982.

158. **McCabe, D.J., Russell, A.A., Karthikeyan, S., Paine, R.T., Ryan, R.R., and Smith, B.,** Crystal and molecular structures of bis(nitrato) [2-(diphenylphosphino)pyridine *N,P*-dioxide]dioxouranium(VI), *Inorg. Chem.*, 26, 1230, 1987.

159. **Yuan, J., Zhao, G., Wang, G., Yuan, W., and Wang, Y.,** Synthesis of cerium(IV) complexes with 1,10-phenanthroline 1-oxide, *Yingyong Huaxue*, 5, 73, 2988; *Chem. Abst.*, 109, 203827s, 1988.

160. **Artz, S.P., De Grandiore, M.P., and Cram, D.J.,** Host-guest complexation 33. Search for new chiral guests, *J. Org. Chem.*, 50, 1486, 1985.

161. **Da Silva, M.L.C.P., Chagas, A.P., and Airoldi, C.,** Heterocyclic *N*-oxide ligands: a thermochemical study of adducts with zinc, cadmium, and mercury chlorides, *J. Chem. Soc., Dalton Trans.*, 2113, 1988.

162. **Carlin, R.L. and de Jongh, L.J.,** Structural and magnetic properties of transition metal complexes of pyridine *N*-oxide, *Chem. Rev.*, 86, 659, 1986.

163. **Furukawa, N., Ogawa, S., Kawai, T., and Oae, S.,** Sulfoxide substituted pyridines as phase-transfer catalysts for nucleophilic displacements and alkylations, *J. Chem. Soc., Perkin Trans. 1*, 1833, 1984.

164. **Titskii, G.D., Shumeiko, A.E., and Litvinenko, L.M.,** Catalytic action of organic bases in the reaction of picryl fluoride with aniline, *Dokl. Akad. Nauk SSSR*, 234, 868, 1977.

165. **Cheshko, F.F. and Shvaika, T.N.,** Synthesis and identification of some mono- and dicarboxylic acid esters, *Zh. Prikl. Khim.*, 44, 1107, 1971.

166. **Zeldin, M., Fife, W.K., Tian, C., and Xu, J.,** Synthesis, characterization, and catalytic properties of poly[methyl(1-oxypyridin-3yl)siloxane], *Organometallics*, 7, 470, 1988.

167. **Savelova, V.A., Belousova, I.A., Simanenko, Y.S., and Titskii, G.D.,** Synergistic effect in catalytic formation of carboxylate and arenesulfonate esters, *Zh. Org. Khim.,* 23, 1571, 1987.

168. **Litvinenko, L.M., Titski, G.D., and Shpan'ko, I.V.,** Catalytic effect of pyridine *N*-oxide on the reaction of benzoyl chloride with primary aromatic amines in benzene, *Zh. Org. Khim.,* 7, 107, 1971.

169. **Litvinenko, L.M., Titski, G.D., and Shpan'ko, I.V.,** Kinetics of pyridine *N*-oxide-catalyzed benzoylation of aniline by benzoyl halides, *Zh. Org. Khim.,* 8, 1007, 1972.

170. **Katritzky, A.R., Duell, B.L., Rasala, D., Knier, B., and Durst, H.D.,** Synthesis and catalytic activity of *N*-oxide surfactant analogs of 4-(dimethylamino)pyridine, *Langmuir,* 4, 1118, 1988.

171. a. **Managle, J.J.,** Conversion of isocyanates to carbon-diimides. Catalyst studies, *J. Org. Chem.,* 27, 3851, 1962; b. **Burbus, J.,** Catalysis of isocyanate reactions. II. Pyridine *N*-oxide catalysis, *J. Org. Chem.,* 27, 373, 1962.

172. a. **Harano, K., Shinohara, I., Murase, M., and Hisano, T.,** Pyridine *N*-oxides as catalysts for thione-thiol rearrangement, *Heterocycles,* 26, 2583, 1987; b. **Pakhomova, G.M., Loktev, S.M., Rezvova, E.A., and Kagan, Y.B.,** Catalysts and conditions for the esterification of higher secondary alcohols by sulfamic acid, *Zh. Vses. Khim. Obshchest.,* 17, 477, 1972; *Chem. Abst.,* 77, 163940k, 1972.

173. **Savelova, V.A., Belousova, I.A., and Litvinenko, L.M.,** Activity of 4-(dimethylamino) pyridine and its *N*-oxide in the phenolysis of arenesulfonic acid derivatives in methylene chloride, *Zh. Org. Khim.,* 22, 133, 1986.

174. **Litvinenko, L.M., Savelova, V.A., Shatskaya, V.A., and Sadovskaya, T.N.,** Kinetic proof of the formation of an intermediate product during nitrogen and oxygen nucleophilic catalysis in a sulfonamide formation reaction, *Dokl. Akad. Nauk SSSR,* 198, 844, 1971.

175. **Savelova, V.A., Solomoichenko, T.N., and Litvinenko, L.M.,** Kinetic principles of noncatalytic and pyridine *N*-oxide catalyzed acylation of *m*-chloroaniline by *p*-toluensulfonyl bromide in a nitrobenzene and cyclohexane mixture, *Zh. Org. Khim.,* 9, 110, 1973.

176. **Litvinenko, L.M., Savelova, V.A., and Belousova, I.A.,** Anomalously high catalytic activity of *N*-oxides of pyridines in acyl transfer reactions in protoinert media, *Zh. Org. Khim.,* 19, 1474, 1983.

177. **Hakimelahi, G.H., Sharghi, H., Zarrinmayeh, H., and Khalafi-Nezhad, A.,** The synthesis and application of novel nitrating and nitrosating agents, *Helv. Chim. Acta,* 67, 906, 1984.

178. **Bao, Y.T., Seltzman, H.H., and Pitt, C.G.,** Polymeric amines and their copper(II) complexes as catalysts for the hydrolysis of organophosphorus esters, *Polym. Prepr.,* 26, 81, 1985.

179. **Efimov, V.A. and Chakhmakhcheva, O.G.,** Use of the oxygen nucleophilic catalysts for the phosphotriester oligonucleotide synthesis, *Chem. Scr.,* 26, 55, 1986.

180. **Hakimelahi, G.H., Prota, Z.A., and Ogilvie, K.K.,** High yield selective 3′-silylation of ribonucleosides, *Tetrahedron Lett.,* 5243, 1981.

181. **Hronec, M. and Vesely, V.,** The effect of aromatic *N*-oxides and *N*-acetylated compounds on the activity of cobalt bromide catalysts, *Collect. Czech. Chem. Commun.,* 43, 728, 1978.

182. **Samsel, E.G., Srinivasan, K., and Kochi, J.K.,** Mechanism of the chromium catalyzed epoxidation of olefins. Role of oxochromium(V) cations, *J. Am. Chem. Soc.,* 107, 7606, 1985.

183. **Srinivasan, K., Michaud, P., and Kochi, J.K.,** Epoxidation of olefins with cationic (salen) manganese(III) complexes. The modulation of catalytic activity by substituents, *J. Am. Chem. Soc.,* 108, 2309, 1986.

184. a. **Bortolini, O.D., Di Furia, F., Modena, G., and Seraglia, R.,** Sulfide oxidation and olefin epoxidation by dilute hydrogen peroxide, catalyzed by molybdenum and tungsten derivatives under phase-transfer conditions, *J. Org. Chem.,* 50, 2688, 1985; b. **Slomp, G. and Johnso, J.L.,** The effect of pyridine on the ozonolysis of 4,22-stigmastadien-3-one, *J. Am. Chem. Soc.,* 80, 915, 1958.

185. a. **Smith, G.B., McDaniel, D.H., Bihl, E., and Hollingsworth, C.A.,** Lithium aluminum amides as catalysts for the reaction of lithium aluminium hydride with 1-hexyne, *J. Am. Chem. Soc.,* 82, 3560, 1960; b. **Wada, H., Watanabe, H., and Hara, Y.,** Ethylene glycol, *Jpn. Kokai* 61/263938 A2, 1986; *Chem. Abst.,* 106, 195880z, 1986.

186. **Chauvin, Y., Commereuc, D., and Hugues, F.,** Hydrocarboxyalkylation of vinyl alkenoates, European patent 205365 A1, 1986; *Chem. Abst.,* 107, 39213z, 1987.

187. a. **Pesa, F. and Haase, T.,** The cobalt-catalyzed hydroesterification of acrylonitrile, *J. Mol. Catal.,* 18, 237, 1983; b. **Ito, K., Kamiyama, N., Nakanishi, S., and Otsuji, Y.,** Selective trimerization of aliphatic aldehydes catalyzed by polynuclear carbonylferrates, *Chem. Lett.,* 657, 1983.

188. a. **Zaugg, H.E., Horrom, B.W., and Borgwardt, S.,** Specific solvent effects in the alkylation of enolate anions. I. The alkylation of sodiomalonic esters with alkyl halides; *J. Am. Chem. Soc.,* 82, 2895, 1960; b. **Zaugg, H.E.,** II. Relationships between structure and physical properties of additives and their catalytic efficiencies, *J. Am. Chem. Soc.,* 82, 2903, 1960.

189. **Lomakin, A.I., Yastrebov, S.I., Gorbunov, Y.A., Samukov, V.V., and Popov, S.G.,** Automated synthesis of oligodeoxyribonucleotides, IV. Use of a new efficient condensing reagent, *Bioorg. Khim.,* 13, 359, 1987.

190. a. **Hudson, M.J. and Glaves, L.R.,** The use of heterocyclic *N*-oxides in the separation of precious metals by solvent extraction, *Chem. Ind.,* 22, 1985; b. **Shatskaya, S.S., Gilbert, E.N., Mikhailov, V.A., and Verevkin, G.V.,** Solvent extraction of platinum and iridium by 2-*n*-nonylpyridine-*N*-oxide, *Zh. Anal. Khim.,* 37, 1089, 1982.

191. **Drew, M.G.B., Glaves, L.R., and Hudson, M.J.,** Solvent extraction of gold and platinum-group metals using 2-nonylpyridine 1-oxide, and the crystal and molecular structure of bis(2-nonylpyridine 1-oxide)hydrogen (1+) tetrachloroaurate (III), *J. Chem. Soc. Dalton Trans.,* 771, 1985.

192. **Bradshaw, J.S, Guynn J.M., Wood, S.G., Krakowiak, K.E., Izatt, R.M., McDaniel, C.W., Wilson, B.E., Dalley, N.K., and Lindh, G.C.,** Synthesis, structural features, and cation transport studies of crown-ethers containing the 4-pyridone *N*-hydroxide subunit cyclic group, *J. Org. Chem.,* 53, 2811, 1988.

193. **Reiner, J.R. and Breister, S.,** Use of heavy metal chelates of 2-mercaptopyridine-*N*-oxide to separate selected precious metals from acidic solutions, U.S. patent 4269621, 1981; *Chem. Abst.,* 95, 118898p, 1981.

194. a. **Ejaz, M.,** Isolation of ionium (thorium-230) from a pitchblende mineral sample, *Radiochim. Acta,* 21, 161, 1974; b. **Ejaz, M. and Carswell, D.J.,** Amine oxides as solvents for uranium, thorium and some fission products, *J. Inorg. Nucl. Chem.,* 37, 233, 1975.

195. a. **Tzeng, W.L., Meloan, C.E., and Hua, D.,** 1-(1′-Oxido-4′-pyridyl)-2-piperidinoethyl 4-nitrobenzoate: a reagent to selectively remove oxidizing dianions, *Anal. Lett.,* 19, 553, 1986; b. **Chumakov, Y.I. and Lopatenko, S.K.,** Extraction of aromatic hydrocarbons from hydrocarbon mixtures, e.g. from medium-distillate petroleum fractions, USSR patent 445680, 1974; *Chem. Abst.,* 82, 173393b, 1975.

196. **Skarzewoki, J.,** Oxidation by metal ions transferred into an organic phase. Reaction of some toluene derivatives with cerium ammonium nitrate in a two-phase system, *J. Chem. Res. (S),* 410, 1980.

197. **Skrzewski, J. and Cichacsz, E.,** The two phase oxidation of some aromatic compounds with cerium ammonium nitrate in the presence of surfactants, *Bull. Chem. Soc. Jpn.,* 57, 271, 1984.

198. **Bortolini, O., Campestrini, S., Di Furia, F., Modena, G., and Valle, G.,** Anionic molibdenum-picolinate-*N*-oxido-peroxo complex: an effective oxidant of primary and secondary alcohols in nonpolar solvents, *J. Org. Chem.,* 52, 5467, 1987.

199. **Bacon, R.G.R. and Stewart, O.J.,** Copper promoted hydrogen transfer from aromatic donors to halides, *J. Chem. Soc. (C),* 301, 1969.

200. a. **Tortorella, V. and Tarzia, G.,** Use of 2-fluoropyridine *N*-oxide in gradual degradation of peptides, *Gazz. Chim. Ital.,* 97, 1479, 1967; b. **Tortorella, V. and Bettoni, G.,** *ibid.,* 97, 1487, 1967.

201. **Tortorella, V. and Bettoni, G.,** Use of 2-fluoropyridine *N*-oxide in *N*-terminal analysis, *Gazz. Chim. Ital.,* 97, 85, 1967.

202. **Sarantakis, D., Sutherland, J.K., Tortorella, C., and Tortorella, V.,** 2-Fluoropyridine *N*-oxide and its reactions with aminoacid derivatives, *J. Chem. Soc. (C),* 72, 1968.

203. **Shrabka-Blotnicka, T.,** Use of 3-fluoro-4-nitropyridine *N*-oxide for the determination of *N*-terminal amino acids in protons and peptides, *Chem. Anal. (Warsaw),* 13, 587, 1968.

204. **Solih, E., Willenbrock, S.J.F., Baines, B.S., and Brocklhurst, K.,** Benzofuroxan as a reactivity probe for the study of dispositions of nucleophilic and acid-base groups in enzyme active centers, *Biochem. Soc. Trans.,* 10, 217, 1982.

205. **Barton, J.R., MacPeek, W.A., and Cohen, W.S.,** Interaction of 2-*n*-heptyl-4-hydroxyquinoline-N-oxide with photosystem II in choroplasts and subchloroplast particles, *J. Bioenerg. Biomembr.* 15, 93, 1983.

206. **Joyeau, R. and Brown, E.,** Synthesis and properties of acrylic monomers bearing an acylating functional group. Preparation of a new acylating copolymer for immobilization of biomolecules, *Bull. Soc. Chim. Fr.,* 391, 1982.

207. **Issidorides, C.H. and Haddadin, M.J.,** Quinoxaline derivatives, U.S. patent. 4343942 A, 1982; *Chem. Abst.,* 98, 4563g, 1983.

208. **Seng, F. and Metzger, K.,** Pharmaceutical quinoxaline di *N*-oxides, *Ger. Offen.,* 2656783, 1978; *Chem. Abst.,* 89, 129545y, 1978.

209. **Sasse, K., Haller, I., Plempel, M., Zeiler, H.J., and Metzger, K.G.,** 3-Sulfonylbenzo-1,2,4-triazines and -benzo-1,2,4-triazine-1-oxides and their use in antimicrobial agents, European patent 16391, 1980; *Chem. Abst.,* 94, 156975y, 1981.

210. **Diel, P.J.,** 3-Imino-1,2,4-benzotriazin-1-oxides, *Ger.Offen,* 2510822, 1975; *Chem. Abst.,* 85, 21474c, 1975.

211. **Glazer, E.A. and Chappel, L.R.,** Pyridoquinoxaline *N*-oxides. 1. A new class of antitrichomonal agents, *J. Med. Chem.,* 25, 868, 1982.

212. **Gullo, J.M.,** Antiinflammatory 2-pyridyl 1-oxide thiocarbamates and dithiocarbamates, European patent 89661 A1, 1983; *Chem. Abst.,* 100, 51455t, 1984.

213. **Danilenko, V.F., Trinus, F.P., Portnyagina, V.A, Ryabukha, T.K., and Klebanov, B.M.,** Preparation and antiphlogistic activity of carboxyphenylamides of nicotinic or isonicotinic acid, USSR patent 539878, 1976; *Chem. Abst.,* 87, 5812x, 1978.

214. **Ramuz, H.,** Imidazole derivatives and their therapeutic use, French patent 2489822 A1, 1982; *Chem. Abst.,* 97, 55809d, 1982.

215. **Pitet, G. and Faure, C.,** 5,6-Diaryl-1,2,4-triazine N-oxides and pharmaceutical compositions containing them, French patent 2485531 A1, 1981; *Chem. Abst.,* 97, 23824p, 1982.

216. **Packman, A.M.,** Ophthalmic and otic composition containing bis(2-pyridyl 1-oxide) disulfide, U.S. patent 4370325 A, 1983, *Chem. Abst.,* 98, 132355e, 1983.

217. **Kunze, B., Hoefle, G., and Reichenbach, H.,** The aurachins, new quinoline antibiotics from myxobacteria: production and physicochemical and biological properties, *J. Antibiot.,* 40, 258, 1987.

218. **Amano, S., Miyamichi, S., Ito, M., Ezaki, N., Niwa, T., Akita, E., and Yamada, Y.,** Mixin, *Jpn. Kokai* 54/86694, 1979; *Chem. Abst.,* 91, 156043f, 1979.

219. **Jackson, R.C., Boritzki, T.J., Besserer, J.A., Hamelehle, K.L., Shillis, J.L., Leopod, W.R., and Fry, D.W.,** Biochemical pharmacology and experimental chemotherapy studies with guanine 7-oxide, a novel purine antibiotic, *Adv. Enzyme Reg.,* 26, 301, 1987.

220. a. **Campese, V.M.,** Minoxidil: a review of its pharmacological properties and therapeutic use, *Drugs,* 22, 257, 1981; b. **Petersen, H.J.,** Substituted pyridyl cyanoguanidine compounds, *Ger. Offen,* 3233380 A1, 1983; *Chem. Abst.,* 98, 198046j, 1983.

221. **Combs, D.W. and Falotico, R.,** Preparation and testing of quinazoline-3-oxides as cardiotonics and bronchodilators, U.S. patent 4745118 A, 1988; *Chem. Abst.,* 109, 93055d, 1988.

222. **Suehiro, H., Kikukawa, K., and Ichino, M.,** 2-Substituted thioadenosine *N*-oxides, *Jpn. Kokai,* 52/65298, 1977; *Chem. Abst.,* 87, 168344K, 1977.

223. **Prasad, R.N.,** N^1-Odixes of adenosine-5′-carboxylates for treating angina, U.S. patent 3928582, 1975; *Chem. Abst,* 84, 165161a, 1976.

224. **Schoenafinger, K., Beyerle, R., Mogliev, A., Bohn, H., Martorana, P., and Nitz, R.E.,** Substituted 1,2,5-oxadiazole 2-oxides as pharmaceuticals for cardiovascular diseases, *Gen.Offen* 3012862, A1, 1981; *Chem. Abst.,* 96, 40909s, 1982.

225. **Pavia, M.R., Taylor, C.P., Hershenson, F.M., Lobbestael, S.J., and Butler, D.E.,** 3-Phenoxypyridine 1-oxides as anticonvulsant agents, *J. Med. Chem.,* 31, 841, 1988.

226. **Nisato, D. and Boveri, S.,** *N*-[(furylmethyl)thio]ethyl]nicotinamide 1-oxide derivatives, European patent 60182 A1, 1982; *Chem. Abst.,* 98, 34508k, 1983.

227. **Crossley, R. and Cliffe, I.A.,** *N*-arylpyridine-4-carboxamides and their *N*-oxides, British patent 2165537 A1 1986; *Chem. Abst.,* 105, 208768a, 1986.

228. **Scherm, A., Hummel, K., Peteri, D., and Schatton, W.,** *N*-oxides of pyridine carboxylic acid esters, *Ger.Offen.* 3315877 A1, 1984; *Chem. Abst.,* 102, 113305w, 1984.

229. **Hashizume, K., Kase, H., and Kitamura, S.,** Quinoline *N*-oxide with lipoxygenase suppressing action, *Ger. Offen.,* 3417573 A2, 1984; *Chem. Abst.,* 102, 84408Z, 1984.

230. a. **Lovisolo, P.P., Briatico-Vangosa, G., Orsini, G., Ronchi, R., Angelucci, R., and Valzelli, G.,** Pharmacological prophile of a new antilipolytic agent, 5-methylpyrazine-2-carboxylic acid 4-oxide (acipimox). I. Mechanism of action, *Pharm. Res. Commun.,* 13, 151, 1981; b. **Cozzi, P., Pillan, A., Bertone, L., Branzoli, U., Lovisolo, P.P., and Chiari, A.,** Antilipolytic activity of new pyrazine *N*-oxides, *Eur. J. Med. Chem., Chim. Ther.,* 20, 241, 1985.

231. **Kontoghiorghes, G.J.,** 2-Hydroxypyridine *N*-oxides; effective new chelators in iron mobilization, *Biochim. Biophys. Acta,* 924, 13, 1987.

232. **Enomoto, S., Immaru, D., Osaka, Y., and Kawasaki, T.,** Antitumor compostition containing a nitrofurylvinylene thiazole *N*-oxide, *Ger. Offen.,* 80, 2920247, 1979; *Chem. Abst.,* 92, 157932c, 1979.

233. **Van der Kleijn, E., Vree, T.B., and Guelen, P.J.M.,** 7-Chloro-1,4-benzodiazepines: diazepam, desmethyldiazepam, oxidiazepam, oxydesmethyldiazepam (oxazepam), and chlorodiazepoxide, *Psychopharmacology (N.Y.),* 2, 997, 1977.

234. **Archer, G.A. and Sternbach, L.M.,** The chemistry of benzodiazepines, *Chem. Rev.,* 68, 747, 1968.

235. **Greve, W., Elben, U., Rudolphi, K., and Schindler, U.,** Aminopyridine *N*-oxides, *Ger. Offen,* 3514073 A1,1986; *Chem. Abst.,* 106, 32862f, 1987.

236. **Marotta, E. and Castronuovo, L.,** Quinoxaline derivatives useful for treatment and prophylaxis of swine dysentery and as animal growth promotants, European patent 192992, A2, 1986; *Chem. Abst.,* 105, 226655r, 1986.

237. **Folz, S.D. and Ursprung, J.J.,** 6-Amino-4-nitro-2-picoline-*N*-oxide compositions and coccidiostat process. U.S. patent 3995035, 1976; *Chem. Abst.,* 86, 89625d.

238. **Matsuno, T. and Imai, K.,** Prophylactic and therapeutic agents for coccidiosis, *Jpn. Kokai,* 54/23133, 1979; *Chem. Abst.,* 91, 13991c, 1979.

239. **Morisawa, Y., Kataoka, M., and Kitano, K.,** Pyridine-3-sulfonamide 1-oxide, *Jpn. Kokai,* 53/56671, 1978; *Chem. Abst.,* 90, 6250m, 1979.

240. **Patchett, A.A. and Wang, C.C.,** Treating coccidiosis with emimycin and its derivatives, U.S. patent 3991185, 1974; *Chem. Abst.,* 86, 34280g, 1977.

241. **Wedig, J.H., Babish, J.G., and Davidson, J.,** Use of selected pyridine *N*-oxide disulfide compounds to treat or prevent bovine mastitis, U.S. patent 4610993 A, 1986; *Chem. Abst.,* 105, 218876y, 1986.

242. **DeGeeter, M.J. and McCall, J.M.,** Animal feed, U.S. patent 4282228, 1981; *Chem. Abst.,* 95, 202320b, 1981.

243. **Menon, G.K. and Sanders, W.J.,** Feed compositions containing copper salts of 2-hydroxypyridine *N*-oxides, European patent 102184Al, 1984; *Chem. Abst.,* 101, 109536j, 1984.

244. **Schmidt, W.,** Quinoxaline di-*N*-oxide derivatives, Swiss patent 630908A, 1982; *Chem. Abst.,* 97, 216233a, 1982.

245. **Seng, F., Ley, K., and Metzger, K.G.,** Veterinary feed additives, U.S. patent 399098, 1976; *Chem. Abst.,* 84, 169665r, 1976.

246. **Niclas, H.J., Hennig, A., and Sukale, S.,** Ergotropes and feed expenditures-decreasing agents for livestock, East German patent 234787 Al, 1986; *Chem. Abst.,* 105, 207994j, 1986.

247. **Menon, G.K. and Sanders, W.J.,** Feed composition containing a (l-oxo-2-pyridyl)disufide, European patent 104836 A1, 1984; *Chem. Abst.,* 101, 37588c, 1984.

248. a. **Seng, F., Scheinpflung, H., and Kraus, P.,** Plant bactericide containing 4-alkylaminopyrimido[4,5-*b*]quinoxaline 5,10-dioxide, *Ger. Offen* 2745007, 1979; *Chem. Abst.,* 91, 1360h, 1979; b. **Creuzburg, D., Elster, S., Kleiner, R., and Klepel, M.,** Plant growth regulator, East German patent 140966, 1980; *Chem. Abst.,* 94, 151893k, 1981.

249. **Lee, K.T.,** Herbicidal pyridine l-oxides, U.S. patent 4201567, l980; *Chem. Abst.,* 94, 11637e, 1981.

250. **Yasu Kagaku Kogyo, K. K.,** Wide-spectrum herbicide compositions, *Jpn. Kokai,* 57/192307 A2, 1982; *Chem. Abst.,* 98, 138989j, 1983.

251. **du Pont de Nemours, E.I., and Co.,** Preparation of *N*-heterocyclyl-*N′*-(pyridylsulfonyl)urea *N*-oxides as herbicides and plant growth regulators, *Int. Appl.* 8803528 A2, 1988; *Chem. Abst.,* 109, 231080m, 1988.

252. **Bell, A.R., Minatelli, J.A., and Doweyko, A.M.P.,** 2-(Ethylsulfonyl)pyridine l-oxide derivatives, European patent 79154 A2, 1983; *Chem. Abst.,* 99, 139781t, 1983.

253. **Sumitomo Chem. Co.,** Benzylsulfonyl pyridine *N*-oxides, *Jpn. Kokai,* 59/167571 A2, 1984; *Chem. Abst.,* 102, 78736p, 1984.

254. **Diehl, R.E. and Walworth, B.L.,** Controlling undesirable plant species using 3-nitropyridines, U.S. patent 3826643, 1974; *Chem. Abst.,* 82, 52661d, 1975.

255. **Johnston, H. and Gulbenk, A.H.,** Aminohalopyridine-*N*-oxides, U.S. patent 3948910, 1976; *Chem. Abst.,* 85, 21132q, 1976.

256. **Tseng, C.P.,** Sulfonylurea *N*-oxides, European patent 57546 A2, 1982; *Chem. Abst.,* 98, 4570g, 1983.

257. **I.C.I.,** Quinoxaline derivatives as herbicides, *Jpn. Kokai,* 57/140771 Al; *Chem. Abst.,* 98, 53933m, 1983.

258. **Serban, A., Farquharson, G.J., Bird, G.J., and Lydiate, J.,** [(1,2,4-Benzotriazin-3-yloxy)phenoxy]alkanoic acid derivatives, their use as herbicides and intermediates, European patent 24932, 1981; *Chem. Abst.,* 95, 169223x, 1981.

259. **Takematsu, T., Tachibana, K., Shimura, M., Suzuki, A., Kai, F., Seki, N. and Sekizawa, T.,** Phenazines as rice paddy herbicide, *Jpn. Kokai,* 80/17341, 1980; *Chem. Abst.,* 92, 210196n, 1980.

260. **Nagano, H., Haga, T., Takase, M., and Sato, M.,** Tetrahydrobenzotriazole l-oxides, *Jpn. Kokai,* 61/43175 A2, 1986; *Chem. Abst.,* 105, 42822m, 1986.

261. **Schoenowski, H., Mildenberger, H., and Sachse, B.,** Fungicidal pyridine *N*-oxides, *Ger. Offen* 3109024 Al,1982; *Chem. Abst.,* 98, 4483f, 1983.

262. **Stanton, D.T., Davis, R.A., and Doweyko, A.M.,** A novel class of fungicides: (2-pyridylthio)methyl benzoate *N*-oxides, *J. Agric. Food Chem.,* 31, 451, 1983.

263. **Goddard, S.J.,** Herbicides, U.S. patent 3883550, 1975; *Chem. Abst.,* 83, 109808j, 1975.

264. **Okano, T.,** Homocysteine derivatives, *Jpn. Kokai,* 75/46672, 1975; *Chem. Abst.,* 83, 114925z, 1975.

265. **Sasamori, H., Wada, T., and Hamada, M.,** 2-Acylthiopyridine 1-oxide derivatives as seed sterilants, *Jpn. Kokai,* 61/186364 A2, 1986; *Chem. Abst.,* 106, 50058c, 1987.

266. **Procter and Gamble Co.,** Fruit abscission enhancement by bispyridine disulfides, *Jpn. Kokai,* 77/54027, 1977; *Chem. Abst.,* 89, 38084f, 1978.

267. **Studeneer, A., Salbeck, G., Emmel, L., and Knauf, W.,** Dimethylaminocabonyloxypyridine l-oxide and -quinoline oxides, insecticides, *Ger. Offen,* 2432635, 1976; *Chem. Abst.,* 84, 150525y, 1976.

268. **Pallos, F.M.,** Derivatives of *N*-oxypyridyl geranyl ethers used in controlling insects, U.S. patent 3928616, 1975; *Chem. Abst.,* 84, 105848v, 1975.

269. **Vladykin, V. I., Zalesov, V.S., Trakhtenberg, S.I ., Markovskii, V.I., and Konova, N.M.,** Raticide, USSR patent 654226, 1979; *Chem. Abst.,* 91, 1380q, 1979.

270. **Mixan, C.E., Goralski, C.T., and Pews, R.G.,** 1,3-Dithiolo[4,5-b]pyrazin-2-ylidenepropanedinitrile 4-oxide, U.S. patent 4039727; *Chem. Abst.,* 87, 153516u, 1977.

271. **Morris, C.E. and Welch C.M.,** Zinc pirithione process to impart antimicrobial properties to textiles, U.S. patent 527894 A0, 1984; *Chem. Abst.,*100, 158193h, 1984.

272. **Weisse, G.K.,** Inhibiting microbial growth in water-based paints, U.S. patent 3892699, l975; *Chem. Abst.,* 83, 133522g, 1975.

273. **Ueno, S., Shimosato, E., and Ohmura, Y.,** Antiseptic composition for food, *Jpn. Kokai,* 49/34810, 1974; *Chem. Abst.,* 82, 138022h, 1975.

274. a. **Grollier, J.F.,** Hair-growth stimulants containing pyridinopyrimidine derivatives and nicotinates, *Ger. Offen,* 3729953 Al, 1988; *Chem. Abst.,* 109, 176083n,1988; b. **Bazzano, G.S.,** Use of retinoids and minoxidil (2,4-diamino-6-piperidinopyrimidine 3-oxide) to increase the rate of growth of human scalp hair and to treat certain types of alopecia, World patent 8302558 A1,1983; *Chem. Abst.,* 99, 181484y, 1983.

275. a. **Saccani, R., Mattac, M., and Saccani, E.,** Preparation of 6-amino-1,2-dihydro-1-hydroxy-2-iminopyrimidine derivatives and cosmetic compositions containing them, French patent 2590897 Al, 1987; *Chem. Abst.,* 109, 93038a, 1988; b. **Kinji, T.,** Hair lotion containing minoxidil, *Jpn. Kokai,* 61/165310 A2, 1986; *Chem. Abst.,* 105, 178218b, 1986.

276. **Wetzel, T.A.,** Antidandruff shampoos containing ethylene glycol fatty acid diesters, European patent 200305, 1986; *Chem. Abst.,* 106, 72692q, 1987.

277. a. **Furia, T.E. and Steinberg, D.H.,** Combatting dandruff with mercaptoquinoline *N*-oxides, U.S. patent, 3961054, 1976; *Chem. Abst.,* 85, 166638u, 1976; b. **Bouillon, C. and Rosenbaum, G.,** Cosmetic and pharmaceuticals containing aluminum tris(N-oxypyridinethiolate), Canadian patent 1130207 A2, 1982; *Chem. Abst.,* 97, 222759w, 1982.

278. **Douglass, M.L.,** 2-Mercaptoquinoxaline 1-oxides, salts and 2-(1-oxoquinoxalinyl)disulfides in detergent compositions, U.S. patent 3971725, 1976; *Chem. Abst.,* 85, 126254j, 1976.

279. **Packman, A.M.,** Antiperspirant-deodorant compositions, U.S. patent 4235873, 1980; *Chem. Abst.,* 94, 151893k, 1981.

280. **Lang, G. and Forestier, S.,** Quinoxaline derivative-based skin tanning composition, *Ger. Offen,* 2719542, 1977; *Chem. Abst.,* 88, 78971d, 1978.

281. **Gaylor, N.G. and Mehta, R.,** Peroxide-catalyzed grafting of maleic anhydride onto molten polyethylene in the presence of polar organic compounds, *J. Polym. Sci. Part A: Polym. Chem.,* 26, 1189, 1988.

282. **Davis, L.H., Sullivan, A.B., and Coran, A.Y.,** New curing system components, *Rubber Chem. Technol.,* 60, 125, 1987.

283. **Crosby, J. and Milner, J.A.,** Modification of polymers, British patent 1586861, 1981; *Chem. Abst.,* 95, 63540h, 1981.

284. **Crosby, J., Rennie, R.A.C., Tanner, J., and Paton, R.M.,** Dinitrile oxides as polymer additives, *Ger. Offen,* 2422764, 1974; *Chem. Abst.,* 82, 141047a, 1975.

285. **Decout, J.L., Lablache-Combier, A., and Loucheux, C.,** Photocrosslinking of polymeric films by amine *N*-oxide initiators, *Photogr. Sci. Eng.,* 24, 255, 1980.

286. **Asani Chem. Ind.,** Catalyst for conjugated olefin polymerization, *Jpn. Kokai,* 59/113003 A2, 1984; *Chem. Abst.,* 102, 7256w, 1984.

287. **Cermak, V. and Mleziva, J.,** Hardening of epoxy resins, Czechoslavakian patent 185416, 1980; *Chem. Abst.,* 95, 98879c, 1981.

288. **Karayannis, N.M. and Lee, S.S.,** Sterospecificity control of propylene polymerization catalysts by sterically hindered alicyclic secondary amines and their nitroxide free radicals, *Makrom. Chem., Rapid Commun.,* 3, 255, 1982.

289. **Hiraoka, H. and Welsch, L.W.,** Deep UV photolithography with composite photoresists containing poly(olefin sulfones), *Org. Coat. Appl. Polym. Sci. Proc.,* 48, 48, 1983.

290. **Ohta, M.,** Nitro derivatives of benzo[*c*]cinnoline-5 (or 6)-oxide, *Jpn. Kokai,* 75/100067, 1975; *Chem. Abst.,* 84, 44108e, 1975.

291. **Ohta, M., Hashimoto, M., and Kojima, A.,** Electrophotographic photoconductors, *Jpn. Kokai,* 77/109939, 1977; *Chem. Abst.,* 88, 129023f, 1978.

292. **Jono, S.,** 2-Phenylbenzotriazole *N*-oxides, *Jpn. Kokai,* 59/172481 A2, 1984; *Chem. Abst.,* 102, 113511k, 1984.

293. **McKellar, J.K. and Warburton, G.G.,** Coating composition for carbon paper, British patent 1395336, 1975; *Chem. Abst.,* 83, 133644v, 1975.

294. **Hamada, M., Kudo, N., Yokoi, J., Shimatani, S., and Oda, R.,** Antifouling coating compositions, *Jpn. Kokai,* 80/36221, 1980; *Chem. Abst.,* 93, 48622b, 1980.

295. **Hamada, M. and Umeno, M.,** Antifouling paints, *Jpn. Kokai,* 79/83035, 1979; *Chem. Abst.,* 92, 24372y, 1980.

296. **Mitamura, H. and Arimatsu, G.,** Preparation of polymerizable (*N*-oxopyridine-2-thio)methylstyrenes as antimold agents, *Jpn. Kokai,* 63/115864 A2, 1988; *Chem. Abst.,* 109, 128843p, 1988.

297. **Kato, T.,** Melt-spinning of polyamide fibers, *Jpn. Kokai,* 78/100253, 1978; *Chem. Abst.,* 88, 171715h, 1978.

298. **Booker, R.A.,** In situ film hardening with pyridine *N*-oxide and aldehyde precursors, U.S. patent 4504578 A, 1985; *Chem. Abst.,* 102, 157941e.

299. a. **Bergthaller, P., Stoltzemburg, R., and Marx, P.,** Color photographic material with cyan dye-releasing compounds, *Ger. Offen,* 3613621 A1, 1987; *Chem. Abst.,* 109, 46029p, 1988; b. **Greenwald, R.B.,** Alkylsulfonylmethyl-substituted pyridine *N*-oxides, U.S. patent 4006150, 1977; *Chem. Abst.,* 87, 60747e, 1977; c. **Bourgeois, G.J., Gaudiana, R.A., and Sahatjian, R.A.,** Photographic products comprising dye developers and *N*-oxides, U.S. patent 4203766, 1980; *Chem. Abst.,* 93, 177209w, 1980.

300. **Ciurca, S.J. and Brault, A.T.,** Photographic color materials with *N*-oxides as oxidant, French patent 2232777, 1975; *Chem. Abst.,* 84, 67800p, 1975.

301. **Kemp, R.J. and Griffiths, D.,** Photographic silver bleach, *Res. Discl.,* 151, 97, 1976.

302. **Norris, W.P. and Spear, R.J.,** Potassium 4-hydroxylamino-5,7-dinitrobenzofurazanide 3-oxide, the first in a series of new primary explosives, *Rep.-Mater. Res. Lab. (Aust.),* MRL-R-870, 1983; *Chem. Abst.,* 99, 40662t, 1983.

303. **Homewood, R.N., Krukonis, V.J., and Loszewski, R.C.,** Safe explosive containing dicyanofuroxan and method, U.S. patent 3832249, 1974; *Chem. Abst.,* 82, 113795z, 1975.

304. **McGuire, R.R.,** Properties of benzotrifuroxan, *Report UCRL*-52353, 1978; *Chem. Abst.,* 90, 4774m, 1979.

305. **Norris, W.P.,** Synthesis of 7-amino-4,6-dinitrobenzofuroxan as an explosive, *Statutory Invent. Regist.,* US 476 H1, 1988; *Chem. Abst.,* 110, 8215j, 1989.

306. **Bell, A.J. and Read, R.W.,** Synthesis of polynitro-substituted 2′-nitrobiphenyl-2-amines, analogs of polynitrobenzo[c]cinnoline oxide derivatives, *Aust. J. Chem.,* 40, 1813, 1987.

307. **Zyss, J.,** New organic molecules crystals with large quadratic nonlinearities, *J. Cryst. Growth,* 79, 681, 1986.

308. **Nicoud, J.F.,** Molecular and crystal engineering for organic nonlinear optical material, *Mol. Cryst. Liq. Cryst.,* 156 Pt A, 257, 1988.

310. **Griffin, A.C., Bhatti, A.M., and Howell, G.A.,** Pyridine *N*-oxides as polymeric nonlinear optical materials, *Mater. Res. Soc. Symp. Proc.,* 109, 115, 25, 1988; *Chem. Abst.,*109, 150111v, 1988.

311. **Chemea, D., Zyss, J., and Nicoud, J.F.,** Use of crystals of 3-methyl-4-nitropyridine l-oxide and of 3,5-dimethyl-4-nitropyridine l-oxide in nonlinear optics, French patent 2472201 Al, 1981; *Chem. Abst.,* 96, 43719r, 1982.

312. **Josse, D., Hierle, R., Ledoux, I., and Zyss, J,** Highly efficient second-harmonic generation of picosecond pulses at 1.32 μm in 3-methyl-4-nitropyridine 1-oxide, *Appl. Phys. Lett.,* 53, 2251, 1988.

313. **Byron, D.J., Lacey, D., and Wilson, R.C.,** Properties of liquid crystals formed by the N-oxides of certain 4-(4′-pyridyl)phenyl-4″-*n*-alkoxybenzoates: comparison of the *N*-oxides and cyano groups as mesogenic terminal substituents, *Mol. Cryst. Liq. Crist.,* 75, 225, 1981.

314. **Mizunoya, K.,** Liquid crystal elements, *Jpn. Kokai,* 61/296086 A2, 1986; *Chem. Abst.,* 107, 15703h, 1987.

315. **Liu, D.,** Resazurin reduction method for toxicity assessment of water soluble and insoluble chemicals, *Toxic Assess.,* 1, 253, 1986.

316. **Janhns, F.D. and Rand, A.G.,** Immobilized enzyme method to asses fish quality, U.S. patent 4105800, 1978; *Chem. Abst.,* 90, 4774m, 1979.

317. **Nakamura, S. and Hamado, K.,** Composition for diagnosis of tooth decay, *Jpn. Kokai,* 79/47700, 1979; *Chem. Abst.,* 91, 62741e, 1979.

318. **Below, H., Kuehn, M., Ohme, R., Rusche, J., and Striegler, C.,** Chromium electroplating from low concentration electrolytes, East German patent 155437 Z, 1982; *Chem Abst.,* 97, 204986p, 1982.

319. **Naito, K. and Ikezaki, T.,** Solid electrolytic capacitor, *Jpn. Kokai,* 62/102512 A2, 1987; *Chem. Abst.,* 107, 107677f, 1987.

320. **Abe, T., Masuda, T., Hashiguchi, M., and Kumaki, K.,** Preparation of benzofurazan-l-oxides as intermediates for dyes, *Jpn. Kokai,* 63/83078 A2, 1988; *Chem. Abst.,* 109, 231030v, 1988.

321. **Sauer, W., Seifert, A., and Binte, H.J.,** Quinoxaline-l-oxide-2-azo dyes, East German patent 101910, 1973; *Chem. Abst.,* 83, 61643x, 1974.

322. **McCapra, F. and Taheri-Kadkoda, M.,** Infrared chemiluminescence from an organic compound, *J. Chem. Soc., Chem. Commun.,* 487, 1988 .

323. **Bouffard, R.A.,** Rust inhibited gasoline compositions, British patent 1249234, 1971; *Chem. Abst.,* 84, 153036a, 1976.

324. **Nolen, G.A. and Dierkman, T.A.,** Reproduction and teratology studies of zinc pyrithione administered orally or topically to rats and rabbits, *Food Cosmet. Toxicol.,*17, 639, 1979.

325. **Nagishi, T., Tanaka, K., and Hayatsu, H.,** Mutagenicity of carbadox and several quinoxaline 1,4-dioxide, *Chem. Pharm. Bull.,* 28, 1347, 1980.

326. **Endo, H. and Kondo, S.,** Molecular aspects of 4-nitroquinoline l-oxide carcinogenesis, *Gann Monogr. Cancer Res.,* 24, 73, 1979.

327. **Takahashi, K., Kawazoe, Y., Tada, M., Ito, N., and Okada, M.,** Carcinogenicity of 4-nitrosoquinoline l-oxide and its possible role in carcinogenesis by 4-nitroquinoline l-oxide, *Gann,* 69, 499, 1978,

INDEX

C